FLAT-PANEL DISPLAYS AND CRTs

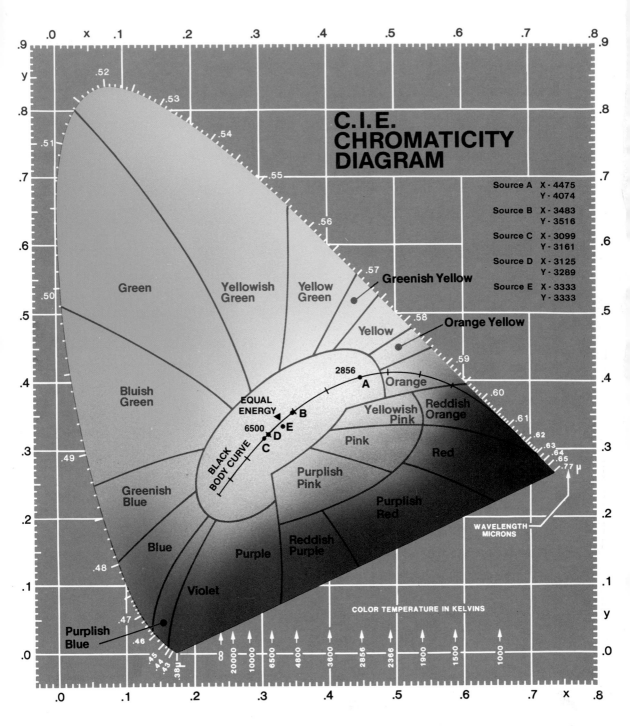

The 1931 CIE (Commission International de l'Eclairge) chromaticity diagram is used throughout the displays industry for engineering and scientific color characterization of emissive and nonemissive displays. The diagram in this form with the color regions and names as derived and revised in 1955 by K. L. Kelly of NBS (National Bureau of Standards) and adapted as such as a standard by EIA (Electronic Industries Association) is called the Kelly Chart. The color rendition is an approximation of the true colors because of the limitations in printers inks. (Courtesy of RCA, New Products Division, Lancaster, PA., and Photo Research, Division of Kollmorgen, Burbank, CA.)

FLAT-PANEL DISPLAYS AND CRTs

Edited by

Lawrence E. Tannas, Jr.

 **VAN NOSTRAND REINHOLD COMPANY**
———————— *New York* ————————

Library of Congress Catalog Card Number: 84-11839
ISBN: 0-442-28250-8

Manufactured in the United States of America

Published by Van Nostrand Reinhold Company Inc.
135 West 50th Street
New York, New York 10020

Van Nostrand Reinhold Company Limited
Molly Millars Lane
Wokingham, Berkshire RG11 2PY, England

Van Nostrand Reinhold
480 Latrobe Street
Melbourne, Victoria 3000, Australia

Macmillan of Canada
Division of Gage Publishing Limited
164 Commander Boulevard
Agincourt, Ontario M1S 3C7, Canada

15 14 13 12 11 10 9 8 7 6 5 4 3 2 1

Library of Congress Cataloging in Publication Data
Main entry under title:

Flat-panel displays and CRTs.

    Includes index.
    1. Information display systems.   2. Cathode-ray tubes.
I. Tannas, Jr., Lawrence E.
TK7882.I6F53 1984   381.3819'532   84-11839
ISBN 0-442-28250-8

PREFACE

Flat-Panel Displays and CRTs, a review of electronic information display devices, is the first systematic and comprehensive coverage of the subject. It is intended to distill our wealth of knowledge of flat-panel displays and CRTs from their beginnings to the present state of the art. Historical perspective, theory of operation, and specific applications are all thoroughly covered.

The field of display engineering is a multidisciplined technical pursuit with the result that its individual disciplines suffer from a lack of communications and limited perspective. Many previously developed standards for, and general understanding of, one technology are often inappropriate for another. Care has been taken here to document the old, incorporate the new, and emphasize commonalities. Criteria for performance have been standardized to enable an expert in one display technology, such as liquid crystals, to compare his device performance with that offered by another technology, such as electroluminescence.

This book has been written with a second purpose in mind, to wit, to be the vehicle by means of which a new scientist or engineer can be introduced into the display society. It is organized to be tutorial for use in instructional situations. The first chapters begin with first principles and definitions; the middle chapters set out requirements and criteria; and the last chapters give a complete description of each major technology.

Flat-Panel Displays and CRTs is primarily concerned with flat-panel displays. It is necessary to discuss CRTs before one can properly understand flat-panel displays, however, and to this end "The Challenge of the Cathode-Ray Tube" chapter has been included.

It is obvious why flat-panel-display technologists are so highly motivated. They all hope to find the ideal replacement for one of the last vestiges of the vacuum tube age, the CRT. This common goal has created the flat-panel display fraternity, and the fever of their enterprise sometimes takes on the air of an olympic challenge. For all that, this book is intended to be a scientific document on flat-panel displays and CRTs rather than a prophecy. In fact, no position is taken as to how or when, if ever, the flat-panel display will become dominant. The reader can make his own prognostications. I, myself, believe that in due time each of the flat-panel display technologies will find a niche of its own in the marketplace alongside the CRT. Perhaps the need for replacing the CRT is no longer the problem, since we continue to find both flat-panel displays and CRTs all around us. If the time should ever come when the flat panel displaces the CRT in all market areas, it will only be when the performance and cost considerations justify it. As of now they don't, but some day they may.

This book is about display devices. The display device includes the image-generating medium, electronic drive, faceplate, bezel, power supplies, and so forth, all of which are parts of the display hardware system. They can be analyzed, characterized, and qualified without regard to the eye-brain system, input signal source, or software. In fact, it makes no difference to the software whether the display is a flat panel or a CRT so long as it can respond to the data presented. Similarly, it makes no difference how the image is generated so long as it meets the output performance requirements. A description of the image-quality-performance measures for quantifying the output performance is given in detail. The chapter on the visual system is included as a prerequisite to the chapter on image quality. It should be noted that the more complex aspects of human factors, pattern recognition, and information theory are not covered.

Each chapter is meant to stand alone, yet as an integral part of the whole, and the whole is intended to be a complete but concise statement of the state of the art of display devices as well as a source for future research and development ideas and study topics. A thorough survey of the published art is presented, along with extensive references. Completeness of coverage is taken to be a more important goal than exhaustive detail, and the identification of classifications and fundamental principles is considered to be more important than the inclusion of all exceptions.

I first conceived of this book in 1976 when it became apparent that Sol Sherr's first book, *Fundamentals of Display System Design*, and H. R. "Lux" Luxenberg and Rudolph Kuehn's book, *Display Systems Engineering*, were becoming outdated. It didn't really get going, however, until 1980 when I formulated the UCLA Extension class called "Flat-Panel Displays," later changed to "Flat-Panel and CRT Display Technologies." Each of the instructors for the class wrote a technology chapter; I wrote the introductory chapters and the chapter on electroluminescence; and Harry Snyder joined us later to write the chapters on the visual system and image quality. The encouragement from senior members of our field such as Irving Reingold and Benjamin Kazan was of great help to me. I wish to further acknowledge the word processing, proofreading, indexing, coordination support, and continued encouragement of my dear wife Carol.

As you read this book, keep in mind that even as it went to press the displays industry was experiencing exciting growth and changing technology. Continuous change of this sort was anticipated at its conception, and hence the emphasis on fundamentals that can be used in analyzing and evaluating new developments. It may be expected that advances in technology will continue to fill the pages of the many publications that serve the field. The real challenge for flat-panel and CRT display technologies is to satisfy this ever-expanding and seemingly insatiable marketplace.

It might be helpful, in conclusion, to give a chapter-by-chapter breakdown of the contents of this book and to acknowledge the efforts of all concerned.

Chapter 1.   INTRODUCTION, by Lawrence E. Tannas, Jr., MS, Displays Consultant, Orange, California.

This chapter provides an historical review of all displays and a detailed technical discussion that categorizes and classifies them. It further attempts to put flat-panel displays and CRTs into relative perspective. It also offers a summary of photometry and colorimetry, includes definitions and nomenclature for all displays, and introduces the concepts of pixel, display array, font, duty factor, pixel contrast ratio, and matrix-addressing.

I would like to express my appreciation for the many helpful suggestions from my fellow chapter authors, associates, and students that were made during the many revisions of this chapter.

Chapter 2.   SYSTEM REQUIREMENTS, by Lawrence E. Tannas, Jr.

An analysis of the display as a part of the overall system is undertaken in this chapter. The overall display requirements are represented in system, installation, and function classifications. The display brightness ratio concept is introduced, and pixel contrast ratio and visual angle are combined to form the universal visual image space with the *display detectability surface*. The chapter also presents a second universal metric called *display ambient performance*, which is a plot of the performance measures of display brightness ratio and pixel contrast ratio, along with the power penalty factor, against the display disturbance factor, ambient illumination. Each display subsystem is discussed, and standards and their sources are reviewed. Modern spectroradiometric equipment appropriate for an up-to-date displays photometry laboratory is shown.

I am indebted to Alan Sobel for his careful reading of, and helpful comments on, a draft of this chapter.

Chapter 3.   THE VISUAL SYSTEM–CAPABILITIES AND LIMITATIONS, by Harry L. Snyder, Ph.D., Department of Industrial Engineering and Operations Research, Virginia Polytechnic Institute and State University, Blacksburg, Virginia.

This chapter gives an overall review of the features of the human eye and the metrics for characterizing the way in which the eye-brain system interprets a display image. It discusses spatial and temporal capabilities in detail and introduces the concepts of modulation, contrast threshold function, and color vision.

Dr. Snyder wishes to acknowledge the valuable review of this chapter by H. Lee Task.

Chapter 4. IMAGE QUALITY: MEASURES AND VISUAL PERFORMANCE, by Harry L. Snyder.

This chapter offers a detailed discussion of the science for evaluating the display image. It introduces the concept of modulation transfer function and uses it in the spatial frequency domain to compare the various image quality metrics.

Dr. Snyder acknowledges the careful reviews of this chapter by H. Lee Task and Robert J. Beaton.

Chapter 5. FLAT-PANEL DISPLAY DESIGN ISSUES, by Lawrence E. Tannas, Jr.

This chapter elaborates on the primary issues of power efficiency, addressability, duty factor, gray scale, color, and cost, all of which affect the performance and ultimate utility of flat-panel displays, and presents a detailed discussion of intrinsic and extrinsic electronic addressing for both emissive and nonemissive displays. It also gives a review of the MOS, TFT, and Si electronic drivers.

I am indebted to Alan Sobel for his careful reading of this chapter and to Thomas Engibous and Andras I. Lakatos for authoring the electronic sections.

Chapter 6. THE CHALLENGE OF THE CATHODE RAY TUBE, by Norman H. Lehrer, MS, Consultant, Cupertino, California.

This chapter provides a complete tutorial review of the fundamentals of the CRT, covering history, basic design, the incorporation of color, electron optics, performance, phosphors, resolution and contrast, life, applications, and drive circuits. It includes an overall summary of CRT performance that can serve as the basis for a comparison of the CRT with flat-panel displays.

Mr. Lehrer would like to express his appreciation to the Watkins-Johnson Co., Inc., for their support in the initial preparation of this chapter, which took place while he was an employee. Grateful acknowledgement is also made for their approval of its publication as well as for the approval of Rank Electronic Tubes, which acquired the cathode-ray tube operation of the Watkins-Johnson Co. Special thanks are due Gus Carroll, who not only read the manuscript but provided several key suggestions for improvements.

Chapter 7. FLAT CATHODE-RAY TUBE DISPLAYS, by Walter F. Goede, MS, Northrop Electronics Division, Northrop, Hawthorne, California.

In this book, the position is taken that the flat-CRT technology is separate from that of the regular CRTs while acknowledging that both technologies have much in common. This chapter discusses the motivations and goals of the flat CRT and then continues with the history, basic designs, cathodes, beam control, luminance enhancement, phosphor screens, vacuum envelope, and examples.

Mr. Goede is indebted to Thomas Credelle, James Smith, and Tadashi Nakamura for the careful review and many valuable suggestions. He would like to thank Northrop Electronics for their support.

Chapter 8. ELECTROLUMINESCENT DISPLAYS, by Lawrence E. Tannas, Jr.

EL displays have had a long and tortuous development cycle and are now finding application. This chapter covers all aspects of the technology starting with a detailed history, theory of operation, performance, fabrication techniques, failure modes, color, and ends with applications for ac, dc, powder, and thin-film EL.

The chapter benefited greatly from the detailed review and comments by Paul Alt and Elliott Schlam, to whom I am indebted. I also wish to thank Aerojet ElectroSystems, my former employer, for releasing the Aerojet data for this endeavor and completing much of the artwork, as well as my former staff at Aerojet who performed the work herein accredited to Aerojet.

Chapter 9. LIGHT-EMITTING DIODE DISPLAYS, by M. George Craford, Ph.D., Optoelectronic Division, Hewlett-Packard, Palo Alto, California.

This chapter is devoted to the single crystal source of electroluminescence, called "LEDs" in the displays industry. All aspects of LED displays are covered, including history, basic technology, display devices, performance, materials, and processes.

Dr. Craford would like to thank Nick Holonyak, Jr., for his careful review of the entire chapter and H. T. Groves, Walter Melnick, Neil A. Obright, Keith Burnette, Roland Haitz, and the Optoelectronic Applications staff at Hewlett-Packard for providing many of the figures.

Chapter 10. PLASMA DISPLAYS, by Larry F. Weber, Ph.D., Computer-based Education Research Laboratory, University of Illinois, Urbana, Illinois.

Plasma displays use the gas discharge light-generation phenomenon applied to a wide range of products from small seven-segment indicators to very large graphic displays. This chapter starts with a thorough treatment of the gas-discharge phenomenon and the history of plasma displays. It then covers the structures, characteristics, fabrication methods and electronic addressing of the full range of ac, dc, and hybrid devices available on the market, along with techniques for achieving gray scale, color, and television.

Dr. Weber is grateful for the support of the Computer-based Education Research Lab of the University of Illinois. The careful review and helpful comments of Thomas C. Maloney are warmly appreciated.

Chapter 11. NONEMISSIVE DISPLAYS by P. Andrew Penz, Ph.D., Central Research Laboratory, Texas Instruments, Dallas, Texas.

The predominant nonemissive display medium is liquid crystals—the primary focus of this chapter. Nevertheless, the nonemissive technologies such as electrochromic, electrophoretic, colloidal, electroactive, and electromechanical are also discussed. The chapter begins with the optical theory of nonemissive displays and continues with history, definitions, and acronyms, as well as all phases of LC displays including construction, alignment, addressing, etc.

Dr. Penz appreciates the helpful suggestions of Allan Kmetz and the careful reading of Milo R. Johnson.

L. E. TANNAS, JR.

CONTENTS

FLAT-PANEL DISPLAYS AND CRTs

1

INTRODUCTION

L. E. TANNAS, JR., *Consultant*

This book surveys the state of the art of flat-panel displays, with discussions on the cathode-ray tube (CRT) as a counterbalance. All of the flat-panel display technologies are at the forefront of science, and as a consequence, engineering advances in some cases have gotten ahead of scientific understanding, slowing further progress until the science catches up. Product demand has at times been immense, forcing advances in all the technologies.

In spite of many difficulties, flat-panel display technologies are now sufficiently mature to support broad product lines, i.e., cathodoluminescence (flat CRTs and vacuum fluorescence), electroluminescence (polycrystalline electroluminescence and light-emitting diodes), gas discharge (plasma panels), and nonemitters (liquid crystals). In the following chapters, each of these technologies is discussed in detail. An historical introduction in each chapter gives insight into the slow evolutionary nature of their development.

All displays have many features in common, such as the overall system requirements, the anatomical requirements of the user, and image quality characterization. Flat panels have shared characteristics because of their flatness. The initial chapters cover these subjects, leading into the detailed technology chapters.

Chapter 6 is devoted to the cathode-ray tube. This chapter serves two purposes:

- It is a tutorial review of the CRT, and
- It serves as a baseline describing what CRTs can display to counterbalance and give perspective to what flat panels can display.

The element of competition between flat-panel engineers and CRT engineers is strong and spir-

ited. Chapter 6 is intended to counterbalance any undue bias that might exist on the part of flat-panel enthusiasts.

1.1 HISTORY OF ELECTRONICS FOR DISPLAYS

Attempts to make flat-panel displays began before most of our professional careers did. A chronology of the most important technical achievements that have affected displays in general are listed below. The immensity of the effort to create light, electronic control, and information imagery forms a background to the flat-panel display evolution.

- Phosphor, a luminescent solid, made by Casciorolo of Bologna, Italy, about 1603.
- Gas discharge light produced in evacuated vessel by agitated mercury, Jean Picard, Paris, France, 1675.
- John William Hittorf first to discover that some type of cathode emission striking the surface of the glass wall of a glow tube caused it to luminesce, Munster, Germany, 1869.
- Glow tube, using gas discharge (Geissler Tube), Henrich Geissler, Germany, 1854.
- Crookes Tube (observable cathode-ray), Sir William Crookes, London, 1875.
- Incandescent filament lamp, Thomas Edison, Orange, N.J., 1879.
- Mercury-arc lamp, L. Arons, Germany, 1892.
- Braun Tube with deflection and fluorescent screen, Karl Ferdinand Braun, Strasbourg, Germany, 1897.
- The electron identified by Sir Joseph John Thomson, Cambridge, England, 1897.

- Photon quantum theory of light, Albert Einstein, Berne, Switzerland, 1905.
- Triode vacuum tube (audion tube) by Lee DeForest, United States, 1906.
- Luminescence from silicon carbide crystals, Capt. H. J. Round, England, 1907.
- Television concept, as known today, A. C. Campbell Swinton, London, 1908 and 1911.
- The Bohr atom model, Niels Bohr, Copenhagen, 1913.
- Neon gas-discharge lamp, D. M. Moore, Schenectady, N.Y., 1917.
- Luminescence from carborundum (Losev's Glow), O. V. Losev, Russia, 1923.
- First demonstration of true TV, J. L. Baird, England, 1926, and color in 1928.
- Sodium-vapor lamp, Europe, 1931.
- Fluorescent lamp, United States, 1934.
- Electroluminescence in polycrystalline phosphor, Georges Destriau, France, 1936.
- Grain of wheat incandescent lamp, United States, 1936.
- Transistor, John Bardeen, Walter H. Brattain and William B. Shockley at Bell Telephone Laboratories, New Jersey, 1947.
- First bistable display storage tube, A. V. Haeff, Naval Research Laboratories, Washington, D.C., 1947.
- First electroluminescent lamp (Panelescent lighting), GTE Sylvania, Salem, Mass., 1948.
- Shadow-mask color television CRT, A. N. Goldsmith and A. C. Schroeder of RCA, New Jersey, 1950.
- First model (single band) of a flat CRT (Thintube), William Ross Aiken, United States, 1951.
- First matrix display patent using electroluminescence, W. W. Piper, United States, filed 1953.
- First patent on flat CRT, Dennis Gabor, Imperial College, London, filed 1953.
- Patent on flat CRT (Kaiser-Aiken Tube), William Ross Aiken, California, filed 1953.
- First color CRT with curved mask and phosphor screen deposited on internal surface of faceplate, Norman Fyler, CBS Hytron, New Jersey, 1953.
- Gas-discharge numeric display (Nixie Tube), Saul Kuchinsky and colleagues at Burroughs, New Jersey, early 1950s.
- Visible injection laser from GaAsP, N. Holonyak and Bevaqua, United States, 1962.
- Flat CRT with area cathode and grid (Digisplay), Jeffries and Hultberg at Northrop, Palos Verdes, Calif., 1962.
- AC thin-film electroluminescent matrix-addressable display using ZnS:Mn phosphor, Edwin J. Soxman, Servomechanisms, Santa Barbara, Calif., 1962.
- AC gas-discharge matrix addressing (plasma panel), Donald L. Bitzer, H. Gene Slottow, and Bob Wilson, University of Illinois, 1964.
- First rare-earth phosphors, Kenneth A. Wickersheim and Robert A. Lefever, GTE-Sylvania Laboratories, Palo Alto, California, 1964.
- Thin-film electroluminescent display with black back layer, E. J. Soxman and G. Steel, Sigmatron, Santa Barbara, Calif., ca. 1965.
- Vacuum-fluorescent display, Tadashi Nakamura, ISE, Japan, 1966.
- Commercially viable ac plasma display (Digivue)., Owens-Illinois, Toledo, Ohio, 1968.
- Commercially viable red LED display, Monsanto and Hewlett-Packard, United States, 1968.
- Liquid-crystal dynamic scattering display, G. H. Heilmeier, L. A. Zanoni, and L. A. Barton, RCA Sarnoff Laboratory, Princeton, N.J., published 1968.
- High performance orange, yellow, and green LED displays by Monsanto, United States, 1970.
- Gas-discharge matrix display (Self-Scan), George Holz and Jim Ogle at Burroughs, New Jersey, 1970.
- First commercial production of a liquid-crystal display (prototype), L. E. Tannas, Jr., A. G. Genovese, and E. T. Fitzgibbons, Rockwell, Anaheim, California, 1970.
- AC plasma shift panel, S. Umedo and T. Hirose, Fujitsu, Japan, 1972.
- Liquid-crystal twisted nematic display, J. L. Fergason, Kent Ohio, and Schott and

Helfric of Hoffman La Roche, Switzerland, 1972.

- Invention of biphenyl liquid crystal material, G. Gray and co-workers, University Hull, England, 1973.
- Thin-film electroluminescence with high brightness, long life, and matrix addressing, Sanai Mito, T. Inoguchi, and colleagues at Sharp Corporation, Japan, published 1974.

It should be clear that all the easy things have been invented. Too many highly motivated and capable people have sifted through the complete composite of human knowledge in search of a better display for anything obvious to have been overlooked. For example, the gas discharge phenomenon has been studied since the Geissler Tube of 1854, and the cathodoluminescence phenomenon since the Crookes Tube of 1879. Further accomplishments will require high technical skills, a sound understanding of the problem, diligence, ingenuity, and further advancements in science.

Invention is not harmoniously connected to science and engineering. Thomas Edison and his staff invented the incandescent lamp before the electron was discovered. A working television was not accomplished until eighteen years after the concept was expounded. The liquid crystallinity phase of materials was observed in 1888, and yet its application to displays was not discovered until around 1963 at RCA Sarnoff Labs (published in 1968). The contemporary flat-panel display challenge originated in the early 1950s with the invention by GTE Sylvania of the AC powder electroluminescent numeric display, the Aiken Thintube,[1] the Gabor Tube, and the Piper electroluminescent matrix addressing patent.

The endurance of the CRT seems to be neverending. In the 1950s it was felt that the new display medium "electroluminescence" (EL) was going to replace CRTs. In the mid 1960s, when matrix addressing was invented for gas discharge (plasma panels,) this technology "definitely was going to replace CRTs." Liquid-crystal and light-emitting diode displays were also a challenge. Still, the CRT lives on.

It is not so much that the new media lack the potential but that CRTs keep getting better and less expensive. Needless to say, the development of flat-panel displays has not been easy. Someone once said that, "If we knew how hard it was going to be in the beginning, we would not have started." Plasma panels are finding a place of their own as large panels and large alphanumeric displays. Liquid crystals are ideal where very little power is available and sunlight readability is required. LEDs are doing extremely well where low cost, small size, and MOS capability are important.

Cathode-ray tube structures were investigated by William Crookes in 1879 and improved upon and first used as a display by Ferdinand Braun in 1897. The CRT is the modern counterpart of the Braun tube and as such will be 100 years old in 1997. The laurels are long. The electron made its presence known by the visible effect coming from the Crookes tube. X-rays were discovered from these electrons. J. J. Thomson first measured the charge-to-mass ratio of the electron using a CRT. The German blockade of Europe in World War II was broken because of the oscilloscope display of radar signal returns. The first digital computers used storage CRTs as memory banks. Nearly all very high-speed measurements in atomic physics were and are made with CRTs. The electronic industry would be helpless without the oscilloscope. And then, of course, there is television.

The CRT is so entrenched in the industrial and commercial world that it may never be replaced. This brash-sounding statement is meant to emphasize that its replacement is not easily predictable. The cost of the CRT display is a key factor guaranteeing its longevity. The trend toward use of more and more high-resolution color displays is another.

Replacement of the CRT, however, is no longer the question. In the 1950s, flat-panel enthusiasts predicted that CRTs would be "out-technologed." In the 1960s, these same enthusiasts predicted that CRTs would be replaced but it was going to take a little longer. The CRT was not replaced by the 1970s, and the credibility of flat-panel enthusiasts was greatly eroded. By the 1980s, it was clear that the CRT may never be replaced, but this fact does not bother flat-panel enthusiasts any more;

they have found applications of their own. Now the question is, "What new products are possible with the flat-panel technologies?" This book is intended to lay the foundation and show the way.

1.2   ELECTRONIC DISPLAYS

1.2.1 Introduction. A display is an electronic component or subsystem used to convert electrical signals into visual imagery in real time suitable for direct interpretation by a human observer. A display is a unique electo-optical device. It must be scaled to human visual and anatomical requirements and yet be as lightweight with as low a profile and using as little power as possible. It serves as the visual interface between user and machine. The visual imagery is pre-processed, composed, and optimized for easy interpretation and minimum reading error. The electronic display is dynamic in that it presents information within a fraction of a second from the time received and continuously holds that information, using refresh or memory techniques, until new information is received. The image is created by electronically making visible contrast patterns.

The use of electronic displays for presentation of graphs, symbols, alphanumerics, and video pictures is doubling every few years. The biggest growth rate in displays is for utilitarian and industrial users. Electronic displays are replacing traditional mechanical and hard-copy (paper) means for presenting information. This change

is due to the increased use of computers, microprocessors, low-cost large-scale integrated (LSI) electronics, modems, and digital mass memories. The success of the electronic calculator is directly attributable to the availability of low-cost LSI electronics and low-cost electronic numeric displays.

Electronic transducers and four-digit (or more) flat-panel displays are replacing the galvanometer movement, thermometer scale, barometer movement, and other classical forms of scientific instrumentation. There are now large signs, arrival and departure announcements, and scoreboards with video imagery using solely electronic means to present changing messages and data.

The computer terminal using a CRT or flat panels is one of the most important industrial applications of electronic displays. The standard computer terminal displays 25 lines of 80 characters, for a total of 2,000 characters. The personal computer with a microprocessor and mass memory is making inroads on the office paper, typewriter, and file cabinet.

The primary information display, aside from small alphanumerics, is the CRT. The primary applications of the CRT are in home entertainment television, scientific and electrical engineering oscilloscopes, radar displays, alphanumeric and graphic display terminals, computer-aided design terminals, industrial machinery displays, and aircraft displays.

The basic elements of the CRT are shown in Fig. 1-1. The viewing screen is coated with a

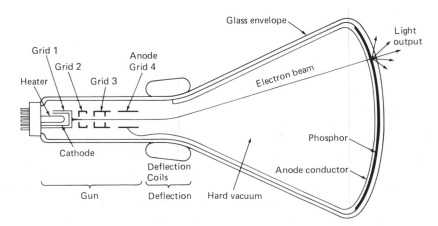

Fig. 1-1. Cathode-ray tube using electrostatic focus and magnetic deflection.

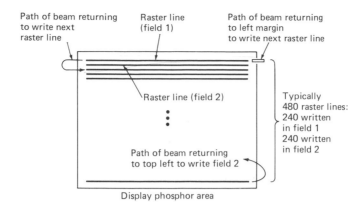

Fig. 1-2. Cathode-ray tube raster for television, using interlace.

phosphor which emits light when struck with a beam of high-energy electrons. The electrons are emitted from the cathode at the rear of the tube in a beam that is focused electrostatically (or magnetically) to a dot or spot on the phosphor screen and positioned in horizontal and vertical coordinates by magnetic (or electrostatic) deflection. The cathode grid and electron-focusing lenses are incorporated into a subassembly called the gun. The beam is accelerated toward the phosphor by a high voltage (20 kV or more) at the anode.

The CRT raster, or image field, is shown in Fig. 1-2. Imagery is created on the raster as it is traced out. The video signal is high-frequency. After amplification, it is applied directly to the cathode gun, and controls the magnitude of the electron beam current and thus the luminous output as it strikes the phosphor on the display surface. The deflection coils steer the beam to trace out the raster. The horizontal-scan deflection signal causes the beam to trace out a horizontal line and then fly back for the next line. The vertical-scan deflection signal causes the beam to be stepped down the raster and then retraced to the top left corner at the beginning of each field. Two fields are typically interlaced with the raster lines of one field traced between those of the other to create a complete picture or frame. This minimizes flicker in the picture.

Because of the depth dimension of the CRT, there has been a concentrated effort to invent a flat-panel display. By the mid 1980s, the electrical phenomena most extensively developed for flat-panel displays are gas discharge (plasma), electroluminescence, light-emitting diodes, cath-

odoluminescence (flat CRTs and vacuum fluorescence), and liquid crystallinity. Flat-panel displays are more expensive than CRTs on a per-character or per-picture element basis.

Flat-panel displays are typically matrix-addressed. A row is enabled to accept display information in parallel via the column lines. The electronics commutate through the rows, serving the same purpose as the vertical deflection amplifier of the CRT. The column data are shifted into the column drivers, and then at the proper time applied to the column lines in parallel.

A typical flat-panel display is shown in Fig. 1-3. The thickness is approximately one-half inch. The row and column lines are spaced typically at 64 lines to the inch. The intersection of each row line with each column line defines a picture element (pixel), the smallest addressable element in an electronic display. The rows and columns in a flat-panel display are analogous to the raster in a CRT.

1.2.2 Display Categories. Electronic displays can be categorized into four classifications, as shown in Table 1-1. Each classification is defined by natural technical boundaries, information density, and cost plateaus. The categorization is useful in visualizing the extent to which electronic displays are used and the breadth of information content.

1.2.3 Display Technique. The essence of electronic displays is their ability to turn on and off individual pixels as shown in the generation of alphanumeric characters in Fig. 1-4. A

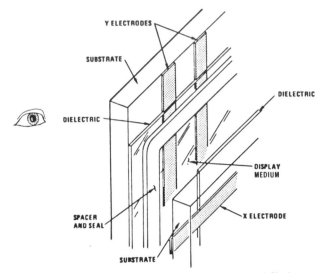

Fig. 1-3. Exploded section of a plasma flat-panel display.

typical high-information-content display will have a quarter-million pixels in an orthogonal array, each under individual control by the electronics. The pixel-to-pixel resolution is normally just at or below the resolving power of the eye at the nominal reading distance. Thus, a good-quality picture can be created from a pattern of activated pixels.

The pixel concept for electronic displays has evolved from the modern flat-panel display tech-

Table 1-1   Direct-View Electronic Display Spectrum of Applications

Classification	Characteristics	Applications	Electronic technologies
Status and Pseudoanalog	Dedicated arrangement of discrete pixels used to present analog or qualitative information	Meterlike presentations, go/no-go messages, legends and alerts, analog-like (watch) dial, on/off, stop/caution/go.	Gas discharge, light-emitting diodes, liquid crystal, incandescent lamps
Alphanumeric	Dedicated alphanumeric pixel font of normally less than 480 characters; most common are 4- and 8-character numeric displays	Digital watches, calculators, digital multimeters, message terminals, games, arrivals and departures	Liquid crystal, light-emitting diodes, vacuum fluorescence, gas discharge, electroluminescence, incandescence
Vectorgraphic	Large orthogonal uniform array of pixels which are addressable at medium to high speeds; normally monochromatic with no gray scale; over 480-character capability and simple graphics	Computer terminals, teletype terminals, scheduling terminals, weather radar, air-traffic control, games, industrial and utilitarian displays	Cathode-ray tube, plasma panels, gas discharge, vacuum fluorescence, electroluminescence, flat CRTs
Video	Large orthogonal array of pixels which are addressable at video rates (30 frames/s); monochromatic with gray scale or full color	Entertainment television, medical electronics, aircraft flight instruments, computer terminals, games, computer-aided design	Cathode-ray tube, other technologies in advanced development

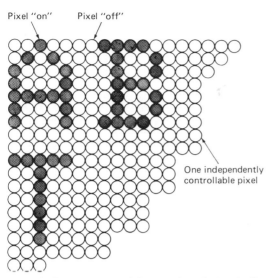

Pixel "on"   Pixel "off"

One independently
controllable pixel

Fig. 1-4. Pixel array used for creating electronic display images.

nologies and digital electronics. It has been extended to the analog-raster-scan CRT in the following way: The electron beam from the gun is deflected magnetically (or electrostatically) so as to sweep across the phosphor and thereby cause a line of modulated luminescence on the face of the CRT. In digitally modulated CRTs, the cathode is modulated by a square wave as the beam is swept across the face of the CRT. Here, instead of a continuously modulated line, a string of dots results. Each dot corresponds to a pixel. Each pixel is on when the beam density is high, and off when the beam is off. The beam is turned off between pixels to make the horizontal and vertical pixel spacing uniform. The pixels are refreshed approximately 30 times per second on a CRT and typically 60 or more times per second on a flat panel.

In a "CRT-digital raster," there are uniformly spaced pixels in all the rows of the CRT raster field. This approach is commonly used in industrial applications and computer terminals, since it is easily interfaced with digital electronics. Home entertainment television uses an analog-raster-scan approach, as do nearly all other video systems. In an analog raster the information content along a row is measured by its modulation transfer function (MTF) rather than by its pixel count.

There are some applications in which a non-raster approach is used to create alphanumeric characters and vectors on CRTs. The electron beam is deflected under control of the deflection amplifiers to stroke out each line of the image. When characters and vectors are generated this way, they are called Lissajous figures, as opposed to (digital or analog) raster characters and vectors. The Lissajous characters and vectors are best suited to large (25-in. or 63.5-cm diagonal) CRTs where there are numerous vectors, straight lines, and curves and not sufficient time to stroke out the raster. Vectors and curves drawn with the raster technique have stair steps when they are at an angle with the raster lines. Lissajous vectors and curves are always smooth and continuous. Lissajous techniques are yielding to the digital raster in newer designs as the cost of digital electronics wanes. There is no analogous Lissajous figure in flat-panel displays.

A hybrid CRT technique combining analog raster and Lissajous vectors is used where the highest of image quality and flexibility is desired. The raster is used for filled-in areas and continuous-tone images. The Lissajous scan is used to write high-quality vectors and alphanumeric characters. The two are usually combined by using the Lissajous mode during the raster field flyback period. The cost penalty for such a configuration is the result of the larger deflection coils required and the additional power needed to drive them.

1.2.4 Font. With flat-panel displays and CRT digital-raster displays, alphanumeric character fonts are created by turning on the appropriate pixels in an array. One standard size is a 5 by 7 array with one or two pixels between characters and two or three pixels between rows, as shown in Fig. 1-4. All of the ASCII (American Standard Code for Information Interchange) letters, numbers, and symbols can be created on this common format array. The 5 by 7 array is considered a minimum for the ASCII letters. Several different combinations of pixels are used to create a letter A. The viewer soon becomes accustomed to the minor variations that are in use. Readers do not read pixels but read letters and words, and therefore the exact detail of the character pixel pattern is of secondary consideration. In general, the more pixels available in the basic array, the more aesthetically pleasing is the character, at the cost of additional electronics to control the extra pixels. A 10 by 10

| 7-bar numeric | 10-bar alphanumeric "double hung window" | 14-bar alphanumeric | 16-bar alphanumeric | 8-bar numeric stylized by LEGI™ | ©Tannas 1977 8-bar numeric "fail-safe font" | ©Tannas 1977 13-bar alphanumeric "fail-safe font" |

Fig. 1-5. Fixed-format alphanumeric matrices.

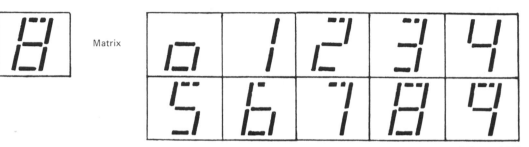

©Tannas 1977

(a)   Fail–Safe Numeric Font Using Eight-Bar Matrix

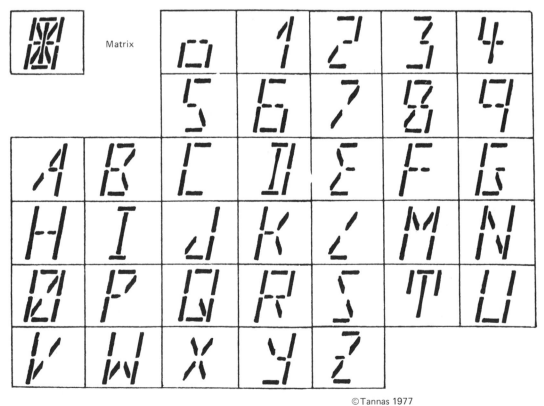

©Tannas 1977

(b)   Fail–Safe Alphanumeric Font Using Thirteen-Bar Matrix

Fig. 1-6. Fail-safe fonts using minimal matrices.

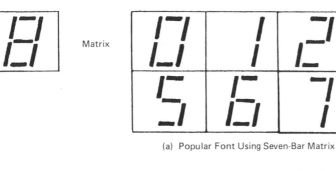

Matrix

(a) Popular Font Using Seven-Bar Matrix

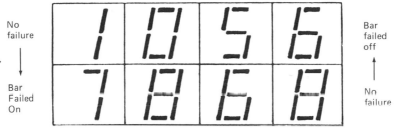

No failure

Bar Failed On

Bar failed off

No failure

(b) Examples of Failures

Fig. 1-7. Seven-bar numeric font with examples of undesirable failure modes.

array is more typically used on CRTs including spaces.

A very efficient and elegant array has evolved for portraying numeric characters only, called the 7-bar matrix. An example of the 7-bar matrix along with other fixed-format matrices are shown in Fig. 1-5. Each bar is a pixel by definition.

The fonts created from these matrices were initially considered crude when compared to the Leroy and Lincoln-Mitre fonts and printer fonts with serifs. They are now universally accepted and appreciated for their simplicity. The 10-bar, 13-bar, 14-bar, and 16-bar matrices are used for alphanumeric characters. There is no analogous form of numeric or alphanumeric character font used on a CRT.

Of particular note are the fail-safe matrices of Fig. 1-5. These fonts, as shown in Fig. 1-6, were developed by this author[2] utilizing a minimum number of bars to avoid the problem of displaying an erroneous character in the event of a failure of any one bar. For example, the 7-bar font can display erroneous characters without providing any clue if one bar should fail on or off, as shown in Fig. 1-7. This can be very critical in aircraft and medical instruments because the display appears to be functioning properly. The 8-bar and 13-bar fonts were coded so as never to display a legitimate char-

acter if one bar fails on or off. This was accomplished simply by making every legitimate character two or more bars different from every other legitimate character. The legitimate character is by definition the intended coded character. The number of bars used in a matrix directly impacts upon the cost of the display. The 8-bar and 13-bar matrices were considered unique in that smaller matrices could not be found to achieve the fail-safe criteria for numerics and alphanumerics, respectively.

The concept of coding the font to make it fail-safe can be carried further by making the font fail-redundant. Fail-redundant means that if one bar fails on or off, and if the user has learned the coding of the legitimate characters, he can deduce the intended character. This is made possible by having each character in the font set at least three bars different from all other characters. Such a font using a minimum size matrix is shown in Fig. 1-8. Now, if one bar fails in a character, the result is only one bar different from the intended character and at least two bars different from all the other legitimate characters in the set. Because of our cultural experience with alphanumeric characters, it is easy to deduce the correct character after only an hour of practice with the alphanumeric character set and even less time with the numeric set.

Matrix

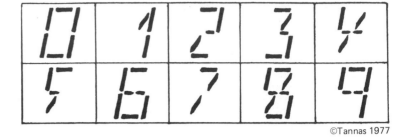

©Tannas 1977

(a) Fail–Redundant Numeric Font Using Ten-Bar Matrix

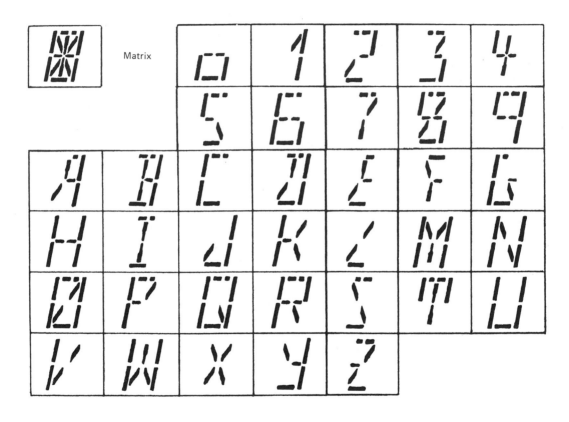

(b) Fail–Redundant Alphanumeric Font Using Fifteen-Bar Matrix

Fig. 1-8. Fail-redundant fonts using minimal matrices.

1.2.5 The CRT Challenge. The invention of flat-panel video display terminals has been a challenge to engineers and scientists since the beginning of the twentieth century. When J. L. Baird invented the first working television in England in 1926, he would have made it with a flat panel if the technology had existed. Instead, he used a CRT which did exist.

A sustained effort since 1950 has been directed at finding a replacement for the CRT. At times it has turned into a "beat-the-CRT" contest. If flat-panel technologists knew how difficult it was going to be then, they would not have started. It is very difficult to get a company or sponsor to fund a program which may take twenty years to replace an existing functional electronic component.

That clumsy old workhorse, the CRT, has turned into a clumsy new workhorse with continuously improving performance at a lower and lower price. As we began the 1980s, the CRT was the only commercially available device that could display full TV video in black and white or color. It is one of just three or four vacuum tubes still widely used in the electronics industry. However, since the first demonstration of true TV, the CRT has undergone over fifty years of development and is still improving. Since the introduction of the shadow-mask color tube in 1950 by RCA, CRTs have increased in brightness by approximately 17% per year while simultaneously increasing in luminous efficiency from 0.3 to 8 lumens/watt. The CRT is also an extremely flexible and versatile device. It is available in a variety of sizes with a wide selection of phosphors. Shadow-mask color CRTs are readily available with 1,200 lines of resolution.

Except for its volume, the CRT is at present superior in all respects to flat-panel displays. Here are a few of the CRT virtues for displays: immunity to high ambient illumination with filtering; low-level red luminance for deep scotopic vision; extremely fast response for scientific measurements; high resolution for intelligence, earth resources, and medical observations; vivid color for entertainment; and small size (one-inch diagonal) and large size (25-inch diagonal) for varied applications. The breadth of application of the CRT is truly amazing. CRTs are now used extensively in high-performance aircraft as the pilot's primary flight instrument. They can be turned down low enough to preserve scotopic vision or up high enough to show color and shades of gray under 10,000 fc of ambient illumination.

By far the most important parameter is cost. The CRT is continually getting cheaper. The cost factor presents the most difficult of all challenges for flat-panel technologies.

With all these laurels, the CRT has one cumbersome feature: it is too deep. The depth dimension is larger than the display screen diagonal dimension. This form factor bars the CRT from many new product applications, such as briefcase portable terminals, wall-mounted displays, and man-packs (hand-held portable terminals). The depth often necessitates structural changes when CRTs are installed in aircraft and vehicles.

The depth is needed in CRTs to focus and scan the image. The simplicity of the CRT focus and scan is at the root of all its virtues

and also its weaknesses. The only insurmountable problem with CRTs is the depth dimension. This factor does justify continued research and development of flat-panel technologies.

Because the CRT is so cumbersome, flat panels are now being used for displaying low-data formats, even though they are more expensive. The penetration of flat panels into the CRT market is complete in the alphanumeric category for up to several hundred characters where gray scale and color are not needed. Several technologies can perform in the vectorgraphic category where high resolution is not needed but low volume is. Flat panels will next grow in complexity into the video category as their technologies improve.

1.2.6 Definition of Flat-Panel Displays. The term "flat-panel display" is more of a concept than a specific entity. It is a display which is flat and light, and does not require a great deal of power. It is not a CRT. Flat is used to mean thin in form, flat as a pancake. It is a foregone conclusion that the viewing surface will also be flat although a slight curvature is acceptable.

A flat-panel display is often defined in terms of the ideal display, that being: thin form, low volume, even surface, having high resolution, high contrast, sunlight readability, color, low power, and being solid-state and lightweight. This is easy to conceive but difficult to deliver. Such a definition is not satisfying because it borders on the science-fiction; but it does serve as a goal, and researchers are pursuing it.

The one common denominator for all non-CRT direct-view displays is flatness (thin form). If flatness is not of primary interest, then the CRT is the obvious display medium of choice (except for the status, pseudoanalog, and alphanumeric classifications and very large direct view displays). Without exception, the objections to the CRT other than flatness can be made tolerable within the one-inch to 25-inch size without a significant increase in cost or reduction in performance.

But is flatness the only virtue? Flatness only has the value of economy of space. How about the other virtues of efficiency of power in an age of power consciousness? How about X-ray radiation, high-voltage, and implosion hazards in a consumer-protection era? How about the cost when over three quarters of the world cannot afford the black-and-white video picture experience?

Flatness is a rallying point in the search for new and better displays. The entire direct-view displays community can be divided into two camps: those dedicated to the promotion and advancement of the CRT display and those dedicated to the discovery of the next-generation display which axiomatically must be flat.

Typically, a CRT has a flatness ratio (depth-to-picture diagonal) of 1:1. It is this author's opinion that a flat-panel display must have a flatness ratio of less than 1:4 (including packaging) so as to render the depth dimension inconsequential. CRT tubes can be made with a flatness ratio of 1:2 with discernible picture aberrations but without any compromise in utility.

1.2.7 Flat CRTs.   The CRT display is defined here as shown in Fig. 1-1: a vacuum envelope with a cathode gun at one end as a source of the electron beam, a phosphor screen at the other end, and in the middle a means for beam focus and deflection. The serial sequence of (1) cathode gun, (2) focus, (3) deflection, and then (4) phosphor screen is what contributes to the depth of the CRT. Numerous attempts have been made to minimize this depth, such as by bending the cathode gun 90° to the screen and using deflection means to deflect the electron beam back to the screen. The Aiken Thintube and the Gabor Tube, patented in the early 1950s, attempted to do this. Since the objective is to make it flat, it is treated here as a flat panel and is in the family of flat CRTs. The Aiken and Gabor Tubes have wide borders in place of depth.

In making flat-CRT displays, a degree of success has been realized by completely replacing the cathode-gun, focus, and deflection mechanisms of the conventional CRT with an area cathode and intermediate electrostatic grids for controlling electron flow to the phosphor screen. This was accomplished by Northrop in the Digisplay® display in black and white, and advanced further by Texas Instruments with the addition of full color.

A high degree of success has been achieved with the vacuum fluorescent display developed by ISE of Japan. This display uses the cathodoluminescent phenomenon in a manner similar to the Digisplay®, and also belongs to the family of flat CRTs.

Flat CRTs are much more difficult to make than they would seem to be. They are so difficult that as of 1983 they are not the obvious alternative to the CRT in achieving flatness. This is despite more than thirty years of large dedicated technical efforts at several major corporations. The flat CRT is considered in this book to be a technology of its own, competing with other flat-panel technologies to become the display of the future. RCA, Sinclair, and Sony, among others, have continuing flat-CRT programs. Chapter 7 is devoted to flat-CRT displays.

A relatively new type of flat CRT has been under development for several years at Siemens AG, Munchen, Germany and Lucitron, Northbrook, Illinois. The novelty is the use of a low pressure gas discharge as the source of electrons which are accelerated by a high anode voltage towards the front phosphor coated faceplate. The process called gas-electron phosphor (GEP) display uses the priming properties of the gas for selective addressing and the high luminous efficiency of phosphors for the light output. Because of the structure, large internally-supported panels are possible.

1.3   DISPLAY CLASSIFICATIONS

Before flat-panel displays and CRTs can be intelligently discussed, a classification must be made of the entire spectrum of electronic information displays. There are many ways of classifying displays—all of them somewhat arbitrary. One useful classification is shown in Fig. 1-9. The entire field can be divided into three categories:

- Projection
- Off-screen
- Direct-view

The different categories in Fig. 1-9 follow the most commonly used phraseology in the

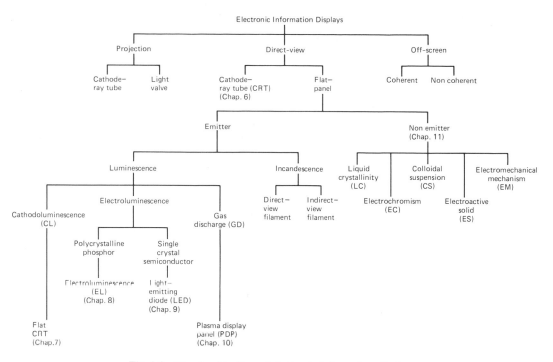

Fig. 1-9. The classification of electronic information displays.

industry. Where discrepancies exist, the names which best describe physical effects are used first, followed at a lower level diagramatically by the term used by the display community. A further categorization of each of the flat-panel display technologies is given in the respective chapters as noted.

The projection display classification encompasses all those displays which utilize a viewing screen separate from the optical image source. As the name implies, the image is projected with appropriate optics onto a screen which serves as the diffusing surface with appropriate screen gain and directionality. The image must be diffused, otherwise it can only be viewed from the specular reflection angle. The diffusing surface is the apparent image source. The viewer cannot see the originating image source. The projection can be either from the front or the rear of the screen. Projection displays can be found in home entertainment centers, management conference rooms, command and control centers, theaters, and sports arenas. The diagonal size can range from approximately three feet (one meter) using one CRT, to 120 feet (40 meters) using three Eidophor projectors.

The off-screen image display category is a minor but important branch of electronic information displays. It encompasses all of those displays where the image is not viewed on a screen. The real or virtual images are perceived to exist in space. Holography falls into this category and is of interest as a means for creating true three-dimensional images. Images focused at infinity in the line of sight of the viewer also fall into this category. Examples are "head-up" and "helmet-mounted" displays used in helicopters and aircraft.

The direct-view classification encompasses all those displays where the image is generated in the immediate proximity of the viewing surface. This is by far the largest category of displays. It is divided into two subcategories: cathode-ray tube displays and flat-panel displays. The size in diameter can range from less than an inch (2.5 centimeters) to 3 feet (1 meter) or more in single panels and 50 feet (15 meters) or more in the diagonal dimension where individual panels are mosaicked together.

By far the predominant display vehicle image source in all these categories is the CRT or kinescope as it is sometimes called. Even when entertainment television is not included, CRT installations outnumber by orders of magnitude the rest of the display devices combined, excluding small numeric displays such as those used in watches, calculators, games, and digital readouts.

1.3.1 Flat-Panel Display Classifications. The flat-panel display subcategory of direct-view information displays is the primary subject of this book. A one-sentence description of the major flat-panel technologies is given in Table 1-2. The relationship between flat panels and CRTs is important, and therefore CRTs and their subcategories are discussed separately in Chap. 6.

Flat-panel displays are logically separated into two groups:

- Emitters
- Nonemitters

Emitters are those displays in which electrical energy is converted to luminous energy as a function of the image signal. Since these are direct-view displays, the final conversion is done at the display-viewing surface. Emissive displays fall into two subgroups—luminescent and incandescent.

Other sources of light could be used in the emissive category of displays; for example, photoluminescence, chemiluminescence, bioluminescence, triboluminescence, and combustion. No realistic model of practical significance has ever been postulated using these forms of luminous energy in a modern display. The problems most often cited with these forms of emissive displays are addressability, reversability, switching time, and luminous efficiency. For example, the physical phenomenon of photoluminescence is not used in displays except as an energy converter from UV to visible because it lacks addressability. It is not inconceivable that these and other emissive techniques would prove feasible in the future.

Incandescent lamps have long been used in flat-panel, direct-view displays. They are primarily used for alphanumeric displays and large outdoor signboard displays where, for example, a 25-watt lamp may be used for each pixel. Incandescence is often used where power is not a major consideration and where sunlight readability is a primary requirement. The two subcategories, direct-view filament and indirect-view filament, are of minor distinction. They

Table 1-2 Flat-Panel Technologies

Technology	Phenomena	Typical Pixel Dwell Time*
Emissive displays		
Gas discharge	Cathode glow from conducting gaseous discharge	15 μsec
Plasma panel	AC capacitively coupled gas discharge	15 μsec
Light-emitting diode	Electron injection in a forward-biased p-n semiconductor junction	10 nsec
Vacuum fluorescence	Electron bombardment of phosphor in hard vacuum under control of a grid	50 nsec
Electroluminescence	Electron conduction in polycrystalline phosphors due to high electric field	20 μsec
Flat cathode-ray tube	Electron bombardment of phosphor in hard vacuum under control of a grid or cathode	50 nsec
Nonemissive displays		
Liquid crystallinity	Electrostatic rotation of organic compounds which exhibit liquid crystallinity	100 msec
Electrochromism	Charging and discharging chemical systems (battery) which exhibit a color change in accordance with Faraday's Law	200 msec
Colloidal suspensions	Electrostatic transport or rotation of light-absorbing particles in a colloidal suspension	200 msec
Electroactive solids	Ferroelectric and ferromagnetic materials with a significant electro-optic effect	10 μsec
Electromechanical	Mechanical motion of elements causing a contrast change	200 msec

*Conventional CRT is 120 nsec.

refer to the way in which the viewer sees the light from the hot filament. In the direct-view filament category, several separate filaments are often used in a single glass envelope, each discernible by the viewer. In the indirect-view filament category, light from the filament is diffused at a surface, normally the glass envelope, or conducted by fiberoptics a short distance to the viewing surface to create a pattern.

The nonemitter category is a very important class of displays. The primary motivation for research and development on nonemissive techniques is sunlight readability at its very low power consumption. The nonemissive displays do not convert electrical energy to luminous energy as a function of the signal intelligence. They merely control ambient light by one or more electrically alterable optical effects such as diffusion, absorption, birefringence, reflection, or refraction. The resulting visual sensation is a brightness or color contrast.

Sometimes the ambient light is supplemented by a light source which is an integral part of the nonemissive display such as in back lighting or edge lighting. The important distinction is that the ambient and supplemental luminance is not switched on and off during the image-creating process. It is necessary that the ambient be supplemented to illuminate nonemissive displays in the dark. The supplemental light is normally from an incandescent filament or electroluminescent phosphors; however, in some cases a converter such as photoluminescent phosphors stimulated by UV or other radiation may be used.

Numerous phenomena are being researched for application to nonemissive displays. The most noted phenomena are liquid crystallinity, electrochromism, electrophoresis, and ferroelectric ceramics.

Liquid-crystal (LC) materials have been used in both the scattering and birefringent modes. Many organic compounds exhibit both liquid and solid properties in a physical phase over a relatively narrow temperature range between the solid and liquid phases. In this phase, they are liquid in that they can flow like water and they are solid in that they possess anisotropic optical and electrical properties like many solid crystals. The liquid-crystal compound is assembled in a thin film approximately one-half mil (12 μm) in thickness between parallel, transparent conductors. When the material is suitably aligned to the surfaces of its container and an electrical field is applied, there is a birefringent or scattering effect. Crossed polarizers are used to see the birefringent displays, and dopants are used to enhance the scattering in a dynamic-scattering display. The cells for each type of display are different in other minor ways. The display must be in the liquid-crystal phase temperature range to exhibit the desired effect. There is no sunlight-readable display material known which can switch images in milliseconds with less power.

Electrochromism (EC) is a change in light absorption as a consequence of a reversible electrochemical reaction. The electrochemical reaction is always directly controlled by electron reaction in accordance with Faraday's Law of electrolysis. Each display picture element can be thought of as a miniature battery. Electrochromic displays have the advantage of exhibiting a vivid color contrast between the on-pixel and off-pixel. Electrochromic displays typically have memory which means they do not need to be continuously refreshed as do LC displays. Power is consumed only when the display is switched. A large category of compounds is classified by chemists as being electrochromic. Electroplating and electrochemical processes using either electronic or ionic conduction are considered to be in this display category. Either the anode or cathode can be made to serve as the viewing screen.

Colloidal suspension covers the class of displays where the movement of charged particles in solution under the influence of an electric field is used to change the absorption or transmission of light. The particles are suspended in a low-viscosity liquid suspendant. The system is assembled in a display structure quite similar to but thicker than an LC cell. An electric field is applied across the cell to cause the particles to either rotate or translate in the suspendant. Colloidal suspension displays can be divided into two subclasses: those with a clear suspendant (sometimes called suspended crystals) and those with an opaque suspendant with electrophoretic particles.

Long dipole particles are used with the clear suspendant. When an ac electric field is applied, the dipoles rotate to align with the field and the system becomes transparent as colloidal dipoles are made with a high aspect ratio and do not absorb or reflect much light in the lowest area profile. Without the electric field, Brownian motion causes the dipoles to be randomly oriented and the system becomes opaque. The dipoles do not translate as the applied field is ac and at a sufficiently high frequency to negate dc effects. The charge mobility in the suspended dipoles is orders of magnitude higher than the dipole mobility in the suspendant.

With an opaque suspendant, particles of a color contrasting with that of the suspendant are used. When a dc field is applied, the particles migrate to the surface of a transparent conductor which acts as the screen as in an electrophoretic process. The surface takes on the color of the particles. When the field is reversed or removed, the particles are dispersed back into the suspendant and the surface takes on the color of the suspendant.

Ferroelectric ceramics are one important form of electroactive solids which exhibit many nonemitting display effects such as the Kerr effect, Pockel's effect, scattering, and birefringence.[3] The electro-optical/ferroelectric ceramic is made of a transparent, doped polycrystalline material. It is characterized by a reversible, net, spontaneous, electrical polarization when subjected to a strong electric field. The spontaneous polarization has its origin in a noncentrosymmetric arrangement of the atoms in the crystal structure. The noncentrosymmetric domains are anisotropic, but are randomly organized so that the bulk properties are isotropic. When the ceramic is subjected to the high electric field, atoms move in response to the electric field in the crystal structure to align domain properties so that the bulk properties approach the anisotropic properties of the domains. The electro-optic effect is realized from the anisotropic properties of the particular material or material mix used. The greatest progress on electro-optic ceramics has been with the hot-pressed, solid-solution system of lead zirconate titanate— $Pb(Zr, La, Ti)O_3$ —modified with lanthanum, commonly denoted PLZT. In thin wafers, PLZT is transparent.

There are many other nonemissive phenom-

ena which produce an electrically alterable or optically alterable and reversible optical effect. Examples are electro-osmosis, electroreflectance (Franz-Keldysh effect), ferromagnetism, photochromism, and microchemical mechanisms (magnetized particles). In general, the optical effect is too small for the power required or the response is too slow for display applications. However, further research may uncover new materials or additional phenomena suited for displays.

1.4 DISPLAY NOMENCLATURE

For historical reasons, the names of display techniques are not consistent. The following paragraphs will clarify the major inconsistencies.

1.4.1 Electroluminescence vs. Luminescence. The term "luminescence" was first defined in 1888 by E. Wiedman as ". . . all those phenomena of light not solely conditioned by the rise in temperature." This was necessary to distinguish those phenomena from incandescence, which comes from a black-body radiator as described by the Stefan-Boltzmann temperature relationship. Ever since then, scientists have been assigning prefixes to luminescence to name some particular phenomenon.

A major problem exists with the classification "electroluminescence." Some authors, mostly outside the display community, take the approach that several luminescent phenomena which are electrically stimulated are a subcategory of electroluminescence. This approach is used inconsistently in varying degrees by encyclopedias and technical dictionaries. There would not be a problem if most agreed upon one consistent set, which is not the case. The extreme and erroneous definition is to include cathodoluminescence, electroluminescence, and gas discharge all under "electroluminescence."

The technical index for the Proceedings of the *Applied Physics Letters* includes both light-emitting diodes (LED) and polycrystalline electroluminescence (EL) under "electroluminescence." Most solid-state physicists and many solid-state physics texts lump LED and EL phenomena under "electroluminescence."

The electronic information displays community of practicing engineers and scientists almost universally use the definitions adopted here. That is, electroluminescence (EL) is used for displays made of polycrystalline phosphors, and light-emitting diode (LED) is used for displays made of single-crystal semiconductors. The LED and EL phenomena, when reduced to practical display devices, are distinct with little in common. An LED display is a low-voltage, single-crystalline, point-source, dc, and low-impedance-current device. An EL display is a high-voltage, polycrystalline, area-source, dc or ac, and high-impedance-voltage device. LEDs are "diodes" in terms of electrical components. The equivalent circuit of an EL display pixel is to a first approximation a leaky capacitor.

In the final analysis, all luminescence phenomena are electric. That is to say, in quantum theory a photon may be emitted only when an electron falls back into a lower energy state. The occurrence and wavelength of the photon depend upon the laws for the conservation of energy and momentum. Photons are emitted in incandescence by the same process. The difference among all of the luminescent and incandescent categories is in how the electron is excited up to the higher energy state.

1.4.2 Gas Discharge, or Plasma Panel.

The adjectives "gas-discharge" and "plasma-panel" are used for the same displays in the display community. Display applications of the gaseous glow discharge phenomenon or simply of gas discharge have been quite successful starting with the Nixie® tube in the mid 1950s, and as neon signs and status indicators before that. The name for this technology has been "gas discharge," and to a lesser extent "glow discharge" following the nomenclature of ionized gas physicists[4] and electrical engineers.[5] The gas-discharge model is consistently carried forth in dictionaries and encyclopedias, with electronic displays as an example of the application of the effect.

In the mid 1960s, with the invention at the University of Illinois of the ac coupled gas-discharge, matrix-addressable display, Bitzer and Slottow[6] coined the name "plasma display panel" or simply "plasma panel." The invention of placing a dielectric between the electrodes and the gas discharge, using the pixel walls to block current flow and ac couple to the gas discharge, was sufficiently evolutionary to warrant a new name. Any gas-discharge display does require a current-limiting device for each pixel. Normally, resistors are used for this purpose and are located in the external circuit. When the walls at each pixel are used as capacitors to limit the current, then it is a unique kind of gas-discharge display and perhaps should be denoted by a new name like "plasma panel" as suggested by its inventors. Similarly, other innovations also deserve special names such as Self-Scan® and Panaplex®.

It is not possible to find a definition for plasma in any technical dictionary or encyclopedia to reflect its present use in displays. The physical phenomenon is gas discharge, for which there is rich and abundant literature.

1.4.3 Cathodoluminescence vs. Vacuum Fluorescence.

The most difficult name to rationalize is that of "vacuum fluorescence" used to describe the display developed at ISE Corporation of Japan. The "Japanese green display," as it is sometimes called, or "vacuum fluorescent display," is a perfect example of cathodoluminescence. That is to say, a phosphor is excited by accelerated electrons in a hard vacuum. The fluorescent lamp is an important light source which uses the glow of phosphor coatings excited by the UV light from a gaseous discharge in a gas-filled envelope. The word "fluorescence" is frequently used to describe the fast decay phase of phosphors when exhibiting cathodoluminescence. Correspondingly, phosphorescence is delayed fluorescence. To a lesser extent, fluorescence is the name applied to the light emitted from a material not normally in a vacuum, when bombarded by electrons, UV, or other radiation.

"Vacuum fluorescence" is to "cathodoluminescence" as "plasma panel" is to "gas discharge." The structure of the vacuum fluorescent display is unique, and the inventors should not be denied their own name. Here again, it is not possible to find a definition for vacuum fluorescence in any technical dictionary or encyclopedia as we presently use it in displays.

1.5   CLASSIFICATION NOMENCLATURE

In any field of endeavor, it is desirable to use standard classification nomenclature. In the display industry, there has been a rush to form new acronyms. Without denying inventors the right to name their new display, a standard generic classification of displays can be used. The outline for such a classification is shown in Table 1-3. The table lists the key words which can be used to categorize any display device. As an example of the use of this scheme, one of the first liquid-crystal displays is classified as:

- Phenomenon: Liquid crystallinity (LC)
- Material: MBBA and EBBA with dopant
- Contrast: Dynamic scattering with back lighting
- Addressing: Direct with ac refresh
- Application: Numeric

This generic classification could be reduced to key words to describe display devices in technical reports, papers, and dissertations.

**Table 1-3   Generic Classification for
Flat-Panel Displays**

Phenomenon	The primary physical phenomenon used to create a particular visual effect. Examples: liquid crystallinity, electrophoresis, light-emitting diodes.
Material	The chemical name and physical state of the display material. Examples: Mn activated ZnS thin-film phosphor, doped MBBA/EBBA.
Contrast	The electrically alterable medium used to create contrast between picture elements. Examples: birefringence, absorption, green luminance on a black background.
Addressing	The method for controlling an array of picture elements. Examples: direct, scan, grid shift, or matrix addressing, with memory or refresh, ac or dc, intrinsic or extrinsic.
Application	The display category most suited. Examples: analog, alphanumeric, vectorgraphic, or video.

If an acronym is needed in the paper, the phenomenon need only be abbreviated, such as LED for light-emitting diode, LC for liquid crystallinity, or GD for gas discharge. The reader has been informed of the other features such as ac or dc, memory or refresh, thin film or powder. Trademarks could be used in place of the abbreviations where it is necessary to refer to specific manufacturer's product or approach.

Further, it makes for easier reading if the letter "D" for display is left out of the acronym. It is easier to read "The EL display uses less power . . ." than "The ELD uses less power . . ." and there is more information in "an EL display" to assist the unfamiliar reader. Additionally, the acronym can be used as an adjective, such as in EL powder, EL light, etc.

1.6   PICTURE ELEMENT OR PIXEL

The basic building block for all displays is the picture element as shown in Fig. 1-10. The noun "pixel" is formed from the contraction of "picture element" and is almost universally used in its place. The pixel is the smallest resolvable spatial-information element as seen by the viewer. There is no spatial information in a display below the resolution of the pixel area.

Some authors, particularly in the word-processing industry use the contraction "pel." By consensus, "pixel" is preferred. Pixel has been directly translated into French, German, Japanese, and other languages with the same meaning.

The pixel may be further subdivided to achieve color (see Fig. 1-11) or gray shades (Fig. 1-12). The key point is that the pixel is the lowest resolvable spatial incremental quantum. The other display dimensions of hue (color), saturation (color purity), luminance (gray shades), and time are all independent of the spatial dimension.

As an example of pixel subdivision, color is often achieved using three dots per pixel, one red, one green, and one blue within the spatial area. A clever geometric arrangement can be used with matrix-addressed displays, as depicted in Fig. 1-11b. Neighboring pixels can share color dots if properly programmed. In this case, the active area is much larger than the pixel spatial area.

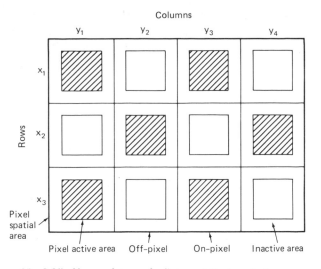

Fig. 1-10. Nomenclature of a flat-panel display pixel array.

Several resolution lines in the rows or columns or both can be used for gray shades. For example, the intersection of two row and two column electric leads in matrix addressing can be used to define a single pixel. The pixel is then made up of four dots, and the excitation of different numbers of dots can be used for gray shades. In Fig. 1-12 the dots are of different size so that fairly uniform steps of sixteen gray shades can be portrayed using all combinations of the four dots.

The pixel spatial dimensions can be defined by their pitch. The resolution is the reciprocal of pitch and is quoted as display lines per inch or millimeter. Display lines do not have spaces

in the sense that optical lines have spaces. For example, it takes a minimum of two display lines to represent an optical line and its space. An optical line space is represented by a display line turned off.

To display 20 optical lines per inch, for example, requires 40 display lines (TV lines) times the Kell factor of 1.4 for a total of 56 display lines per inch. The Kell factor is used to determine the number of TV raster lines needed to reproduce a resolution test chart. Other methods are used today, as discussed in Chap. 4.

The active area may be less than the pixel area, as shown in Fig. 1-10. The checkerboard pattern in Fig. 1-10 is useful as a test pattern

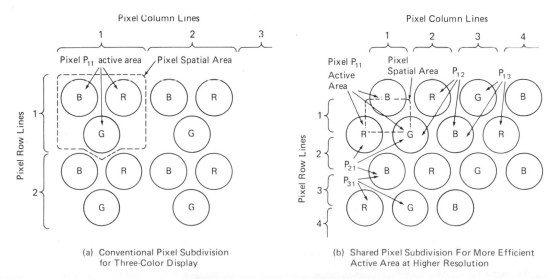

(a) Conventional Pixel Subdivision for Three-Color Display

(b) Shared Pixel Subdivision For More Efficient Active Area at Higher Resolution

Fig. 1-11. Pixel subdivision for color.

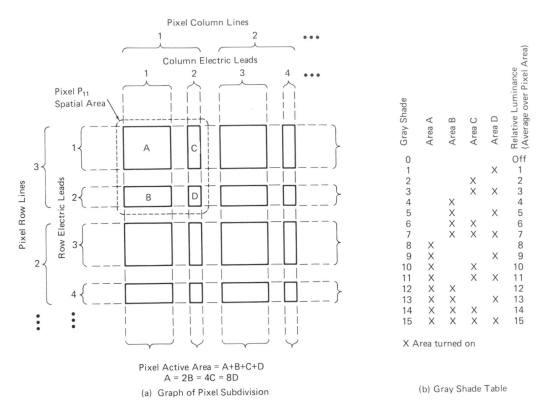

Gray Shade	Area A	Area B	Area C	Area D	Relative Luminance (Average over Pixel Area)
0					Off
1				X	1
2			X		2
3			X	X	3
4		X			4
5		X		X	5
6		X	X		6
7		X	X	X	7
8	X				8
9	X			X	9
10	X		X		10
11	X		X	X	11
12	X	X			12
13	X	X		X	13
14	X	X	X		14
15	X	X	X	X	15

X Area turned on

Pixel Active Area = A+B+C+D
A = 2B = 4C = 8D

(a) Graph of Pixel Subdivision.

(b) Gray Shade Table

Fig. 1-12. Pixel subdivision for gray shades.

for measuring contrast ratio between on-pixel and off-pixel. Using this test pattern would lead to a conservative measure of contrast ratio, since each off-pixel is surrounded by on-pixels which can indirectly contribute to the brightness of the off-pixel by light piping or scattering. Studies have shown that the readability of an emissive display is not adversely affected with pixel-active areas as low as 50% or less. Diffusers can be used to spread out the pixel luminance to fill the pixel spatial area.

Except for CRTs and some flat CRTs, the invention of a display begins with the pixel. It must be electronically alterable and reversible. The contrast ratio must be high, the speed of response fast, and the power consumption low. Additionally, it must be operable over a wide temperature range with long life. It is not important whether it is an emissive or nonemissive technique. There are thousands of ways to create a pixel. Only a few are worthy of pursuit. In general, the pixel must have the following properties to be suitable for a general-purpose display (500 rows by 500 columns):

- Resolution: 64 lines/in.
- Pixel contrast ratio: 10:1 (over a wide ambient illumination range)
- Directionality: Lambertian over 60-deg cone from normal
- Operating temperature range: $-40°C$ to $80°C$
- Life (maintenance): Memory—10^6 cycles
  Refresh—10^4 hours
- Dwell time: 20 μsec
- Power (less than 25 W total): 0.1 mW per pixel
- Duty cycle: Memory—100%
  Refresh (100 Hz)—0.2%
- Discrimination ratio (luminance of on-pixel to off-pixel at one-third voltage): 10^4:1

It is necessary but not sufficient that a pixel have all these properties. If it does not, then it may be suitable as a smaller display or special-purpose display depending upon the limitations. There are no liquid-crystal materials which possess all these pixel properties, and there are no successful 500 by 500-line LC displays either.

Yet there are many successful liquid-crystal display applications primarily because of the low power and sunlight readability properties.

1.7 DISPLAY ARRAY

A display is simply an array of independently controllable pixels. The number of pixels needed depends upon the application. A minimum of seven pixels is needed for a numeric character. A four-digit watch display would require four times seven plus one for the colon, or 29 pixels. The colon is only one pixel if the two colon dots are not independent. NTSC (National Television System Committee) standard commercial television requires approximately 150,000 pixels (480 rows by 320 columns) when displayed in a pixel array (digital equivalent).

The total pixel count is independent of refresh time, color, etc. The pixel count limits the total instantaneous information content of the picture.

The array is normally organized in rows and columns. The address of a pixel is defined by its row number and column number, normally counting from the upper left-hand corner as shown in Fig. 1-10. The electronic drive controls the state of the pixels according to their address.

In matrix-addressed displays, all the columns are normally addressed in parallel to save time. The complete array is addressed one row at a time, and the time to address the complete array is then set by the number of rows multiplied by the time per row. The rows are kept to a minimum in order to minimize frame time and maximize duty factor. The duty factor is the percent of the total time spent on each row or pixel and is therefore the reciprocal of the number of rows, a key point. The duty factor is very critical for slowly responding display material such as in liquid-crystal displays. Liquid-crystal displays are so slow that only 64 to 128 rows can be addressed per frame directly. However, the display can be made to have more display rows than it has addressed rows. Techniques for doing this are shown in Fig. 1-13. In example Fig. 1-13d, every set of four addressed column lines are rotated to appear as row lines. This complicates the data signal and electrode lines, but it increases the useful display row lines by a factor of 4 or more. Since the columns are addressed in parallel, the number of addressed column lines is of no consequence, time-wise.

A flat-panel display array is normally a fixed-digital set due to its construction; the exceptions are the flat-CRT Aiken Tube, the Gabor Tube, and others that use the scan addressing techniques which will be discussed in the next section. The number of rows and columns is constrained by the electrode configuration. The number of rows and columns is discrete. To make the best use of electronic counters in the electronic drive, and of memory locations in a picture memory map, the number or rows and columns is often a power of two, such as 256 rows by 512 columns.

The CRT, Aiken Tube, and Gabor Tube are analog displays. The pixel locations are defined by an analog voltage on the vertical and horizontal deflection amplifiers. The display array is called a raster scanning pattern or Lissajous pattern. The pixel size, called a spot size (diameter at 50% luminance), is principally defined by the electron beam focus, the video amplifier bandwidth, and the phosphor light diffusion. The number of rows and columns possible in a CRT corresponds to the raster vertical dimension divided by the spot vertical dimension and the raster horizontal dimension divided by the spot horizontal dimension, respectively.

The CRT display phosphor screen is usually continuous and without electrode definition. The deflection and video amplifiers are external to the tube itself. These analog amplifiers can be adjusted or modified to greatly alter the pixel size and raster size of a CRT display without changing the tube. This capability is used in the zoom feature of some commercial TV sets.

The number of row lines greatly impacts upon the display media performance requirements in the following way:

1. Refresh duty factor is inversely proportional to the number of electronic row lines, and
2. Pixel contrast ratio is inversely proportional to the number of electronic row lines.

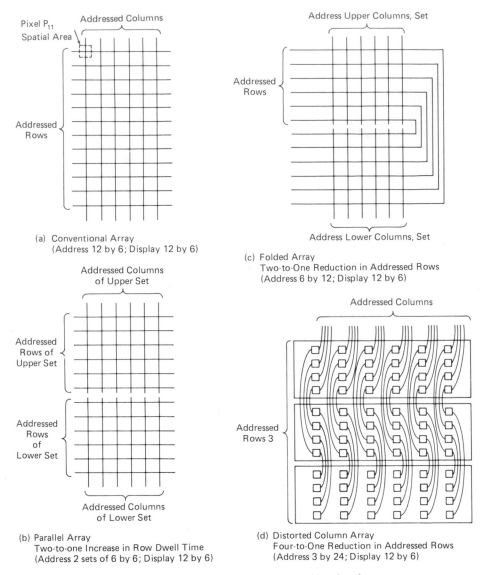

Fig. 1-13. Techniques for reducing row addressing time.

The requirements on the display media can be reduced if the electronically addressed row lines can be reduced. The first and obvious thing is to simply skew the presentation by increasing the column lines and reducing the row lines. This distorts the display presentation and can only be used within limits. The next step is to reduce the electronic rows without reducing the displayed rows by using techniques such as those shown in Fig. 1-13.

1.7.1 Duty Factor. A display is addressed or strobed sequentially. The time spent on each pixel is inversely proportional to the number of pixels and is called the *duty factor*. For an array of 500 by 500 the duty factor becomes 4 parts per million, or .0004%, which is quite small. Only CRT phosphors and LEDs can operate satisfactorily with a duty factor this small.

The duty factor can be greatly increased by using line-at-a-time addressing. That is to say, a complete row is addressed in parallel, and the rows are then commutated sequentially. For an array of 500 by 500 the duty factor becomes 2 parts per thousand, or .2%, which is a significant improvement. This is why line-at-a-time addressing is used whenever it is at all practical and why the number of addressed rows is kept to a minimum. How hard a display can be driven is inversely proportional to the duty factor.

A second consideration is the pixel dwell-time. This is the time required to turn a pixel on to its required luminance or contrast. The dwell time is the product of the duty factor and the frame time. For a display array of 500 by 500 addressed or refreshed at 60 frames per second, the frame time is 16 msec, and, using line-at-a-time addressing, the duty factor is 0.2%, which computes to a dwell time of 32 μsec.

Each technology has its characteristic required dwell time, as shown in Table 1-2. The response of most CRT phosphors is sufficient for a 120 nsec dwell time. Obviously, the dwell time requirements can limit the size of the array or the refresh rate. This is one of the reasons why the techniques of Fig. 1-13 are used to reduce the number of addressed rows, as would be required, for example, with liquid-crystal displays. The CRT cannot utilize line-at-a-time addressing, and the nsec response of the CRT type scan addressing is necessary for 500 by 500 arrays.

1.7.2 Pixel Contrast Ratio. The Pixel Contrast Ratio is simply the ratio of the luminance of the on pixel to the off pixel in its ambient environment. The influence of the ambient environment will be covered in Chapter 5. We are left simply with the emitted portion affecting the pixel contrast ratio; this will be used to demonstrate the significance of the number of row lines on the pixel contrast ratio. The pixel contrast ratio for refreshed displays using line-at-a-time addressing is:

$$PCR = \frac{L_{on} + (M - 1)L_{off}}{ML_{off}}$$

where PCR is pixel contrast ratio; L_{on} is the luminance of the pixel when strobed on; L_{off} is the luminance of a pixel intended to be off when another pixel is strobed on; and M is the number of electronically addressed rows.

The derivation of the equation can be done heuristically. Consider the fact that the display row lines must be strobed sequentially when refreshing the display image. Now, depending upon the display addressing technique when a

row is being addressed, the pixel in that row being addressed will have a luminance of L_{on}, and all the pixels intended to be off will experience a partial signal and will have a luminance L_{off}. When the next row is addressed, the pixel turned on in the previous row will experience the partial signal and will be stimulated to the L_{off} value, and so on through the remaining (M - 1) rows. The pixel intended to be off will again experience the partial signal, and so on for every addressed row. Over an entire frame, the total luminance is the sum of the individual light pulses; therefore, the on pixel has a luminance of $L_{on} + (M - 1)L_{off}$, and the off pixel has a luminance of ML_{off} for M addressed rows in the array. This is conceptual, and the specific application will depend on the specifics of the particular display configuration being analyzed.

It is appropriate to introduce at this point a metric called *discrimination ratio*, which is used to characterize a display material nonlinearity. Discrimination ratio, D, is defined as:

$$D = L_{on}/L_{off}$$

where L_{on} and L_{off} are the same as before. The full signal causes the display material to emit at L_{on}, and the partial signal causes it to emit at L_{off}. The magnitude of the partial signal is dependent upon the addressing technique, as will be discussed in Chapter 5. For example, with matrix addressing, the very best that can be achieved due to internal sneak circuits is for the partial signal to be one-third the full signal.

Without regard to the value of the partial signal, an important observation can be made relating to the required discrimination ratio to the number of rows in the array. Substituting D into the PCR equation for L_{on}/L_{off}, we have

$$PCR = D/M + (M - 1)/M$$

and for more than ten rows, a good approximation is

$$PCR = D/M + 1$$

Therefore, for a pixel contrast ratio of greater than 10, the discrimination ratio must be greater than 9 times the number of addressed rows.

Most display technologies in combination with the addressing techniques have a limited discrimination ratio.

1.8 ADDRESSING

For the uninitiated, the least understood and most underestimated task in flat-panel displays is that of addressing the hundreds of thousands of pixels. It is the single most difficult problem. The solution chosen has a great impact on the display cost. The problem is to convert a serial electrical data sequence into a rectilinear pixel array in real time.

The success of the CRT is directly attributable to the simplicity of its scan-addressing technique used to generate the raster or Lissajous patterns. Scan addressing is possible because of highly efficient, fast-responding cathodoluminescent phosphors. Scan addressing is further unique and ideal in that it can directly accept data in real time at video speeds from a single serial data channel without the need for intermediate data storage or shift registers.

In flat-panel displays, the addressing problem is similar in many ways to that in randomly addressable digital memories and solid-state imaging devices. The cross-coupling anomaly of a partial selection of nonselected pixels is identical. However, with displays the array must be of a size appropriate for viewing. It must be planar, with the proper linearity, size, shape, and percent active area. The viewing signal-to-noise ratio and power must be appropriate for human interpretation. There is no opportunity for noise filtering or error correction once the information is displayed.

All electronic displays are addressed by one of five basic techniques as summarized in Table 1-4. Normally, power is applied at the same time the pixels are addressed. In some flat-panel displays, the information is applied by one addressing scheme and the power is applied by another. The pixels in the array must effectively have three or four electrical leads when information and power are applied separately.

Addressing becomes more and more difficult as the number of pixels in the array becomes larger and larger. A successful addressing solution for an array of 64 rows by 128 columns may not work at all for an array of 128 rows by 128 columns using the same display media. The

reason for this is that the requirements of discrimination ratio (nonlinearity), speed of response, duty factor, dwell time, and power increase in proportion to the number of rows in the array. Typically, one axis is used for timing and the other is used for data input. It does not matter which one is used for which, except that there is some economy in the overall device if the smaller of the two is used for timing. For purposes of this book, the timing axis lines are called the rows and the data axis lines are called the columns. The addressing techniques are discussed in more detail in Chap. 5.

1.8.1 Direct Addressing. Direct addressing applies to the hard wiring of each pixel to a driver amplifier. It is only used with discretes and a few alphanumeric characters. Direct addressing becomes unacceptable for five or more numeric character displays. For example, for five seven-segment numeric displays with decimal, direct addressing would require 40 leads plus power return (8 for the seven segments plus decimal times 5 characters). If an integrated circuit were used to drive the display, 40 pins in the package would be dedicated to drive the display which greatly impacts the cost of the integrated circuit. Additional pins would be required for data-in, clock, and power. With matrix addressing, the number of leads could be reduced to 8 for the seven segments and decimal plus 5 for the five characters, making 8 column lines plus 5 row lines for a total of 13 leads. The array can be visualized as five rows and eight columns, and would be wired accordingly for matrix addressing. Direct addressing is sometimes used on large signs where there is adequate space and large amounts of power are required.

1.8.2 Scan Addressing. The addressing problem can best be understood by studying an array of pixels such as would be required for a commercial TV picture. There are approximately 480 rows (controlled by the raster sync signals) and 320 columns (limited by the video amplifier bandwidth) for a total of 153,600 usable addressable pixels in NTSC standard TV. To use an individual lead as in the direct-addressing technique would require 153,600 leads. This is virtually impossible except for a direct-view display the size of a billboard. Also, the

Table 1-4 Classification of All Known Addressing Techniques

Addressing Technique Name	Typical Pixel Electrode	Number of Amplifiers	Display Applications
Direct	One lead to each pixel with a common signal return for power or a pair of leads to each pixel	Number of rows multiplied by the number of columns	Four or fewer alphanumeric characters
Scan	Each pixel defined by beam size focused on continuous screen of pixel media	One for horizontal scan deflection; one for vertical scan deflection; one for beam intensity control	Cathode-ray tube and some flat CRT's
Grid	Each pixel defined by the grid hole geometry; one to four grids typically	Variable, dependent upon number of grids and subdivision of each grid but fewer than in matrix addressing	Vacuum fluorescence, some flat CRT's, and some gas discharge technqiues
Shift	Each pixel electrically connected between one row channel and a pair of column leads	Number of rows plus number of columns divided typically by four for shift articulation (assume shift is along the rows only)	Uniquely used with some gas discharge and plasma panels
Matrix	Each pixel electronically connected between one row lead and one column lead	Number of rows plus number of columns	Possible with all flat-panel technologies with high discrimination ratio (large nonlinearity)

cost of completing all the connections, routing all the wires, and assembling all the amplifiers is prohibitive. The best technique is scan addressing, which unfortunately adds depth to the display. Scan addressing is uniquely integral to the CRT and to date has not been successfully applied to a flat-panel display except in flat CRTs.

1.8.3 Grid Addressing. The number of line drivers in matrix-addressing display arrays is a significant cost factor. Grid addressing is used to reduce the number of line drivers even further at the cost of increasing the physical structural complexity. The grids must all be electroded and constructed with holes. The display media need not possess a nonlinearity when grid addressing is used. However, the media must be sensitive to charged particles which can be controlled with a grid, such as gas discharge priming particles, electrons, or ions. Grid addressing does not have the partial-selection problem that matrix addressing has. Each grid effectively adds another electrode to each pixel. However, the array must be addressed pixel-at-a-time or row-at-a-time to prevent direct

cross-coupling as with matrix addressing. The name "grid" is used because the grid structure function is analogous to the grid of a triode vacuum tube.

1.8.4 Shift Addressing. Another way to reduce the number of line drivers even further is with the shift-addressing technique. Typically, the data are introduced in parallel in all the columns of one row at the left side of the display and shifted to the right. The gas-discharge-controlled switching characteristics are particularly suited for this approach. It is not the nonlinearity but the gas switching or priming properties that permit shifting of the input information. Channels may be constructed along each column to contain the discharge and prevent column-to-column crosstalk. After a gas discharge has been started in a gas-filled column channel, priming particles drift and diffuse along the column (data) channel to the next row (timing), lowering its firing voltage. When the scanning voltage is applied to a row, the pixel that has been primed turns on, and the pixel that has not been primed remains off. In this manner, the gas discharge can be articulated along the channel by means of four inter-laced and interconnected row-line drivers. With the shift-addressing technique, additional time is needed to shift the data into the display. Minimizing columns saves on the number of amplifiers—minimizing rows saves addressing time. Here the rows run vertically and the columns run horizontally.

1.8.5 Matrix Addressing. Matrix addressing is the most commonly used method for flat-panel displays. In comparison with direct addressing, the number of leads is reduced from the product of the number of rows and columns to the sum of the number of rows and columns. The over-all physical construction is quite simple. How-ever, the display media must possess a strong nonlinear characteristic to prevent partial selec-tion of the nonaddressed pixels, and the array must be addressed pixel-at-a-time or line-at-a-time to prevent direct cross-coupling. Examples of display media that inherently possess a strong nonlinearity found to be most suitable for matrix addressing are light-emitting diodes, ac thin-film electroluminescence, and gas discharge.

If the media does not inherently possess sufficient intrinsic nonlinearity, as for example with liquid crystallinity and electrochromism, active electronic components must be added at each pixel. This has been successfully accom-plished using transistors and/or diodes. How-ever, complexity has now been added at each pixel. The necessary nonlinearity has also been achieved by introducing another material such as nonlinear resistor films of ZnO or ferroelec-tric wafers to act in concert with the display media to render a net nonlinear response. Mild nonlinear characteristics are realizable with liquid-crystal materials, permitting matrix ad-dressing for small arrays.

When something is added to the display media to achieve matrix addressability, it provides "extrinsic matrix addressing." Otherwise it is simply matrix addressing or intrinsic matrix addressing. The adjectives "extrinsic" and "intrinsic" are analogous to extrinsic and intrin-sic semiconductor materials. An extrinsic semi-conductor is one where a dopant has been added to get the desired semiconductor properties. An intrinsic semiconductor inherently has the desired properties without adding a dopant.

1.9 DISPLAY DEVICE DEVELOPMENT

Display device development requires a sustained effort of advanced technical talent over an ex-tended period of time. The effort requires the appropriate balance of research, engineering, manufacturing, marketing, finance, and manage-ment. A single organization with a grasp of the total technical problem, the stability for a pro-longed development cycle, and the proper blend of corporate functions is indeed rare.

Moreover, a display is only a component. Production orders for displays come only after production of the product using the display has been committed to. After the display prototype is released for product application, it takes three years at a minimum until production quantities are ordered. The reasons for this are imbedded in corporate annual planning and budgeting procedures. There are three phases in new-product development, each of which takes a significant increase in commitment of company resources and financial support. The first year or phase is spent by the company, or organization responsible for the product, in eval-uating the display and designing the product and making breadboards of the product or critical parts of the product. The second year or phase is spent making several engineering prototypes meeting functional requirements and evaluating them internally. The third year or phase is spent making pilot line units to form, fit, and function for limited sales and evaluation by product cus-tomers and users, planning production, and making marketing evaluations for prodution recommendations. Production could commence the fourth year. The production order for the display component comes only after the com-mitment to production and sales have been made.

The three-or-more-year delay from display prototype delivery to receipt of production order is a real negative aspect in any business plan for display development or in any present-value analysis or return-on-investment analysis. If a large investment is needed during the display prototype phase, the aspect of not getting pro-duction orders until three years after the first prototypes are delivered turns most investors away to greener pastures.

The development of a new display technology is often very long and financially and technically

Table 1-5 Display Technology Development Cycle.

	LC Dynamic Scattering	AC Plasma Panel	AC Thin-Film EL
Display phenomenon	1963 RCA	1964 Univ. of Illinois	1962 Servomechanisms
Prototype	1969 Rockwell	1969 Owens-Illinois	1965 Sigmatron
Testing of prototype	1970 Rockwell	1970 Sperry	1974 Sharp
Pilot plant	1971 Rockwell	1970 Owens-Illinois	1978 Sharp
Production	1972 Rockwell	1974 Electro-plasma	1983 Sharp

hazardous. It usually requires ten to twenty years from the time of the research breadboard demonstration of an alphanumeric character or small array to the beginning of full production. One institution may do basic materials research, a second may pick it up with some engineering effort in hopes of entering a new market, and then a third may continue into manufacturing because it has a product need. This type of staggered development occurs for any one of several reasons. Most frequently, the starting company does not initially grasp the magnitude of all of the technical problems and the corresponding cost and risk. As a consequence, as the costs mount up and schedules slip, the venture dies for lack of internal management support. Other causes of truncated development include changes in corporate management personnel or in corporate objectives, recession cutbacks, loss of key technical personnel, inability to overcome technical problems, lack of technical breadth, poor market timing, or inadequate return from present value analysis.

Frequently, a researcher or engineer finds a new twist, or more likely rediscovers an old one. He demonstrates it to management, which offers some seed money. The initial market analysis is clear and simple: "It could lead to a display appropriate for a wristwatch (100,000,000 of which are sold per year) or a TV set (10,000,000 of which are sold per year)." The project is sustained by additional seed money, interested customers, and perhaps a small contract or two. A patent is issued, and the project is reported at a technical conference and in the literature.

Quanta of individualized effort as just de-scribed make important contributions to our industry. Many individual quanta of research and engineering are needed to advance the technology from the concept to the product. Once a technology is in production, there is a cash flow necessary to sustain advanced engineering, and the evolutionary process sets in for bigger and better displays using the same technology.

The display research, design, and development cycle is outlined in Table 1-5. Three actual examples are outlined to emphasize the development cycle time for a new technology.

1.10 MULTiDISCIPLINE

Electronic display device development is an extremely multidisciplined venture. First of all, displays have a human eye/brain optical interface which must be accomplished over a wide environmental envelope. Second, the display media itself can involve any one of many and diverse scientific disciplines—organic chemistry, solid-state physics, etc. Third, the input signal is electrical; addressing, formatting, and control of the display is entirely in the domain of the electronic circuit design engineer. Fourth, the fabrication of the display is normally not along the lines of other electronic devices. A display is an electro-optic device on a macro scale. The device may require thick and thin films deposited on glass substrates with hermetic seals, polarizers, and antireflective coatings involving several manufacturing disciplines.

The breadth and multidisciplined nature of display device development was noted, among others, by Dr. Ruth M. Davis in her keynote address at the Society for Information Display (SID) Third National Symposium on Information Displays in San Diego in February 1964. Dr. Davis, then Director of DDR&E (Directorate of Defense Research and Engineering) made the following lead-off statement:

"A good catch-phrase with which to start a discussion on displays might be 'There is something in it for everyone.' Certainly, the interdisciplinary nature of display development attests to the truth of this phrase. Contributing disciplines include optics, psychology, physiology, engineering, the behavioral sciences, and the computing sciences. Such an interdisciplinary approach is both a strength and a weakness to display development. Its strength lies simply in the multiplicity of disciplines upon which it may feed and expand; its weakness lies in the difficulty of effecting parallel advance on all the essential fronts typifying any display development."

This statement is as true today as it was in 1964. There are so many diversities that typically a company will get a reputation for expertise in one display technology or medium. Any extended effort using one display media will result in staffing experts and facilities which will perpetuate a company's interest. There is so much inertia that companies need to develop a display technical capability anticipating a future market opportunity.

1.11 TECHNOLOGY IMPETUS

Display device development seems always to lag behind product needs and electronic capabilities, and this gap is greater now than ever before. Displays may always lag behind the development of other components because of the high technical risk and large capital investment required during their development. The accelerated demand for displays has evolved for three reasons:

- The accelerated development of the overall miniaturization of all integrated circuits,
- Technology advances in microprocessors, and

- Continuing cost reductions leading to expanding applications and production volume.

The most important impetus for displays in the 1980s has been new developments in drivers. High-voltage and low-voltage custom LSI drivers with appropriate shift registers and latches are available for all the display technologies. The last devices to fill out the spectrum of drivers has been for high-voltage displays such as plasma, vacuum fluorescent, and electroluminescent displays, for which up to 200 volts peak at one amp peak are required. The high voltage is achieved using a MOS technology developed for the power-switching industry using double diffusion (DMOS) to control the souce-to-drain gap.

There are numerous new products that are awaiting suitable cost-effective flat-panel displays. A tabulation of the most sought-after display-oriented products is given in Table 1-6.

Other, more specialized new products include military displays for command and control, hand-held display and control units, helmet-mounted displays, etc. Aircraft displays and controls are perhaps the most complex as a set, with similar requirements for military and commercial aviation. The requirement for sophisticated, compact, lightweight, wide-temperature-range, sunlight-readable, colored, and low-powered displays all apply to aircraft. The same holds for shipboard applications, with more emphasis on corrosion resistance and moisture proofing and less on weight and power. Toy and game products require low-cost, rugged, low-power displays that are appealing and colorful, which get the attention of the buyer (parents) and survive the rigors of the user (children). Educational toys are a very important new product area where cost is a key factor.

The electronic panel meters and multimeters with numeric displays cannot compete directly with instruments using galvanometers, aneroid movements, or drag-cup speedometers. These transducer-display devices that have developed over decades are very low-cost, simple, and rugged. Electronic displays require a processor of some sort and an electronic analog to digital transducer. The new display products must excel in performance to overcome the higher cost.

Table 1-6 Product Examples That Benefit from Flat-Panel Displays

Generic Product	Typical Information Source	Approximate Resolution	Nominal Size	Special Feature
Entertainment:				
High-quality video	Video memory	1000 rows, 1000 columns	48-in. diagonal	Natural color
Wall-mount TV	Broadcast	480 rows, 500 columns	36-in. diagonal	Natural color
Portable TV	Broadcast	240 rows, 360 columns	12-in. diagonal	Natural color or black and white with gray shades
Graphics	Memory	4000 rows, 4000 columns	25-in. diagonal	Synthetic color
Computer:				
Terminal	Computer	2000 characters	14-in. diagonal	Low cost, high quality
Personal	modem	1000 to 2000 characters	12-in. diagonal	Small space
Portable	memory	2000 characters	9-in. diagonal	Light weight, low power
Picture phone	Telephone	256 × 512 columns or more	12-in. diagonal	Phone power, memory display
Oscilloscope	Memory transducer	512 rows, 512 columns	8-in. diagonal	Low power, portable, light weight
Departure and arrival	Computer	16 to 64 rows of 64 characters and larger	Large character	Computer power or phone power, memory display
Text editor	Keyboard	2 to 32 rows of 80 characters	$\frac{3}{16}$-in. character	Typewriter power, low cost, high quality
Tickertape, news, or weather	Telephone or broadcast	1 to 4 rows of 80 characters	$\frac{3}{16}$-in. character	Memory and scroll
Point of sale	Cash register Gas pumps	2 to 4 rows 8 numerics each	1 in.	Sunlight-readable
Calculator	Microprocessor	8 numerics with plus and minus and flags	$\frac{3}{16}$ in.	Sunlight-readable
Panel meter	Transducer	4 to 8 characters with plus and minus and flags	$\frac{1}{2}$-in. characters	Transducer-powered, sunlight-readable
Watch	Oscillator	Time and date	$\frac{3}{16}$ in.	Aesthetically pleasing, sunlight-readable, extreme low power
Clock	Oscillator	Time and date	1 in.	Aesthetically pleasing
Counter	Transducer	Numeric digits	Approx. $\frac{1}{2}$ in.	Nondestruct memory, tamper-proof
Automobile panel	Transducers	40 to 100 pixels	Stylized	Sunlight-readable, wide temp. range
Aircraft primary instrument	Transducer computer radio	512 × 512 pixels	8-in. diagonal	Sunlight-readable, synthetic color

Selling more performance for more money can be difficult. More performance can only be achieved with the application of large-scale integrated electronics and microprocessors.

Flat-panel displays can help fill many product gaps. There is a large gap between a galvanometric meter and a CRT in information displayed. There are many products that need more digits than an 8-digit calculator and fewer than a 2,000-alphanumeric-character CRT.

A digital presentation offers a performance advantage in itself. Digital displays can be read faster and with less error than analog displays where interpolation is required. Exceptions exist with rapidly changing parameters such as with speedometers and altimeters. The least significant digit required to be read may be a blur. Another exception exists with nondecimal systems such as time which is to the base 60. The display of ": 50" may be read as half-past the hour instead of 50 minutes.

Some products may not be complex enough to justify the cost of an electronic implementation. An example of this may be the automobile panel. In its simplest form, there are only six items of data: odometer (17 bits) and speedometer (8 bits), engine temperature (1 bit), oil (1 bit), alternator (1 bit), seatbelt (1 bit), turnlights and high beam (2 bits). The speedometer could be represented by a pseudo analog scale with 6 bits for an accuracy of 1 part in 64. The odometer presents a special problem in that it must not be volatile to power-off and it must be tamper-proof. The total data word is low and the update rate is slow—at 5 times per second, maximum. A very low-cost display technology or marketing consideration may overrule this argument. Further, the cost of electronic displays would be less significant to rally car drivers and sports car enthusiasts.

The most difficult problem with the new display products is cost. The most expensive component is often the display itself. Within the display component, the most costly part is the display addressing and power-drive electronics. Also, the display often requires different voltages than are otherwise available; thus additional power supply requirements. For military display products, performance can be more important than cost. That is never the case with commercial products, with the possible exception of medical and aircraft electronics. The cost problem is compounded by low production volume in some markets. This is best exemplified in aircraft display products. The total production of new aircraft, including military, commercial, and general aviation, numbers in the thousands per year. Sophisticated, high-performance display products are justified in aircraft. However, the production volume is so low that it is not feasible to justify a large development effort and capital investment aimed solely at this market.

1.12 CONCLUSION

Display and electronic processing evolve together, although not always in complete synchronization. Being a part of the display evolution is the most exciting thing that can happen to an engineer, marketeer, businessman, or entrepreneur. We now have Dick Tracy's wristwatch TV/radio, Captain Kirk's command and control flight-deck display, and the briefcase personal computer. What will be next?

REFERENCES

1. Aiken, Wm. Ross private communications, Kihei, Maui, Hawaii, February, 1983.
2. Tannas, Jr., L. E. "Fail-Safe Matrix Fonts." *SID 77 Digest*, p. 54, 1977.
3. Land, C. E.; Thacher, P. D.; and Haertling, G. H. *Electro-optic Ceramics*. Applied Solid State Science, Advances in Material and Device Research, Vol. 4. New York: Academic Press.
4. Loeb, L. B. *The Nature of a Gas*. New York: John Wiley & Sons, Inc., 1931.
5. Ryder, John D. *Electronic Engineering Principles*. Englewood Cliffs, N. J.: Prentice-Hall, Inc., 1956.
6. Slottow, Hiram Gene. "Plasma Displays." *IEEE Transactions on Electron Devices*. Vol. 23, No. 7, July 1976.

2

SYSTEM REQUIREMENTS

L. E. TANNAS, JR., *Consultant*

2.1 INTRODUCTION

All display requirements must flow down from those of the primary system and the overall objective. System performance is paced by the display performance. Since the display is often the most expensive subsystem of a display-oriented product, the system performance is often tailored to minimize the display cost impact. On the other hand, the display features are among the most important selling features of a display-oriented product.

Functionally, the display itself is a system made of several parts or subsystems as outlined in Fig. 2-1 and discussed in Section 2.6. However, the design objectives must come from a higher system level which would include the mission objectives and functional performance requirements. Before a display system can be designed, certain requirements must be given which presumably are needed to achieve the mission objective or product capability. These requirements might include:

- Detection criteria and readability requirements
- Operating scenarios
- Display image repertoire
- Modulation transfer function
- User interface
- Ambient illumination
- Environment
- Weight and volume
- Operating power
- Life and utilization cycle
- Maintenance and repair objectives
- Cost—initial and lifetime

The system constraints are further discussed in Section 2.5.

The display system designer takes the above requirements and derives detailed requirements such as:

- Resolution, luminance, and contrast
- Array size
- Refresh, memory, and processing requirements
- Faceplate requirements
- Bezel requirements
- Interactor selection
- Electronic interface
- Packaging

Once the display system requirements are defined, the display device designer can get down to work. The primary goals for flat-panel displays boil down to selecting a display medium and designing for proper performance in terms of the following:

- Resolution and array size
- Pixel contrast ratio (see Section 2.5.1)
- Display brightness ratio (see Section 2.5.1)
- Ambient illumination
- Temperature range
- Life
- Power
- Cost

The system requirements cause the display to be placed into one of four system classifications (Section 2.2), one of four installation classifications (Section 2.3), and one of four functional classifications (Section 2.4). The functional and

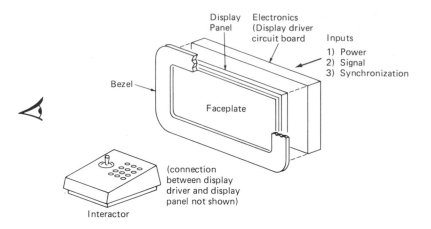

Fig. 2-1. Schematic of display subsystems.

installation classifications impact data rate and power respectively. This is an arbitrary but consistent set of classifications intended to give one an overview of the major factors affecting the display performance requirements.

2.2   SYSTEM CLASSIFICATION

All display products fall into one of four classifications (Table 2-1) depending upon their purpose. The distinction among system classifications is one of emphasis. The status display is designed for error-free reading on a sample basis. The data display is designed for utilitarian presentations over prolonged periods. The interactive display is designed to be compatible with the sytem dynamics and manual and visual interface. The entertainment system is designed to emphasize the aesthetic quality of the presentation.

Table 2-1   System Classification

Classification	Characteristics	Display Subsystem Impact
Status	A display used to answer a single question such as: Safe, go/no·go, speed, voltage, time, cost, magnitude	1) Color for status emphasis 2) Memory display an advantage 3) Low data rate 4) Bright and easily read
Data	A continuous flow of alphanumeric or pictorial data such as: News, stock quotations, teletype, weather report, flight plan, medical scan, surveillance of earth resources	1) Utilitarian 2) Memory display 3) Medium data rate 4) Hard-copy accessory desirable 5) Scroll 6) Playback
Interaction	A feedback display with the operator controlling the outcome such as: Electronic games, text editing, command and control, interactive graphics, design, scheduling, flight control	1) Interactive control 2) Refreshed display 3) Real time (min. delay) 4) Microprocessor 5) Medium to high data rate
Entertainment	A visual experience such as entertainment commercial television	1) Color desirable 2) Refreshed display 3) High data rate 4) Real time rate, delay OK 5) Wide dynamic range 6) High MTF

Table 2-2   Display Installation Classification

Classification	Characteristics	Display Subsystem Impact
Personal	A hand-held or carried display with power pack, operated while being carried by means of its own power. Examples are watch, calculator, radio communicators, TV at the beach.	1) Very lightweight 2) Very low power 3) Sunlight- or shade-readable 4) Readable in the dark 5) Wide environmental extremes for operating
Portable	A person-portable display which need not operate in transit and need not have its own power pack. Operates in controlled environment. Examples are portable TV, portable computer terminal.	1) Lightweight 2) Sealed rugged packaging 3) Wide environmental extremes for storage
Vehicular	A display mounted in a vehicle and using vehicle power. Examples are aircraft flight director, horizontal situation display.	1) Sunlight-readable 2) Wide environmental extremes operating and storage 3) Lightweight (aircraft) 4) Low power 5) Rugged packaging
Stationary	A display mounted and operated in an environmentally controlled, permanent installation using, facility power. Examples are computer terminal, TV.	1) Minimum 2) Controlled environment for operating 3) Shipping environment for nonoperating and storage

2.3   DISPLAY INSTALLATION CLASSIFICATION

All display products are designed for and installed in one of four environments. The installation is independent of the system classification. The different installations are outlined in Table 2-2. Each classification evokes certain environmental requirements which cannot be compromised without compromisimg the performance of the system. The most taxing and difficult environments are operating temperature range and sunlight readability. Both of these requirements may simultaneously impact the vehicular installation classification.

2.4   DISPLAY FUNCTIONAL CLASSIFICATION

All displays can be placed into four functional categories, which differ in complexity in terms of their pixel array size and data rate. The functions are tabulated in Table 2-3. The distinction among functions is the display complexity. There are natural display technology barriers which reinforce these categories. The barriers are primarily due to increasing levels of difficulty and complexity in addressing the display array.

2.5   SYSTEMS CONSTRAINTS

Ninety percent of everything we learn comes through our eyes. A display is unique among the myriad of electronic devices in that it is the primary data channel to the human user and operator. Alphanumerics and symbols must be chosen or designed so that they are recognizable to a user. New symbols cannot be introduced without the possibility of misinterpretation due to similarity with an old symbol. The display format and polarity conventions of an aircraft instrument cannot be changed without the permission of the entire pilot community. If they are changed, then the pilot population must be trained or retrained accordingly. The display designer *cannot unilaterally change* display symbols, conventions, formats, scales, speed of response, or other parameters without consulting with the affected user community.

Anatomically, a display must fit the human race. It must be bright enough to be seen in its

Table 2-3   Display Functional Classification

Classification	Characteristics	Display Subsystem Impact
Annunciator	A discrete indicator indicating "on" or "off." Examples are go/no-go, off, talkback, flag, exit, pictorial (no-smoking picture, seat belt picture).	1) Hardwired 2) Fixed array 3) Dedicated electronics 4) Lowest data rate
Alphanumeric	A fixed, formatted display with a font array suitable only for the presentation of alphanumeric characters, bar array, and special symbols. Examples are calculator, watch, digital panel meter, analog-type bargraph.	1) Fixed array 2) Stylized font 3) Dedicated electronics 4) Low data rate
Vectorgraphic	An orthogonal array of pixels which are randomly addressable for alphanumeric and graphical presentations. Examples are computer terminals and graphic displays.	1) Limited gray shade 2) Limited color 3) Full matrix array or raster 4) Medium to high data rate
Video	An orthogonal array of pixels which are suitable for video presentations in gray scale and color. Example is commercial TV.	1) Gray scale 2) Color 3) Refreshed 4) Video rate 5) Highest data rate

environment. It must have the required resolution at the viewing distance. It must be refreshed without objectionable smear or flicker. The human anatomy cannot be changed, and therefore the display must be designed to satisfy the pecularities of the eye-brain-hand system.

A display must be designed to meet the off-nominal design conditions. The display must be operationally functional in the environmental situation in which the operator will use it. The user may be of any age. He can have degraded visual capabilities such as lack of visual accommodation, color aptitude, visual acuity, or monocular vision. The human can also be fatigued, ill, stressed, or despondent—or all four. Even with the compounding of all these conditions, the objective is that reading errors not be due to deficiencies in the display.

2.5.1 Performance. From the system standpoint, a display's performance can be characterized by many major parameters. The qualitative parameter readability is perhaps the most important functional requirement. The degree of readability depends upon visual image detection, training, and motivation. Training and motivation will not be covered here as these parameters are of concern at a higher system level. Visual image detection will be discussed as it is the

most important system parameter which the display's engineer must provide. Visual image detection threshold at a given error probability is quantitative; it is used for greater precision in place of readability at the hardware level. The parameters of luminance, contrast, and visual angle, as a set, determine the threshold of detectability. It can be represented as a surface in three-dimensional space, as shown in Fig. 2-2. Human-factors engineers use brightness, modulation, and visual angle in their analysis and testing for visual threshold criteria. In display-device performance characterization, the display brightness ratio, pixel contrast ratio, and visual angle are easier to use. They are determined by the hardware and can be measured without regard to its application. The display brightness ratio (DBR) and pixel contrast ratio (PCR) are new parameters not too different from brightness and modulation; however, they are defined in a specific way to better characterize the display hardware as a function of the ambient illumination. The display brightness ratio (DBR) is defined as:

$$DBR = \frac{\text{Average luminance of the display}}{\text{Average luminance of the surround}}$$

The term "display brightness ratio" is used to

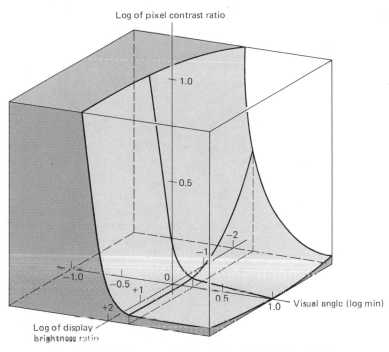

Log of pixel contrast ratio

Visual angle (log min)

Log of display
brightness ratio

Fig. 2-2. Heuristic example of a display detectability surface.

describe the performance requirement on the display luminance relative to the illumination environment in which it will be used. The luminance of the display surround controls the eye adaptation, and therefore the display area will either be brighter or dimmer or equal to the surround, which profoundly affects readubility and eye fatigue.

The nominal DBR is 1, corresponding to a uniformly illuminated display and surround. Contrast threshold increases when departing from unity by making the display brighter or dimmer. However, the departure is not linear or symmetrical, as determined by Ireland, et al., 1967.[1] They used a test setup exactly as required for the DBR definition here and measured the detectable contrast threshold shift using a Landolt ring target in the photopic visual range. The contrast threshold decreases by less than 10 percent as DBR increases by an order of magnitude. There is no problem in resolving bright images, as would be expected. Conversely, as the DBR decreases the contrast threshold increases significantly from that when DBR is equal to 1. At DBR of 0.1, the contrast threshold increased to 40 percent and for DBR of 0.01, the contrast threshold increased to 270

percent, as shown in Fig. 2-3. These figures quantitatively describe one of the problems that pilots have when they try to read their panel instruments with a bright sky around them, or that skiers have when they look into the shade. For optimal reading, the display should be as bright as or brighter than the surround. Normally, values of 0.5 to 10 are acceptable for DBR, whereas 2 to 3 is the desired nominal for minimum fatigue.

As of this writing, the display detectability surface of Fig. 2-2 has not been measured and plotted for a digital type flat-panel display. It is presented for its conceptual value to show tradeoff between these three parameters. A similar surface was presented by Lukiesh and Moss (1943)[2] using brightness in place of display brightness ratio as used here. The surface of Fig. 2-2 was constructed using the data from Figs. 2-3a and b. The data of Fig. 2-3a were used for the slope in the display brightness ratio and pixel contrast ratio plane. The data of Fig. 2-3b from Blackwell[3] were used for the slope in the pixel contrast ratio and visual angle plane. This could not be done rigorously, and Fig. 2-2 should only be considered qualitative. It could not be done rigorously because the Blackwell data abscissa

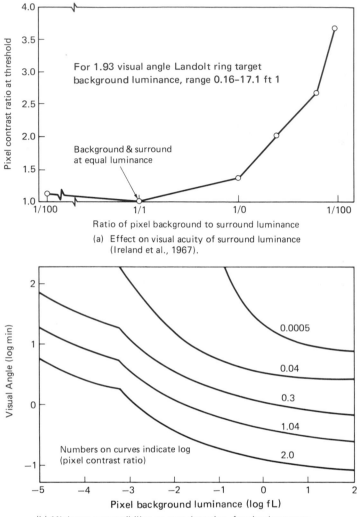

(a) Effect on visual acuity of surround luminance (Ireland et al., 1967).

(b) Minimum perceptibility, or spot detection, for circular targets as a function of contrast, and background luminance (Blackwell, 1946)

Fig. 2-3. Basic data showing the relationship between contrast, luminance, and visual angle.

was completed for (pixel) background luminance and not display brightness ratio. It remains for the reader to create the specific surface for his application.

The pixel contrast ratio (PCR) is defined as

$$PCR = \frac{\text{Average luminance of the on-pixel}}{\text{Average luminance of the off-pixel}}$$

The term "pixel contrast ratio" is used to describe the performance requirement on the display spatial image as a function of ambient illumination. Normally, values of 3 to 40 are acceptable for PCR, whereas 5 to 10 is the desired nominal.

The literature abounds with contrast studies using various definitions that are all mathematically interrelatable. One of the more common terms along with PCR is contrast modulation (M). Contrast modulation is used in Chap. 4. The relationship between PCR and M is as follows:

$$PCR = (1 + M)/(1 - M)$$

and

$$M = (PCR - 1)/(PCR + 1)$$

where

$$M = (L_{on} - L_{off})/(L_{on} + L_{off})$$

$$PCR = L_{on}/L_{off}$$

The luminance, L, in all these equations refers to the surface luminance as measured by a spot photometer. The luminance can be due to several sources of luminous energy such as reflected energy from ambient illumination, supplemental illumination, transillumination (edge lighting), and photon emission.

PCR is a measure of the ratio of luminance between two pixels. The more advanced approach is to use both chrominance and luminance contrast,[4] where the net PCR is the root mean square of chrominance and luminance contrast. This has been successfully demonstrated by several proponents of this approach.[5,6] The 1960 CIE (Commission International d'Eclairage) chromaticity diagram is used as the graphical representation of color differences. A normalized chrominance index is derived that corresponds to a just-perceivable luminance difference.[4] This has been experimentally determined to be:

Pixel chrominance contrast ratio

$$\frac{(\Delta u^2 + \Delta v^2)^{1/2}}{0.027}$$

where Δu and Δv are the difference in chrominance between the on and off pixel as plotted on the CIE 1960 chromaticity diagram, and 0.027 is the experimentally determined normalizing factor. The chrominance and luminance contrast ratio can be combined as the root mean square of the two, as follows:

PCR = [(Pixel chrominance contrast ratio)^2

+ (Pixel luminance contrast ratio)^2]$^{1/2}$

Most color measurements and discussions are in terms of the famous CIE 1931 x, y chromaticity coordinates.[7] However, color sensitivity differences are not uniform, as pointed out by MacAdam in 1943[8] and as discussed in such books as Bellmeyer and Saltzman.[9] This led to the CIE 1960 u, v chromaticity diagram to make the color space sensitivity more uniform, a diagram that was further improved upon by the CIE 1976, u', v' chromaticity diagram. In 1976, the CIE committee recommended the use of color difference equations denoted CIELUV (pronounced see-love) (L*, U*, V*) whenever precision in perceived color differences of emitted light is needed.

It is appropriate that displays engineers use the CIELUV system in the PCR equations when color differences are included.

All the CIE systems are related to the basic X, Y, Z photometric measurements (see Section 2.8).

For the CIE 1931 x, y system,

$$x = \frac{X}{X + Y + Z}$$

and

$$y = \frac{Y}{X + Y + Z}$$

For the CIE 1960 u, v system,

$$u = \frac{4X}{X + 15Y + 3Z} = \frac{4x}{-2x + 12y + 3}$$

and

$$v = \frac{6Y}{X + 15Y + 3Z} = \frac{6y}{-2x + 12y + 3}$$

For the CIE 1976 metric chromaticity coordinates, u', v',

$$u' = u$$

$$v' = 1.5v$$

For the CIELUV color space L*, U*, V*,

$$L* = 116 \left(\frac{Y}{Y_n}\right)^{1/3} - 16$$

$$u* = 13L*(u' - u'_n)$$

$$v* = 13L*(v' - v'_n)$$

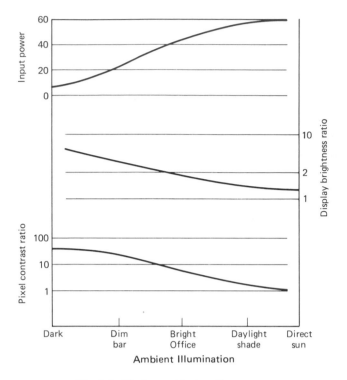

Fig. 2-4. Display ambient performance.

where u_n' and v_n' refer to the reference white or light source.

DBR is independent of color unless there is a big variance between the color of the display and the color of the surround—in which case, more studies would need to be performed.

These two parameters, DBR and PCR, are both a function of the display ambient illumination and display power. The ultimate *display ambient performance* is its DBR and PCR and associated power requirements all plotted simultaneously as a function of ambient illumination. Here, such a plot is called the *Display Ambient Performance*; it is developed in detail in Section 5.2.5. The plot for a typical high-quality display is shown in Fig. 2-4. The reason for making the effort to plot the display's ambient performance is that this plot will give a system designer the ability to determine rigorously the detectability of displayed information in any ambient illumination environment from any reading distance without further detailed knowledge or testing of the display. In addition, with the display power utilization curve, the display system designer can determine the power penalty to achieve a given level of detectability. This ap-

proach is directly suitable for both emitters and nonemitters. The display ambient performance plot (Fig. 2-4) completely characterizes the display hardware capability. The display detectability surface (Fig. 2-2) gives the means to present the threshold of detectability as a function of DBR, PCR, and visual angle—the major system requirements.

The display ambient performance plot is also a design synthesis tool. Higher or lower values in DBR and PCR can be obtained from measurements of the display using correspondingly more or less power. The plot shows immediately when the ambient illumination overcomes the display-generated luminance. The methods for making the display basic measurements and computing the display ambient performance plot are given in Section 5.2.5.

Another important system display-characterizing parameter is its spatial frequency response called "modulation transfer function"(MTF). Spatial frequency is in lines (line pairs) or minimum/maximum intensity pairs per unit distance. The MTF is used as a performance measure of many optical systems. One advantage of such a parameter is that the MTF of a series of optical operations is the product of the individual subsystem MTFs. The MTF cutoff frequency of a flat-panel display can never be any better than its pixel line-pair spatial frequency divided by 1.4 (the Kell factor). Display characterizing measures and performance measures are discussed more fully in Chap. 4.

Other important overall performance parameters are physical size, display lines per inch or millimeter, color, gray shades, pixel-active area, refresh rate, and memory.

The system designer needs all of these parameters plus electronic interface and power interface in order to determine the suitability of a display to perform a given system function.

2.5.2 System Design. The design of display systems is a very subtle discipline, particularly in the design of crew stations for high-performance aircraft and spacecraft. Here again, the past experience of the pilot must be taken into account. The most difficult task is to make the reading of displays error-free when the crew member is fatigued and under stress.

Many design conventions have been established for the control and monitoring of machines. A significant amount of anthropometric data have been tabulated for establishing workspace and worktasks. An important compilation of such data created for the design of controls and displays is:

• *MIL-STD-1472 Military Standard, Human Engineering Design Criteria for Military Systems, Equipment and Facilities.*

Another significant publication in this regard is:

• *Human Engineering Guide to Equipment Design* edited by Harold P. Van Colt and Robert G. Kinkade, 1972, sponsored by Joint Army–Navy–Air Force Steering Committee, available from the Superintendent of Documents, U.S. Government Printing Office.

The latter book is an updating and revision of a book of the same title by Wesley F. Woodson and Donald W. Conover, University of California Press, 1966. *MIL-STD-1472* and *Human Enginering Guide to Equipment Design* represent the state of the art in human-factors considerations for system equipment design. The

former is an active document with a formal revision procedure in effect. Both are as well suited to commercial products as they are to military products.

Two other documents should be consulted when designing systems including displays and controls, as follows:

• *Analysis of Human Factors Data for Electronic Flight Display Systems*, Technical Report AFFDL-TR-70-174, April 1971, sponsored by Air Force Flight Dynamics Laboratory, Wright-Patterson Air Force Base, Ohio 45433.

• *NAVSHIPS Display Illumination Design Guide*, July 1973, Technical Report NELC/TD223, sponsored by Naval Electronics Laboratory Center, San Diego, CA 92152.

The design from an electronics standpoint is specified by several documents.

• *MIL-STD-454: Standard General Requirements for Electronic Equipment*

is used extensively by the government. Other documents on design specification can be obtained from ANSI, IEEE, UL, and EIA. See Section 2.5.4 for addresses.

2.5.3 Environmental Testing. A display can never be considered to be any more than "in development" until it has passed the rigors of environmental testing conducted on certified test equipment under the supervision of the quality assurance department.

The environmental requirements for a display are ultimately defined by the test specifications. No requirement is definitive until both the qualifing test and the test procedure are defined. The most thorough, consistent, and rigorous tests are contained in military specifications. These specifications and supporting test equipment form an excellent foundation for environmentally qualifying a display. The major tests are listed in Table 2-4. One of the hazards in using these tests is that a test category may be selected which is more severe than the environment to which the display will ever be exposed. Electronic components can be overtested. The testing should be no more severe than the expected environmental exposure.

All electronic components essentially have the same environmental exposure as the system

Table 2-4   Military Display Environmental Tests

Test	Test Specification
High temperature	MIL-STD-810
	MIL-E-5272
Low temperature	MIL-STD-310
	MIL-E-5272
Altitude	MIL-E-5400
	MIL-E-5272
	MIL-STD-810
Temperature-Altitude	MIL-E-5400
	MIL-E-5272
	MIL-STD-810
Humidity	MIL-E-5272
	MIL-STD-810
Vibration	MIL-STD-167
	MIL-E-5272
	MIL-STD-810
	MIL-T-4807
Mechanical Shock	MIL-S-901
	MIL-E-5272
	MIL-STD-810
	MIL-T-4807
Acoustical Noise	MIL-STD-810
	MIL-STD-740
Acceleration	MIL-E-5272
	MIL-STD-810
Salt Spray	MIL-STD-810
	FED-STD-151
Sand and Dust	MIL-E-5272
	MIL-STD-810
Fungus	MIL-STD-810
Explosive Atmosphere	MIL-STD-810

MIL-E-16400 and MIL-E-5400 cover all tests at higher system level than display system.

that contains them. Temperature exposure is an important exception in that internal heating dictates higher operating temperatures than those to which the external surfaces are exposed. Also, systems may be sealed to some degree to protect internal components from exposure to sand, dust, moisture, and salt spray. Internal parts can be mounted to provide some degree of shock, vibration, and acoustic noise isolation.

The sources of standard tests are as follows:

- EIA Standard RS-186: Standard Test Methods for Electronic Component Parts
- IEC Publication 68: Basic Environmental Testing Procedures for Electronic Components and Electronic Equipment (published in several parts)
- MIL-STD-810: Military Standard Environmental Test Methods
- ASTM Standard Test Methods--primarily applicable to materials used in electronic components

The general military specifications which invoke electronic requirements and test methods such as those in MIL-STD-202 and 820 are contained in:

- MIL-E-16400: *Military Specifications, Electronics Equipment, Naval Ship and Shore, General Specifications*
- MIL-E-5400: *Military Specification, Electronic Equipment, Airborne, General Specification*

The qualification of a display is of no consequence unless the future displays are manufactured by the same methods and using the same materials as used in the qualification units. A manufacturer must have a written quality assurance program if he is to achieve uniformity in the product quality. A good example of a quality assurance program is defined in:

- MIL-Q-9858, *Military Specification, Quality Program Requirements.*

2.5.4 Standards.   Except for MIL-SPEC, writing standards in the United States is strictly a voluntary operation. The standards are usually prepared by a working group in one of the professional societies or trade association groups. The central clearinghouse for approval and distribution of standards in the United States is the American National Standards Institute (ANSI). ANSI is the coordinating agency which represents the American position at the International Standards Organization (ISO) located in France. Groups representing other countries at ISO include the British Standards Institute (BSI), the Deutsches Institut fur Normung (DIN), and the Association Française de Normalisation (ANFOR). ANSI issues the Catalog of American National Standards, which is available upon request from their New York Office.

Military standards are developed by DOD

agencies in cooperation with industry. The MIL-SPEC catalog is called DOD Index of Specifications and Standards (DODISS) and can be ordered from the Commanding Officer, Naval Publications and Forms Center, 5801 Tabor Avenue, Philadelphia, PA, 19120, (215) 697-2000.

Numerous professional societies and agencies maintain standards, test specifications, and perform tests. Several are listed here:

- Aeronautical Radio, Inc. (ARINC)
  2551 Riva Road
  Annapolis, MD 21401
  (301) 266-4000
- American National Standard Institute, Inc. (ANSI)
  1430 Broadway
  New York, NY 10018
  (212) 354-3300
- American Society of Mechanical Engineers (ASME)
  345 East 47th Street
  New York, NY 10017
  (212) 705-7722
- American Society for Testing and Materials (ASTM)
  1916 Race Street
  Philadelphia, PA 19103
  (215) 299-5400
- Electronic Industries Association (EIA)
  2001 Eye Street, NW
  Washington, DC 20006
  (202) 457-4900
- Illuminating Engineering Society of North America (IES)
  345 East 47th Street
  New York, NY 10017
  (212) 705-7926
- The Institute of Electrical and Electronic Engineers (IEEE)
  Standards Department
  345 East 47th Street
  New York, NY 10017
  (212) 705-7960
- National Associaton of Photographic Manufacturers (NAPM)
  600 Mamaroneck Avenue
  Harrison, NY 10528
  (914) 698-7603

- National Bureau of Standards (NBS)
  National Center for Standards & Certification
  Technical Building, Rm. B-166
  Washington, D.C. 20234
  (303) 497-3000
- Optical Manufacturers Association (OMA)
  6055-A Arlington Boulevard
  Falls Church, VA 22044
  (703) 237-8433
- SAE (Society of Automotive Engineers)
  400 Commonwealth Drive
  Warrendale, PA 15096
  (412) 776-4841
- Underwriters Laboratories (UL)
  333 Pfingston
  Northbrook, IL 60062
  (312) 272-8000
- Bureau Central De La CIE
  52, Boulevard Malesherbes 75008
  Paris, France

2.6 DISPLAY SUBSYSTEMS

A display can be broken down into several essential parts or subsystems, each worthy of elaboration. The display system in its most complete form is as diagramed in Fig. 2-1. The heart of the system is the panel itself. The various technologies used to achieve an image on the flat panel or CRT are discussed in Chaps. 6 through 11.

2.6.1 Faceplate. The faceplate may be an integral part of the front surface of the display, or a separate transparent plate adhered to and optically coupled to the display. The faceplate performs several functions:

- Structural and protective shield
- Radiation shield
- Electromagnetic interference (EMI) shield
- Front surface etching to scatter reflections
- Antireflective (AR) coating
- Optical filtering
- Optical focusing

The physical arrangement of the faceplate is typically as shown in Fig. 2-5. Faceplate design has been developed for high-performance CRTs

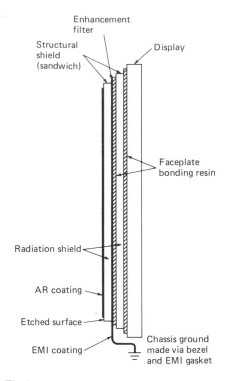

Fig. 2-5. Cross section of faceplate structure.

used in aircraft.[10] All of these features may not be necessary, depending on the display technology and display system requirements. For example, surface etch and AR coating may only be justified on a critical display required to achieve sunlight readability or low-fatigue, error-free reading. Structural shield would normally be required on portable equipment or equipment installed in a high-vibration environment.

The faceplate performs several vital functions, and when properly designed enhances the readability of the display by increasing its contrast ratio and brightness ratio. The different functions performed by the faceplate require different materials of differing thicknesses and optical properties. The optical transmittance, absorptance, and reflectance of the total faceplate sandwich must be considered when satisfying the individual requirements. That is to say, additional optical coupling thin films may need to be added to minimize intralayer reflections and interference effects.

Personal and portable equipment must survive a standard drop test. It is reasonable to assume that at some time during the life of portable equipment, it will be dropped from carrying height or table top height. Well-designed equipment will survive a drop onto a hard surface landing on any side or corner from a height of 3 to 6 ft (1 to 2 m). The case may be marred, but the display and electronics must continue to operate without any degradation in performance or life. The structural shield aspects of the faceplate must protect the display during such a drop.

Structural Shield. Structurally, the faceplate must protect the operator from high voltage and protect the display panel from breakage. The most severe test to which a display faceplate may be subjected is the shatter test. In this test, a calibrated steel ball is dropped from a prescribed height upon the display, face-up. Typical values are a 2-oz. ball from six feet. This test is primarily intended to ensure that the operator in aircraft, ships, vehicles, and elsewhere will be protected from flying debris, sharp edges, and exposure to electrical and material hazards. It is normally not necessary that the display operate after the test, but only that the operator not be subjected to additional hazards. Flying glass is usually the greatest concern.

In the event of internal explosion, implosion, or damage of any foreseeable kind, the faceplate is to contain the parts, and thus protect the operator from possible injury. Again, it is not necessary that the display continue to operate.

For the greatest protection, the shield can be made as a sandwich of two sheets of tempered glass bonded together with an adhesive sheet of Polyvinyl Butyrate (PVB) between. With time, exposed plastic shields tend to become scratched. The sandwiched plastic sheet can also serve as an optical filter if impregnated with the proper absorbing dyes.

Radiation Shield. Flat-panel display technologies normally do not present any X-ray or UV radiation hazards. Some UV radiation and EMI is emitted from gas-discharge displays. Such radiation is normally sufficiently absorbed in the display panel glass itself. Most glasses are good UV absorbers. When high voltage (>20 kV) is used, X-rays may be a problem. Moderate amounts of barium or lead in the glass will absorb the X-rays if the source cannot be reduced or eliminated. Barium is preferred over lead as lead tends to darken the glass with age.

The UV and EMI source may not be within the display itself. However, since the display presents a hole in the panel with respect to electromagnetic radiation, the faceplate must prevent its passage either into or out of the display. The EMI may be externally generated and interfere with the display system by penetrating through the display panel window. The sun is rich in UV, which may have an aging or other detrimental effect on the display media materials.

Military displays must be nuclear-radiation-hardened to a wide spectrum of higher energy radiation than is normally required for civilian displays. Since it is assumed that a display is to accompany a human operator, it need not survive an exposure any longer than the operator can survive. These levels are typically 500 to 5,000 rads total dosage, which is the maximum survivability level for a tactical environment.

Electromagnetic Interference Shield. An electromagnetic interference (EMI) shield, if needed, must completely cover the display opening and be connected to chassis ground potential through a low-resistance path. The shield is used to prevent EMI or static from radiating into or out of the display opening. At some level, the electromagnetic energy will interfere with the operation of the display electronics, or the display can interfere with other electronic equipment. Also, gas-discharge displays emit some radio frequency interference. In some applications it is necessary to prevent detection and interception of display signals. The EMI shielding is usually accomplished with a transparent conductor deposited over the entire faceplate. It conducts the energy to ground potential via an EMI conductive gasket. The faceplate is covered with a semitransparent thin film of metal approximately 0.01 μm thick or with transparent thin-film conductors of tin oxide or indium-tin oxide of approximately 0.1 to 0.2 μm. The allowed resistivity is typically 100 ohms per square; the lower the better, until it degrades the display's readability. The gasket may be a metal-impregnated rubber or a metal mesh that is compressible. The gasket permits disassembly

and reassembly of the display in the field and prevents EMI from leaking out between the faceplate and metal chassis.

First-Surface Etching. An improvement in display readability can almost always be realized by etching the first surface to cause a slight scattering of the specular reflections. This has the effect of diffusing or fogging the reflected image from the first surface or any of the internal surfaces. The operator is unable to focus on the reflected image, and its impact is therefore diminished. The fogging of the display image is minimal due to its proximity to the etched surface. The minor amount of fogging or blurring of the image is usually beneficial as it tends to diffuse the sharp pixel boundaries and give a more uniform quality. The etching is done before the first-surface antireflective coating is applied. Etching and AR coatings complement each other. Both are beneficial regardless of the ambient illumination.

The optimum amount of etching is subjective. Etching should normally be used to the maximum degree until it begins to degrade the resolution of the display. If it is acceptable to lose resolution by diffusion from an etched faceplate, the display resolution requirements should be challenged. An overall savings may be realized by reducing the resolution of the display panel. The degree of etching can be measured and specified with a standard resolution chart such as EIA Resolution Chart 1956 or U.S. Air Force Chart AF 1951. The chart is placed directly behind the faceplate in place of the display media. The resulting readable resolution, as read from the viewside of the faceplate, is then a direct measure of the degree of diffusion and therefore etching.

Antireflective Coating. The first surface presented to the air is a source of severe reflections due to the unavoidable mismatch in index of refraction between air and faceplate. The index for air is approximately 1 and for the faceplate glass it is approximately 1.5. From Fresnel's formula, the reflected light is approximately 4% of the incident light from each air-glass surface interface. The reflected image of the operator and surrounding background area is easily seen if the operator focuses his eyes on it. This extraneous image only serves to clutter the display. A good multilayer antireflective coating can reduce this to 0.1% over a broad band. For a balanced design, all the optical interfaces must be considered. In multilayer faceplates, there will be several material interfaces. For example, an EMI shield of indium-tin oxide has an index of refraction of 1.7 to 2.0, which will cause reflections at its interface surfaces.

One of the more common antireflective coatings for glass-to-air interface is a thin-film coating of MgF_2 at one-quarter wavelength for the primary color. MgF_2 is easy to deposit, and it is hard and chemically resistant. It is a good antireflective coating, reducing reflectivity to approximately 1%. Reducing reflections to below 1% requires multiple thin films with graded indexes of refraction.

Antireflective (AR) coating requires special care in cleaning. If the AR coating cannot be kept clean, its use will be more detrimental than beneficial. Simple wiping is not satisfactory as it abrades the AR film and will not remove fingerprints, salt spray, or oils. A cleaning agent of ammonia and water is required to remove fingerprints. AR coatings should not be used on the first surface if the display must be used while exposed to rain, fog, dust, or fingerprints. The AR coating makes the contamination stand out. First-surface etching greatly reduces the adverse effects of fingerprints.

Filtering. Optical filtering will always improve the contrast ratio of an emitting display under ambient illumination. However, it does so at the cost of reduced brightness. Filtering has no advantage with nonemitters, as will be shown using the pixel-contrast-ratio equation. The pixel contrast ratio (PCR) is the ratio of luminance of an on-pixel, L_{on}, to that of an off-pixel, L_{off}, plus the effect of ambient illumination, I, reflected from the display by the reflectivity coefficient R_{on} for the on-pixel and R_{off} for the off-pixel. From Section 2.5.1 (Performance):

$$PCR = \frac{L_{on} + IR_{on}}{L_{off} + IR_{off}} \quad (Eq.\ 2\text{-}1)$$

Internal reflectivity effects and display media reflectivity are lumped into one overall reflectivity coefficient. The reflectivity of the display media is not directly alterable without changing the display material itself, changing the internal structure, or adding antireflection coatings.

A neutral-density filter in front of a display will reduce the emitted light onefold and the reflected ambient illumination twofold. The ambient illumination must pass through the neutral filter twice, once inbound and once after reflection back out toward the viewer. First-surface reflections from the filter are required to be made negligible by the use of antireflection coatings on the filter and from the second surface by optically coupling the filter to the display. To show the advantage of filtering in equation form, let the filter transmittance be T where T is greater than zero (a transmissivity of zero is the trivial case of a completely occluded display):

$$PCR = \frac{TL_{on} + T^2 IR_{on}}{TL_{off} + T^2 IR_{off}} \quad (Eq.\ 2\text{-}2)$$

where the R_{on} and R_{off} are not changed with the addition of the filter. The magnitude of the reflected ambient illumination is reduced by T, and thus the argument for neutral-density filtering for emitting-type displays. Filtering has no advantage with nonemitting displays, since all of the light is originally ambient illumination. Thus for nonemitters there is no improvement in PCR and always a reduction in display brightness ratio.

The PCR for emitters is optimum when T approaches zero, according to Eq. 2-2. As T approaches zero, the average brightness of the display would be proportionally reduced regardless of the ambient illumination. For the same detectability, the PCR must be increased if the DBR is decreased.

DBR is a new expression to account for the detrimental aspects of filtering, to yield filter design criteria. Over the years, this term has been used in different forms by human-factors engineers. The display brightness ratio (DBR) is the macroscopic requirement that the display has in order to be easily read by the eye adapted to ambient light. It serves to set the range over which the pixel contrast ratio (PCR) has a useful meaning. The PCR is a microscopic parameter relating to spatial information detectability. For example, a PCR of 10 is almost meaningless if it is for an average display brightness 100 times below that to which the eye is adapted. The display brightness ratio is simply the average luminance of the display area divided by the average luminance of the area surrounding the display within the cone of vision which affects the light adaptation of the eye. From Section 2.5.1 (Performance):

$$DBR = \frac{K(L_{on} + IR_{on}) + (1 - K)(L_{off} + IR_{off})}{IR_s}$$

(Eq. 2-3)

where K is the average fraction of pixels that are on and R_s is the average reflectivity of the "surround."

The minimum magnitude of T for an emitting display is constrained by the requirement for display brightness. For easy reading, a display should be the same brightness as the surrounding area, or slightly brighter. This is generally true for any workspace such as desktop, workbench, or countertop. An equation for T as a function of display reflectance can be derived showing the interaction.

When a filter is added to a display, the display brightness ratio is reduced, as can be seen by rewriting Eq. 2-3 with the proper addition of T for filter transmittance and observing that R_{off} is equal to R_{on} for an emitting display and can be represented by R_d.

$$DBR = [KTL_{on} + (1 - K)TL_{off} + T^2IR_d]/IR_s$$

(Eq. 2-4)

A general solution can be found by solving Eqs. 2-2 and 2-4 for T using the following display design criteria for numerical values for DBR and PCR:

1. Under normal reading conditions, an average brightness ratio between display and surround should be 2:1.[2]
2. Similarly, it is desirable to have a display contrast ratio of 10:1[2] between on-pixel and off-pixel. Therefore, Eq. 2-2 reduces to:

$$L_{on} = 10 L_{off} + 9 TIR_d \quad \text{(Eq. 2-5)}$$

and from Eq. 2-4:

$$2 IR_s = KTL_{on} + (1 - K)TL_{off} + T^2IR_d$$

(Eq. 2-6)

Combining Eqs. 2-5 and 2-6 results in a quadratic in T. Solving for T, and noting from physical considerations that T is greater than zero and less than 1, and letting C^2 equal $8/(9K +1)$:

$$T = \frac{-L_{off} + \sqrt{L_{off}^2 + C^2I^2R_dR_s}}{2 IR_d} \quad \text{(Eq. 2-7)}$$

An interesting solution results if L_{off} can be taken as equal to zero, as would be the case in many displays.

$$T = \sqrt{2/(9K + 1)} \sqrt{R_s/R_d} \quad \text{(Eq. 2-8)}$$

Note that the filter value is no longer a function of the ambient illumination.

Narrow-band filtering is beneficial if the emitted light is radiated over a narrow band (50 nm wide or less at half power). So far, we have assumed that the emitted and illuminating luminance is independent of wavelength, which is an idealized case. LEDs tend to be yellow, green, or red. Gas discharge light is from a neon/argon gas mixture (Penning Mixture) that is orange. EL light is from terbium- or manganese-activated zinc sulfide, which emits in the yellow-green and yellowish-orange range, respectively.

Equations 2-2 and 2-4 must be rewritten as a function of wavelength and integrated over the visual spectrum to properly represent brightness as perceived by the eye. In general, the transmittance is made low at wavelengths where the display does not emit light to improve the PCR, and made high where the display does emit light to improve the DBR.

A combination of narrow-band filtering and neutral-density (gray) filtering has been used to improve luminance contrast and yield color contrast to achieve sunlight readability with light-emitting diode displays.[5] Color contrast is an important second dimension for improving PCR.

A third type of filter used for displays is a circular polarizer. Circularly polarized light is shifted 180 degrees in phase when it is reflected off a conductor. Ambient illumination reflected from metals is therefore absorbed when passing back through a circular polarizer placed in front of a display. This type of polarizer is very effective in eliminating annoying reflections from metallic conductors used in the display-active area. If metal surfaces are not used in the display, circular polarizers will be of no advantage.

Circular polarization is achieved by using a quarter wave retarder plate in conjunction with and after a linear polarizer. As a consequence of the linear polarizer, at least 50% of the unpolarized light is absorbed when it passes through the filter. Polarizers of any type reduce the useful viewing angle of a display to approximately 45 degrees off the normal.

On emitter displays that are isotropic radiators in a transparent medium, such as with EL displays, the loss in a circular polarizer can be compensated for. A nondepolarizing metallic reflector can be placed behind the emitter media thus doubling the light emitted forward to the viewer. Half of the emitted light will still be absorbed by the polarizer, but now there is twice as much. All the ambient illumination will still be absorbed by the circular polarizer. Metals can depolarize light as a result of scattering from surface roughness.

The circular polarizer can be located anywhere after the first surface and before the emitting-surface metal electrodes and metal structure. Neutral-density filtering and color filtering may be used as before. The first-surface etch and antireflective coatings may still be used as before.

Optical Focusing. Optical focusing may also be included in the faceplate. Individual pixel and character-enlarging lenses have been successfully used to enhance the readability of small monolithic LED displays. The reflectivity effects of the lens surface must be taken into account.

High-gain faceplates are sometimes used in high ambient illumination when the viewer's head position is restricted, such as is the case with aircraft pilots. These take on the form of optical fiber faceplates, louvers, micromesh, and honeycomb structures. The objective is to deflect the emitted light back towards the viewer that would otherwise be lost off to the side. The granularity of the mesh or optical fiber must be greater than the pixel spatial resolution. The depth of the mesh or optical fiber

prevents the sunlight which enters at the side from reflecting off the display.

2.6.2 Bezel. The bezel is the display picture frame. In addition to framing the display aesthetically, it serves several other functions:

- Provides means for attaching display to chassis or covering attachment hardware and covering gap between display faceplate and chassis;
- Completes circuit for EMI shielding and path for conducting current to chassis and ground;
- Provides for water seal and gas seal between display and chassis;
- Provides mechanical structure for holding switch overlay and bezel-mounted switches.

2.6.3 Interactors. An integral part of a display subsystem is the means for operator interaction. There are two approaches:

- Electronically coupling through the processor;
- Coupling through the display emitter.

The operator is not aware of any difference. Electronic coupling is more universally used, as it does not place any additional requirements upon the display. The components used are:

1. Keyboard
2. Bezel keys
3. Display overlay
   (a) Ultrasonic
   (b) LED with photocells
   (c) Pressure-sensitive
   (d) Resistance detection
   (e) Capacitance detection
   (f) Insulating displacement
   (g) Surface acoustic waves
4. Joy stick
5. Trackball
6. Tablet
7. Mouse
8. Light pen

All of these are completely independent of the display panel except the light pen.

The light pen may be a photo cell which detects the pixel emission during the addressing sequence. When this signal is synchronized with the addressing timing signal, the exact row and column location of the light pen is known, assuming pixel-at-a-time addressing. This technique is used with raster-addressed CRTs.

Coupling through the display may be done with a light pen. The light pen emits UV at the appropriate wavelength which can be made to trigger the operation of pixels which had been excited up to just below operating threshold. This has been demonstrated with EL and GD displays. The electronics detect the extra current that flows in the corresponding row and column lines due to the new operating pixel.

2.6.4 Electronics. The electronics are generally organized into several parts, as shown in Fig. 2-6, all dedicated to the operation of the display. The electronics contribute a larger share to the cost of the display system than does the display panel. Because of the numbers and power involved, the set of row and column line drivers is the largest cost element.

The heart of the electronic drive is the processor or microprocessor which is programmed to manipulate the data bus, buffer, and memory. The timing circuit is used to generate a basic clock signal for all the display operations and synchronize the enabling of the row drivers. The memory is used to hold the data in coded alphanumeric form or in pixel-bit map form. The data format converts the data out of the memory and prepares it in parallel form with the proper timing for the column drivers. The buffer interfaces the memory to the data line and strips out any synchronization signal. The row and column drivers apply a voltage to the display row and column electrodes.

This description is for a matrix-addressed panel display. Variations are used for other types of addressing. For large panels with 512 by 512 lines, the electronic drive will require 16 MOS chips where each chip can drive 64 lines. For small displays such as eight-digit numerics, it may all be done in one custom LSI circuit in one MOS chip.

In general, matrix-addressed displays have a

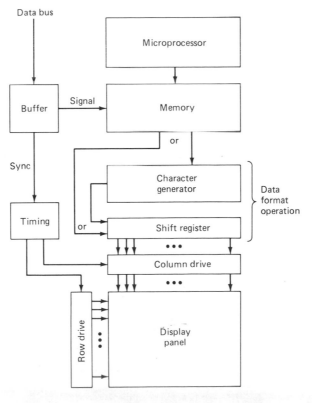

Fig. 2-6. Electronics for control and drive of displays.

large number of row and column lines which require a large number of pinouts, one for each line. This necessitates a large number of leads between the drive electronics and the display panel which in turn dictates the mounting of the display drivers next to the display panel as shown in Fig. 2-7. A small wiring harness is then used to connect this assembly to the remainder of the electronics.

The connection of all the drivers to all the row and column lines presents an electronic packaging problem which has been solved by one of several approaches, each of which is expensive and cumbersome. One technique is to simply mount the driver chips right on the display and seal them with the display. This technique makes repair and rework very difficult. A second approach is to mount the driver chips in a conventional package, solder the package

to a flexible polyimide-type cable with the appropriately etched circuit fanout which registers with the display panel electrical leads, and attach the polyimide flexible circuit to the display panel by soldering or spring-clamping. A third approach is to use an elastomeric connector to connect a conventional circuitboard with drivers to the display panel. The circuitboard is aligned and pressed against the display panel with the elastomeric connector in between.

The elastomeric connector is made of conductive layers of carbon or silver-impregnated polymer separated by layers of dielectric polymer on approximately 5 mil (127 μm) centers. The conductive layers are arranged to be parallel with the electric leads on the display panel and circuitboard with the conductive plane perpendicular to the surface. The metal on elastomer (MOE) connector performs in the same

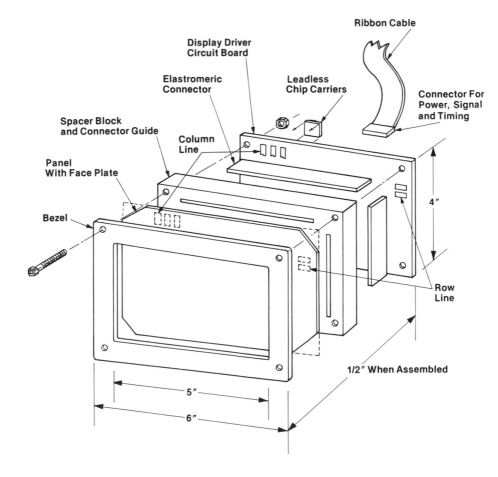

Fig. 2-7. Exploded view of display assembly.

way that the elastomeric connector does but is constructed differently.

The display panel and circuitboard leads must be aligned to ensure proper interconnection. The elastomeric connector need not be aligned. The conductor/dielectric layer centers of the connector are nominally two to three times smaller than the pitch of the leads on the display panel and circuitboard. By this means, one is assured of at least one dielectric polymer layer between adjacent leads.

This latter method is the most repairable technique of the three and is used in the packaging concept of Fig. 2-7.

2.7 TRANSILLUMINATION

Transillumination is used to give brightness or to light up a nonemissive display. "Transillumination" is synonomous with "supplementary lighting." Transillumination is more commonly used in the aircraft industry to describe the lighting of nonemitting displays. The transillumination is built into the display, as opposed to flood lighting or post lighting. Transillumination is only used during night operations or in low-light-level situations. The techniques used for transillumination are as follows:

- Edge lighting
- Back lighting
- Wedge lighting
- Internal lighting
- Louver-controlled back lighting

The term "transillumination" is used in MIL-STD-1472 to describe supplementally illuminated displays.

Transillumination can always be used to make a nonemissive display as bright as an emissive display.

2.8 PHOTOMETRY

2.8.1 Photometric Measurements. The two primary measurements used in display characterization are luminance and spectral radiance. Luminance is the quantitative measure of brightness and is quoted in English units in footlamberts (fL) and in SI units in candela/m² (nit). The spectral radiance is the radiation power as a function of wavelength and is quoted in units

of microwatts/cm² × nm of radiation × steradian or (μW/(cm² × nm × sr)). These two measurements are needed spatially and temporally in performing detailed measurements on display panels and CRTs in characterizing the performance of display materials, phosphors, faceplates, and geometries.

The PCR and DBR are computed based on luminance values measured with a spot photometer. The ambient illumination is normally measured using a standard reflectance block and and the same spot photometer. The reflectance block, typically made of pressed barium sulfate powder, is placed where the ambient illumination is to be measured, and a spot photometer is then used to measure the luminance of the reflectance block. If the reflectance block is a perfect Lambertian reflector, a luminance of 1 footlambert (fL) will be measured if the block is illuminated by 1 footcandle (fc). A perfect Lambertian reflector is a surface that reflects all the incident energy according to a cosine distribution about the normal without regard to the angle of incidence. Matte white paper is a good approximation to a perfect Lambertian reflector.

If color is a factor, spectral radiance measurements must be made with a spectral radiometer to obtain the energy per unit wavelength. Once these measurements are made, they are mathematically converted to the desired chromaticity coordinates by using the photopic spectral luminous efficiency tables, V(λ), as published by the CIE committee for the standard observer. Modern photometers and spectral radiometers with built-in, integrated microprocessors for making these calculations are shown in Figs. 2-8 and 2-9.

The microprocessor has revolutionized photometry, colorimetry, and spectroradiometry. Desktop computers are now regularly used, both to control measurements and reduce data.

The addition of computers to light measurement systems is particularly important in spectroradiometry. Now when the raw spectroradiometric data is taken, the computer can apply correction factors wavelength by wavelength and produce absolute data in terms of spectral irradiance (μW/(cm² × nm)) or spectral radiance (μW/(cm² × nm × sr)). The computer is used to develop the correction factors by scan-

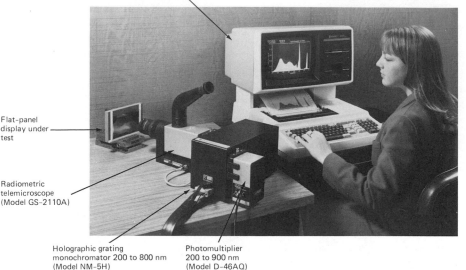

C-9 Color graphic computer
with integral hardcopy printer
(shown outputting hardcopy page)

Model GS4100
microprocessor board (controller
for calibration, interface, ranging, dark
current suppression, etc. is located
in the consolette below the computer)

Flat-panel
display under
test

Radiometric
telemicroscope
(Model GS-2110A)

Holographic grating
monochromator 200 to 800 nm
(Model NM-5H)

Photomultiplier
200 to 900 nm
(Model D-46AQ)

Fig. 2-8. Spectroradiometer. (Courtesy of EG&G Gamma Scientific, San Diego, CA)

ning a calibrated source and dividing the raw reading at each wavelength into the absolute value of the calibrated source at the same wavelength to produce the correction factor at each wavelength. These correction factors are then put into working memory, when measuring an unknown source, and used to multiply the un-known source raw data to produce corrected data in the same units as the calibrated source.

The computer may also be used to provide the measurements with correction factors applied to both the wavelength scale (wavelength correction) and linearity scale (amplitude correction).

The spectroradiometric data, once reduced

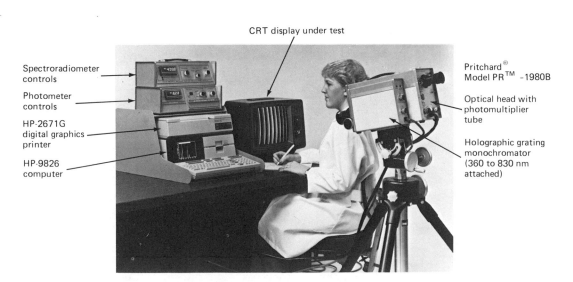

CRT display under test

Spectroradiometer
controls

Photometer
controls

HP-2671G
digital graphics
printer

HP-9826
computer

Pritchard®
Model PR™ -1980B

Optical head with
photomultiplier
tube

Holographic grating
monochromator
(360 to 830 nm
attached)

Fig. 2-9. Spectroradiometer. (Courtesy of Photo Research, Div. of Kollmorgen, Burbank, CA)

to absolute terms, may then be converted by the desktop computer into tristimulus values (X, Y, Z) and then chromaticity coordinates (x, y) or into universal color coordinates (u', v'), correlated color temperature, radiance (irradiance), Munsell hue, value, and chroma, or any units of luminance or illuminance. The beauty of these mathematical transformations is that they may all be done with the full CIE-table values so that no error is introduced in the conversion. This improves the accuracy substantially over even the best filter colorimeters or photometers. It is particularly important when measuring sources with sharp emission lines such as modern, high-pressure gas lamps or color cathode-ray tubes using rare earth phosphors.

Recently, radiometers have been introduced to the market which include microprocessors. These instruments take much of the routine calculation burden away from the desktop computer. They perform zero correction, background subtraction, wavelength scale correction, reliability correction, as well as signal averaging, detector sensitivity setting, and time constant selection. All of these functions are performed in machine language. The time needed to exe-

cute them is substantially reduced over doing them in the desktop computer in higher-level language.

2.8.2 Photometric Units. The interpretation and conversion of photometric units is sometimes confusing. They are one level more complex than radiometric units in that the $V(\lambda)$ convolution is added. Further, the conversion from luminance to illuminance involves a geometric description which is often not achievable or must be idealized. The most often used idealization is the Lambertian (cosine law) surface for reflection and emission.

A graphical pictorialization connecting luminous flux to luminous intensity to luminance and illuminance is given in Figs. 2-10 and 2-11. Unit conversions are given in Tables 2-5 and 2-6.

The final authority on photometry as used in engineering in the U.S. is Publication RP-16, ANSI/IES RP-16-1980 revision of ANSI Z7.1-1967 (R1973), *Nomenclature and Definitions for Illuminating Engineering*, sponsored and published by the Illuminating Engineering Society of North America.

A higher level authority on photometry and colorimetry is the CIE Publication No. 15

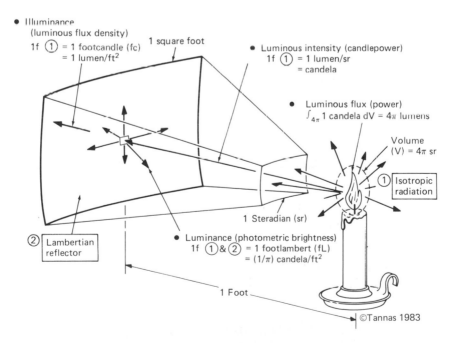

Fig. 2-10. Photometric English units.

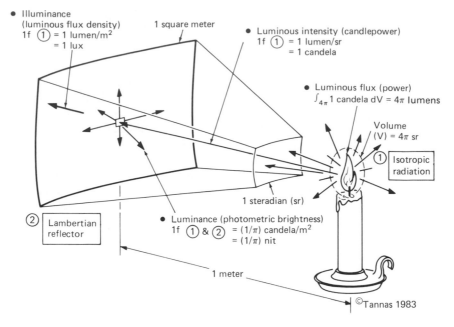

Fig. 2-11. Photometric SI units.

(E-1.3.1, 1971), *Colorimetry*, with Supplement No. 2 (1976), *Recommendations on the Uniform Color Spaces*, *Color Difference Equations*, *Psychometric Color Terms*, CIE, Paris. Both are official recommendations of the International Commission on Illumination made with representation from the U.S. and other interested technical parties (see Section 2.5.4 for addresses.)

It should be noted that the photometric unit most used by engineers, the footlambert (fL), is depricated in all the new standards. The nit, which is equal to one candela/m², or 3.142 fL, should be used in its place.

Table 2-5 Luminance Conversion Factors*
(Basic units are lumens/steradian × unit area)

Number of → Multiplied by ↓ Equals Number of ↓	Foot-lambert	Candela/m²†	Milli-lambert	Candela/in²	Candela/ft²	Stilb
Footlambert	1	0.2919	0.929	452	3.142	2,919
Candela/m² (nit)†	3.426	1	3.183	1,550	10.76	10,000
Millilambert	1.076	0.3142	1	487	3.382	3,142
Candela/in²	0.00221	0.000645	0.00205	1	0.00694	6.45
Candela/ft²	0.3183	0.0929	0.2957	144	1	929
Stilb	0.00034	0.0001	0.00032	0.155	0.00108	1

*Reprinted from RP-16, *Nomenclature and Definitions for Illuminating Engineering*, with permission from the Illuminating Engineering Society of North America.
†International System (SI) unit. Example: 1 candela/ft² = 3.142 footlambert. Moreover:

1 nit = 1 candela/m²
1 stilb = 1 candela/cm²
1 apostilb (international) = 0.1 millilambert = 1 blondel
1 apostilb (German Hefner) = 0.09 millilambert
1 lambert = 1000 millilamberts

Table 2-6 Illuminance Conversion Factors*
(Basic units are lumens/unit area)

Number of → Multiplied by ↘ Equals Number of ↓	Foot- candles	Lux†	Phots	Milli- phots
Footcandles	1	0.0929	929	0.929
Lux†	10.76	1	10,000	10
Phot	0.00108	0.0001	1	0.001
Milliphot	1.076	0.1	1,000	1

*Reprinted from RP-16, *Nomenclature and Definitions for Illuminating Engineering*, with permission from the Illuminating Engineering Society of North America.
†The International System (SI) unit. Example: 1 lux = 0.0929 footcandles. Moreover:

$$1 \text{ lumen} = 1/683 \text{ lightwatt}$$
$$1 \text{ lumen-hour} = 60 \text{ lumen-minutes}$$
$$1 \text{ footcandle} = 1 \text{ lumen/ft}^2$$
$$1 \text{ watt-second} = 1 \text{ joule} = 10^7 \text{ ergs}$$
$$1 \text{ phot} = 1 \text{ lumen/cm}^2$$
$$1 \text{ lux} = 1 \text{ lumen/m}^2$$

REFERENCES

1. Ireland, F. H., et al., "Experimental Study of the Effects of Surround Brightness and Size on Visual Performance." *Report No. AMIRL-TR-67-203* (1967), Aerospace Medical Research Lab.,WPAFB, Ohio.

2. Luckiesh, M., and Moss, F. K. *The Science of Seeing.* New York: D. Van Nostrand Co., p. 125, 1943.

3. Blackwell, H. R. "Contrast Thresholds of the Human Eye." *J. Opt. Soc. Am.* Vol. 36, No. 11, pp. 624–642, Nov. 1946.

4. Galves, Jean-Pierre, and Brun, Jean. "Color and Brightness Requirements for Cockpit Displays: Proposal to Evaluate Their Characteristics." Twenty-ninth AGARD Avionics Panel Technical Meeting, Paris, 1975.

5. Christiansen, Peggy. "Design Considerations for Sunlight Viewable Displays." *SID '82 Digest*, p. 194, May 1982.

6. Merrifield, Robin M., and Silverstein, Louis D. "Color Selection for Airborne CRT Displays." *SID '82 Digest*, p. 196, May 1982.

7. Wyszecki, G., and Stiles, W. S. *Color Science: Concepts and Methods.* New York: John Wiley and Sons, 1967 (rev. 1982).

8. MacAdam, D. L. "Specification of Small Chromaticity Differences." *J. Opt. Soc. Am.* Vol. 33, pp. 18–26, 1943.

9. Bellmeyer, Jr., Fred W., and Saltzman, Max. *Principles of Color Technology*, 2nd ed. New York: John Wiley and Sons, 1981.

10. Carroll, Gus F. "Display System Design." Seminar lecture notes cosponsored by SID and MIT, SID 1977 International Symposium.

3

THE VISUAL SYSTEM: CAPABILITIES AND LIMITATIONS

HARRY L. SNYDER, *Virginia Polytechnic Institute and State University*

3.1 INTRODUCTION

While the main substance of this book is a detailed compendium and evaluation of the design features and merits of the various display technologies, it is critical for the display designer and systems engineer to remember at all times that the ultimate purpose of any visual display is to provide useful and appropriate information to the person(s) using the display. In most operating environments, the display user needs to receive the desired information accurately and promptly; hence, two of the main evaluation criteria of display efficacy are speed and accuracy of information reception from the display. These criteria for display evaluation will be discussed in Chap. 4, which summarizes the visually important display parameters and the effects of these parameters upon human visual performance with the display.

In this chapter, we describe the characteristics of the human visual system which impact display design and the reception of displayed information. First, the overall organization of the visual system is discussed, including pertinent characteristics of the anatomy of the visual system which have direct functional importance for display design and selection. Next, we introduce contemporary and useful concepts of measuring spatial, temporal, and chromatic visual performance. These concepts have been used extensively in laboratory experiments and can be shown to be directly applicable to describing the visual performance with and image quality of various displays.

3.2 ANATOMY OF THE VISUAL SYSTEM

3.2.1 Overall Organization. In general, we regard the visual system as the two eyes, the neural network and pathways which connect the eyes ultimately to the brain, and that portion of the brain largely concerned with visual sensation, i.e., the visual cortex. It is beyond our purposes in this text to describe the neural pathways and information-processing levels of the visual system except to indicate that subcortical levels have been shown to provide significant information processing prior to the termination of the sensory information at the cortex.

What is important for our purposes, however, is the organization of the eye itself, its primary anatomical characteristics, and the relation between these anatomical characteristics and the functional capabilities of the visual system. Disregard for these characteristics can lead (and has often led) to poor display design and low utility of otherwise good engineering of the display device.

A horizontal section through the eye is shown in Fig. 3.1. As light enters the eye, the larger amount of refraction is provided by the cornea, the front hard surface of the eye. The pupil, or hole in the iris, controls the amount of light entering the inside of the eye. The diameter of the pupil is important in that a larger diameter admits light rays to the edges of the lens and thereby reduces the quality of the optical image. The eye's lens provides a lesser

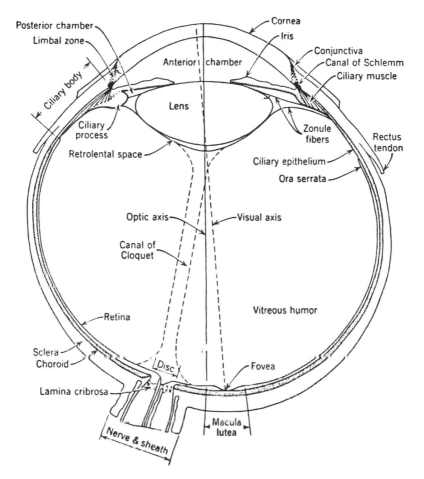

Fig. 3-1. Horizontal section through the eye.

amount of focusing of incoming light rays, permitting the viewer to alter focus distance ("accommodate") to objects at varying distances. The light rays then pass through the internal fluid of the eye, ultimately reaching the retina, or rear surface, at which the neural imaging process begins. It is at the retinal level that the anatomical organization of the visual system becomes critical to the display designer.

3.2.2 Rod and Cone Characteristics. The human eye is said to have a duplex visual system in the sense that the retina is composed of two distinctly different photoreceptors—rods and cones. These receptors are distributed in different parts of the retina, have different sensitivities to light energy, and are connected differently to higher neural pathways. For these

reasons, the duplex retina must be understood in order to make maximum use of the close relationship between the display design and the visual system's capabilities (and limitations).

As illustrated in Fig. 3.2, the cones are concentrated in the center of the eye, or the fovea centralis. For most practical purposes, the fovea is regarded as the central two degrees of the visual field or about one degree either side of the visual axis.

Rods, virtually absent in the fovea, increase in density with distance from the fovea, reaching a maximum density about 18–20 degrees from the fovea. Rods continue out to about 80 degrees from the visual axis but with nearly linearly decreasing density.

The area between 15 and 18 degrees nasal from the fovea and on the horizontal meridian

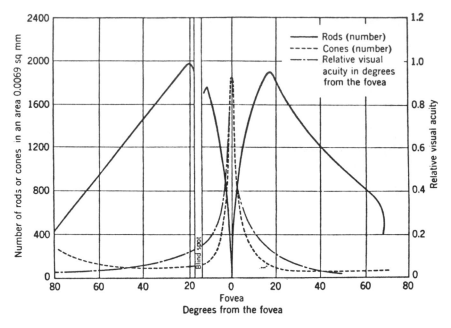

Fig. 3-2. Rod and cone distributions.

of the eye is totally devoid of both rods and cones. This area, about 3 degrees in diameter, is termed the "blind spot" and is the point at which the neural fibers from the rods and cones exit the eye. Any image falling on the blind spot is not seen by the visual system. (The blind spot can be demonstrated easily by closing one eye and moving a small object inward from the periphery while fixating on a point straight ahead. When the object is imaged on the blind spot, it will visually disappear.) Fortunately, the blind spots on the two eyes are bilaterally symmetric, with the result that an object imaged on one blind spot is "seen" by the other eye. We are usually totally unaware of the partial loss of vision in the two blind spots and early in life become totally accustomed to this anatomical restriction.

There are four fundamental differences between rods and cones that are vitally important to information reception from visual displays:

1. Rods have much greater sensitivity to light than do cones.
2. Cones are differentially sensitive to the wavelength of light (i.e., they can "see" colors) whereas rods can only "see" shades of gray.
3. Cones have greater resolving power than do rods, and hence provide greater discrimination of details.
4. Rods are more sensitive to temporal changes in luminance levels and are therefore more likely to "see" flickering displays.

Figure 3.3 illustrates the greater sensitivity of rods to light energy. It should be noted that there are several orders of magnitude between the lowest intensity that the dark-adapted rod can "see" and that which the cone can "see." Further, the rod is insensitive to the longer wavelengths in the visual spectrum, those wavelengths typically perceived as red and above 670 nanometers. Traditionally, the area above the cone threshold curve is referred to as *photopic vision* and represents the combinations of intensity and wavelength to which the cones are sensitive. In this photopic region, the eye is sensitive to color and sees with the cone photoreceptors.

The area between the rod and cone curves is the *scotopic* region of vision, the combinations of wavelength and intensity to which only the rods are sensitive. Intensities in this region are seen only by the rods, and are achromatic, i.e.,

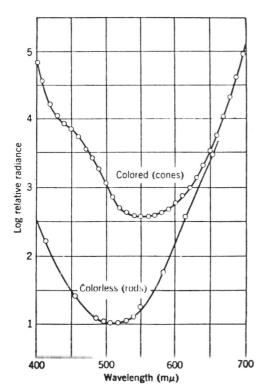

Fig. 3-3. Rod and cone threshold curves.

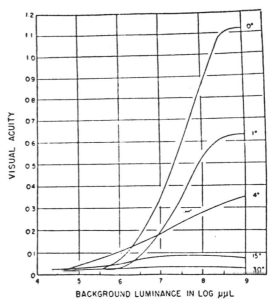

Fig. 3-4. Acuity as a function of off-axis angle.

shades of gray, even though they may be composed of various wavelengths which, were they more intense, would be seen to have color by the cones.

One of the more interesting and important visual phenomena is the gradual transition between the photopic and scotopic sensitivity regions, such as that which occurs when a multicolor display is gradually reduced in intensity or when a multicolor illustration is viewed under a light source of gradually decreasing intensity. As the intensity (or luminance) of the displayed image is reduced from the full-color photopic region into the scotopic region, colors "drop out" sequentially. The first colors to shift to gray are the blues and reds, followed by the cyans and yellow-oranges, and finally by the greens. This differential shifting of colors to gray is termed the Purkinje shift and is obviously critical to the use of color coding under low luminance levels.

While the density of rods and cones in their most dense regions is approximately the same (see Fig. 3.2), the interconnections between the receptors and the neural connections to higher neural centers are quite different. The cones are essentially connected "one-to-one" with the exiting neural fibers, while the rods are matrixed together in a local summation network. This summation permits the rods to have greater sensitivity, thereby increasing their equivalent signal-to-noise ratio and permitting the detection of lower-intensity stimulation as shown in Fig. 3.3. At the same time, the spatial resolution of the rods is diminished since the exiting neural signal cannot distinguish between the stimulation of one rod and its neighbor. Figure 3.4 illustrates the decrease in spatial resolution or acuity with increasing distance from the fovea. This decrease in resolution is due simply to the gradual shift from cones to rods as the photoreceptors. Thus, information which must be resolved with great detail must be imaged on the cones, that is, on the fovea. Functionally, this means that the observer must look directly at the source of information. Conversely, information which is of very low intensity can be seen only in the visual periphery (e.g., 20 degrees off axis) and should be viewed by looking "to the side."

The functional tradeoff in the above change in direction of fixation is that the directly fixated object, if sufficiently intense, can be seen with greater detail and in color, whereas the peripherally fixated object will be seen with less detail and only in shades of gray. However, the improvement in sensitivity with off-axis viewing is the tradeoff for loss of color and resolution, a tradeoff which has withstood the test of time in the evolutionary process.

3.3 SPATIAL VISION

Perhaps the most important characteristic of any visual display is the information density and resolution of the display. Resolution is referred to in many ways in the literature, some of the ways describing the number of display elements per unit distance at the display and some referring to the ability of the observer to discriminate among or between adjacent elements. It is clearly necessary to differentiate between these two classes of meanings. In this chapter, we shall refer to resolution in the sense of the discriminatory power of the visual system. In the following chapter, we shall relate the visual system's ability to resolve detail to the detail present on the display, in which case the concept of "resolution" shall be broadened to include the interaction between the information density on the display (in elements per unit distance) and the resolving capability of the visual system (also in elements, or cycles, per unit distance).

The ability of the visual system to resolve (or detect) at a 50% probability criterion (termed the "absolute threshold") depends on many attributes of the display. Among these attributes are the size of the object of interest, the luminance of the object, the luminance of the background, the luminance of the surround of the display, the shape of the object, the colors (hue and saturation) of the object and the background, and the adaptive state of the visual system. It is beyond our purposes to summarize or review this voluminous set of data; the interested reader is referred to other secondary sources, such as Graham (1965) and Kling and Riggs (1971).

Traditionally, the ability of the visual system to resolve detail has been specified as the visual "acuity" under a given set of circumstances. While there is some interest in the various types of acuity, the most important spatial type of

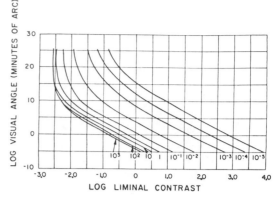

Fig. 3-5. Resolvable target size as a function of background luminance and contrast, from Blackwell (1946). The parameter is background luminance in foot-lamberts.

acuity is the "minimum separable" acuity, that minimum visual angle required for the observer to distinguish between two objects at a 50% probability level. The minimum separable acuity has been related to many of the display variables listed above; Fig. 3.5 is but one classical example of acuity, in this case as a function of both the contrast between the object and its background and the adapting (background) luminance level.

The type of data illustrated in Fig. 3.5 shows that the human observer can see small discs at varying luminance contrasts which depend upon the size of the disc and the adapting luminance level. Further, other data could have been cited to indicate that the threshold for small square objects or rectangular bars also behave in a reasonably similar manner but lead to slightly different performance levels due to the changes in shape. In fact, one can define many variations of this type of experiment which will show small (or, in some cases, large) changes in threshold levels. Thus, the person attempting to apply these data to a real-world display design problem is faced with a substantial dilemma. The application of these basic visual threshold data requires the use of a "field factor," that factor by which the threshold value is multiplied to account for differences in the complexity of the displayed information and the nature of the viewing situation. Such field factors vary from 2.5 to over 10, depending on the circumstances. It is often difficult, if not impossible, to estimate the magnitude of the field factor, and persons typically find that they can only do so by after-the-fact curve fitting techniques rather than by pre-experimental analysis.

On the other hand, one cannot simply perform the virtually infinite set of experiments which would be necessary to identify thresholds for objects of varying sizes, shapes, contrasts, adapting luminance levels, etc. Fortunately, in recent years the application of linear systems analysis has been applied to research in visual psychophysics with the result that the very powerful techniques of Fourier analysis can be used to abbreviate greatly the need for such a myriad of experiments.

Specifically, linear systems analysis, applied to the visual system, permits us to analyze any displayed pattern into its component frequencies, amplitudes, and phase relationships. It assumes that the visual system behaves as a Fourier analyzer, decomposing complex patterns into the frequency/amplitude/phase combinations and responding to each of the component frequencies independently. Given that a complex display can be Fourier analyzed and given also that the visual system behaves as a Fourier analyzer, then it is a tractable analytical matter to determine which of the frequencies contained in the display are "visible," i.e., above the visual threshold.

Numerous experiments have carefully demonstrated that the visual system, in fact, behaves as a Fourier analyzer in the spatial domain, at least to an adequate first approximation. As a result, we can indicate the sensitivity of the visual system to a standard pattern which is used in linear systems analysis and then compare this sensitivity to the frequency spectrum of the displayed information to determine the sensitivity of the visual system to that information. The standard pattern used for this purpose is the sine-wave grating, an example of which is illustrated in Fig. 3.6. In visual threshold experiments using this approach, the sine-wave grating is varied in spatial frequency (cycles per unit display distance or, more usefully, cycles per visual degree of angular subtense). The observer adjusts the luminance contrast (modulation) of the grating to a threshold criterion. This process is repeated many times for each spatial frequency over the range of spatial frequencies from very low (e.g., 0.5 cycle/degree) to either the upper limit of

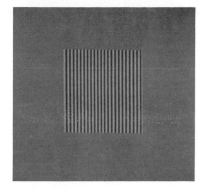

Fig. 3-6. Sine-wave grating.

vision (about 50 cycles/degree) or the upper frequency limit of the display device. Assuming that the displayed modulation is uniform, the modulation needed to reach a threshold response is then an indication of the sensitivity of the observer to that spatial frequency. Modulation, M, is defined as

$$M = \frac{L_{max} - L_{min}}{L_{max} + L_{min}},$$

in which

L_{max} = luminance of the lighter grating half-cycle, and

L_{min} = luminance of the darker grating half-cycle.

When plotted as threshold contrast as a function of spatial frequency, the resulting function is termed the Contrast Threshold Function or CTF. Often, one finds the inverse of this function, the Contrast Sensitivity Function (CSF), in the literature. For purposes of display design and evaluation, it is more useful to plot the CTF for reasons to be given in Chap. 4.

The typical CTF has a minimum in the region of 3-5 cyc/deg, with increasing modulation required to reach a threshold response at both higher and lower spatial frequencies. Figure 3.7 illustrates a typical CTF for normal, healthy, corrected adult eyes. Also illustrated is the estimated deviation from this typical curve for 90% of the population. What is interesting is not the fact that more modulation is required as the spatial frequency increases (i.e., the bars get smaller), but rather that more modulation is also

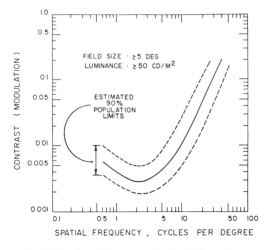

Fig. 3-7. Spatial contrast threshold function.

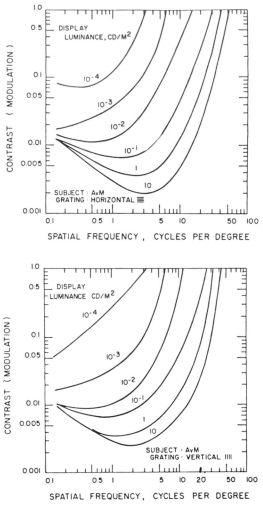

Fig. 3-8. Effect of luminance on the CTF.

required as the spatial frequency *decreases* below 3 cyc/deg. That is, as the bars get larger, the visual system becomes less efficient in integrating over distance, and greater modulation is required. For this reason, the eye is conceptualized by some to behave as a "passband filter," tuned to a frequency about 3 cyc/deg. Its d-c response in the spatial domain is near zero. While this analogy is attractive and permits generalization from other technical areas, one must be cautious in extending the analogy too far, in that the concept really applies only to the threshold response of the visual system and not to the sensitivity of the system well above threshold.

The CTF has become the basis for the quantitative analysis of display quality, as will be discussed in the next chapter. It is important to realize, however, that the CTF is altered by various display/surround conditions pertinent to operational display design and usage. For example, as illustrated in Fig. 3.8, decreases in the space average display luminance cause the CTF to both increase in threshold modulation and to shift toward the lower spatial frequencies. *That is, lower luminance displays require greater contrast for equal discriminability.* Further, smaller details become more difficult to see on lower luminance displays. With a display having an average luminance of 10 cd/m², the eye can resolve about 50 cyc/deg at 100% modulation; however, with an average luminance of 1 cd/m², this same observer can only resolve about 40 cyc/deg at 100% modulation and 20 cyc/deg at a more typical 10% modulation.

Also shown in Fig. 3.8 is the effect of grating orientation. The sensitivity for a horizontal grating is slightly greater than it is for a vertically oriented grating, but the difference is quite small. Gratings oriented at 45 deg to the horizontal result in noticeably less sensitivity (Campbell, Kulikowski, and Levinson, 1966). This relationship has prompted several researchers to suggest that dot matrices and rasters be oriented diagonally to reduce their visibility. It might be noted that the halftone dots of most newspaper screening processes follow this 45-deg convention, apparently for reduced visibility.

Figure 3.9 illustrates one of the more important effects upon the CTF. As viewing distance increases, the optical quality of the eye improves, as does that of most lenses. As a result, the CTF shifts slightly to the upper frequency end as the grating is moved farther away. This effect is of

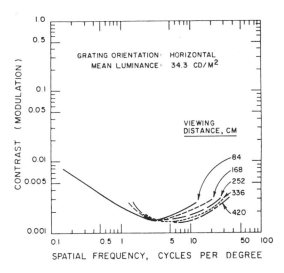

Fig. 3-9. Effect of viewing distance on the CTF, from Watanabe et al. (1968).

greatest importance at viewing distances less than 0.5 meter, for within this small distance the optical image-forming capability of the eye has been shown to decrease rapidly as illustrated in Fig. 3.10. It might be noted that the optical quality of the visual image for the typical uncorrected eye continues to improve out to 3 meters, and probably beyond, which suggests that larger displays viewed at greater distances can be more visually compatible than are smaller displays which subtend the same visual angle.

Numerous experiments have tested the notion that the visual system behaves as a spatial Fourier analyzer, decomposing a complex image into its component spatial frequencies with each frequency having its particular amplitude and phase

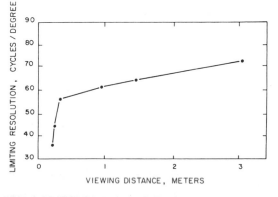

Fig. 3-10. Effect of viewing distance on the limiting spatial frequency of the sine-wave grating, from De Palma and Lowry (1962).

relationship to other frequencies. Without reviewing this literature in detail, it is fair to conclude that there is a strong balance of evidence that supports the Fourier hypothesis. For this reason, it has become accepted to use the Fourier decomposition approach in determining what content of a complex display can be seen by the visual system and what content is below the visual threshold. Following this logic, several models of image quality have been developed to assess display designs and configurations. These models will be discussed in Chap. 4.

3.4 TEMPORAL VISION

For many display technologies, there exists a need to refresh the display periodically to maintain adequate luminance and to avoid the perception of flicker. Because the determination of the minimum acceptable refresh rate is often critical to the bandwidth requirements of the device, one needs to be able to predict the refresh rate at which an image ceases to flicker and is seen as nonflickering or fused.

There is a large volume of literature dealing with research on the sensitivity of the visual system to pulses of light. Much of this research has used a stimulus or display which is sequentially turned on and off in rapid succession, with the observer asked to determine the minimum rate at which flicker is no longer perceived. This minimum rate is termed the critical fusion frequency or CFF. Plots of the CFF vs. other display parameters, such as display mean luminance or off-axis angle, abound in the literature. Brown (1965) has written an excellent review of this literature.

Unfortunately, the data are not of direct use to the display designer because the luminous sources used for the experiments are typically fast-response devices, such as glow modulator tubes. Thus, in these experiments the luminous intensity over time is not characteristic of many display devices which have noticeable and important persistence.

Fortunately, the approach used in the previous section to characterize the spatial response of the visual system can also be used meaningfully and usefully to characterize the temporal response, which is logically termed the *temporal contrast threshold function*, or temporal CTF.

3.4.1 Temporal CTF.

One can take a repetitive waveform of varying luminance and Fourier analyze it in the same fashion that one does a spatially varying signal. In fact, the Fourier analysis of a temporally varying signal is the more traditional application of this analysis form. The results of the analysis are, of course, a series of sine-wave frequencies, each having its own amplitude and phase shift.

The utility of a Fourier analysis approach to temporal vision was demonstrated in 1958 by deLange, who conducted an experiment to determine the effect of various waveforms upon the CFF at several luminance levels. His results, shown in Fig. 3.11, indicate that the sensitivity of the visual system to temporally varying stimulation can be predicted by the frequency and modulation of the Fourier fundamental, irrespective of the waveform itself. He also noted that the sensitivity to flicker is increased with increasing luminance, and that the typical bandpass response shifts to a higher center frequency with increasing luminance, not unlike the response of the visual system in the spatial domain. DeLange's (1958) results cover the frequency range from 1 to over 50 Hz, while the

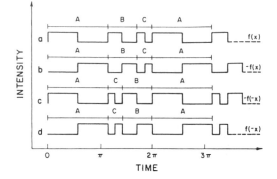

Fig. 3-12. Stimulus trains used by Forsyth (1960).

research of Almagor, Farley, and Snyder (1979) extended these data to a lower frequency of 0.01 Hz. They found that the sensitivity continues to decrease (i.e., the threshold modulation continues to increase) monotonically below 1 Hz.

Further support for the Fourier analysis of temporally varying stimuli by the visual system was provided by Forsyth (1960), who investigated a variety of square-wave pulse trains and predicted the CFF from the Fourier fundamental of each. Forsyth's waveforms are shown in Fig. 3.12 and his results, for six observers, are shown in Fig. 3.13. In this latter figure, all combinations of frequency and amplitude above the curves are seen as flickering and all combinations below the curves are seen as fused.

Numerous other experiments have been conducted since those of Forsyth and of deLange, all of which show with good consistency that the visual system behaves essentially as a Fourier analyzer in the temporal domain. Fortunately, this means that we can take the Fourier fundamental and the first several harmonics of the waveform, compare those coefficients with known threshold curves, and estimate whether or not the display exhibiting that waveform will be seen as flickering or fused. The only difficult part of this process, other than measuring the waveform, is to know which visual threshold curves to use for the purposes of making the comparison. Ample research exists to define temporal CTFs for a variety of conditions, including field size, target size, and mean luminance. See Snyder (1980) for a review of this approach and for references. The application of this approach to representative displays will be demonstrated in Chap. 4.

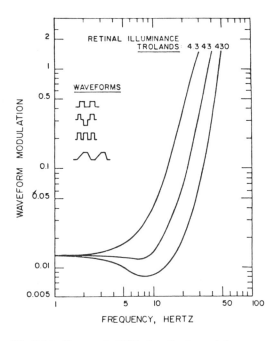

Fig. 3-11. Temporal CTF for fundamental component of four waveforms, from de Lange (1958).

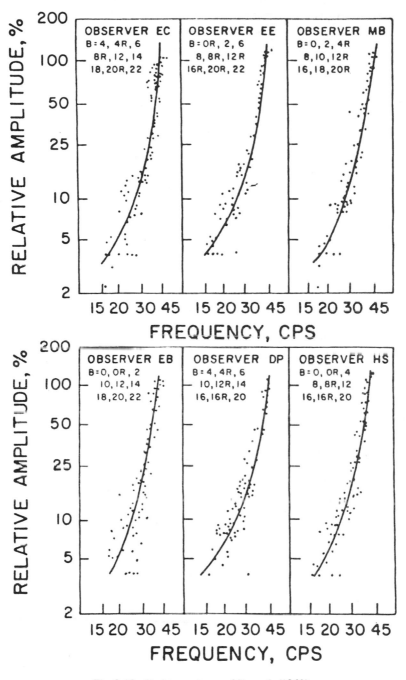

Fig. 3-13. Fusion contours of Forsyth (1960).

3.4.2 Other Temporal Psychophysical Data. In addition to the CTF, there are other temporal psychophysical data which are directly pertinent to the design of visual displays. One of the most critical, yet most ignored by display designers, is the concept of the integration time of the eye. As early as 1885, Bloch showed that the visual threshold is constant in total energy if the stimulus duration is less than some upper limit, t. That is, the eye integrates light energy over the stimulus duration, and the threshold is a direct function of this integrated energy rather than of the momentary energy level itself. Several experiments (e.g., Herrick, 1956) have demonstrated that this upper limit for foveal vision is about 100 ms; further, it appears that the upper limit for peripheral vision is approximately the same, at least out to 20 deg off axis (Baumgardt, 1972). The results of Barlow (1958) further support the upper limit of temporal integration as 0.1 s, with no difference in this limit for fields between 0.111 deg^2 and 27.6 deg^2.

This integration time has been shown to vary significantly with differences in luminance of the flickering field (Almagor et al., 1979). Increases in luminance result in decreases in integration time, as measured by both CFFs and the duration of the temporal Mach band, an overshoot perception of continuing luminance change at the end of a ramp function of luminance change. Fig. 3.14 illustrates the stimulus luminance function and the related subjective brightness experience.

The importance of this varying integration time lies in the viewing of dynamically noisy displays, such as low S/N television displays of the low-light-level sensor type. Images imbedded in dynamic, uncorrelated noise should be viewed in a dim environment and with low display luminance to increase the integration time and thereby permit the visual system to filter out more noise. Because the eye trades off integration time for integration space, the acuity remains reasonably high while the perceived noise is reduced through increased temporal integration. Conversely, images of moving objects should be viewed under higher display luminance conditions to reduce integration time and therefore

reduce image blur induced by the integration time.

3.5 COLOR VISION

Our subjective visual experience permits us to classify visual scenes into various parameters or dimensions, among them brightness, contrast, texture, and color. The visual object can be measured physically in terms of luminance and luminance contrast, chrominance, and texture. However, our visual experiences are not linearly related to the physical world. Thus, we perceive *brightness* as a stimulus characteristic which is determined by the combination of luminance (the physical measure of intensity) and chrominance. Chrominance is the physical combination of dominant wavelength and purity, reduced purity being caused by increases in the amount of achromatic mixture added. Correlated to the chrominance domain are the perceptual dimensions of *hue*, which is related to dominant wavelength, and *saturation*, which is related to purity. Thus, it is important to distinguish between the physical quantities of wavelength, purity, and luminance, and the perceptual quantities of hue, saturation, and brightness. The physical quantities are orthogonal while the perceptual quantities are not, changes in one often causing changes in the others. As a result, one can see that it is most inappropriate to speak of "brightness" in luminance terms, such as foot-Lamberts or candelas per meter squared. Similarly, it is inappropriate to speak of hue or saturation in physical terms such as wavelength or percent.

The visual system is not equally sensitive to all wavelengths of visible light energy. Accordingly, it is necessary to understand the visual sensitivity in the chromatic domain to be able to compare the emission spectra of various display technologies to the likelihood of seeing the display. The starting point for this type of comparison is to determine the radiant or luminous emission of a display as a function of wavelength. If the emission is measured in radiance terms, such as watts per square centimeter per unit solid angle, then the radiance can be converted directly to luminance by the formula

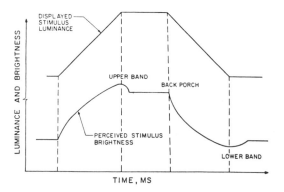

Fig. 3-14. Comparison of physical stimulus and perceived brightness, from Almagor, Farley, and Snyder (1979).

$$L = 683 \int_{350\,nm}^{750\,nm} P(\lambda)\, V(\lambda)\, d\lambda,$$

in which

L = luminous flux, in lumens,
$P(\lambda)$ = radiance, in watts, and
$V(\lambda)$ = the value of the photopic luminosity function at wavelength λ.

The conversion of 680 lumens/watt applies to photopic (cone) vision, while the conversion constant changes to 1745 lumens/watt for scotopic vision.

There are various devices available with which luminance can be measured, and many of the devices employ photomultiplier tubes coupled with filters to give an integrated response output proportional to luminance. While these are generally accurate within 5% for broad spectrum colors such as white, narrow band emission spectra can often be measured with much greater error. In addition, such devices typically use X, Y, and Z filters to estimate the amounts of the three color primaries in the CIE system (to be described below) from which can be derived the dominant wavelength, purity, and luminance. Again, such integrating filters can be in substantial error for narrow band emission spectra, such as those obtained with the blue P22 phosphor or the emission of an LED display. It is not uncommon to "measure" a nonvisible color from a standard P22 television CRT even with such devices in current calibration. For this reason, careful colorimetric work requires the use of radiance measurement as a function of wavelength and subsequent conversion to luminance or chrominance values. A carefully calibrated radiometer and standard radiance source are needed, but such exist commercially on the market. For a discussion of radiometric and photometric measurement see Walsh (1965) or other standard references on the subject.

3.5.1 Color Systems and Measurement.

The color spectrum is best described and measured in terms of radiant energy as a function of wavelength over the range to which the eye is sensitive, typically between 380 and 720 nm. Very narrow spectral bandwidth sources produce "pure" (highly saturated) colors. Colors having a broader spectrum are still seen by the eye as having a dominant wavelength (single, but less pure color) that can be calculated directly from the intensity distribution. The dominant wavelength or hue of a broad spectrum color is not seen to be any different than is the same wavelength/saturation combination of a pure, monochromatic color. This property of the visual system permits the specification of a color measurement system which can handle, with equal accuracy, both broad band and monochromatic narrow band colors.

The most often used color measurement system is that specified by the CIE, which is summarized in Fig. 3.15. This system is based upon three artificial primaries, X, Y, and Z, which are approximately equal to red, green, and blue, respectively. However, they are artificial in the sense that the specification of the primaries places them outside the visual spectrum; that is, they are nonvisible "colors." The proportions of each of the primaries in a real color is given by:

$$x = \frac{X}{X + Y + Z},$$

$$y = \frac{Y}{X + Y + Z}, \text{ and}$$

$$z = \frac{Z}{X + Y + Z}.$$

From these equations, one can see that

$$x + y + z = 1.$$

Thus, it is necessary to plot only two of these coefficients to characterize the color, as is done is Fig. 3.15. Conventionally x and y are plotted, and one can view y as the amount of the green primary and x the amount of the red primary. The solid line is the spectrum locus, which bounds all visible colors. Dominant wavelengths are indicated around the spectrum locus, as are the typical subjective names which define these color locations. White (or black or gray) is located in the center. The purity of a color is the proportional distance of that (x, y) color from the spectrum locus to the center (0.33, 0.33) point, expressed as a percent. The "purple" line, which connects 750 nm to 350 nm, is a combination of blue and red, and is expressed in

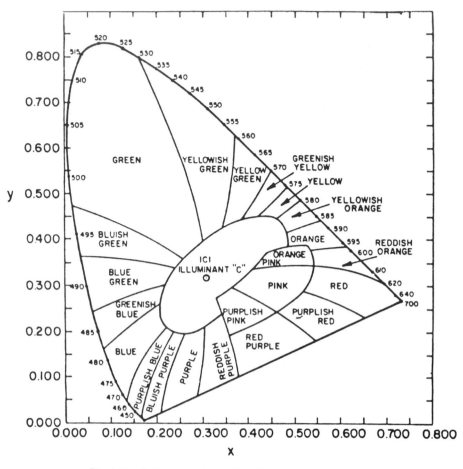

Fig. 3-15. Subjective colors within the chromaticity diagram.

terms of its complementary wavelength, the intersection of a line drawn from the color's x, y position through the center to the dominant wavelength on the spectrum locus.

The CIE system is a means by which visible colors can be specified, measured, and reproduced. Further, the CIE coordinates can be converted directly into other physical system values, such as those of the Munsell system. It is absolutely critical to note, however, that distances in the CIE system are not proportional to perceptual space. That is, distances in x, y space are not Euclidean in our perceptual world. For this reason, attempts have been made to transform the x, y units into equal perceptual distances. One series of transforms has been based upon the discrimination ellipses of MacAdam (1949), shown in Fig. 3.16. Each axis of each ellipse is ten times the distance required for threshold color discrimination, *based upon a*

single observer. Other transforms have also been suggested, based in part on visual sensitivity data and upon theoretical surface transformations designed to convert CIE space into perceptually Euclidean space.

Since perceptual distances in color space are important for the design of color displays, some recent effort has been made to relate color contrast perception to various color space transformations, the desired end product of which would be a mathematical prediction of perceived color distance. Post, Costanza, and Lippert (1982) concluded that the best available metric of suprathreshold color distance, for either equal or unequal brightness stimuli, is the 1976 CIE $L*u*v*$ distance between the stimuli, specified as

$$dL*u*v* = (dL*^2 + du*^2 + dv*^2)^{1/2},$$

in which

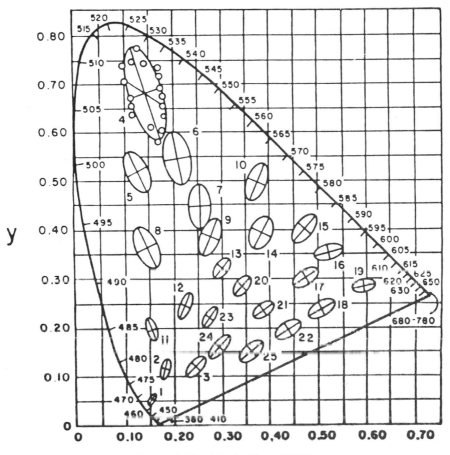

Fig. 3-16. MacAdam's ellipses (1949).

$dL^*u^*v^*$ = distance between the colors,

dL^* = distance between the colors' L^* values,

du^*, dv^* = distances between the colors' u^* and v^* values, respectively,

$L^* = 25(100\ Y/Y_0)^{1/3} - 16$,

$u^* = 13\ L^*(u')$,

$v^* = 13\ L^*(v')$,

$u' = 2x(6y - x + 1.5)$,

$v' = 3y(6y - x + 1.5)$, and

Y_0 = Y value of the illuminant, $1 < Y_0 < 100$.

Post et al. found the correlation between this distance measure $dL^*u^*v^*$ and perceived color contrast to predict about 73% of the variance among colors, a percentage which is quite acceptable for prediction of human perceptual performance. While there is obviously some room for improvement in this prediction, it serves at the present time as our best estimate of perceived color contrast between any two visible colors. Of course, this correlation is based upon the average response of several color-normal individuals. Differences among individuals and color deficiencies of particular individuals will result in exceptions to this rule.

3.5.2 The Chromatic CTF. The previously introduced concept of the contrast threshold function can be applied to chromatically varying stimuli as well as to temporally varying stimuli. The results of van der Horst (1969) serve as a good example of the visual system's CTF for chromatic variation. He used a CRT to vary a sinusoidal grating with one "bar" of the grating one color and the other "bar" another color, as indicated in Fig. 3.17. Sinusoidal variation was presented along the axis BEyy, located at point E, or along the axis bgER, but also at point E. E is the equal energy point in the spectrum, or white. His results are shown in Fig. 3.18, which

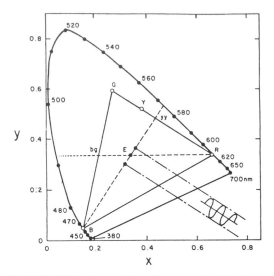

Fig. 3-17. CIE chromaticity diagram stimuli of van der Horst (1969).

indicates that the threshold (measured in percent purity) varies with both the waveform of the grating (sine vs. square vs. triangular) and the spatial frequency of the grating. The CTF is a low-pass filter function, with the threshold increasing beyond about 4 cyc/deg.

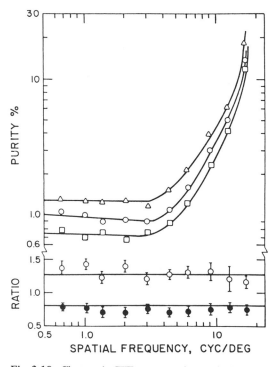

Fig. 3-18. Chromatic CTFs: square, sine, and triangular wave gratings, from van der Horst (1969). Purity is in percent for a dominant wavelength of 492 nm.

Of equal interest is the verification of the Fourier processing capability of the visual system in the chromatic domain, also plotted in this figure. If one computes the first harmonic of the sine, square, and triangular waveforms, the amplitudes of the three fundamentals are in the ratio 1:1.27:0.81. At the bottom of Fig. 3.18 are plotted the actual ratios of sine-wave to square-wave, and sine-wave to triangular-wave thresholds. These ratios are in good agreement with the actual ratios of the amplitudes of the fundamentals. Thus, the visual sensitivity to chromatic gratings is also well predicted by the Fourier analysis approach. Granger and Heurtley (1973) have extended this result to gratings lower in spatial frequency, and have concluded that there is no further change in sensitivity between 0.5 and 0.1 cyc/deg. Thus, the visual system is essentially a low-pass system for the chromatic CTF, whereas it behaves as a bandpass system for determining thresholds in the monochromatic spatial and temporal domains.

In addition, it is important to note that the upper frequency cutofff of the chromatic CTF is less than 20 cyc/deg. That is, beyond this frequency, the visual system ceases to distinguish color and responds only to luminance changes, up to about 50 cyc/deg. This distinction is important for color coding problems, as it clearly points out that single colored elements, no matter how intense, will not have any perceived color if they are smaller than 1.50 arcmin. (A single point can be considered as a half-cycle of a grating; a half-cycle at 20 cyc/deg subtends 1.50 arcmin.)

3.6 SUMMARY

This chapter has attempted to present key psychophysical concepts of vision as they relate to the design and evaluation of visual displays. One can find many other visual system functional characteristics in the literature; in fact, there are literally hundreds of texts and thousands of journal articles describing research performed on the visual system's capabilities and limitations. It is beyond our purpose and intent to summarize all of these, or even a small subset. Rather, we have presented a coherent overview of the measures currently used by visual scientists to characterize the visual system in the spatial, temporal, and chromatic domains. These measures are directly useful and compatible with the measurements important to the characterization of visual displays. Because of this direct compatibility, one can arrive at meaningful predictions of human performance with various displays and estimates of perceived image quality of various displays. These approaches are the subjects of Chap. 4.

REFERENCES

1. Almagor, M., Farley, W. W., and Snyder, H. L. "Spatiotemporal Integration in the Visual System." Wright-Patterson AFB, OH: Aerospace Medical Research Laboratory *Technical Report AMRL-TR-78-126*, February 1979.

2. Barlow, H. B. "Temporal and Spatial Summation in Human Vision at Different Background Intensities." *Journal of Physiology*, *141*, 337–350, 1958.

3. Blackwell, H. R. "Contrast Thresholds of the Human Eye." *Journal of the Optical Society of America*, *36*, 624–643, 1946.

4. Bloch, A. M. "Experiences sur la vision," *C. R. Societe de Biologie (Paris)*, 37, 493–495, 1885.

5. Brown, J. L. "Flicker and Intermittent Stimulation." In *Vision and Visual Perception*, ed. C. H. Graham. New York: Wiley, 1965.

6. Campbell, F. W.; Kulikowski, J. J.; and Levinson, J. "The Effect of Orientation on the Visual Modulation of Gratings," *Journal of Physiology*, *187*, 427–436, 1966.

7. De Palma, J. J., and Lowry, E. M. "Sine-Wave Response of the Visual System." II. "Sine-Wave and Square-Wave Contrast Sensitivity," *Journal of the Optical Society of America*, *52*, 328–335, 1962.

8. Forsyth, D. M. "Use of the Fourier Model in Describing the Fusion of Complex Visual Stimuli."

Journal of the Optical Society of America, *50*, 334–341, 1960.

9. Graham, C. H. *Vision and Visual Perception*. New York: Wiley, 1965.

10. Granger, E. M., and Heurtley, J. C. "Visual Chromaticity-Modulation Transfer Function." *Journal of the Optical Society of America*, *63*, 1173–1174, 1973.

11. Herrick, R. M. "Foveal Luminance Discrimination as a Function of the Duration of the Decrement or Increment in Luminance." *Journal of Comparative and Physiological Psychology*, *49*, 437–443, 1956.

12. Kling, J. W., and Riggs, L. A. *Experimental Psychology*. New York: Holt, Rinehart, and Winston, 1971.

13. de Lange, H. "Research into the Dynamic Nature of the Human Forvea-Cortex Systems with Intermittent and Modulated Light: I. Attenuation Characteristics with White and Colored Light." *Journal of the Optical Society of America*, *48*, 777–784, 1958.

14. MacAdam, D. L. "Color Discrimination and the Influence of Color Contrast on Visual Acuity." *Oblique Physiologique*, *Coleurs*, *28*, 161–173, 1949.

15. Post, D. L.; Costanza, E. B.; and Lippert, T. M. "Expressions of Color Contrast As Equivalent Achromatic Contrast." Santa Monica, CA: *Proceedings of the 26th Annual Meeting of the Human Factors Society*, 1982.

16. Snyder, H. L. "Human Visual Performance and Flat-Panel Display Image Quality." Blacksburg, VA: *VPI&SU Technical Report HFL-80-1*, 1980.

17. van der Horst, G. J. C. "Fourier Analysis and Color Discrimination." *Journal of the Optical Society of America*, *59*, 1482–1488, 1969.

18. Walsh, J. W. T. *Photometry*. New York: Dover, 1965.

19. Watanabe, A.; Mori, T.; Nagata, S.; and Hiwatashi, K. "Spatial Sine-Wave Responses of the Human Visual System." *Vision Research*, *8*, 1245–1263, 1968.

4

IMAGE QUALITY: MEASURES AND VISUAL PERFORMANCE

HARRY L. SNYDER, *Virginia Polytechnic Institute and State University*

4.1 INTRODUCTION

Analysis, research, and recommendations regarding the image quality of displays have come from a variety of disparate sources, intended for a wide variety of applications and uses. These sources of data include those which mathematically describe images for purposes of determining analog system requirements, as in commercial and closed-circuit television; the encoding studies which are designed to minimize transmission and storage bandwidth; and the psychophysical studies which attempt to relate physical measures of the image to what the user or observer believes to be "image quality" in a utility sense. It is not surprising that persons with academically and scientifically different backgrounds have performed these many and diverse studies, and it is similarly not surprising that there is a general lack of familiarity of results from one segment of the research and engineering community to the next.

In this chapter we attempt to put bounds on the meaning of "image quality," provide useful operational definitions, and summarize the current state of data and knowledge which are helpful to the display researcher and designer in his quest for achieving an optimal display for a given objective. Alternative concepts of image quality are offered, mathematical definitions of the various image quality metrics are stated, and results that relate these mathematical quantities to the performance of the user are summarized in an effort to discern objectively which image quality descriptors or models are valid and meaningful.

4.1.1 Image Measurement and Specification.
Monochromatic images can be represented as a two-dimensional light emission (or reflection) function, $f(x, y)$, where x and y denote spatial coordinates in the horizontal and vertical directions, respectively. The value of the function is proportional to the intensity (or reflection) at each x, y location. If the image is approximately continuous, as in a photograph, then x and y can be considered to be continuous variables. On the other hand, if one is referring to most flat-panel displays or digitally addressed CRTs, then x and y are discrete variables. *Digital* images are discrete in intensity as well as in spatial coordinates. The x, y elements of a digital image are called image elements, picture elements, pixels, or pels. To be compatible with digital computer programs and storage conventions, digital image sizes are usually a power of 2 such as 512×512 or $1,024 \times 1,024$, with 8, 16, 32, or 64 intensity levels (3, 4, 5, or 6 bits, respectively).

Regardless of whether the displayed image is continuous (analog) or discrete (digital), the intensity dimension must ultimately be measured (or calculated) in luminance units in order to relate the appearance of the image to the visual system of the observer. As noted in Chap. 3, appropriate units are candelas per square meter (cd/m^2) in the SI system or foot-Lamberts (fL) in the English system.

All useful measures of image quality can be calculated from the $f(x, y)$ data measured from the image. The procedure and hardware/software used for making these measurements are beyond the scope of this book, and the reader is referred to the literature on microdensitometry and image digitizers for these techniques.

For reasons indicated below, the analytically most convenient pattern used in making measurements for the calibration of a display or an image is a sine-wave grating of the form illustrated in Chap. 3. Data resulting from one-dimensional sine-wave grating scans are usually stored in a one-dimensional computer file, while intensity measurements from two-dimensional patterns are stored in x, y arrays. The outputs from two different one-dimensional microphotometric scans are illustrated in Fig. 4.1a and 4.1b. These scans were made with a 25-by-2,500-micrometer aperture with the long dimension oriented parallel to the sine-wave pattern imaged on a raster scan CRT. The pattern measured with the sine-wave bars perpendicular to the raster lines shows a fairly uniform sine-wave output smoothed somewhat by space averaging over the length of the 2,500-micrometer aperture. The scan made with the sine-wave pattern parallel to the raster clearly indicates the effect of the discontinuous raster on the display of the sine-wave intensity grating. In both cases, the nonlinearity of the CRT is seen by the peaking of the high-intensity part of the sine wave and the flattening of the low-intensity part. This nonlinearity in the CRT output is typical of most displays, although it is rarely recognized in image quality models of CRT images.

The term "image quality" has been used in two general contexts: (1) that dealing with physical measures of the image itself and with little or no regard for the ability of the observer to obtain information from the image; and (2) that dealing with perceived or measured quality from the human observer, sometimes with little regard for the physical characteristics of the image. Both concepts of the meaning of image quality are critical for an understanding of the literature and the importance of clearly defining image quality in any design context.

4.1.2 Physical Measures of Image Quality. Physical measures of image quality attempt to define or describe pertinent image statistics relative to a baseline or ideal image. For example, some physical image metrics relate intensity distributions of an image to assumed ideal intensity distributions or relate an original image to a degraded version of the image such that the differences in the statistical intensity distributions are a measure of the degradation or the reduced image quality. This type of metric is usually described as a "pixel error" metric and will be discussed later in this chapter.

Another set of physical metrics of image quality is based on the modulation transfer function (MTF), a measure of the displayed contrast in an image as a function of the size of objects in the image. The MTF will be defined and discussed below.

An important distinction with regard to the physical measures of image quality is that they do not attempt to relate the measured quantities directly and empirically to the visual performance of the user of the image. That is, the emphasis is placed solely upon the physical measurement, and the validity of the measurement, calculation, or construct is assumed from some statistical characterization of the user or the visual system rather than upon a direct experimental test of the perceived quality of the image with a user population or sample.

4.1.3 Behaviorally Validated Measures of Image Quality. Behaviorally validated measures of image quality, on the other hand, strongly emphasize the visual performance or perception of the user and relate this performance empirically to physical characteristics of the image. Most of the behavioral measures of image quality are simpler in form than are the purely physical measures, perhaps because of the less sophisticated level of mathematical interest of the researchers performing psychophysical experiments or perhaps more because of the awareness of the limited precision of perceptual measurement and the desire to maintain a comparable level of sophistication in the measurements of both the physical image and the perceptual response to the image.

A careful evaluation of the literature of image quality convinces the reviewer that it is both meaningful and useful to conduct and apply research that relates physical measures of image quality to user performance. The real issues center around the nature of the assumptions of the visual system that are important in the con-

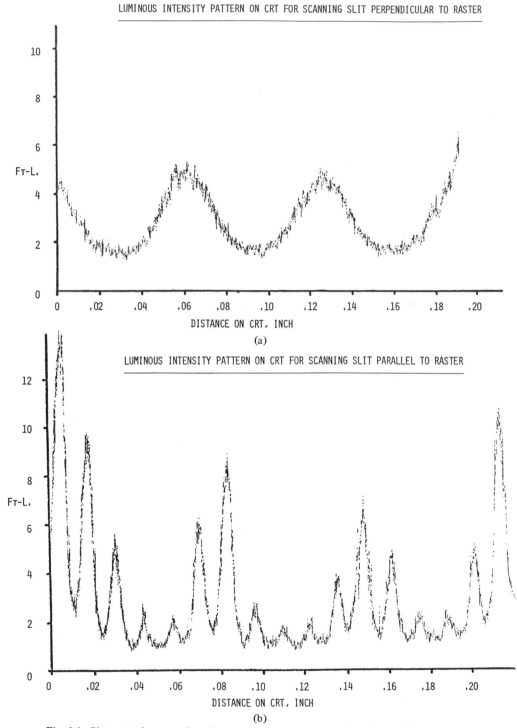

Fig. 4-1. Photometric scans of gratings perpendicular to raster (a), and parallel to raster (b).

text of image quality and the selection of the physical characteristics of the image that influence user visual performance. In subsequent discussions of the extant image quality concepts, the reader will find a large range of emphasis, running from a strong weighting on the physical side to a strong emphasis on the behavioral side. Since work in this general research area is still receiving heavy emphasis, it seems inappropriate at this time to suggest that a definite position toward either extreme is the more suitable. Rather, it is logically apparent that useful measures of image quality *must* contain both repeatable and design-relevant physical descriptors of the image as well as suitable behavioral measures by which to assess the validity of the metric.

4.2 THE MODULATION TRANSFER FUNCTION

4.2.1 Concept and Measurement. In recent years, a strong popularity for image quality metrics (and other types of measurements) based upon the Modulation Transfer Function (MTF) has developed. Using the concept of modulation as defined in Chap. 3, the MTF relates the relative attenuation of modulation to the spatial frequency input to a device. The device of interest can be a complete imaging system, a subsystem (e.g., a display), or a component, such as a lens or film. Irrespective of the device being measured by the MTF, the resulting function relates the modulation output of the device to the modulation input to the device. The measurement is made for a sine-wave input of known amplitude or modulation.

Because most imaging systems or devices are not perfect or possessing of infinite bandwidth, the output modulation of the sine wave is typically reduced by some amount due to losses within the system (limited bandwidth, aberrations, diffraction limits, spot size, etc.). Generally, the loss of output modulation relative to input modulation increases with increasing frequency of the sine-wave input. The extent of this loss is denoted by the ratio

$$\text{Modulation transfer factor} = \frac{M_{\text{out}}}{M_{\text{in}}}.$$

The plot of the modulation transfer factors for all spatial frequencies is the MTF, which is shown in Fig. 4.2.

The MTF is a representation of imaging system capacity in the spatial domain and is based upon the theory of linear systems analysis and the mathematics of Fourier transformations. Because of the widely ranging applications of Fourier analysis to a variety of physical systems, the MTF has been used as the basis of a

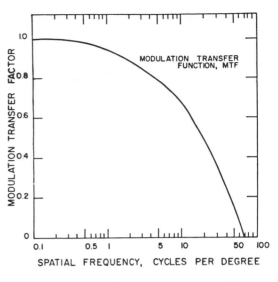

Fig. 4-2. Modulation transfer function (MTF).

number of image quality metrics. It has been shown to be a very valuable tool in this regard and therefore deserves careful consideration and emphasis. Its fundamental importance lies in the fact that, through the Fourier theorem, any waveform (or object intensity distribution) may be broken down into a series of sine-wave components, each having a unique amplitude and phase relationship to the other components. For a discrete (digital) image, the Fourier transform is given by

$$\text{FT}(w, v) = \sum_{x=0}^{N-1} \sum_{y=0}^{N-1} L(x, y)$$

$$* \exp\left[-j2\pi\left(\frac{wx + vy}{N}\right)\right],$$

where

$L(x, y)$ = image intensity at spatial location x, y in rectangular coordinates,

$\text{FT}(w, v)$ = Fourier transform coefficient at spatial frequency w, v, and

N = number of discrete image samples along one spatial dimension ($N = N_x = N_y$).

Generally, $\text{FT}(w, v)$ is a complex function consisting of a real part and an imaginary part indicated as

$$\text{FT}(w, v) = R(w, v) - jI(w, v),$$

where $j = (-1)^{0.5}$. The amplitude of the Fourier coefficient is given by

$$A(w, v) = [R^2(w, v) - jI^2(w, v)]^{0.5}$$

and the phase angle is given by

$$P(w, v) = \arctan \frac{I(w, v)}{R(w, v)}.$$

The two-dimensional modulation spectrum (MTF) of a digital image is computed from the normalized Fourier amplitude spectrum,

$$M(w, v) = \frac{A(w, v)}{A(0, 0)},$$

where the zero frequency component $A(0, 0)$ is equal to the average intensity level of the image or $[L_{max} + L_{min}]/2$.

It may be helpful to elaborate further on the MTF and its relationship to other display concepts and measurements, particularly to measurements in the space domain. For example, the MTF is the normalized Fourier transform of the line spread function often used in photographic image analysis for analog images (Dainty and Shaw, 1974; Gaskill, 1978). The line spread function is the spread of an image of an infinitely narrow line input. When the image of the narrow line is formed, the measured image is no longer a sharp line but has "rounded" edges— that is, the intensity profile is "spread" or "blurred" by the imaging device. The line spread function defines the profile of the resulting image and can be obtained by measuring the luminance distribution directly. Alternatively, the luminance or intensity distribution can be measured across a displayed step ("knife") edge and differentiated. The differentiated edge is the line spread function.

The line spread function and the normalized MTF are inverses of one another. In addition, the width of the line spread function is inversely proportional to the passband of the MTF. Thus, either concept may be used to characterize the physical performance of an imaging device or component in one dimension. This mathematical similarity is used as the basis of a number of proposed metrics of image quality. The point spread function (PSF) is a two-dimensional function that can be transformed into a two-dimensional MTF. For most systems, the two-dimensional MTF is more useful.

The MTF has several advantages which have contributed to its popularity in characterizing image quality. For example, the composite MTF of an imaging system can be determined simply from the cascading of the MTFs associated with the several components of the system. That is,

$$F_{system}(w, v) = F_1(w, v) * F_2(w, v) \ldots F_N(w, v)$$

where $F_{system}(w, v)$ is the system modulation transfer factor at spatial frequency (w, v) and the other terms are the modulation transfer factors for the components of the system. Careful analysis of these component MTFs can be very useful in localizing the major source(s) of image degradation in a system.

The mathematical definitions given above describe the MTF concept for a digital image rather than an analog one. This is deliberate in that all flat-panel displays receiving emphasis in this book are digitally addressed and are composed of discrete pixels rather than of continuous image information. The exception to this generalization is the ever-popular CRT image, particularly the raster-scanned image in which the along-raster dimension is a continuous image and the across-raster image is discrete. As will be seen later, this distinction is not critical to an appreciation of the MTF-based measures of image quality; rather, the distinction is important only in the calculation of the MTF. For subtleties and approaches to the calculational differences, the reader is referred to Gaskill (1978).

4.2.2 Relation to Vision. Chapter 3 contains a discussion of the spatial Contrast Threshold Function (CTF). This function, when inverted, is often (inappropriately) considered to be the spatial frequency equivalent of the MTF of the visual system. It is inappropriate to term this inverse function an MTF, however, because the MTF of a system is noise independent and the CSF is decidedly noise dependent. Other distinctions also need to be made at this point, however, in that the MTF concept assumes certain characteristics derived from linear systems analysis. First, linear systems analysis requires that the system be linear—that is, that the output/input relationship be independent of the amplitude of the input. Such is not the case for the visual system (Bryngdahl, 1966). Secondly, linear systems analysis as applied to the visual system requires homogeneity with regard to both axial orientation and retinal location. These requirements are not rigidly met, in that the CTF is lower for gratings oriented vertically and horizontally than for gratings at oblique angles (Campbell, Kulikowski, and Levinson, 1966). However, the increases in the CTF with obliquity of the grating are very small and probably insignificant in the real-world range of image quality variation. Similarly, the variations in CTF with retinal location, while pronounced as the image is moved outside the fovea, are of little importance because most displays are foveally fixated when information is critical to the observer, at least for displays of substantial image quality.

What is of importance, of course, is the extent to which the visual system can be approximated to a useful degree by linear systems analysis. While the ultimate acceptance of the notion will be left to the reader, it is this author's opinion that sufficient research has been conducted to justify this approach (e.g., Campbell and Robson, 1968) as a very useful first approximation and that variance in visual behavior unexplained by this approach is small relative to that which is predicted by the powerful analytical techniques made available through linear systems analysis and Fourier theory (Snyder, 1980).

4.2.3 Design Utility of the MTF. The paramount advantage of the MTF as a basis for image quality measurements is that it (1) describes a large amount of what is generally considered to be critical to an image, specifically the effect of spatial frequency on contrast or modulation, and (2) can be used to describe, in similar units, the capabilities of the human observer using the display. Thus, it provides us with an analytical tool which permits the common evaluation of the display and the user in the same units, units which are quantitative, well understood, and most important of all, predictable in advance

before much of the hardware is built. This last advantage is economically important in that the acceptance of the MTF-based approach lets the designer simulate the effect of a variety of design perturbations on the display system MTF and then estimate quantitatively the effect of the resultant MTF change on user performance. Research supporting this contention is summarized below. However, before reviewing this literature, it is important that another approach to image quality, based on statistical pixel error distributions, be described.

4.3 PIXEL ERROR MEASURES

An alternative approach to image quality has achieved some attention, particularly among researchers using images containing several levels of intensity. These measures are based on an error or variance concept in which the extent of the difference in intensity levels, averaged in some fashion across pixels, is taken as a measure of the quality of degradation of an image between the original image and the image whose quality is measured. This error measure can be made in either the intensity distributions or in the modulation spectra of the respective images. These error measures of image quality have seen the greatest popularity in the field of digital image processing, largely because of the ease of their calculations for images already stored in digital array files. For other purposes, however, the calculations can become burdensome and of perhaps less utility. Nonetheless, because they offer a different and popular look at the concept of image quality, they are reviewed and summarized here.

4.3.1 Measurement of Pixel Errors. All of the pixel error metrics of interest perform similar calculations on the x, y image arrays, essentially determining the differences between corresponding pixels in the original and the to-be-evaluated image. These differences are then treated mathematically in one fashion or another, resulting in a summed or multiplied term which serves as an overall index of quality. The mathematical definitions are defined later in this chapter when the research evaluating these metrics is described.

4.3.2 Relation to Vision. Pixel error metrics of image quality are less supported by empirical vision research then are the MTF-based metrics. While some of the authors of pixel error metrics claim a "good physical and theoretical basis" to vision (Granrath, 1981), it can be argued that the correspondence is not well substantiated, at least to the satisfaction of the visual science community. It has been suggested that the appeal of these measures lies more in their computational simplicity than in their inherent or demonstrated correspondence to visual performance and knowledge of the visual system's capabilities, even though the phrase "visually optimized" appears frequently in this literature. In the opinion of this author, the pixel error metrics should be viewed cautiously and used as the basis for research evaluation rather than as the basis for display system design and optimization.

4.4 MTF-BASED MEASURES OF IMAGE QUALITY

4.4.1 Equivalent Passband, N_e. The emphasis on MTF-based measures of image quality probably originated with the Equivalent Passband (N_e) concept suggested by Schade (1953). Conceived as a means to describe the quality of a television signal, N_e is the equivalent bandwidth of a rectangular MTF (a perfect filter) which contains the same total sine-wave power as does the actual MTF of a system; that is, it is the

cut-off frequency of the perfect filter passing the same power. This concept is illustrated in Fig. 4.3.

The notion of an equivalent passband is not unique to imaging systems or to the television industry but rather is derived from the statistical concept of the variance of a distribution. Specifically,

$$VAR = A^2(0) * \sum F_s^2(w),$$

in which

> VAR = variance of the distribution,
> $A(0)$ = average intensity level of the image, and
> $F_s(w)$ = system modulation transfer factor at spatial frequency w.

For an MTF of unity modulation from $w = 0$ to $w = N_e$, this expression reduces to

$$VAR = A^2(0) * N_e.$$

Rearranging this equation and expressing it in terms of the system MTF,

$$N_e = \sum F_s^2(w).$$

While this last equation describes the one-dimensional case, the imaging system is assumed to have circular symmetry. Thus, the Equivalent Passband is a measure of the blur or "sharpness" in an image and certainly relates impor-

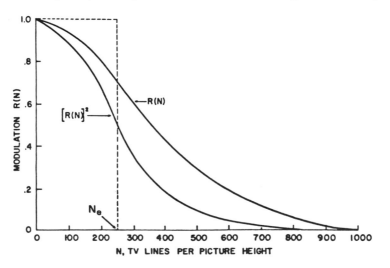

Fig. 4-3. Noise equivalent passband (N_e).

tantly to the perceived quality of the image. At the same time, this metric is severely limited in that it does not take into account the "error" data in an image caused by correlated or uncorrelated noise from any source. Since, by definition, the MTF is measured in the absence of noise, any metric such as N_e based solely upon the MTF cannot account for changes in image quality resulting from the introduction of noise into the image. Unfortunately, many degraded images are in fact degraded solely by the introduction of uncorrelated noise which is unrelated to image content. For this reason, the N_e concept has been rejected in recent years by researchers of concepts of image quality (Snyder, Keesee, Beamon, and Aschenbach, 1974; Task, 1979).

4.4.2 Strehl Intensity Ratio.

While the Equivalent Passband concept has its limitations, it has influenced the thinking of many image quality researchers in the development of measures based on weighted MTFs or integrated, weighted MTFs. Hufnagel (1968) suggested several alternative weighting schemes for the MTF calculated directly from the line spread function. For example, weighting the line spread function by a Dirac delta function results in the maximum intensity of the spread function. The ratio of the maximum spread function values for an imaging system to that of an equivalent aberration-free system is known as the Strehl Intensity Ratio (Linfoot, 1960). Beaton (1984) has suggested an alternate, simpler computational approach to the Strehl Ratio for digitized images. Of course, the Strehl Intensity Ratio is no more useful than is the N_e concept in evaluating the quality of images containing noise and for that reason is unsuitable for many image quality determinations.

4.4.3 Modulation Transfer Function Area (MTFA).

One of the most researched metrics of image quality takes into account the MTF of the imaging system or display as well as the CTF of the visual system. The concept was originally suggested by Charman and Olin (1965) and subsequently evaluated experimentally or analytically by a variety of both laboratory experimentalists and theorists (e.g., Blumenthal

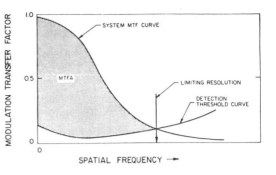

Fig. 4-4. MTFA concept.

and Campana, 1981; Borough, Fallis, Warnock, and Britt, 1967; Snyder, 1973, 1974, 1976; Snyder et al., 1974; Task, 1979).

The MTFA is illustrated in Fig. 4.4, and is defined in the one-dimensional case as the area between the MTF and the CTF, between zero spatial frequency and the crossover frequency of the two curves. Thus, it is often conceptualized as a "signal minus noise" integrated over all usable spatial frequencies. Furthermore, the crossover spatial frequency is the "limiting resolution" of the imaging device. Mathematically, the MTFA is defined as

$$\text{MTFA} = \int_{w=0}^{F_c} [F_s(w) - T(w)]\, dw,$$

in which

F_c = cross over frequency, or limiting resolution,

$F_s(w)$ = value of the MTF at spatial frequency w, and

$T(w)$ = observer's CTF.

The rationale behind the MTFA is simple. It summarizes the excess of signal (MTF) over the threshold requirement (CTF) of the visual system over all usable spatial frequencies. It further assumes that the area is homogeneous in image quality; i.e., that the excess of MTF over CTF is uniformly important or isotropic for all spatial frequencies and for all amounts of modulation above the threshold requirement. As will be noted below, this assumption of homogeneity of the area has been questioned and tested experimentally but with no substantial and consistent improvement in the concept.

As originally proposed by Charman and Olin and as used by subsequent researchers, the MTF is measured for a given system in the traditional fashion. The CTF is determined either experimentally or analytically. The CTF is used to account for differences in viewing conditions, gamma of the display or imaging system, and noise content of the display. In general, as gamma increases or as noise decreases, the CTF is lowered to provide a larger MTFA value. For the rationale and quantitative approach to these manipulations, see Snyder (1973; 1980).

Numerous studies have been conducted to evaluate the MTFA concept to predict target recognition with video displays. Variables that have been included in the research include target size, target type, background clutter, display magnification, rate of dynamic image motion, and others. In general, the correlations between MTFA and various observer performance measures range from .60 to .97, with the latter being obtained for static photographic imagery (Borough et al., 1967). Nearly all the studies have resulted in statistically significant correlations, some higher than others. For examples of this research, see Snyder et al. (1974), Snyder (1973), Snyder (1974), and Beamon and Snyder (1975).

Gutmann, Snyder, Farley, and Evans (1979) tested the isotropic assumption of the MTFA and found that the assumption was unsupported for systems having atypical MTFs. For systems having similarly shaped MTFs, the correlations between MTFA and observer performance are typically quite high (on the order of .80 and above). However, when high-pass filtering is done to the image or when the image MTF is rolled off substantially, the correlations are considerably reduced. Thus, the MTFA is decidedly anisotropic. Beamon and Snyder (1975) have suggested that the area immediately above the CTF is of greater importance to the observer than is the area well above the CTF. Stated differently, it is critical to have adequate signal (modulation) above that minimally required for detection (CTF), but additional increases in this excess of MTF over CTF are of less value in most real-world tasks. The next quality measure is an effort to overcome this problem in the MTFA.

4.4.4 Gray Shade Frequency Product (GSFP).

Task and Verona (1976) proposed a nonlinear transform of the MTFA to weight the area near the CTF more heavily than the area well above the CTF. This transformation uses as its logical basis the assumption that the visual system can be modeled as a logarithmic amplifier which sees modulation proportional to the logarithm of the modulation. Accordingly, they transformed the modulation axis into "just noticeable differences" or "shades of gray" (G), by the formula

$$G = 1 + \frac{\log\left[(1 + M)/(1 - M)\right]}{\log 1.414},$$

where the numerator is the modulation and the denominator is the approximation of the modulation difference between successive shades of gray (but see Chap. 3).

Tests of the GSFP have demonstrated slightly greater correlations between observer performance measures and GSFP than between MTFA and performance. The differences were small, however, and the manipulation of image quality was only by MTF differences, not by differences in the noise content of the image.

4.4.5 Integrated Contrast Sensitivity (ICS).

Van Meeteren (1973) proposed another approach to "perceptually weighting" the system or display MTF and then cascading it with the visual system MTF or CTF. In this approach, ICS is defined as

$$ICS = \sum F_s(w, v) * CTF^{-1}.$$

van Meeteren suggested that the ICS is more sensitive to small changes in the shape of either the MTF of the system or the CTF than would be the MTFA and is therefore more sensitive to small changes in image quality. In fact, Task (1979) found slightly higher correlations between observer performance and ICS than between observer performance and MTFA under three different experimental conditions.

4.4.6 Visual Capacity (VC).

Cohen and Gorog (1974) took yet another approach to the modification of the MTF concept and built upon Schade's N_e metric, extending it to a more modern knowledge of visual perception. In this approach, VC is given by

$$VC = B\left[\sum \sum F_s^2(w,v) * F_e^2(w,v)\right],$$

in which B is the area of the display device.

The rationale behind this metric is that N_e is related to the width of edge transitions (sharpness) in the image field and that VC must therefore express the perceptual width of these edge transitions. Normalizing the summed quantity (perceived edge transitions) by the area of the display is suggested as a means of expressing the maximum number of perceived edge transitions within the image from which the name "capacity" is taken.

4.4.7 Discriminable Difference Diagrams (DDD).

Subsequent work by Carlson and Cohen (1978) built upon the earlier RCA activity, and developed a model to predict the just noticeable differences in contrast discrimination for sine-wave gratings. Using the concept of independent spatial frequency channels in the visual system, these researchers have developed a series of DDDs which correspond to a variety of display conditions. A DDD indicates the increases in modulation necessary to achieve a just noticeable difference in modulation as a function of spatial frequency. Vertical lines are centered at each spatial frequency channel, and small tick

4.4.8 Displayed Signal-to-Noise Ratio (SNR_D).

Using the analyses of Schade (1953) as a background, Rosell (1971) developed an approach for analyzing television systems which takes into account the temporal and spatial integration capability of the visual system. Rosell's approach is to relate all system parameters to the analytically derived SNR_D. Assuming that the human observer required an SNR_D of approximately 2.8 for a 50% probability of detection, system tradeoffs can be made to achieve this or some other level of detection through the relationship between detection probability and SNR_D. Rosell and his colleagues have performed many laboratory studies to establish the probability of detection as a function of size for geometric figures and single tactical vehicles. Observer confidence levels, task loading, ambient environments, dynamic scenes, target textural characteristics, and other factors have not been investigated; however, this metric and the MTFA have probably generated more research data than have any other metrics discussed in this chapter.

There are many variants of the SNR_D concept, depending on whether one assumes the limitations in the line-scan system to be photon limited, preamplifier limited, display limited, etc. For purposes of discussion, however, an elementary calculational formula is given by Rosell and Wilson (1973):

$$SNR_D = [at\Delta f_V/A]^{0.5} * [Ci_{max}]/[[(2-C) e\Delta f_V i_{max}]^{0.5}$$
$$= [(a/A)t\Delta f_V]^{0.5} * SNR_V,$$

marks indicate the increments at each just noticeable difference. The number of just noticeable differences reflects the perceptual extent of image structure in each spatial frequency channel, limited only by the MTF of the display system at that channel. Thus, an image quality metric derived from this approach is the sum of the just noticeable differences under the MTF, given by

$$JND = \sum_{i=1}^{n} J(i),$$

in which $J(i)$ indicates the number of just noticeable differences at center frequency i.

in which

SNR_D = displayed signal-to-noise ratio,

a = area subtended by the target at display,

A = total display area,

t = integration time of the eye, assumed to be constant at 0.2 s,

Δf_V = video bandwidth, in MHz,

C = target contrast,

i_{max} = maximum photocurrent,

e = charge of an electron, and

SNR_V = signal-to-noise ratio in the video, defined as peak-to-peak signal divided by RMS noise.

The key to the SNR_D concept lies in the

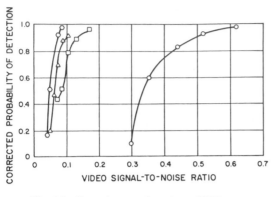

Fig. 4-5. Detection as a function of SNR_V

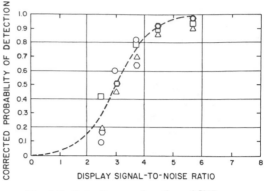

Fig. 4-6. Detection as a function of SNR_D.

bracketed term in this equation. Essentially, this term provides for both spatial and temporal integration of the signal, and reflects the visual system's spatial and temporal integration capabilities. The larger the portion of the display subtended by the target, the greater the signal, with signal strength directly proportional to the square root of the target area, a. In addition, the signal is integrated over the integration time of the visual system, t, which is assumed to be a constant, 0.2 s. More recently, Almagor, Farley, and Snyder (1979) have shown that the integration time is decidedly not constant and varies greatly with adapting luminance, individual observer differences, and the noise level on the display. In fact, these investigators have shown that the visual system typically trades off spatial integration with temporal integration to obtain an optimum visual image.

Analysis of the SNR_D equation indicates that SNR_D varies directly with the square root of a and with SNR_V. Thus, the same SRN_D and hence the same detectability of a target will exist with reciprocal covariation in SNR_V and $a^{0.5}$. Figures 4.5 and 4.6 illustrate this relationship. Rectangles of varying size were displayed on a television monitor and the observer was asked to respond if he could detect the target. Variation in SNR_V resulted in consistent variation in detection probability, as illustrated in Fig. 4.5 for several target sizes. When the SNR_V was converted to SNR_D, taking into account target size, the results show that probability of detection is directly related to SNR_D, as pre-dicted by the model. The data should be fit by

an ogive, as drawn in Fig. 4.6. It is clear, however, that a negatively accelerating exponential curve would be a better fit, thus casting some doubt on the model. No statistical evaluations of goodness of fit were offered by the authors.

Rosell and Wilson have conducted several other experiments to evaluate SNR_D with conditions of clutter and tactical targets. They also have tested the model for recognition and identification responses. In general, the higher the cognitive processing required or the more complex the scene, the poorer the prediction of the model. Unfortunately, no statistical evaluations of the model are given in any of the reports, so the reader is left to draw his own conclusions as to the validity of the predictions. As with the case of other metrics presented above, the concept is heuristically desirable, and further empirical testing with statistical analyses would be desirable.

4.4.9 Visual Efficiency (VE). Overington (1976, 1982) has developed a sophisticated mathematical model of human visual performance for simple and complex visual environments, basing much of the approach upon basic mechanisms in visual perception. In developing the model, Overington assumes that the illuminance gradients between retinal photoreceptors provides important information for target detection, and uses the derivative of the edge (line) spread function (or the Fourier transform) to obtain the following metric, which assumes that the photoreceptor spacing is 25 arcmin:

$$VE = \frac{\sum\sum[F_s w, v) * F_e(w, v)] * \cos[2\pi(wx/N)] * \cos[2\pi(vy/N)]}{\sum\sum F_e(w, v) * \cos[2\pi(wx/N)] * \cos[2\pi)vy/N)]},$$

in which $x = y = 25$ arcmin.

When VE > 1, the perception of image detail is limited by the optics of the eye, whereas when VE < 1, edge transitions are limited by the sharpness of the image. Overington (1975) suggests that, in the absence of empirical performance data, the VE metric contains the same fundamental information as the MTFA-type metric and therfore should yield similar correlations with performance.

4.4.10 Information Content (IC).

The concept of information theory (Shannon and Weaver, 1949) has had a noticeable impact upon developments in image quality metrics, as it has in other technical areas. As applied to images, the amount of information (in bits) is an image is:

$$IC = N * \log_2 (L),$$

in which

IC = image information, in bits,
N = number of pixels in image, and
L = number of response levels.

Schindler (1976, 1978) has considered in detail the application of IC to pictorial displays and has derived an equivalent spatial frequency expression for information content, given by

$$IC = \sum \sum \log_2 \left[1 + \frac{F_s(w,v)}{F_d(w,v)} \right],$$

in which $F_d(w, v)$ refers to the "just detectable" response level of the imaging system. Experimental tests of this metric have not been very encouraging in predicting observer performance with real images.

4.4.11 Summary of MTF-Based Metrics.

The preceding metrics have been proposed by various persons, some of whom have had the capability and inclination to test the predictive ability of the metrics in a controlled experiment. As indicated, some of the correlations have been acceptably good, a couple have been excellent, and several have been disappointing. In many cases, the metrics proposed have never been subjected to experimental test and have remained only the offerings of theorists. One major research effort evaluated a variety of these metrics to predict the results of three separate experiments. Task (1979) determined the correlation between metric values and observer performance in three target detection/recognition studies in which image quality was varied by changing the system MTF. No variation in noise content of the image was introduced in the studies. The results of Task's research and analyses are shown in Table 4-1.

Table 4-1   Task's (1979) Correlations

Quality Metric	Target Recognition (TV Study)	Target Detection (TV Study)		Target Recognition (Film Study)
		POL 1000	POL 2000	
log BLMTFA	−.948	.931	.878	−.977
JNDA-log (0.5 cpd)	−.937	.928	.853	−.983
log S.T. Res	−.909	.923	.864	−.969
log JNDA	−.906	.902	.838	−.984
JNDA-log (2 cpd)	−.896	.902	.835	−.983
S.T. Res	−.888	.900	.834	−.973
log MTFA	−.878	.866	.777	−.950
JNDA	−.876	.880	.802	−.983
log SSMTFA	−.869	.857	.766	−.978
GFP-log	−.847	.869	.760	−.858
log IC	−.820	.880	.842	−.971
ICS	−.818	.837	.724	−.978
MTFA	−.811	.829	.717	−.912
Info Dens	−.795	.824	.766	−.974
log Lim Res	−.783	.795	.709	−.885
SQF	−.781	.803	.702	−.979
GFP	−.781	.798	.670	—
Lim Res	−.764	.778	.695	−.851
N_e	−.726	.761	.618	−.729

As indicated in Table 4-1, there is considerable variation in the predictive capability of the metrics. In general, the MTFA and just-noticeable-difference (JND) type metrics performed well. Minor modifications to the MTFA and JND formulae improved predictions to some extent. Clearly, these metrics support the notion that some integration of excess modulation over a threshold level, when integrated in some (perhaps weighted) fashion, produces a numerical value which is generally proportional to the capability of the observer to obtain information from the display. Refinements of these concepts are possible, and there is some room for improvement. Nonetheless, all researchers in this general area are likely to agree that over half of the variance in performance across various displays can be predicted from such an approach, and that this level of prediction is far superior to the level of prediction that existed before the development of these metrics. As is typically the case, additional research can contribute significantly to reducing the remaining unpredicted variance.

4.5   PIXEL ERROR MEASURES OF IMAGE QUALITY

Most of the pixel error metrics have been devised as a means to evaluate digital image quality and to assess the effects of various processing algorithms on image quality. Because most of these metrics have been suggested by persons whose primary training and interests are in mathematical digital image description and modification, there is a noticeable lack of behavioral data accompanying the metrics. Therefore validation of the metrics has often taken the form of examples of "before-and-after" illustrations and demonstrations rather than the preferred empirical testing under controlled conditions. Nevertheless these metrics have considerable heuristic value and need to be considered in the context of image evaluation. As in the previous section, each of the metrics is described mathematicaly and any evaluation data are summarized.

4.5.1 Normalized Mean Square Error (MSE).
This metric is the basic quantity from which most of the other pixel-error metrics are derived or borrowed. It is defined as

$$\text{MSE} = \frac{\sum \sum [M_o(w,v) - M_m(w,v)]^2}{\sum \sum M_o^2(w,v)},$$

in which $M_o(w, v)$ and $M_m(w, v)$ refer to the modulation spectra of the original and modified images, respectively.

This equation, in its basic form, is simply the sum of the normalized squared deviations between the two images, with the summation unweighted over all pixels. Variations of this general concept have been created by the application of different weighting functions (Pratt, 1978). Four of these weighted approaches follow.

4.5.2 Point Squared Error (PSE).
The PSE normalizes the squared deviations with respect to the maximum value of the original image distribution, as given by

$$\text{PSE} = \frac{\sum \sum [M_o(w,v) - M_m(w,v)]^2}{\max [M_o^2(w,v)]}.$$

No evaluation data are known to exist for this metric.

4.5.3 Perceptual MSE (PMSE).
This metric weights the deviations in the MSE by the MTF of the visual system, and is given by

$$\text{PMSE} = \frac{\sum \sum M_e^2(w,v) * [M_o^2(w,v) - M_m^2(w,v)]}{\sum \sum M_e^2(w,v) * M_o^2(w,v)}.$$

As for the MSE, no known empirical data exist to test the validity of this measure against observer performance.

4.5.4 Image Fidelity (IF).
Linfoot (1960) suggested that MSE, with appropriate normalization, may be interpreted as a fidelity deficit in the modified image as compared to the original image. Following this logic, he defined the IF metric as unity minus the fidelity deficit, or

$$\text{IF} = 1 - \frac{\sum \sum [M_o(w,v) - M_m(w,v)]^2}{\sum \sum M_o^2(w,v)}.$$

He also suggested two other variants of MSE which use different normalization values. As might be expected, none of the three metrics has been investigated experimentally. These other two metrics follow.

4.5.5 Structural Content (SC).
Structrual Content is defined as

$$SC = \frac{\sum \sum M_o^2(w,v)}{\sum \sum M_m^2(w,v)}.$$

4.5.6 Correlational Quality (CQ).

The CQ metric is defined by the following equation

$$CQ = \frac{\sum \sum M_o(w,v) * M_m(w,v)}{\sum \sum M_o(w,v)}.$$

As pointed out by Beaton (1984), there are some interesting relationships among the various metrics. For example, SC may be interpreted as a normalized equivalent of N_e. Since N_e is related to the width of edge transitions, the SC metric expresses the width of edge transitions in the modified image normalized with respect to the original image. In addition, SC retains the basic form of the Strehl Intensity Ratio if the original image is assumed to be the equivalent of an aberration-free image. The QC metric can be interpreted as the cross correlation of the original image with the modified image, normalized to the original image.

In many respects, these pixel error concepts are simply discrete calculational formulae, for digitized images, of the continuous image concepts advanced under the MTF-based measures. For that reason, it is not surprising that Task (1979) found similar correlations for the various measures with observer performance.

4.6 AN EMPIRICAL IMAGE QUALITY MODEL

All of the above metrics of image quality have one thing in common—they are based upon some theoretical approach to the notion of image quality and the quantification of the visual system, and lead directly to a model of image quality based upon that theoretical approach. A totally different approach is to offer no pet theory or concept and simply determine empirically which concepts best predict observer performance, letting the resulting pool of predictors define quantitatively what is meant by "image quality." Such an approach was taken by Snyder and Maddox (1978) for digitally addressed displays.

Using three different tasks, they performed experiments which varied the structure of the display in terms of pixel size, shape, contrast, spacing, and the like. They collected observer performance on two different search tasks and a reading task and correlated these performance measures with a variety of physical measurements of geometric and photometric characteristics of the image. Table 4-2 lists the predictor variables which were tested in a stepwise, linear multiple regression approach. In this statistical approach, all known variables are permitted to enter into a linear prediction equation, and the computed result is a "model" that defines best predictive combination of any or all of the variables. The resulting R^2 value gives the percent of the variation among the various display conditions that can be predicted by the model.

Table 4-3 indicates the resulting prediction equations from the Snyder and Maddox (1978) experiments for two of the tasks, the reading task and the structured visual search task. It

Table 4-2.   Pool of Predictor Variables

Vertical	Horizontal	Description
VFREQ	HFREQ	Fundamental spatial frequency (c/deg)
VFLOG	HFLOG	Base 10 log of fundamental spatial frequency
VSQR	HSQR	Square of (fundamental spatial frequency minus 14.0)
VMOD	HMOD	Modulation of fundamental spatial frequency
VDIV	HDIV	Fundamental spatial frequency divided by its modulation
VLOG	HLOG	Base 10 log of VDIV and HDIV
VMTFA	HMTFA	Pseudo-MTFA
VMLOG	HMLOG	Base 10 log of VMTFA and HMTFA
MCROS	HCROS	Spatial frequency at which MTF crosses CTF
VRANG	HRANG	Crossover frequency minus fundamental frequency

Snyder and Maddox (1978)

Table 4-3   Predictive Equations (Snyder and Maddox, 1978)

Task	Predictive Equation and Related Information
Reading Task	Time (s) = 5.74 + 0.3111 HFREQ + 2.479 HMOD + 4.365 HLOG − 14.973 HFLOG + 1.112 VMLOG. $R = 0.72$ ($R^2 = 0.525$) Asymptotic $R^2 = 0.637$
Search Task	Time (s) = 7.27 + 0.027 HDIV + 2.159 HLOG + 5.916 VFLOG − 0.339 VMTFA − 0.054 VRANG + 5.487 VMLOG $R = 0.71$ ($R^2 = 0.500$) Asymptotic $R^2 = 0.575$

can be seen that this empirical model, which has subsequently been cross-validated, predicts 50% of the variance for the search task and 52% of the variance in the reading task. Of perhaps more interest are the combinations of variables which entered into the prediction equations. These predictor variables are almost entirely modulation and MTFA type measures, and generally support the results which have been previously obtained for these types of image quality measures.

As noted by Snyder and Maddox (1978), the equations in Table 4-3 represent the best empirically derived measures of image quality for digital displays, for the purpose of display design specification. They do not deal directly with the recommended dynamic range of a given image or any other image-specific parameters as do some of the measures described above. Thus, these equations are useful by the designer to optimize displays particularly for the presentation of alphanumeric information. They are of no help in describing or suggesting processing algorithms to make a literal image more interpretable.

4.7   PROBLEMS IN IMAGE QUALITY MEASUREMENT

The preceding discussion has summarized some of the more pertinent concepts of image quality and mathematical quantities suggested by researchers for the specification of image quality. In some cases, research has been conducted to evaluate the predictive capability of the metric, and in some cases the metric has only face validity. At the present time, it is not possible to single out a single metric as "the best," and no responsible researcher would do so. At the same time, several of the measures are interrelated and obviously share some common predictive capability. The reader is left to choose among these metrics to suit his or her individual purposes, but with the often happy realization that many of the metrics are similarly derived and result in similar prediction of visual performance (Task, 1979).

Irrespective of the metrics to be used, one is often confronted with terms and quantities in the literature that relate to these measures and to display design in general. Some of these quantities have been used incorrectly, and many are ambiguously defined. The following section is intended to clarify some of these terms and relate them to approaches that can be used for image quantification for metric calculation.

4.7.1 Lines, Line Pairs, and TV Lines.   Historically, an optical line pair was defined as a pair of adjacent lines in an optical test pattern, one line being light and the other dark. In most optical publications, the term "optical line" is the same as one line pair, that is, a light line plus a dark line.

Unfortunately, the television industry adopted a slightly different definition, using the term "TV line" to mean *either* a light line or a dark line on a TV test pattern. Thus, one optical line pair is equal to two TV lines. This distinction is critical in that some measures of display performance require the specification of the number of visible TV lines. Since this number will differ by a factor of two from the number of visible optical line pairs, one must be careful in definitions. More than once, this calculational error has been made in evaluating the capability of imaging systems (Biberman, 1973).

It is critical that the distinction be made between TV lines or optical lines and the scanning lines on a TV display, which are called "raster lines." While it is not possible to have greater resolution in the vertical direction (assuming a horizontal raster) than one has raster lines, the effective resolution is often less than the number of raster lines, due to misregistration, contrast, noise level, and the like. The optical line, line pair, or TV line refers to the resolvability of a test pattern of either square-wave or sine-wave modulation which is imaged on the display. It does not refer to the raster lines which are used to paint the picture in TV format.

4.7.2 The Measurement System.

Photometric measurement of emissive or reflective displays can be made with either a photometer or a radiometer equipped with appropriate optics and scanning capability. If a photometer is used, then the output is directly recorded in luminance units, and no correction is necessary for the spectral sensitivity of the visual system. The photometer is usually equipped with a spectral filter which, when cascaded with the spectral sensitivity function of the photodetector, produces a response proportional to the photopic luminosity function for all visible wavelengths of light.

On the other hand, one can use a radiometer and then calculate the equivalent luminance from the radiance measure by the relationship

$$L_v = K_m \int_{380}^{720} L_{e,\lambda}\, V(\lambda)\, d\lambda,$$

in which

L_v = luminance, in candelas/square meter,
K_m = 683 lumens/watt,
$L_{e,\lambda}$ = radiance at wavelength λ, in watts,
$V(\lambda)$ = value of photopic luminous efficiency function at wavelength λ, and
λ = wavelength, in nanometers.

In general, greater accuracy is obtainable from the radiometric approach than is typically obtained using an integrated photometer. For a detailed discussion of these approaches and the physics of measurement, see Walsh (1965) or Wyszecki and Stiles (1982).

It is usually helpful to plot unidimensional functions of luminance vs. distance on the display to understand the influence of various display parameters. Such an example for a sine-wave grating was illustrated in Chap. 3. Storage of the one- or two-dimensional radiometric or photometric data in a digital computer facilitates subsequent calculations of image quality metrics and other values of interest.

One of the major decisions in photometric or radiometric measurement of a display is that of the aperture size over which the energy is to be integrated. In general, edge gradient and MTF-related measurements require an aperture size considerably smaller than a single pixel. In practice, aperture widths equal to or smaller than 5% of the pixel width are adequate for most applications. Corrections for the MTF of the aperture can be made analytically (Wyszecki and Stiles, 1982). If the aperture selected is to be very small, then there is a premium on the sensitivity of the detector, which in turn can lead to higher noise levels.

Edge gradients are best measured with elongated apertures to average out noise along the major axis while the aperture is scanned in the direction of the minor axis. A 100:1 aspect ratio for the aperture is common and works quite well. Scanning can be accomplished either optically or physically. One convenient technique is to use a larger X, Y stage driven by stepping motors in both axes. Under computer control, the stage can be moved to each new position and the photometer/radiometer output can be recorded prior to the next step. A typical system for this type of measurement can take over 200 samples per second with a distance between samples as small as 0.001 inch. It is not uncommon to calibrate a display by taking 6,000 × 6,000 samples in an area of 6 × 6 inches. Typical times for scanning and recording this number of pixel intensities and recording the 3.6×10^7 bytes are about 12 minutes.

Overall display luminance measures can be made either photometrically or radiometrically by changing the optics on the detector to subtend a greater area of the display. For field work, there are several battery-powered luminance meters that have an acceptable accuracy (e.g., about 10%). Greater accuracy is obtainable in the laboratory with controlled, calibrated luminance or radiance sources, although these also are subject to drift.

4.7.3 What to Measure and Calculate.

While there are no accepted standards for the quantification of visual displays for either modulation or spatial capabilities, there have been developed common practices used by several laboratories and manufacturers. In general, the following are recommended.

First, overall luminance of the display can best be measured by integrating the radiance (or luminance) over a large number of "on" pixels, with the number of pixels subtended at least 100. Care must be taken to avoid lens flare in the optical collection of the energy.

Second, edge or line spread measurements can most easily be made by scanning an aperture across a line or knife edge written on the display, as by turning on a large area of pixels. The edge gradient can be differentiated to obtain the spread function, and then Fourier transformed to obtain the MTF. This process works quite well with carefully controlled and calibrated apparatus, but requires an on-line computer. Alternatively, the line spread function can be measured directly by turning on a row or column of pixels and scanning across the row or column with a slit aperture. Fourier transforming the data yields the MTF. A third alternative is to write either a square-wave or sine-wave grating on the display at a number of spatial frequencies and scan across this grating with a slit aperture. The output is then recorded and analyzed. In the case of the sine-wave grating, the modulation can be calculated directly, typically with some averaging over a number of cycles. A plot of the modulation at the various spatial frequencies of the grating is the MTF. If a square wave is used, the output should either be corrected to get the sine-wave fundamental equivalent modulation (by Fourier transforming the square-wave data) or the square-wave modulation can be used directly, although it should be noted that this does not produce an MTF.

Third, spectral emission (or reflectance) can be obtained by radiometric measurement using a scanning monochromator to obtain radiance per unit bandwidth. Corrections for the slit width of the monochromator can be made to refine the accuracy of the measurement (Wyszecki and Stiles, 1982).

4.7.4 Test Pattern Selection.

It is generally best to use a sine-wave intensity grating to avoid problems in Fourier transforming the data. On the other hand, discrete pixel displays will not permit sine-wave gratings at high spatial frequencies, and even at lower spatial frequencies the stepwise increments distort the analytical results. For such displays, a square-wave pattern is typically most useful, with the recorded data subsequently Fourier transformed to yield the sine-wave fundamental frequency modulation.

Most physical displays are anisotropic, with different MTFs in the vertical and horizontal directions. Accordingly, it is strongly recommended that the spatial measurements be done separately in both dimensions. Note that differences in these measured characteristics in the two dimensions can result in a significant contribution to observer performance (e.g., Table 4-3).

4.8 CONCEPTS RELATED TO IMAGE QUALITY

The term image quality has taken on a variety of meanings to different people over the last decade. In some situations it has held a very specific meaning, using for example the quality metrics defined above. In other situations, the term is used more generally and suggests various characteristics of displays or pictures that are less than ideal but for which no metric has been suggested or for which no quantitative concept exists. To avoid further misconceptions, some of these issues are discussed below.

4.8.1 Uniformity: Large and Small Area.

The uniformity of a display can best be defined in its absence, or by nonuniformity. Tannas and Goede (1978) described three types of nonuniformity—large-area nonuniformity, small-area nonuniformity, and edge discontinuities. Large-area nonuniformity is a gradual change in luminance (or chrominance) from one area of a display to another, such as center-to-edge. In general, many displays exhibit such nonuniformities, such as TV displays which have typical luminance gradients as large as 50% from the center to the edge. Similarly, photographic systems, lenses, projectors, etc., drop off from center to edge in both luminance and MTF. Because of the very gradual nature of these gradients (very low spatial frequency), they are largely not recognized. But they become important in some instances, as when the display is dimmed to its minimum intensity and the darker areas become invisible. No adequate criteria exist for acceptable limits of large-area nonuniformity, nor do any good experimental data exist on this problem.

Small-area nonuniformity pertains to pixel-to-pixel changes in luninance or chrominance, and are usually described by pixel-to-pixel contrast ratio measures. The detectability of such nonuniformities can usually be predicted by a Fourier transform of the gradients and comparison of the Fourier coefficients with the visual system CTF. Coefficients that exceed the CTF values indicate detectable nonuniformity at that spatial frequency. This technique has been experimentally evaluated by Snyder (1980) and appears to work satisfactorily.

Edge discontinuities are changes in luminance or chrominance which extend over an extended boundary. Detection of these nonuniformities can also be predicted with Fourier analysis and comparison with the visual system CTF.

4.8.2 Shades of Gray.

An often used, but typically misunderstood, concept in the literature is "shades of gray." While there is an image quality measure based upon this notion (GSFP), the concept has a basic underlying assumption that is incorrect—namely, that a just detectable increment in luminance is an increase by the square root of two. It is not clear where this misconception originated, but the many studies of visual perception dealing with difference thresholds clearly indicate that the just noticeable increment in luminance varies with the original luminance level, the size and shape of the target, the number of targets, the adapting luminance, and a variety of other variables. To use a single number is overly simplistic. In fact, as indicated in Chap. 3, the CTF is a 50% detectability contrast (discrimination of one half of the sine-wave cycle over the other) for luminance differences. It is far more meaningful, and therefore less confusing, in the visual sense to specify the dynamic range of the display (maximum minus minimum luminance) than the shades of gray, no matter what means is used to "measure" the shades of gray. On the other hand, purely for purposes of defining the dynamic range of the display, there is no harm in using square root of two increments to arrive at a number of "gray shades," which is merely another way to indicate dynamic range.

4.8.3 Resolution.

The number of pixels that can be displayed is constrained by the size of the display, the pixel size, the spacing between pixels, and occasionally the pixel shape. On an analog display, such as a CRT, there is also some overlap between addressed pixels which can cause heterogeneity of pixel sizes and shapes. For analog displays it is more meaningful to speak of the general concept of "resolution" than to speak of pixel density. Conversely, pixel density is more meaningful for discretely addressed displays.

"Resolution" is often used imprecisely and inappropriately. In some contexts it is used to mean the number of pixels per unit display distance, in which case it is equivalent to pixel density. More typically, it is used to refer to the number of "resolvable" elements per unit dimension, where the resolvable element is measured either physically or subjectively. Because there are at least eight or nine measures of resolution in the literature, the term has become an opportunity for "specsmanship" among manufacturers, with various persons using different definitions to obtain the most favorable numbers. Some discussion of these alternate definitions may be helpful.

The most often used measure of resolution for a television image is that of television limiting resolution. It is typically measured with a test pattern such as the "Indian Head" or RETMA pattern. The limiting resolution is simply the spatial frequency (in TV lines per picture width or height) at which the observer can no longer discriminate the light and dark bars of the (square-wave) image. It is tacitly assumed that the observer's MTF exceeds that of the display system, perhaps an unwarranted assumption. Generally, the limiting resolution is taken to be about 3% contrast of the square-wave pattern (RCA, 1974, p. 119). "Contrast" in this text, however, is defined as the luminance difference between dark and light bars divided by twice the (nonzero) background luminance. Thus, TV limiting resolution is equal to the spatial frequency at which the CTF is 0.024, correcting for square-wave to sine-wave transformation.

If a CRT image is formed with a Gaussian spot intensity distribution, then two adjacent "white" lines will be barely discernible from an interspersed black line if the line spacings are 1.18 standard deviation units. Fig. 4.7 illustrates this relationship at TV limiting resolution.

For a Gaussian spot, the luminance of the spot is equal to 60.7% of its maximum intensity at ±1σ. At one-half its maximum intensity, the width of the spot is ±1.18σ, or 2.35σ wide. This is the so-called "full-width half-amplitude" value of spot size that is used in the image processing literature. Resolution measurements using this half-amplitude approach are called 50% amplitude measurements. This separation is also shown in Fig. 4.7.

Several other measures of resolution are also used with CRTs. Assuming a Gaussian spot intensity distribution, it is possible to convert from one to another of these measures, as illustrated in Fig. 4.8. For convenience, all measures have been related to the display MTF in normalized units (Slocum, 1967).

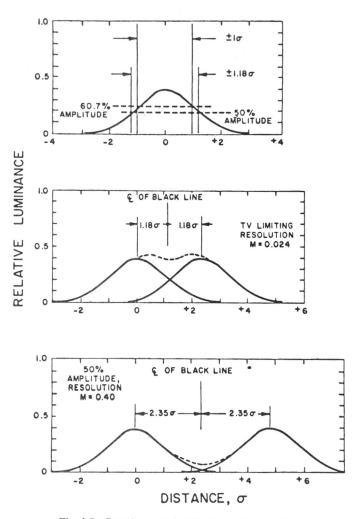

Fig. 4-7. Gaussian spot distribution and resolution.

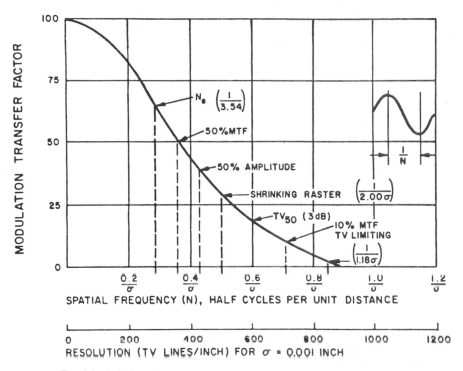

Fig. 4-8. Relationships among measures of resolution for Gaussian spot.

REFERENCES

1. Almagor, M., Farley, W. W., and Snyder, H. L. "Spatiotemporal Integration in the Visual System." Air Force Aerospace Medical Research Laboratory *Technical Report AMRL-TR-78-126*, 1979.

2. Beamon, W. S. and Snyder, H. L. "An Experimental Evaluation of the Spot Wobble Method of Suppressing Raster Structure Visibility." Air Force Aerospace Medical Research Laboratory *Technical Report AMRL-TR-75-63*, 1975.

3. Beaton, R. J. "A Human-Performance Based Evaluation of Quality Metrics for Hard-Copy and Soft-Copy Digital Imagery Systems." Unpublished Ph.D. dissertation, Virginia Polytechnic Institute and State University, 1984.

4. Biberman, L. M. *Perception of Displayed Information.* New York: Plenum Press, 1973.

5. Blumenthal, A. H., and Campana, S. B. "An Improved Electro-optical Image Quality Summary Measure. *Proceedings of the Society of Photographic Instrumentation Engineers, Image Quality, 30,* 1981.

6. Borough, H. C.; Fallis, R. F.; Warnock, R. H.; and Britt, J. H. "Quantitative Determination of Image Quality." Boeing Company *Technical Report D2-114058-1,* 1967.

7. Bryngdahl, O. "Characteristics of the Visual System: Psychophysical Measurement of the Response to Sine-wave Stimuli in the Photopic Region." *Journal of the Optical Society of America, 56,* 811–821, 1966.

8. Campbell, F. W., and Robson, J. G. "Application of Fourier Analysis to the Visibility of Gratings," *Journal of Physiology, 197,* 551–566, 1968.

9. Campbell, F. W.; Kulikowski, J. J.; and Levinson, J. "The Effects of Orientation on the Visual Modulation of Gratings." *Journal of Physiology, 187,* 427–436, 1966.

10. Carlson, C. R., and Cohen, R. W. "Visibility of Displayed Information: Image Descriptors for Displays." Arlington, VA: Office of Naval Research, *Technical Report ONR-CR213-120-4F,* 1978.

11. Charman, W. N., and Olin, A. "Tutorial: Image Quality Criteria for Aerial Camera Systems." *Photographic Science and Engineering, 9,* 385–397, 1965.

12. Cohen, R. W., and Gorog, I. "Visual Capacity—An Image Quality Descriptor for Display Evaluation," *Proceedings of the Society for Information Display, 15,* 53, 1974.

13. Dainty, J. C., and Shaw, R. *Image Science.* New York: Academic Press, 1974.

14. Gaskill, J. D. *Linear Systems, Fourier Transforms, and Optics.* New York: Wiley, 1978.

15. Granrath, D. J. "The Role of Human Visual Models in Image Processing," *Proceedings of the IEEE, 69,* 552–561, 1981.

16. Gutmann, J. C.; Snyder, H. L.; Farley, W. W.; and

Evans, J. E., III. An Experimental Determination of the Effect of Image Quality on Eye Movements and Search for Static and Dynamic Targets." Air Force Aerospace Medical Research Laboratory *Technical Report AMRL-TR-79-51*, 1979.

17. Hufnagel, R. E. "Significance of the Phase of Optical Transfer Functions." *Journal of the Optical Society of America*, *58*, 1505–1506, 1968.

18. Linfoot, E. H. *Fourier Methods in Optical Image Evaluation*. London: Focal Press, 1960.

19. Overington, I. "Visual Efficiency: A New Figure of Merit for Optical Quality." British Aircraft Corporation Study Note No. 5, 1973.

20. Overington, I. "Visual Efficiency: A Means of Bridgeing the Gap Between Subjective and Objective Quality," *Proceedings of the Society of Photographic Instrumentation Engineers, Image Assessment and Specification*, *46*, 93, 1974.

21. Overington, I. "Some Considerations of the Role of the Eye As a Component of an Imaging System." *Optica Acta*, *22*, 365–374, 1975.

22. Overington, I. *Vision and Acquisition*. London: Pentech Press, 1976.

23. Overington, I. "Towards a Complete Model of Photopic Visual Threshold Performance." *Optical Engineering*, *21*, 2–13, 1982.

24. Pratt, W. K. *Digital Image Precessing*. New York: Wiley, 1978.

25. Radio Corporation of America. *Electro-optics Handbook*. Camden, NJ: RCA, 1974.

26. Rosell, F. A. "Analysis of Electro-optical Imaging Sensors." Westinghouse Electric *Technical Report ADM 105*, 1971.

27. Rosell, F. A. and Wilson, R. H. "Recent Psychophysical Experiments and the Display Signal-to-Noise Ratio Concept." In *Perception of Displayed Information*, ed. L. M. Biberman. New York: Plenum Press, 1973.

28. Schade, O. H. "Image Gradation, Graininess, and Sharpness in Television and Motion-Picture Systems. Part III: The Grain Structure of Television Images." *Journal of the Society of Motion Picture and Television Engineers*, *61*, 97–164, 1953.

29. Schindler, R. A. "Optical Power Spectrum Analysis of Display Imagery. Phase I: Concept Validity." Air Force Aerospace Medical Research Laboratory *Technical Report AMRL-TR-76-96*, 1976.

30. Schindler, R. A. "Optical Power Spectrum Analysis of Display Imagery." Air Force Aerospace Medical Research Laboratory *Technical Report AMRL-TR-78-50*, 1978.

31. Shannon, C. E., and Weaver, V. *The Mathematical Theory of Communication*. Urbana, Ill.: University of Illinois Press, 1949.

32. Slocum, G. K. "Airborne Sensor Display Requirements and Approaches." Hughes Aircraft Company *Technical Report TM-888*, 1967.

33. Snyder, H. L. "Image Quality and Observer Performance." In *Perception of Displayed Information*, ed. L. M. Biberman. New York: Plenum Press, 1973.

34. Snyder, H. L. "Image Quality and Face Recognition on a Television Display." *Human Factors*, *16*, 300–307, 1974.

35. Snyder, H. L. "Visual Search and Image Quality: Final Report." Air Force Aerospace Medical Research Laboratory *Technical Report AMRL-TR-76-89*, 1976.

36. Snyder, H. L. "Human Visual Performance and Flat Panel Display Image Quality." Virginia Polytechic Institute and State University *Technical Report HFL-80-1*, 1980.

37. Snyder, H. L.; Keesee, R. L.; Beamon, W. S.; and Aschenbach, J. R. "Visual Search and Image Quality," Air Force Aerospace Medical Research Laboratory *Technical Report AMRL-TR-73-114*, 1974.

38. Snyder, H. L., and Maddox, M. E. "Information Transfer from Computer-Generated, Dot-Matrix Displays." Virginia Polytechnic Institute and State University *Technical Report HFL-78-3*, 1978.

39. Tannas, L. E., Jr., and Goede, W. F. "Flat-Panel Displays: Six Major Problems and Some Solutions." Stanford/SID Seminar, San Francisco, 1978.

40. Task, H. L. "An Evaluation and Comparison of Several Measures of Image Quality for Television Displays." Air Force Aerospace Medical Research Laboratory *Technical Report AMRL-TR-79-7*, 1979.

41. Task, H. L. and Verona, R. W. "A New Measure of Television Display Quality Relatable to Observer Performance." Air Force Aerospace Medical Research Laboratory *Technical Report AMRL-TR-76-73*, 1976.

42. van Meeteren, A. "Visual Aspects of Image Intensification." Soesterberg, The Netherlands: Institute for Perception TNO, 1973.

43. Walsh, J. W. T. *Photometry*. New York: Dover, 1965.

44. Wyszecki, G., and Stiles, W. S. *Color Science*, 2nd ed. New York: Wiley Interscience, 1982.

5

FLAT-PANEL DISPLAY DESIGN ISSUES

L. E. TANNAS, JR., *Consultant*

5.1 INTRODUCTION

Flat-panel displays are the wave of the future. They make it possible to capitalize on the size and power advantages of microelectronics and microprocessors and will lead to new products previously only imagined in science fiction. We are just beginning to realize the potential of the flat-panel technologies and their promise of providing a dramatic increase in performance with dramatic reductions in cost. Flat-panel displays have many common features, and this book is devoted to categorizing and clarifying their complexities. Although the CRT is still the high-performance king and will be for some time, flat panels add to the realm of display options and are being applied to many new products for which the CRT is not suitable.

The display medium must have certain inherent performance capabilities. These have been described as the "six basic problems,"[1] as follows:

- Power efficiency
- Addressability
- Duty factor
- Gray scale
- Color
- Cost

The first three—power efficiency, addressability, and duty factor—are primary. Lack of performance with any of these three fundamentally limits the functional classifications to which the technology can be applied. Without gray scale and full color, a display technology will nerver penetrate the video market. Cost is the bottom line. Most of our display inventions have been too expensive to manufacture and sell. Flatness will sell but not at any price.

There are many other flat-panel design issues. The above six are usually the most critical to flat-panel displays. Other considerations are discussed in Chap. 2, System Requirements. Storage and operation life are important issues. Storage and operating temperature ranges often present difficult requirements for the new technologies to meet. Sunlight readability is difficult but necessary for outdoor and vehicular applications.

5.2 POWER EFFICIENCY

Luminous efficiency[1] is one of the most useful parameters in evaluating the practicality of the light-emitting display technologies. Low luminous efficiency often leads to one or more of the following undesirable characteristics:

- High power
- Low luminance
- High temperature
- Short life
- Bulky and expensive electronics
- Limited resolution
- Opaque row and column conductors
- High current and/or voltage
- Elaborate cooling means

Unfortunately, no standard exists for calculating and reporting luminous efficiency because researchers do not include the power for the same levels of supporting electronics. Also, re-

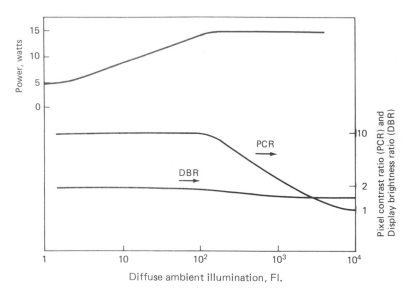

Fig. 5-1. Presentation of display ambient performance based on display system efficiency data.

searchers do not consistently and accurately measure the useful luminous flux output. For example, much of the emission is often absorbed internally in the display panel materials, and much is lost by radiating in directions outside the useful viewing envelope.

The following sections attempt to resolve this dilemma through the introduction of two new terms:

- Display system efficiency
- Display ambient performance

The efficiency of a display is a measure of its conversion of electrical power into an inter-

pretable image. There is no single number to characterize the overall efficiency, primarily because of the influence of the ambient illumination. A graphic presentation is used to characterize a display in this regard and illustrates the performance in terms of input power, pixel contrast ratio, and display brightness ratio as a function of the ambient illumination. The hypothetical example shown in Fig. 5-1 is called the "display ambient performance." The uses of this diagram were discussed in Chap. 2.

Figure 5-2 is a schematic diagram of the major elements which convert power and the information signal into a readable image, with the

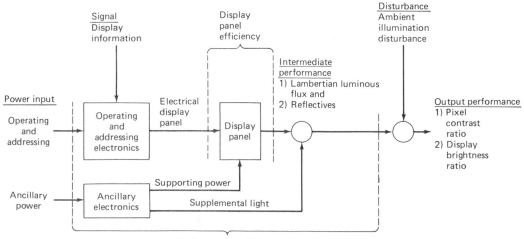

Fig. 5-2. Schematic diagram of display process (emitters and nonemitters).

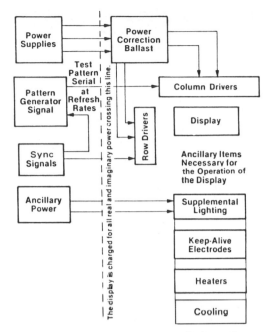

Fig. 5-3. Definition of power into display system.

ambient illumination as a disturbance. This diagram is useful in conceptually characterizing the major conversions necessary to create a usable image of the information signal. A more detailed block diagram of the display system is given in Fig. 5-3. The electronic conversion represents all the operations necessary to create the electrical impulse that stimulates the display panel. The display panel includes the line loss in the panel, the real and imaginary load, elec-

trical-to-luminance conversion, and optics of the panel materials.

A schematic diagram of the conversion at the pixel level from the power input to display panel is given in Fig. 5-4. The efficiency of a display emitter material is governed by the basic quantum efficiency, photon wavelength, and optical path out of the pixel cell toward the viewer. Display material efficiency is computed using real-power input at terminals A and B. Display material efficiency is only a small part of the display system as can be seen from Figs. 5-2 and 5-4. The display material efficiency is easy to measure experimentally. A single pixel is constructed, excited with laboratory power supply, and measured with laboratory spot photometer and power measurement instrumentation. In this context, the display is usually not charged for the imaginary component of power. To help emphasize that it is only a small part of the total process, it is called the "display material efficiency." If the pixel is not constructed to represent the panel geometry, or if all the luminous flux is measured with an integrating sphere, then it is simply called "material efficiency." The "material efficiency" is always much higher than "display material efficiency" as there is always large internal light piping and electrode blockage. For example, with ac thin-film EL, the material efficiency is of the order of 20 lm/W and the corresponding display material efficiency is on the order of 4 lm/W.

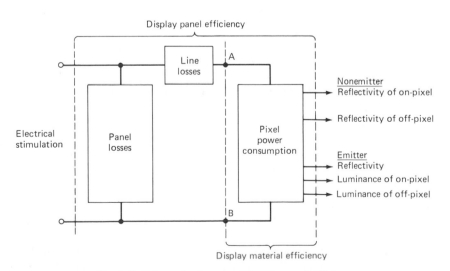

Fig. 5-4. Schematic diagram of display panel efficiency.

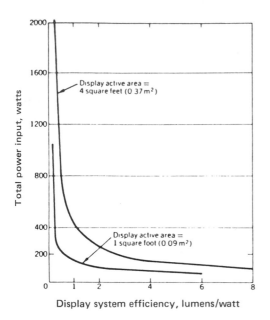

Fig. 5-5. Display system power efficiency. When a display is used for TV, the total input power required varies with the display system efficiency, as shown. The conditions under which the curves apply are as follows: power input calculated assuming 50 fL averaged luminance; display power plus 60 watts added for receiver functions, power supply inefficiency, logic, and timing.

Figure 5-5 shows the total power input required vs. "display system efficiency" for a display used in a TV receiver.[2] Below about 1 to 2 lm/W display system efficiency, power requirements quickly become untenable. Unfortunately, most configurations of the flat-panel display emitter fall into this category. Table 5-1 gives typical display material efficiencies for flat-panel technologies. The display system's system efficiency can be a factor of 10 below the display panel efficiency, which can be a factor of 2 below display material efficiency as shown in Figs. 5-2 and 5-4, and outlined in Table 5-2. Most of the flat-panel technologies have display system efficiencies below 1 lm/W.

If the display system efficiency is 0.1 lm/W, for example, we can calculate that if one were to build a display array one foot square and operate it at 30 fL average luminance, the power dissipated in the panel and electronics would be 300 watts. These calculations are based on the assumption that the directional luminous intensity of the display falls off in accordance with the cosine law. The total luminous flux F emit-

ted from a flat surface of area A and luminance L can be found by integrating over the hemisphere into which the pixel radiates. The result is as follows:

$$F = LA$$

where L is in fL and A is in ft^2. Since

$$F = \sigma P$$

where P is the input electrical power and σ is the display power efficiency, then

$$P = LA/\sigma$$

Another important principle is that of how the power is delivered to the emitting element. Many researchers consider low-voltage operation to be a desirable attribute because more driving circuits are available for low-voltage displays. However, continuing the 0.1 lm/W example, a panel requiring 200-volt excitation would have peak row conductor currents of approximately 500 milliamperes if addressed line at a time, whereas the same efficiency requiring 10 volt excitation would require peak row currents of 100 amperes.

Even if two devices have identical display system efficiency, there may be substantial viewability differences. For example, both CRT and LED materials have high reflectivity coefficients, which reduce the pixel contrast ratio in moderate to high ambient illumination. Transparent thin-film phosphor materials have very low reflectivity coefficients and need not emit the same light to achieve the same pixel contrast ratio. The "display ambient performance" diagram allows the effects of ambient illumination to be accurately accounted for in both of these cases.

The nonemissive display technologies possess a definite inherent advantage over the light emitters when considering power and immunity to high ambient illumination. The power advantage of nonemitters over emitters is measured in orders of magnitude in medium to high-ambient-illumination applications. This advantage is sufficient to justify further research and development despite serious limitations such as low

Table 5-1 Display Material Efficiencies

Display Type	Color	Reflection	Emission	Efficiency Commercial, lm/W	Efficiency Experimental, lm/W	Comments
Flat Panels						
Gas-UV phosphor	White	80% broadband	Lambertian		7.8	
Gas discharge, dc	Orange	5% plus electrode	Lambertian	0.1 to 0.5		Ref. Weber, U. of Illinois
Gas discharge (plasma panel), ac	Orange	5% plus electrode	Skewed	0.3	1.5	
Gas-electron phosphor	P 4, white	80% broadband	Lambertian		2 to 4	Ref. Sobel, Lucitron
Grid-addressed flat CRT	P 31, green	80% broadband	Lambertian		1 to 5	Ref. Goede, Northrop
VF	ZnO, green	80% broadband	Lambertian	1 to 3		
Thin-film EL, ac	ZnS: Mn, yellowish-orange	Transparent phosphor	Lambertian	3	4 to 6	Ref. Tannas, Aerojet
Thin-film EL, ac	ZnS: Tb, green	Transparent phosphor	Lambertian		1 to 2	
Powder EL, dc	Green	80%	Lambertian		0.5 to 1	
Red LED	GaAlAs, 655 nm	Package dependent	Lambertian	2 to 3	11.0	High brightness indicators
Green LED	GaP: N, 565 nm	Package dependent	Lambertian	0.5 to 2.5	4.5	Lamp, alphanumeric displays
Yellow LED	GaAsP: N, 585 nm	Package dependent	Lambertian	0.5 to 1.5	2.0	Lamp, alphanumeric displays
Blue LED	GaN, 490 nm	Package dependent	Lambertian	0.015	0.8	Lamp
CRTs						
P 1 phosphor	Yellowish-green	80% broadband	Lambertian	30		
P 31 phosphor	Green	80% broadband	Lambertian	40 to 60		
P 39 phosphor	Yellowish-green	80% broadband	Lambertian	20		
P 43 phosphor	Yellowish-green	80% broadband	Lambertian	40		
P 45 phosphor	White	80% broadband	Lambertian	22		Ref. Seats, Thomas Electronics
P 53 phosphor	Yellowish-green	80% broadband	Lambertian	30		
P 22 color family	Blue at 450 nm	80% broadband	Lambertian	9		
	Green at 535 nm	80% broadband	Lambertian	40 to 70		
	Red at 626 nm	80% broadband	Lambertian	18		
Lamps without ballast						
60 watt incandescent	White	—	Isotropic	20		Low cost production lamp
Powder EL, ac	Green	—	Lambertian	20		Lighting films
Mercury Vapor	Bluish-white	—	Isotropic	63		Bulb
Fluorescent	White	—	Isotropic	82		Tube
High pressure sodium vapor	Yellowish-white	—	Isotropic	140		High power lighting
Low pressure sodium vapor	Yellowish-white	—	Isotropic	183		High power lighting

Table 5-2   Summary of Luminance Losses in a Typical Emitting Flat-Panel Display System

Display Process	Conversion Losses	Incremental Efficiency	Process Efficiency
Quantum Efficiency	Electron to Photon	10%	10%
Material Efficiency	Photon to Luminance	25%	2.5%
Display Material Efficiency	Pixel Optics	25%	0.63%
Display Panel Efficiency	Panel Optics and Lead Resistance	50%	0.31%
Display System Efficiency	Electronics	50%	0.16%

speed of response and addressability. The low display brightness ratio of nonemitters can be corrected by using supplemental lighting.

The display system efficiency is a set of curves which can be used for characterizing both emitters and nonemitters. The data for the curves are based upon laboratory measurements using inexpensive instrumentation. The display system efficiency characterizes the luminous output of the completed display as a function of input power as shown in Fig. 5-6. It is an end-to-end test measurement.

Researchers doing display device development should make estimates about a new approach before a complete display is fabricated and measured. The following is a procedure for generating the display system efficiency diagram.

The ultimate performance of a display is measured by its pixel contrast ratio and display brightness ratio in its intended environment. This measurement is called the "display ambient performance" (Fig. 5-1) and can be computed from the "display system efficiency" curves of Fig. 5-6.

The following discussions describe the problems and sources of power losses and light losses internal to the display. The various losses of power and light and conversion efficiencies are so interrelated and complex that only end-to-end testing can be relied upon to give accurate display system efficiency data. Finally, there is no substitute for testing a display in the ultimate user environment to determine the power required to give the desired performance.

5.2.1 Emitters and Nonemitters.   Displays must be bright and contrasty so they can be easily read without error and fatigue. On the basis of

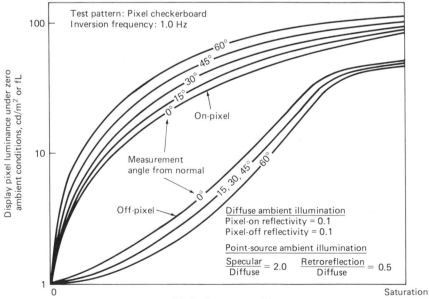

Fig. 5-6. Display system efficiency diagram.

display brightness ratio and pixel contrast performance, nonemitters can be directly compared with emitters. The common factors to which both are subjected are the ambient illumination and the need for input power. The advantages of emitters vis-a-vis nonemitters are subtle and worthy of further elaboration.

When discussing the origin of the light emanating from a display surface, the following distinction among emitted luminance, supplemental light, and reflected illuminance is used: Emitted luminance is light created at the viewed surface from electrical power, and reflected illuminance is from ambient illumination falling upon the viewed surface and reflected back toward the viewer. Supplemental light is a constant light source created within a display package for the express purpose of indirectly increasing its surface brightness. Supplemental light contains no intelligence. The supplemental light is used with nonemitters. The light is transformed by birefringence, reflectance, absorptance, or transmittance by the nonemissive display. Instrumentation measuring the luminance of a display surface cannot distinguish between the different contributing sources of luminous flux—*nor can the human viewer.*

The off- and on-pixels have the same reflectivities for the emitter and different reflectivities for the nonemitter. With an emissive display, the generated luminance is primarily created by the

Table 5-3   Summary of the Effects of Ambient Illumination, Supplemental Illumination, and Luminance upon Display Performance

	Pixel contrast ratio	Display brightness ratio
Emitter with increasing ambient illumination	Decreasing	Decreasing
Emitter with increasing luminance	Increasing	Increasing
Nonemitter with increasing ambient illumination	No change	No change
Nonemitter with increasing supplemental illuminance	No change	Increasing

on-pixel. For nonemitters, the supplemental light falls upon both on- and off-pixels equally; however, the on- and off-pixels transform the supplemental light unequally. A summary of these effects is given in Table 5-3.

A comparison of the readability of emissive and nonemissive techniques is given in Fig. 5-7. Nonemitters cannot be as contrasty as emitters at low and medium ambient illumination, even with supplemental lighting. This is due to the

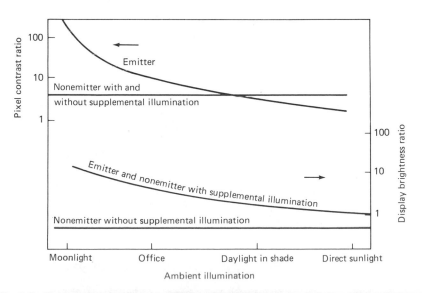

Fig. 5-7. Comparative significance of PCR and DBR for emissive and nonemissive displays.

fundamental optical property of nonemissive displays, which is that their pixel contrast ratio is independent of the ambient illumination (as can be seen from the defining equation for pixel contrast ratio in Eq. 5-2). On the other hand, the pixel contrast ratio of emitters will always wash out at high ambient illumination (as can be seen from the pixel contrast-ratio in Eq. 5-1). At high ambient illumination, the pixel contrast ratio equation asymptotically converges toward one. Supplemental light can always be added to nonemitters to bring up their brightness and make them comparable in brightness to emitters. Supplemental light can be created by the most efficient means independent of the display effect and addressing technique.

The pixel contrast ratio equations written to emphaize these points are as follows:

Emitter:

$$PCR = \frac{\text{Emitted luminance from on-pixel} + \text{Reflectivity} \times \text{ambient illumination}}{\text{Emitted luminance from off-pixel} + \text{Reflectivity} \times \text{ambient illumination}}$$

(Eq. 5-1)

Nonemitter:

$$PCR = \frac{\begin{array}{c}\text{Supplemental light transformed} \\ \text{by on-pixel} + \text{On-pixel reflectivity} \\ \times \text{ambient illumination}\end{array}}{\begin{array}{c}\text{Supplemental light transformed} \\ \text{by off-pixel} + \text{Off-pixel reflectivity} \\ \times \text{ambient illumination}\end{array}}$$

(Eq. 5-2)

Similarly, the display brightness ratio equations are as follows:

Emitter:

$$DBR = \frac{\begin{array}{c}\text{Emitted luminance from display} \\ \text{with 50\% of the pixels on} \\ + \text{Reflectivity} \times \text{ambient illumination}\end{array}}{\begin{array}{c}\text{Reflected ambient illumination} \\ \text{from the surround}\end{array}}$$

(Eq. 5-3)

Nonemitter:

$$DBR = \frac{\begin{array}{c}\text{Supplemental light transformed by} \\ \text{display with 50\% of the pixels on} \\ + \text{Reflectivity with 50\% of the pixels on} \\ \times \text{ambient illumination}\end{array}}{\begin{array}{c}\text{Reflected ambient illumination} \\ \text{from the surround}\end{array}}$$

(Eq. 5-4)

Note that all of these measurements are average over at least one frame time.

The actual reflected light that exists in the display surround installation should be used in the DBR equations. If the display setting is not known, then a reflectivity of 18% may be used. The value of 18% is the value that photographers use based on experience from taking black-and-white photographs of varied scenes. The supplemental light is modified by the on- and off-pixels in accordance with the geometric arrangement of the light source and display panel. The term "transformed by" is used in Eqs. 5-2 and 5-4 to generalize how the on- and off-pixels affect the supplemental light source.

The PCR measurements are averaged over one pixel. The DBR measurements are averaged over several pixels in a test pattern and are to include a typical number of on-pixels and off-pixels corresponding to actual display image content. A conservative approach is to include an equal number of on- and off-pixels.

5.2.2 Ambient Illumination. The ambient illumination and display reflectivity are very important in computing display ambient performance and the display system efficiency. For example, consider two EL display panels, identical in all respects, including display material luminous efficiency, except that one uses a transparent phosphor with a black, light-absorbing back layer and the other an opaque white phosphor. They are both isotropic radiators. Both phosphors have a high index of refraction. One phosphor is opaque because of the grain structure which causes multiple reflecting surfaces. The other phosphor is a thin film of amorphous or small-grain polycrystralline struc-

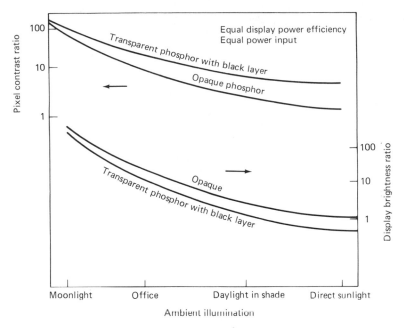

Fig. 5-8. Qualitative significance of display surface reflectivity on display performance.

ture, rendering it transparent with a very low reflectivity, and has a black, light-absorbing back layer. The qualitative performance of each, for the same power input, is shown as a function of ambient illumination in Fig. 5-8. As the ambient illumination is increased, the opaque phosphor display washes out (PCR = 1) long before the display with transparent phosphor and black layer.

5.2.3 Light Losses.   Other factors affect pixel contrast ratio and display brightness ratio separate from the power input. The means by which the emitted luminance gets out and away from the light center is significantly different for dif-

ferent technologies. In gallium arsenide LEDs, most of the light is internally absorbed. In gallium phosphide, the light is piped to the next pixel, degrading the contrast. Once a photon is free from the emitting junction, it must pass through the substrate air interface out of the front of the display. Typically, LED materials have high indices of refraction, and as a consequence (in accordance with Snell's Law) a majority of the light is reflected internally. Similarly, EL phosphors have high indexes of refraction, and as can be seen in Fig. 5-9, only a small cone (approximately 18 degrees) of the isotropic radiation escapes. Internal reflections occur at the phosphor-glass interface and at the glass-air

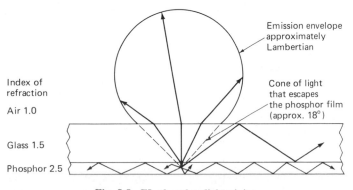

Fig. 5-9. EL phosphor light piping.

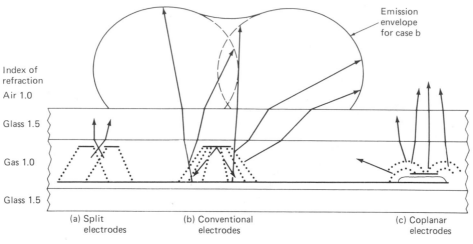

Fig. 5-10. GD plasma-panel electrode blockage.

interface. GD displays do not suffer from total internal reflections because the index of refraction of the gas is essentially identical to that of air. A larger cone of light (approximately 48 degrees) can pass through the glass faceplate. However, for the larger GD panels, an opaque conductor is needed to conduct the row and column current.[3] As shown in Fig. 5-10, these conductors block light from passing out of the front of the panel. Split electrodes and coplanar electrodes are used to minimize this loss, but not without other complications.

5.2.4 Power Loss. A significant amount of power is lost due to ohmic heating in the ballast of LED displays and addressing leads of EL. For EL, the electrical impedance at each pixel is highly reactive. As a consequence, several times more current is conducted in and out of the display than is actually used as power to create light.

Either power is lost in transparent leads or light is blocked by opaque ones. The best transparent conductors have resistivities of approximately 1 to 10 ohms/square. Indium-tin-oxide (ITO) thin-film transparent conductors can be made with 90% optical transmittance and 10 ohms/square resistivity. An eight-inch column line in a 60-line/inch display will have 960 squares or 9.6 kilohm resistance from end to end, assuming equal lead and space width. Metal lines can be made two or three orders of magnitude lower in resistance but with similarly reduced optical transmittance.

5.2.5 Display System Efficiency Diagram.
The "display system efficiency" (Fig. 5-6) is an end-to-end measurement characterizing the power performance of the display. The desired ultimate performance is a pixel contrast ratio and a display brightness ratio in a given ambient. The ultimate performance is presented in a "display ambient performance" diagram (Fig. 5-1). The display ambient performance is computed using Eqs. 5-1 and 5-3 for emitters or Eqs. 5-2 and 5-4 for nonemitters and the display system efficiency diagram.

The basic measurements needed for constructing the display system efficiency diagram are as follows:

Emitters:
1. On- and off-pixel luminance as functions of viewing angle and input power under zero ambient illumination (see Fig. 5-11).
2. Display panel luminance as a function of viewing angle under an arbitrary, extended-source ambient illumination at arbitrary intensity with display power off (used to compute reflectivity for extended source) (see Fig. 5-12).
3. Same as (2) above except with point-source ambient illumination (used to compute reflectivity for point source) (see Fig. 5-13).

Nonemitters:
1. On- and off-pixel luminance as functions of viewing angle and display power under

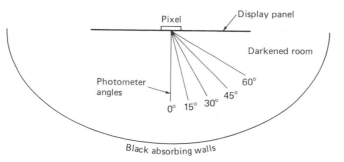

Fig. 5-11. Test setup for display luminance measurements.

zero ambient illumination with supplemental light power on. This test cannot be done if supplemental light is not used, as there will be no light to measure (see Fig. 5-11).

2. On- and off-pixel luminance as a function of viewing angle under an arbitrary, extended-source ambient illumination with display power on and with supplemental light power (if used) off (see Fig. 5-12).

3. Same as (2) above with point-source ambient illumination (used to compute reflectivity for point source) (see Fig. 5-13).

A consistent set of ground rules for making measurements is needed to permit comparisons of different display technologies by different laboratories. A checkerboard test pattern as shown in Fig. 1-2 of Chap. 1 is recommended. The luminance should be averaged over the entire pixel spatial area. In order to quantify the temporal factor for comparison between refresh- and memory-type displays, the pattern should be inverted at a specified rate while measuring

input power. A test pattern change of once per second is recommended.

The power input should include all power required for the display itself and display supporting electronics as outlined in Fig. 5-3. The display system measurements are summarized in Fig. 5-6.

In the test geometries of Figs. 5-11, 5-12, and 5-13, the display luminance measurements are made at different angles, using a conventional spot photometer. The performance of a display becomes exceedingly difficult to characterize (and utilize) if any of these measurements from Figs. 5-11 and 5-12 are very angle-dependent. Extra effort should be taken in order to minimize the viewing directional dependence in order to improve the display's utility in varied lighting installations. A diffusing faceplate and other optical techniques should be added in order to make the display independent of viewing angle as much as possible over a 60-degree cone angle from the normal. If the luminance is not uniform, then the display should be rated by the lowest luminance reading throughout a

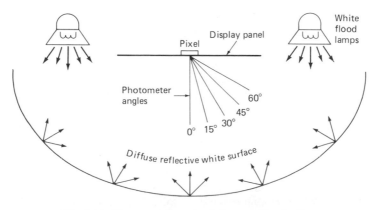

Fig. 5-12. Test setup for extended-source illumination.

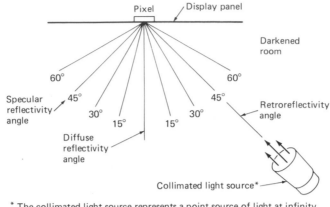

* The collimated light source represents a point source of light at infinity (the sun). The collimated beam of light is necessary to give a suitably large illuminated area for measurements with a photometer at different locations on the display

Fig. 5-13. Test setup for point-source illumination.

45-degree cone angle from the normal. This is an appropriate penalty for its nonuniformity. The specular reflection associated with the collimated source of Fig. 5-13 is particularly difficult to minimize. This is usually handled procedurally. When the sun is the source never view the display in the specular reflection direction. The performance is quoted in terms of the ratio of the specular-to-diffuse reflectivity and retroreflectivity-to-diffuse reflectivity as measured along the axes shown in Fig. 5-13. These ratios are used to quantitatively describe the display's point-source reflection properties.

In Fig. 5-6 a sample set of measurement data is presented for a high-quality display. The performance of the display is computed from the appropriate Eqs. 5-1 through 5-4 and is plotted in Fig. 5-1. Note the power requirement as a function of ambient illumination. The power saturates at 15 watts. Lower power is used at the low ambient illumination due to excess performance capability. This performance characteristic is very typical of emitting displays because of the reduction in performance at high ambient illumination. The display performance of Fig. 5-1 is computed for diffuse ambient illumination. If it is done for a point-source ambient illumination, then the geometry of the source and viewer must be specified.

5.2.6 Quantum Efficiency. The computation of PCR and DBR is based on laboratory measurements taken of a complete operating device

under appropriate ambient illumination and frame rate. These measurements are not possible until a completed display can be made and tested. Until a complete device can be made, PCR and DBR must be calculated from more basic measurements. For emitters, the most basic property is the material quantum efficiency. The quantum efficiency is the ratio of the number of photons emitted from the material to the number of electrons injected into the material.

5.2.7 Material Efficiency. The material luminous efficiency of an electric lamp or display is its primary figure of merit. It is still true for electric lamps or luminaires but more complex for displays with improved optical design and with the introduction of nonemitters. The material efficiency is a material property that is useful in appraising the potential merits of a display approach. It can be computed from the quantum efficiency. Assuming everything else is the same, the ultimate display performances can be ranked based on quantum efficiencies. This is particularly useful in comparing one material with another within one technology, such as two thin-film EL materials.

The material efficiency can be measured directly using an integrating sphere. Illuminating engineers do this routinely in order to determine the performance of luminaires. The integrating sphere collects all the emitted photons, and the instrumentation is calibrated in lumens. Because all directional properties are lost in the integrating sphere, these measurements have little value for displays except for comparison purposes, as with quantum efficiency.

5.2.8 Estimating Display Performance. A reasonably good estimate of the display system efficiency can be made knowing the material efficiency and general optical and electrical properties of the display material. Such calculations are appropriate to justify further work on a new material. The addressing technique must be selected and designed to the point where power requirements can be computed. An optical model of a pixel similar to Figs. 5-9 or 5-10 must be derived. The losses can then be calculated based on the index of refraction of the material and the nature of the material packaging. The material efficiency, the pixel structure efficiency, and the addressing efficiency can then be used to derive the display system efficiency curves of Fig. 5-6. The luminous intensity of the off-pixel is computed from internal light piping and addressing cross-coupling. Finally, knowing the reflectivity of the display surface, the presentation of display ambient performance, Fig. 5-1, can be computed. *Such calculations always tend to be optimistic* because all loss effects are usually not included and the detrimental effects of the ambient illumination are underestimated.

A similar procedure can be used for nonemitting display materials.

5.3 ADDRESSABILITY

The most underestimated problem in flat-panel displays is that of addressing a large array of pixels. All known addressing techniques fall into one of five classifications, as summarized in Chap. 1. The most useful addressing technique for flat-panel displays has been the matrix-addressing technique using display materials which are intrinsically matrix addressable. Extrinsic matrix addressing is also very useful, where active and passive electronic switches and nonlinearities are added at each pixel. Each of the addressing techniques will be briefly discussed. The matrix-addressing technique will be discussed in detail.

5.3.1 Direct Addressing. Direct addressing is the most fundamental technique for addressing a pixel array. A single lead is used between the control switch in the external display circuit and each pixel. "One-third-select" or "cross-coupling-select" cannot occur in direct addressing. By definition, each pixel circuit is electrically isolated. The display material need not have intrinsic nonlinear properties. Direct addressing is the oldest form of addressing a small array of pixels but becomes inpractical for large arrays. Nixie® tubes and incandescent lamp displays have always used direct addressing. Many liquid crystal displays use direct addressing. The essence of direct addressing is shown in Fig. 5-14.

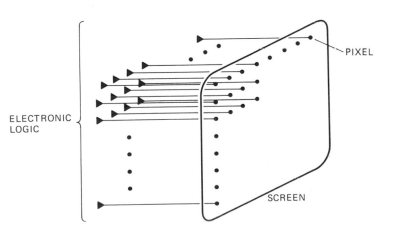

Fig. 5-14. Direct addressing.

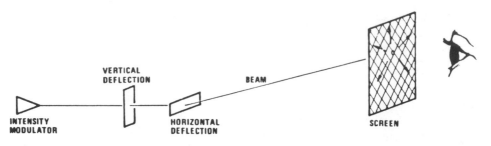

Fig. 5-15. Scan addressing.

5.3.2 Scan Addressing.

Scan addressing is used in CRTs. A beam of electrons serially sweeps out a raster or patterns under the control of two deflection amplifiers and an intensity modulator. When the x and y deflection is harmonic motion, the resulting patterns are called Lissajous figures. The essence of the scan-addressing technique is shown in Fig. 5-15. Unfortunately, the deflection requires a large depth dimension, because of practical limits on electron-beam trajectory curvature. The flat-CRT Kaiser-Aiken and Gabor tubes used scan addressing where the depth dimension was folded parallel with the screen which increased the screen perimeter area. Scan addressing as applied to CRTs is discussed in detail in Chaps. 6 and 7.

5.3.3 Grid Addressing.

Grid addressing is entirely different from the other forms of addressing[4]. The excitation of the display media is by the gating of an electron or other particle through a physical hole at each pixel. In grid addressing, the pixel is defined by the hole locations. The display medium may be a uniform film. There are no x and y matrix lines, and therefore no sneak circuit problems. However, cross-coupling can occur under some circumstances. The primary benefit of the grid-addressing technique is that it uses considerably fewer amplifiers than matrix addressing. However, the need for the individual grids with a hole for each pixel greatly increases the cost of the display physical structure.

A diagram of the grid-addressing process as applied to cathodoluminescence is shown in Fig. 5-16[5]. Imagine a uniform area of electrons emitted from a series of filament wires (cathodes) and a front screen completely electroded as the anode. The electrons will be attracted toward the anode according to the electrostatic forces through the holes in the grids. The grids are electroded in sections, and if negatively charged by their individual amplifiers, the grid

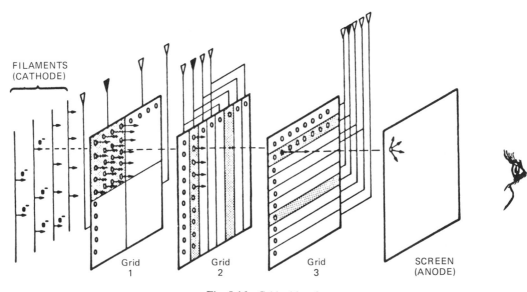

Fig. 5-16. Grid addressing.

will repel electrons and prevent their passage through their respective holes. By the process of elimination through a series of grids, one unique pixel may be excited.

Successful grid-addressed devices using cathodoluminescence[5,6] and gas discharge[7] have been fabricated. In the gas-discharge approach, a reservoir of metastable particles, ions, electrons, and UV photons is generated at the back of the display. The particles drift and diffuse through holes in the grids under control of the grid electrode potentials. When a metastable particle or ion passes all the way to the front section of the display, they lower the firing voltage at that particular pixel location and a sustainable gas discharge occurs.

The name *grid addressing* is chosen because the grid element used in these displays is the same functionally as the grid in a vacuum triode or pentode tube. Grid addressing does not suffer from the one-third select problems associated with matrix addressing.

5.3.4 Shift-Addressing Technique.

Shift-addressing[8] is quite simple, but is perhaps limited to the gas-discharge display technology. A typical physical/electrode structure is shown in Fig. 5-17. The gas discharge is initiated with the I-row of amplifiers at the left side, one column at a time. Amplifiers A, B, C, and D are then used to articulate the discharge to the right along its channel, first between A and B, then B and C, then C and D, then D and A, etc. If a discharge is not initiated in a channel, then there are no priming particles to diffuse along that channel. If a discharge in a channel does exist, then the A and B amplifiers can sustain the discharge and generate more priming particles to perpetuate it. A discharge at one AB combination of

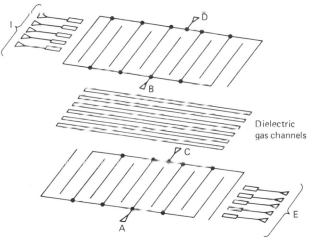

Dielectric
gas channels

(a) Exploded view showing channels and electrodes

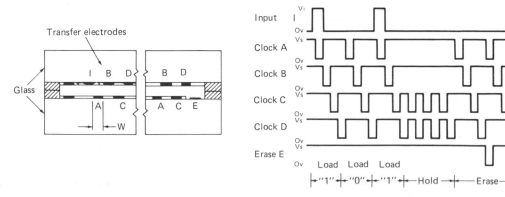

(b) Cross-sectional view of a single-channel

(c) Voltage pulses for input, shift, hold, and erase

Fig. 5-17. Shift-addressing.

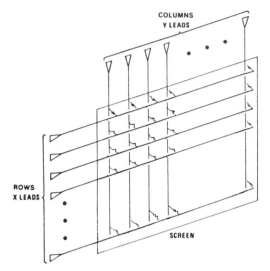

COLUMNS
Y LEADS

Fig. 5-18. Matrix addressing.

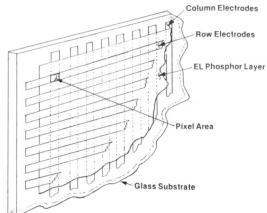

Fig. 5-19. Typical flat-panel display construction.

electrodes will not skip to the next AB electrodes, because the priming particles cannot diffuse that far down the channel before they recombine. The priming particles are physically blocked from diffusing from one row to the next. The role and nature of the priming particle is discussed in Chap. 10.

The shift-addressing technique uses fewer amplifiers than the matrix-addressing technique. It is simple in construction and ideally suited to small-to-medium-size numeric gas-discharge displays. Time is required to shift the image into the display. Except in scroll mode, the image cannot be read while it is being written, as it can with matrix addressing. In the scroll mode, the data is shifted in slowly. Scrolling of data is ideal for such applications as tickertape data.

The Self-Scan^{TM} display, formerly manufactured by Burroughs, uses the shift technique to enable each column, and the grid technique to activate each row. The Self-Scan^{TM} and other gas-discharge configurations using these basic techniques are discussed in Chap. 10.

5.3.5 Matrix Addressing. The simplest display panel structure in flat-panel displays has been achieved using the matrix technique. The essence of the matrix-addressing structure is a set of electrically isolated leads arranged orthogonally, with a pixel at each intersection. The horizontal set is called the *rows* and the vertical set is called the *columns*, as shown in Fig. 5-18. A typical

matrix-addressed display structure is shown in Fig. 5-19. This particular structure is used in electroluminescent displays where the medium is a phosphor.

One-Third-Select. A major difficulty with matrix addressing is the sneak circuit, which causes a voltage to be applied across all the pixels when only one is addressed.[9, 10, 11] This phenomenon will be called *one-third-select* for reasons which will become apparent later. An equivalent circuit for the matirx-addressed display is shown in Fig. 5-20. The display material electrical impedance is schematically shown as a resistor. The impedance may be linear or any conceivable form of nonlinearity such as hysteresis, diode, or threshold effect. The voltage drop across each pixel in the array can be computed from Kirchhoff's Voltage and Current Laws for solving electrical network problems.

Before proceeding, it will be appropriate to briefly review Kirchhoff's Laws.[12] The Voltage Law, simply stated, is that the sum of the instantaneous voltage drops around any closed loop of branches from node to node is equal to zero. The Current Law, simply stated, is that the sum of the current into any node is zero. In Fig. 5-21, the nodes are a, b, c, d, and e, and the branches are 1, 2, 3, 4, 5, 6, 7, and 8. The arrows are arbitrarily assigned to define the polarity. The voltage drop between nodes a and b is V_1. If voltage V_1 is equal to a positive numerical value, then V_a is more positive than V_b. Furthermore, the current flow is postive in each branch when

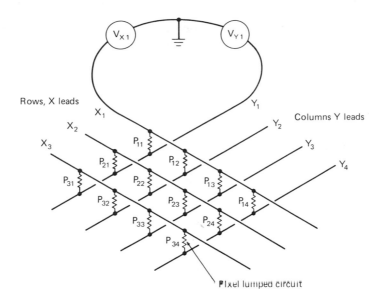

Fig. 5-20. Electrical network for matrix-addressed flat-panel display array of pixels.

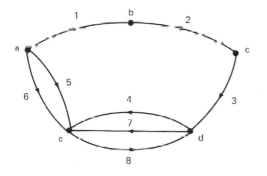

Fig. 5-21. Electrical circuit network.

it flows in the direction of the arrow. All of the circuit elements of a pixel are lumped into a branch. The node corresponds to an electric terminal or display line.

Example voltage and current equations are:

Loop 1, 2, 3, 7, 5, (1)

$$V_1 + V_2 + V_3 + V_7 - V_5 = 0 \qquad \text{(Eq. 5-5)}$$

Loop 4, 8. (4)

$$V_4 + V_8 = 0 \qquad \text{(Eq. 5-6)}$$

Node b

$$I_1 - I_2 = 0 \qquad \text{(Eq 5-7)}$$

Node e

$$I_4 + I_3 + I_0 + I_7 - I_8 = 0 \qquad \text{(Eq. 5-8)}$$

The network of Fig. 5-21 has eight branches ($b = 8$) and five nodes ($n = 5$). It can be logically deduced [12] that there are $j = n - 1$ independent Current Law equations and $k = b - n + 1$ independent Voltage Law equations.

We can now set out to analyze the electrical network for the matrix-addressed display. First, we draw the circuit in terms of branches and nodes. Each lead is really a node neglecting the lead resistance and any parasitic capacitance. The addressed leads terminate at an amplifier represented by a branch connected to a ground node. Each pixel is a branch connecting a row node to a column node.

The electrical circuit network can now be drawn and is shown in Fig. 5-22. Note that there are 14 branches and 8 nodes. Seven current equations and seven voltage equations are required for solution of this network. The solution is not trivial. A computer is required for a 512 × 512 array. For example, such a matrix would have 1,025 nodes (rows + columns + ground) and 262,146 branches (number of pixels + row drive branch + column drive branch). There would be 1,024 current equations and 261,122 voltage equations requiring simultaneous solution. The direct approach for solving the matrix in order to find the individual pixel voltages is not practical.

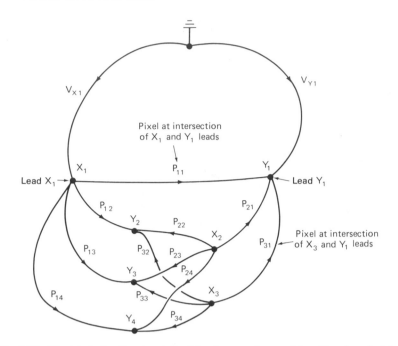

Fig. 5-22. Electrical circuit network for 3 by 4 array when matrix-addressing pixel 1, 1.

The problem can readily be solved by noting some of the physical peculiarities of an electronic display:

- Observation 1: All the nonaddressed row and column leads are electrically connected to amplifiers, which are in turn connected to ground. This must be so because any pixel must be able to be turned on.
- Observation 2: All the nonaddressed row amplifiers could be at one potential and all the nonaddressed column amplifiers could be at one potential. This must be so, as the row and column drivers will be set to the most desirable voltage. Due to the symmetry of the array, what is best for one non-addressed row will be best for all the non-addressed rows. Similarly, what is best for one nonaddressed column will be best for all nonaddressed columns.

The network of Fig. 5-22 now reduces to that shown in Fig. 5-23.

It is now possible to make a very important deduction; that is, that the minimum voltage drop across each of the nonselected (off) pixels is one-third the voltage drop across the selected (on) pixel.[11] Further, this can only be achieved if the nonselected columns and nonselected rows

are driven to the one-third and two-thirds intermediate voltages. That is to say:

$$P_{11} = V_{Y1} - V_{X1} \qquad \text{(Eq. 5-9)}$$

$$-P_{11} = -P_{12} + P_{22} - P_{21} \qquad \text{(Eq. 5-10)}$$

The voltage across all the nonselected pixels must divide into three voltage steps because there are three branches between the selected row X_1 and column Y_1 as shown in Fig. 5-23. The number of pixels between X_1 and Y_{off}, Y_{off} and X_{off}, and X_{off} and Y_1 will not be equal (except for the trivial case of a 2 X 2 matrix array). Therefore, the voltages will not divide equally between the nodes if branches V_{Yoff} and V_{Xoff} are made open circuits.

The complex equivalent circuit network for the 3 by 4 array of Fig. 5-22 has now been reduced to the simple case of Fig. 5-23 using Observations 1 and 2 for a matrix-addressed display. The *minimum* sneak circuit voltage across any nonselected pixel is one-third the voltage across the selected pixel; thus the name *one-third-select*.

The voltage across any one nonselected pixel could be less than one-third the voltage across the selected pixel, but only at the cost of causing the voltage across some other nonselected pixel

Where Y_{off} is the combined
node for Y_2, Y_3, & Y_4

X_{off} is the combined
node for X_2, X_3

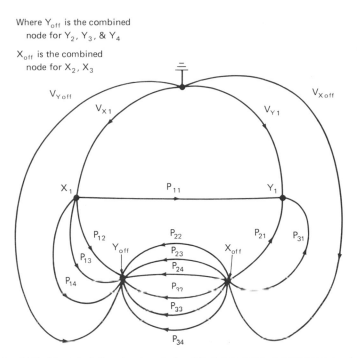

Fig. 5-23. Electrical circuit network simplified as a result of display symmetry.

to be greater than one-third. This last statement can be deduced by inspection of Fig. 5-23 and observing that there are always at least three pixel branches between the nodes of the addressed pixel. Using Kirchhoff's Voltage Law, it is apparent that if one of the nonaddressed pixel branches in any loop including the branch P_{11} is less than one-third the voltage of P_{11}, then at least one of the remaining two unaddressed pixel branches must be greater than one-third the voltage of P_{11}.

The one-third-select phenomenon applies if two or more pixels are addressed in the same row or same column. This can be seen from a reconstruction of Fig. 5-22 using Observations 1 and 2 where pixels p_{11} and p_{12} are addressed as shown in Fig. 5-24. This is called *line-at-a-time addressing* when all the pixels in one row are actually being addressed at the same time. For example, in this case, pixels P_{11} and P_{12} are commanded "on" and pixels P_{13} and P_{14} are commanded "off." The sneak circuits can still be limited to the one-third-select voltage. The proof is that all the nonaddressed pixels are in a section of the loop three branches long between the addressed nodes. If the voltages of V_Y and V_X are set to satisfy Eqs. 5-9 and 5-10, then the nonaddressed pixels will be at one-third-

select voltage. The same holds true if two pixels are selected in a single column.

Cross-Coupling-Select. If two pixels are selected in different rows and different columns, we now have what may be called *cross-coupling-select.* The result is that four pixels are selected when only two are intentionally selected. The four are at the corners of a rectangle, with the intentionally selected pixels at opposite corners.

Cross-coupling selection is obvious from inspection of the diagram of Fig. 5-20. It is not obvious what happens to the nonaddressed pixels. In attempting to turn on pixels P_{11} and P_{33} only, it is obvious that pixels P_{31} and P_{13} cannot be prevented from coming on at full voltage. The remaining pixels would experience one-third-select voltages as before, as can be seen from Fig. 5-24.

Due to the cross-coupling-select phenomenon, only line-at-a-time or pixel-at-a-time addressing can be used with the matrix-addressing technique.

Voltage Reversal. The set of nonselected pixels in the center branch has its applied voltage reversed from that applied across all the other pixels. This can be seen from inspection of Eq. 5-10 and Fig. 5-23. This can be very beneficial

Where X_{off} is the combined node for X_2 and X_3
Y_{off} is the combined node for Y_3 and Y_4

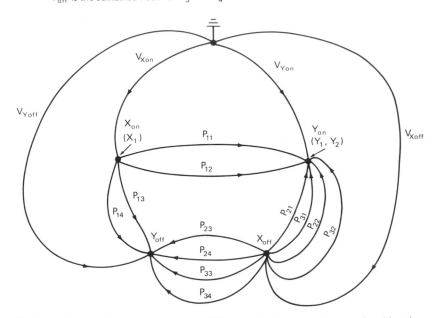

Fig. 5-24. Electrical circuit network simplified showing line-at-a-time matrix addressing.

when using a display medium with a nonsymmetrical current vs. voltage characteristic, such as with LEDs.

This makes LEDs intrinsically matrix addressable even though they respond to one-third-select voltage in the forward direction. The reversed-biased diode prevents any current from flowing in the nonselected pixel branches except for reverse bias leakage current. The leakage current must be less than that which would turn on a diode when applied in the forward direction. One bad diode could cause all of the "nonselected row" and "nonselected column" diodes to emit light.

Four-Case Model. The electrical circuit network for matrix addressing can readily be visualized and analyzed by reducing the network to its four cases. The four-case model is shown in Table 5-4. Each of the four cases represents a set of branches between the four nodes of Fig. 5-24. Pixel-at-a-time addressing is a special case of row-at-a-time addressing with only one or no columns turned on. The entire network rigorously and explicitly reduces to four cases using Observations 1 and 2. If all of these conditions

Table 5-4   Four-Case Model for Pixel Excitation Analysis

	On-column leads $V_{Y\text{on}}$	Off-column leads $V_{Y\text{off}}$
On-row leads $V_{X\text{on}}$	1. Selected pixels	1. Row-selected pixels
	2. On-pixels	2. Off-pixels
	3. $V_{Y\text{on}} - V_{X\text{on}}$	3. $V_{Y\text{off}} - V_{X\text{on}}$
	4. Postive	4. Positive
	5. None	5. Row-streak
	6. $1 \times n$	6. $1 \times (N - n)$
Off-row leads $V_{X\text{off}}$	1. Column-selected pixel	1. Nonselected pixel
	2. Off-pixels	2. Off-pixels
	3. $V_{Y\text{on}} - V_{X\text{off}}$	3. $V_{Y\text{off}} - V_{X\text{off}}$
	4. Positive	4. Negative
	5. Column streak	5. Complementary image
	6. $(M-1) \times n$	6. $(M-1) \times (N - n)$

Legend;
1. Case name
2. Pixel status
3. Pixel voltage
4. Instantaneous polarity
5. Anomaly
6. Number of pixels in case
M Number of rows in array
N Number of columns in array
n Number of on-columns

do not hold, the problem defies analysis without the aid of matrix algebra and a digital computer.

This four-case model is very powerful in analyzing the status and transients of all the pixels in the display array. Even if the conditions to make it rigorously correct do not hold, it can still be used to analyze the array. Conceptually, the array reduces to the four different cases. Each case must be analyzed to determine the voltage of the off-pixel and whether the on-pixel will come on. If the lead resistance and the line driver source inpedance are not negligible, then a qualitative analysis can be completed. The qualitative analysis is completed by analyzing each of the four cases as a circuit with the appropriate loading impedance.

The four-case analysis can be used for numerous matrix-driving configurations. It may not always be advantageous to reduce the off-pixels to one-third the voltage of the on-pixel. The four-case model can be used to analyze the impact of the various voltage levels and the time sequence for turning on and off the row and column amplifiers. For some display materials, to achieve one-third voltages on all the off-pixels at all times during the addressing sequence requires a three-level voltage driver on the column leads and a two-level driver on the row leads.

The generalized circuit can now be drawn for M rows and N columns using the four case model of Table 5-4 and the electrical circuit network of Fig. 5-24. Such a circuit is shown in Fig. 5-25. This circuit can be easily solved for impedances in all branches between the principle nodes. The worst cases are usually for n equal to zero or N.

5.3.6 Intrinsic Matrix Addressing. The display media must have inherent *nonlinearities* to be directly matrix-addressed using a simple flat-panel structure such as shown in Fig. 5-19. Examples of such materials with nonlinear characteristics are shown in Fig. 5-26.

Diode Nonlinearity. A diode such as an LED has an excellent nonlinearity for a pixel. There are two problem areas:

- Reverse diode current
- Diode impedance

The diode (pixel) with reverse bias in the nonselected pixel condition will have some reverse bias current. The sum of the reverse bias currents must flow through the diodes of the row-selected condition and (in series) the column-selected condition. This will lead to some of the row-selected or column-selected diodes (pixels) turning on due to low diode conduction threshold and excessive reverse bias diode current. Second, diodes are current devices, and will self-destruct from overheating if the current is not limited. Current limiting is usually done with a resistor in series with the diode.

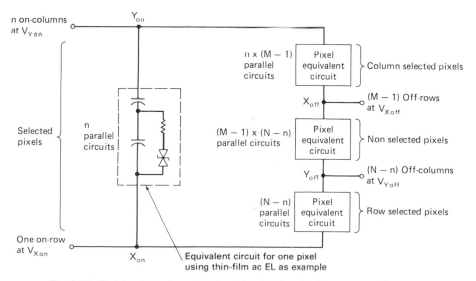

Fig. 5-25. Matrix addressing equivalent circuit using the four-case model.

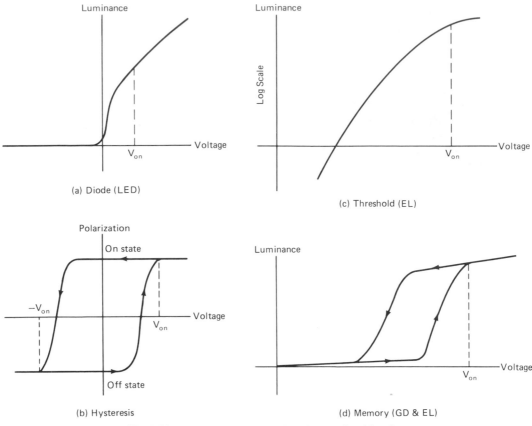

Fig. 5-26. Nonlinearities appropriate for matrix addressing.

Hysteresis. Any hysteresis effect is an excellent pixel nonlinearity. There are many materials which can be used for a pixel hysteresis nonlinearity. Gas discharge is the most commonly used in displays. Ferroelectric and ferromagnetic materials are also used. Ferroelectric hysteresis loops exhibit hysteresis creep when partially traversed many times as would occur at an unselected pixel at one-third-select voltage. Hysteresis creep results in the slow collapse of the switched state after numerous partial traversals.

The wall charge in a gas-discharge cell of a plasma panel contributes to a hysteresis effect when driven by an ac voltage.[13]

Threshold. Any significant threshold can be used as a pixel nonlinearity. The best example of display material with a good threshold is thin-film electroluminescence.

Any nonlinear property may provide a means for matrix addressing. The four-case model can be used to analyze all possible pixel conditions

that can result, given the row and column-lead voltage (or current) signals. The *analysis* of pixel status has been made easy. The *synthesis* of the best driving voltage signals requires ingenuity.

5.3.7 Extrinsic Matrix Addressing. If the display medium is not inherently nonlinear or is only mildly nonlinear, other materials or circuits must be added at each pixel to render it sufficiently nonlinear for matrix addressing.

The name "extrinsic matrix addressing" is suggested over "active matrix" or other terms for several reasons: "Extrinsic" and "intrinsic" are used analogously in semiconductor materials to denote adding a dopant or not adding a dopant; "active" is sometimes used to describe an emitting display; "active" does not suggest the addition of circuits or materials; extrinsic matrix addressing can be achieved using passive circuit elements as well as active circuit elements; "active" does not have a logical or descriptive opposite like "intrinsic" is to "extrinsic."

The oldest and most common way to achieve the required nonlinearity is to add a diode at each pixel. This has been done successfully with incandescent display arrays, but the cost is high and the size is limited due to the cumulative diode reverse bias current.

The most extensively studied approach to extrinsic matrix addressing is the use of polycrystalline thin-film transistors (TFTs) as proposed by P. K. Weimer,[14] T. P. Brody,[15] G. Kramer,[16] A. G. Fischer,[17] J. C. Anderson,[18] and others. The first successful polycrystalline TFTs were made by Weimer using CdS. Their evolution to a successful product has been slow.

During the 1960s, the advances in the single-crystal metal-oxide-semiconductor (MOS) transistor were much more rapid since its characteristics were more reproducible than those of the polycrystalline TFT. As a consequence, the interest in TFTs was diverted to MOS. During the 1970s there was a strong push to use TFTs in large-scale arrays to matrix-address displays. This was promoted and supported by ERADCOM of the U.S. Army at Ft. Monmouth, N.J. The technical effort was primarily carried out by Westinghouse Research Center, Pittsburgh, Pa., and to a lesser extent by RCA Sarnoff Laboratory, Princeton, N.J., and by Aerojet Electro-Systems, Azusa, Calif. Although numerous laboratories around the world continued to do research on TFTs as electronic circuit devices, overall development diminished further as a consequence of the spectacular success in MOS technology. The 1980s started with MOS LSI and VLSI, MOS microprocessors, 256 K bit MOS memories, etc., while the TFT technologists were still perfecting the basic TFT transistor.

During the 1970s, technical interest in TFTs was heightened with the introduction of such nonemissive technologies as liquid crystallinity and electrochromism. There are two overriding features about the nonemitters: They require extremely low power because they do not emit light, and they are not intrinsicaly matrix-addressable because they lack adequate nonlinearity. The TFT with its potential low cost in large arrays is a natural for matrix-addressing nonemitters. MOS has been successfully used to address[19] LC displays. However, MOS wafers are limited in size to 4 or 5 inches in diameter, and single-crystal wafers are expensive ($50.00 per processed wafer in 1980).

In summary, the 1980s began with a few points made vividly clear by hindsight:

- Thin-film polycrystalline transistors (TFTs) will never compete with metal-oxide semiconductor (MOS) single-crystal field-effect transistors (FETs) in large-scale integrated circuits (LSI).
- There is a continuing need for large-size display arrays and small-size display arrays of low-cost transistors for extrinsic matrix addressing of flat-panel displays.
- TFTs, as reported by Weimer in 1962, still hold the lead for extrinsic matrix addressing of large display arrays (greater than 5-inch diagonal).
- MOS arrays are still best for extrinsic matrix addressing of small arrays (less than 5-inch diagonal).
- The application of recrystallized, amorphous, and small-grained silicon TFTs is a promising new approach.
- Passive approaches such as ZnO varistors and PLZT ceramics are still plausible solutions for enhancing the nonlinearity or adding storage.

Examples of nonlinear arrays applied to matrix addressing of displays are shown in Figs. 5-27 and 5-28. In Fig. 5-27, the capacitor is the liquid-crystal pixel charged at a refresh rate of 30 times per second using line-at-a-time addressing. Each liquid-crystal capacitor discharges through the LC material during the balance of the frame time after being addressed. When addressed, the PLZT controls the charge level. The electrical parameters of the pixel are adjusted so that the LC can be charged in one line time and will discharge through the resistance of the LC material for a period of one frame time or more.

Liquid-crystal displays have been made using MOS for extrinsic matrix addressing. The first examples were constructed by Hughes Aircraft Company with the basic structure shown in Fig. 5-28. During the addressing period, the capacitor is charged to a level corresponding to the desired gray shade excitation of the LC pixel. The actual excitation occurs during the

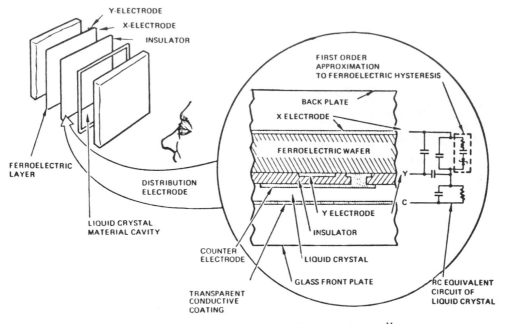

Fig. 5-27. LC/PLZT extrinsic matrix-addressing technique.[11]

balance of the frame time. The slow speed of response of LC material (30 to 100 ms) is of no consequence in this type of addressing. The display exhibits neither smear nor flicker at 30 frames per second, even for the faster responding LC materials.

5.4 DUTY FACTOR

There are two types of duty factors of interest to the display designer and user:

- Addressing duty factor
- Pixel dwell time

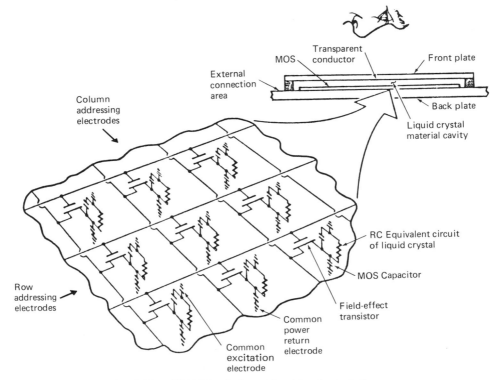

Fig. 5-28. LC/MOS extrinsic matrix-addressing technique.

Addressing duty factor is the percent time that is available to address each pixel. Pixel dwell time is the fraction of frame time that is available to excite each pixel. For most applications, pixel dwell time is the frame time multiplied by the duty factor. Activation and persistence make up the actual time that a display effect is exhibited. Frame time is the total time available to address the entire display array. Frames are sometimes separated into two or more fields where alternate rows are sequentially excited during each field. The purpose of using two or more fields to make up a frame is to decrease the frame time while minimizing flicker. The significance of both of these duty factors will be elaborated further.

A display can be addressed with one of three possible sequential techniques:

- Random Sequence—Possible with memory displays or addressing with refresh using Lissajous technique.
- Line-at-a-time Sequence—Possible with all addressing techniques except scan addressing. Very advantageous with matrix addressing. Limited by display medium speed of response. Yields the longest possible duty factor except for random sequence.
- Pixel-at-a-time Sequence—Possible with all addressing techniques. Very advantageous with cathodoluminescent phosphor display media and scan addressing. Suitable with any display medium with memory. Disadvantage is the short duty factor time for each pixel.

A display is excited at one of two possible times:

- Excitation When Addressing—Simultaneously applied with addressing sequences. Limited time for excitation.
- Excitation During Sustaining—Possible with all displays that are bistable and use a sustaining signal.

An example of the addressing pixel dwell time per frame is as follows:

- Line-at-a-time; 512 columns × 256 rows @ 60 frames per second,

$$\text{Pixel time} = \frac{\text{frame time}}{\text{No. of rows}} = 65 \ \mu s$$

- Pixel-at-a-time; 512 columns × 256 rows @ 60 frames per second,

$$\text{Pixel time} = \frac{\text{frame time}}{\substack{\text{No. of columns} \\ \times \text{ no. of rows}}} = 127 \ ns$$

The addressing time is very short for large arrays. In fact, it is inversely proportional to the size of the array. At some point, as the array becomes very large, the addressing duty factor becomes so short that all the pixels cannot be addressed during one refresh frame time because of the required pixel dwell time, and a memory approach, interlace, or random addressing must be resorted to in order to gain more dwell time. As discussed in Section 1.7, the array can be rearranged to increase dwell time.

The pixel dwell time is further complicated when gray scale or color is required. The pixel dwell time period must be divided up for pulse width modulated gray scale control if amplitude modulation is not used. The pixel time period must also be divided up for color excitation. Amplitude modulation for color selection is uncommon, though it is used for some two-color LED displays and penetration phosphors.

The excitation duty factor is a very critical parameter. The percent time allowed to excite a pixel may place a direct limit on the average brightness of the panel. For most refreshed displays, the addressing time and excitation time are one and the same. Often the display medium does not respond sufficiently to give the required average brightness. The average power is limited by the instantaneous power that the display materials, electronics, leads, power supply, and cooling can tolerate. Some display materials, such as liquid crystal, do not respond to short electrical pulses, only rms voltages. Some display emissive materials such as some phosphors and LEDs are very tolerant of high power pulses where some phosphors actually become more efficient at high power. Limitations other than high current eventually predominate. One in particular is the resistance of the row and column conductors to current pulses and pixel loads requiring fast rise and decay voltage waveforms.

5.4.1 Pixel Dwell Time.

There are inherent limits in the display medium dictating a minimum excitation time. Typical values are outlined in Table 1-2. At some point as the array becomes larger and the available dwell time becomes too short, either a display medium with memory or extrinsic matrix addressing must be resorted to. Extrinsic matrix addressing of the type shown in Figs. 5-27 and 5-28 completely decouple the addressing duty factor from the pixel dwell time by the use of four electrical terminals to each pixel.

Four terminal pixels are used to separate the addressing signal from the excitation power. There are several advantages in this arrangement:

- The excitation timing and voltages are separated from the addressing timing and voltages.
- Resonant power supplies can be easily used to save all the reactive power component.
- Voltage amplification can be achieved if transistors are used at each pixel.
- Very low or random addressing rates can be used.
- Up to 100% duty factor can be used to achieve high luminance. No limitation is placed on pixel dwell time.
- Excitation frequency can be altered to minimize flicker.
- Brightness control can be independent of the addressing signal.
- Common display pixel electrodes can be used for the power circuit.

The display luminance or contrast is controlled during the dwell time. Additional dwell time above the minimums cited in Table 1.2 or higher frame rates above the critical flicker frequency may be needed to improve the pixel contrast ratio and display brightness ratio.

5.4.2 Summary of Dwell Time Effects.

The required display brightness ratio and pixel contrast ratio are predominant display performance requirements. The pixel array arrangement, addressing technique, and size can be adjusted to improve the display performance by relieving the dwell time requirements in a variety of ways:

- Provide the optimum power pulse when addressing a pixel. This normally means providing the maximum current and voltage that the display can tolerate and that is compatible with the light-emission properties of the display and driving capabilities of the electronics.
- Use line-at-a-time addressing.
- For line-at-a-time refreshed displays, widen the array to use as few rows and as many columns as possible.
- Increase the frame time period.
- Minimize background luminance caused by sneak optical paths and electrical circuits.
- Increase the pixel-active area to obtain more luminous flux for a given dwell dime.
- Optimize the faceplate optical design.
- With nonemitters, incorporate supplemental lighting.
- With emitters, improve the material luminous efficiency.
- Reduce the electrically addressed rows without reducing the displayed rows by the methods outlined in Fig. 1-13.
- Improve material persistance and phosphorescence.

5.5 GRAY SCALE

To achieve video or TV performance, a display must possess gray scale. To exhibit gray scale, it must have good uniformity. There are basically four types of nonuniformities which can degrade display performance:

- Large-area nonuniformity
- Small-area nonuniformity
- Edge discontinuities
- Viewing-angle irregularities

These nonuniformities can manifest themselves in the form of changes in brightness, contrast, color saturation, hue, and noticeable positional offsets.

The degree of uniformity required of a display device varies considerably from one application to another, and is primarily a function of the end use of the display. For TV displays, large-area nonuniformities are the least objectionable because the eye is very poor at detecting gradual changes in luminance. Two-to-one variations from center to edges of a display generally go unnoticed.

Small-area nonuniformities in gray scale or color are troublesome, since the eye is much better at detecting abrupt or localized intensity variations. Adjacent pixels should vary no more than about 10%. Small numbers of inoperative elements need not affect the utility of an array if they are randomly located, though even a single missing pixel is aesthetically objectionable. Edge discontinuities are much more noticeable and objectionable than a missing pixel. This type of nonuniformity generally results when large displays are made up of many small modules edge-butted together or when a defect occurs in processing (such as an error in a photomask). For this type of long-line defect, the eye is remarkably sensitive. The eye-brain system can detect lines where the width of the line is below the resolving power of the eye. A missing line is never acceptable.

At a viewing distance of 24 inches (61 cm), the eye can detect positional changes of less than 12 micrometers and contrast changes of less than 1% in a display array. The extreme eye sensitivity to edge discontinuties is what makes the mosaic or modular approach extremely difficult to implement in displays for continuous-tone imagery. Viewing-angle irregularities result when a display's viewing characteristic (brightness or contrast) changes abnormally with the observer's location. Ideally, a display's perceived brightness at all wavelengths should have a Lambertian intensity profile.

The achievement of good gray scale performance remains a critical problem area for most flat-panel displays and is essential for a TV display. For good gray scale performance, a pixel contrast ratio of approximately 20:1 is desirable with the luminance either continuously variable, or, if quantized, controllable into at least 16 logarithmically spaced steps. With fewer than 16 shades, modeling or contouring at shade changes in large, continuous tone-changing areas becomes objectionable. Sixty-four shades are desired for good, aesthetic picture quality.

Analog techniques of achieving gray scale are generally both visually superior and simpler to implement. Most flat-panel displays cannot use analog control for a variety of reasons. For example, emitters may become nonuniform at low power, or the devices may be basically incapable of analog amplitude modulation. These devices usually use one of the following techniques:

- Pulse-width modulation
- Multiple display cells per pixel
- Changing the element's duty factor on a frame-time basis,
- Dither, etc.

Of these, the pulse-width modulation approach is probably the most common. Its primary shortcoming is the higher speeds required of the display drivers. The pixel addressing duty factor for full brightness in a TV display driven line-at-a-time is approximately 65 μs. If 16 shades are required, then this time period must be divided into 16 subparts when using pulse-width modulation. The factor driving speed increases power dissipation and increases the driver cost. Also, except for LEDs and CRTs, display materials are not sensitive to 1 μs variations.

It is also highly desirable that a display be capable of being dimmed while maintaining gray scale range. Even though the required dimming range may not be large, it is still a useful feature in home TV applications. It is mandatory for military and other specialized industrial applications that a display have large dimmable range of over 10:1, particularly for displays that must operate in day and night ambient lighting. For emissive displays that use digital rather than analog methods to control luminance, further demands are placed on the required contrast ratio and switching speeds. Here, the nonemitters have an advantage in that they achieve automatic dimming control via the surrounding ambient. At very low room ambient, some form of illumination is required.

It is most plausible to use amplitude modulation for overall brightness control, pulse-width modulation for gray scale, and multiple-celled pixels for color.

5.6   COLOR

Although many display applications do not require multicolor capability, it is mandatory for entertainment TV. One exception is a portable TV, where customers will tolerate monochrome as long as it is either substantially less expensive than a color version or is the best choice when considering the consequences of weight, size, and power. Most of the flat-panel displays have trouble achieving color for one or more of the following reasons:

- No way of achieving multiple emission wavelengths
- Limited range of available colors to form a chromaticity triangle
- Low efficiency in one or more colors
- Poor color purity
- Insufficient resolution to allow the use of the triad technique
- Short life of one or more of the color materials

Research continues, and each year progress is reported on new techniques and improved materials for color displays. As of 1984, no commercializable flat-panel display has demonstrated good color performance at full TV resolutions. Exciting new developments are being made using back-lit LC-controlled color filter triads with extrinsic matrix addressing by Suwa Seikosha (1982)[20] and Sanyo Electric (1983).[21]

Flat-CRT color displays have been built both by Northrop (1972)[22] using beam-penetration phosphors and by Texas Instruments[6] using phosphor triads (1978). Performance was good (limited color range with beam-penetration phosphors), but device complexity and cost limited their military and commercial applications. A flat-panel color TV using a gas discharge cathode has been developed and demonstrated at Siemens AG and reported by A. Schauer (1982).[23] Both powder and thin-film EL materials have been developed with color capability at limited luminances. LEDs have good color capability except for blue. Liquid crystals have demonstrated limited color performance in small arrays using optical birefringence and pleochroic dyes. Electrochromics and electrophoretics have also provided a variety of color ranges for small arrays. As of the early 1980s, however, material problems such as short life, low luminance, low efficiency, and low color saturation have prevented commercial implementation of multicolor flat panels.

5.7   COST

Cost is the most important factor in flat-panel technologies. The addressing problem is solvable in many ways but always at high cost. Contrast and brightness ratios can be improved with polarizers, antireflection coatings, filters, or more power, but at high cost. The electronic circuits to drive flat-panel displays are usually more expensive than the panels themselves.

Orthogonal arrays of pixels are cost-competitive with CRTs in the medium alphanumeric size (480 characters). As of the mid-1980s, larger sizes cannot compete with CRTs.

Each display technology has its own cost problems. LEDs are constrained by the cost of the diodes themselves. Diodes suitable for a quality matrix display cost from ten cents to a dollar per pixel, assembled. Electroluminescent and plasma displays require high-voltage line drivers which sold in 1980 for over twenty-five cents per line driver and projected to 1985 at six cents per line driver. A driver is usually required for every column and some fraction of the rows, depending on the addressing scheme. TFTs and MOS FETs will always be expensive for display applications because of the large area, huge numbers, and low yields.

All flat panels have a similar "pinout" requirement, which can cost approximately one to ten cents or more per connection, times the number of row-plus-column connections. The line drivers must be independently connected to each line, and that ultimately leads to numerous driver packages being mounted on one or two large printed circuitboards. One possible solution to reducing pinout and circuitboard costs is to place the LSI chips in chip carriers and place the chip carriers on the display substrate itself with the interconnections integrally fabricated with the substrate. Placing LSI chips directly on the display substrate has been devel-

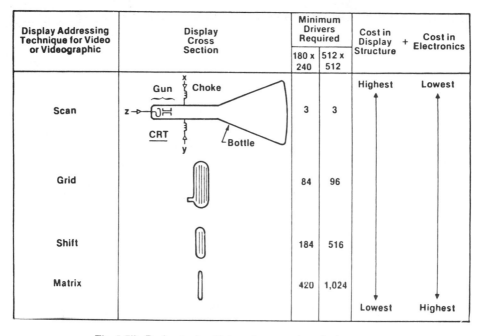

Display Addressing Technique for Video or Videographic	Display Cross Section	Minimum Drivers Required		Cost in Display Structure	+ Cost in Electronics
		180 x 240	512 x 512		
Scan		3	3	Highest	Lowest
Grid		84	96		
Shift		184	516		
Matrix		420	1,024	Lowest	Highest

Fig. 5-29. Design trade-offs to reduce cost in a display system.

oped in the electronic watch and calculator industry, but is more difficult in larger arrays. Pretesting, rework, and repair are not as easy when the LSI chips are mounted integrally with the display panel.

Complex panel designs, tailored to reduce the number of line drivers, have been successfully used. These include the grid- and shift-addressing approaches. The number of line drivers and connections is reduced at the expense of increased cost of display panel fabrication. In the future, as the cost and availability of electronic drivers in LSI packages continue to improve, the advantage will swing to the matrix-addressed configurations. The relationship between display panel cost and electronic cost is shown in Fig. 5-29.

After years of experience, the production cost of CRTs has come down to an extremely low level. It is obvious that the CRT is a complex physical structure. When comparing the complexity of the CRT with many of our flat-panel approaches, everything else being equal, the flat-panel approaches should be even less expensive. The difference is the production volume and experience. The cost of flat-panel production should be reducible to less than that of the CRT. The problem is to get the initial cost low enough so that the display is attractive enough to start large-volume production.

REFERENCES FOR SECTIONS 5.1–5.7

1. Tannas, Jr., L. E., and Goede, W. F. "Flat-Panel Displays: A Critique." *IEEE Spectrum*, pp. 26–32, July 1978.
2. Markin, Joe. "Some Factors Affecting the Development of Flat-Panel Video Displays." *SID Proceedings*, Vol. 17, No. 1, p. 2, 1976.
3. Burke, R. W.; Hoehn, H. J.; and Fein, M. E. "Optical Characteristics of AC Plasma Panels," *SID Symposium Digest*, p. 104, 1974.
4. Landrum, B. L., and Jeffries, L. A. "A Novel Digitally Addressed Electron Beam Scanner for Advanced Sensor and Display Applications." Presented at the SID Symposium, Image Display and Recording, April 8, 1969.
5. Goede, W. F.; Gunther, J. E.; and Lang, J. E. "512-Character Alphanumeric Display Panel." Presented at the 1972 SID International Symposium, June 1972.
6. Scott, Warner C., et al., "Flat Cathode-Ray-Tube Display," *SID 1978 International Symposium Digest*.
7. Lustig, Claude D., and Watts, Geoffrey P. "The Multilayer Gas-Discharge Display Panel." *SID 1974 Digest*, p. 128.
8. Coleman, W. E., and Craycraft, D. G. "A Serial Input Plasma Charge Transfer Display Device." *SID 1975 Digest*, p. 114.

9. Ivey, H. F., and Thornton, W. A. "Preparation and Properties of Electroluminescent Phosphors for Display Devices." *IRE Transactions on Electron Devices*, pp. 265, July 1961.

10. Sobel, A. "Selection Limits in Matrix Displays." *IEEE Proceedings, Conference Display Devices*, December 1970.

11. Tannas, Jr., Lawrence, E. "One-Third Selection Scheme For Addressing a Ferroelectric Matrix Arrangement," U.S. Patent No. 4,169, 258, Sept. 25, 1979.

12. Guillemin, Ernst A. *Introductory Circuit Analysis*. New York: John Wiley & Sons, 1955.

13. Bitzer, D. L., and Slottow, H. G. "The Plasma Display Panel—A Digitally-Addressable Display with Inherent Memory." *FJCC Proceedings*, November 1966.

14. Weimer, P. K. "The TFT—A New Thin Film Transistor," *IRE Proceedings*, Vol. 50, p. 1462, 1962.

15. Brody, T. P., et al. "Operational Characteristics of a 6″ × 6″ TFT Matrix Array Liquid Crystal Display." *SID 1974 Digest*.

16. Kramer, G. "Thin-Film Switching Matrix for Flat Panel Displays." *IEEE Record of Conference on Display Devices*, 1974.

17. Fischer, A. G., et al. "Design of a Liquid Crystal Color TV Panel." *IEEE Record of Conference on Display Devices*, New York, 1972.

18. Anderson, J. C. "Thin Film Transistors—Past, Present and Future." *Thin Solid Films*, Vol. 50, pp. 25–32, 1978.

19. Ernstoff, M. N., et al. "Liquid Crystal Pictorial Display," *IEDM Technical Digest*, p. 548, 1973.

20. Morozumi, S., et al. "B/W and Color LC Video Displays Addressed by PolySi TFTs." *SID 1983 Digest*, p. 156.

21. Yamano, M., et al. "A-Si TFT Active-Matrix Full Color LC TV." *Japan Display 1983 Digest*, p. 214.

22. Goede, W. F. "A Digitally-Addressed Flat Panel CRT-Review." Northrop Corp., January 1973.

23. Schauer, A., "Flat Color TV Panel Using a Plasma Cathode." *Japan Display 1983 Digest*, p. 234.

5.8 INTRINSIC ELECTRONIC DISPLAY DRIVE

5.8.1 IC Technology for High-Voltage Display Drivers (by Tom Engibous, Texas Instruments).

Up until the mid 1970s, the mainstream of the semiconductor industry virtually ignored the market potential of flat-panel displays. Almost no high-voltage IC development was reported until 1977. The cost of flat-panel displays was such that the vast majority of applications were military. The high-volume-conscious semiconductor industry was not willing to commit resources to such a market. What has happened in the late 1970s and early 1980s in the area of high-voltage IC development is an explosion of technology development equaled only in the development of dynamic RAMs. The result today is that several semiconductor manufacturers have high-voltage IC-processing technologies in production and several more have announced plans.[3,5,6,8]

There are three distinct requirements of high-voltage flat-panel display drivers: high-voltage capability, high-level integration, and superior reliability.

The first moderate-volume ICs designed for high-voltage applications were developed in the mid 1960s. The technology utilized was a standard bipolar junction-isolated process. The high-voltage component was an NPN transistor. The typical capability of this technology was under 50 volts, with some bold claims of up to 100-volt capability. High-voltage bipolar technologies are still in use today mainly in the area of interface circuits, designed to drive small relays and solenoids. Reliable operation limits the voltage capability to less than 70 volts for most applications.[1] The need for higher voltage capability drove a major effort in developing a manufacturable dielectrically isolated IC technology. In this technology, the high-voltage component is again an NPN transistor but it is capable of much higher breakdown voltage than in the junction-isolated process. However, isolating components is a mechanical process and has resulted in significantly higher wafer cost than a junction-isolated process. The large boom of the late 1970s in high-voltage IC development led to a common conclusion on a superior high-voltage component. However different the various technologies appear, all are centered around the DMOS or double diffused MOS transistor. The major reasons for this commonality are the high-voltage performance of the DMOS transistor and its compatibility with other processes. The DMOS transistor can be built using the lower-cost junction-isolation process and can be easily integrated with standard bipolar and MOS components.[7]

As can be seen in Fig. 5-30, the DMOS appears very similar to a standard NPN bipolar transistor. The main difference is in the opera-

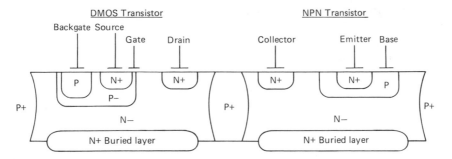

Fig. 5-30. Comparison of DMOS and NPN transistors.

tion of the devices. In the case of the DMOS transistor, the P and N+ doped regions making up the backgate and source are shorted together to eliminate any vertical bipolar current gain. Eliminating all bipolar current gain results in the most significant advantage of the DMOS transistor. It no longer suffers from the secondary breakdown phenomenon which is a characteristic of all NPN bipolar transistors. This is illustrated in Fig. 5-31, where the breakdown characteristics of typical NPN and DMOS transistors are shown. Although the BVDSS voltage of the DMOS and the BVCES voltage of the NPN are very similar, the NPN also possesses the secondary breakdown or snapback characteristics described by the BVCEO voltage. If, in operation, the loadline of the NPN transistor should cross outside of the shaded area of the NPN transistor, secondary breakdown will occur and in many cases destroy the transistor. Consequently, if an NPN transistor is operated above its BVCEO voltage, reliable operation is very dependent on the load it is driving. Further complicating reliable operation, many of the

parameters which determine the onset of the snapback condition vary with stress and age. The conclusion among most semiconductor technologists is that optimum reliability can only be obtained when the operating voltage is limited to below the BVCEO for NPN transistors. Increasing the BVCEO breakdown voltage is dependent on several parameters, the main one being epitaxial layer thickness. Manufacturability limits the epitaxial layer thickness to approximately 0.7 mils, producing a worst-case BVCEO of 75 volts.

The conclusion is simply that a bipolar NPN transistor produced with standard junction-isolation techniques cannot be operated above 75 volts without sacrificing some degree of reliability performance, while breakdown in the DMOS transistor involves simple avalanching which is nondestructive by nature and does not snap back. If properly field-plated, the BVDSS of the DMOS does not vary with stress and age. Consequently, the reliability of the DMOS transistor is virtually independent of the external load.

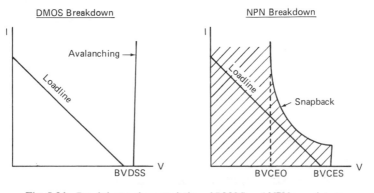

Fig. 5-31. Breakdown characteristics of DMOS and NPN transistors.

In addition to operating voltage advantages, the DMOS transistor also exhibits all the advantages of MOS components. Turn-on and turn-off times are in the tens of nanoseconds, significantly less than bipolar counterparts since they are majority carrier devices. The most significant disadvantage of the DMOS transistor is in its current-sinking capability per unit area of silicon. Compared to an NPN transistor, the DMOS cannot handle as high current densities and also has a much higher drain-to-source ON resistance. As a result, the DMOS transistor occupies approximately five times the area of a similar-performance NPN. This tranlates directly to cost and is the single reason why bipolar NPN transistors are still used for operating voltages below 70 volts.

The ability to integrate additional low-voltage logic on a high-voltage IC is necessary to even approach the cost requirements of flat-panel ICs. Of the high-voltage DMOS technologies available today, all integrate standard metal-gate or silicon-gate CMOS with the DMOS transistors to obtain low-voltage logic circuitry. Two of these technologies go even further in integrating standard bipolar components along with the CMOS logic and DMOS transistors. Figures 5-32 and 5-33 illustrate a cross section and a chip photo of the various components available on a single chip using this unique technology. In the past, the amount of logic placed on a flat-panel IC has been limited to shift registers and latches, for the purpose of making a serial-to-parallel interface. Putting more intelligence on the high-voltage driver chips has significant potential in further reducing the cost of driving flat-panel displays.

The output-circuit configuration for high-voltage flat-panel drivers fall into one of two categories: open-drain and totem-pole, as shown in Fig. 5-34.[2,4] From a cost standpoint, open-drain outputs have approximately a 30% cost advantage over a similar-performance totem-pole output when integrated in a common serial-in parallel-out logic configuration. For this reason, electroluminescent row drivers are priced below the totem-pole circuits required for ac plasma displays. A further advantage of open-drain outputs is that they are capable of much higher operating voltages. In order to implement a totem-pole circuit, it is necessary to pass high-voltage leads over isolating junctions, which

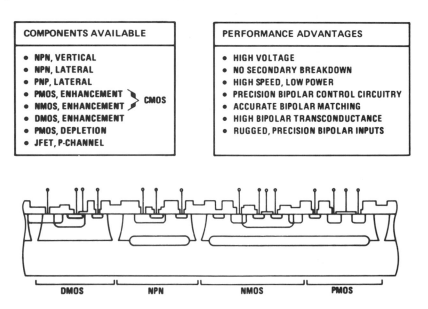

Fig. 5-32. BID-FET process technology.

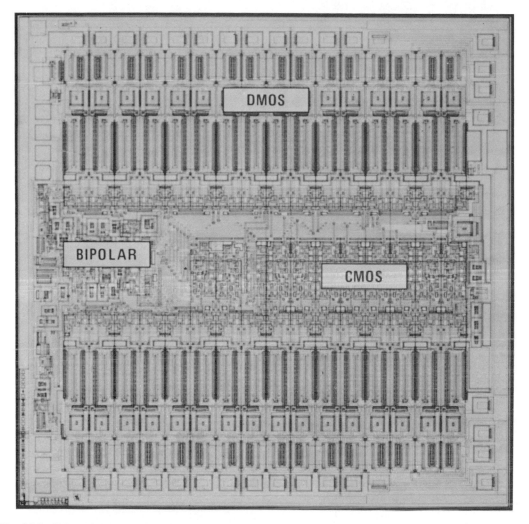

Fig. 5-33. Chip with integrated bipolar components, CMOS logic, and DMOS transistors for high-voltage displays.

causes premature surface breakdown. This limits breakdown to as low as 120 volts if special precautions are not observed. In the open-drain configuration, operating voltages over 400 volts are easily obtained since no high-voltage leads must pass over isolating regions. Unfortunately, most flat-panel displays require the higher cost and complexity of totem-pole outputs. Figure 5-34b summarizes the cost tradeoffs which must be made when increased voltage capability is required. The step increase in cost at 20 volts results from requiring NPN transistor outputs rather than self isolating CMOS. The switch to DMOS transistors at 70 volts results in the largest jump in cost. The step increase at 100 volts

is due to increased wafer processing steps needed to limit the maximum electric field at the surface to below avalanche levels.

IC packaging not only affects the actual cost of the IC but also can have a much more significant impact on overall system cost. The method of interconnecting the driver ICs to the panel glass is usually a very costly process which also has great effects on the final form factor. Most driver ICs today are packaged in dual in-line packages and mounted on standard printed circuitboards. This configuration is quickly being replaced with leaded chip carriers mounted on a flexible printed-circuit film which is attached directly to the glass panel. The ultimate

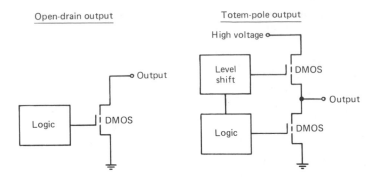

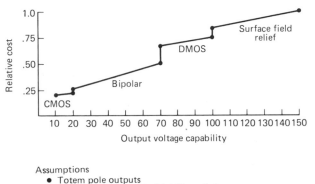

Fig. 5-34a. Output circuit configurations for high-voltage flat-panel drivers.

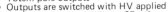

Assumptions
- Totem pole outputs
- Outputs are switched with HV applied
- Output current capability = 25 mA

Fig. 5-34b. Driver IC output voltage vs cost.

goal is to mount the chip carriers directly on the glass substrate. Although this appears to be the most cost-effective method, the issue of serviceability is yet to be resolved.

Several flat-panel developers have suggested mounting unpackaged driver circuits directly on the glass substrate. This method would preclude any repairability and would make the panel manufacturer responsible for the reliability of the driver circuitry. At this point in time, the optimum configuration appears to be mounting leaded chip carriers directly on the glass substrate.

In an extremely short period of time, high-voltage IC technology has moved from very basic voltage buffers to multiple output circuits with MSI logic complexity. The volume introduction of flat-panel displays to the commercial market in the mid 1980s will further fuel semiconductor technology development in high-voltage ICs. Higher logic complexity, higher voltage capability, and larger pin count packages will be the major advances in reducing the cost per line of high voltage ICs which is so critical to the success of flat-panel displays.

REFERENCES FOR SECTION 5.8

1. Spencer, J. "The High Voltage IC and Its Future." *SID 1982 Digest.*
2. Graves, S. "Electronics for Addressing High Resolution Thin Film EL T.V. Display." *SID 1980 Digest.*
3. Gielow, T. "Monolithic Driver Chips for Matrixed Gray-Shaded TFEL Displays." *SID 1981 Digest.*
4. Smalter, M. "Integrated AC Plasma Display Panels," *SID 1982 Digest.*
5. Blanchard, R. "High Voltage IC's for Electroluminescent Panels." *SID 1982 Digest.*
6. Fields, S. "Circuit Design Boosts 15-V CMOS Process to Drive 60-V Displays." *Electronics*, p. 37, May 5, 1982.
7. Schreler, P. "Components." *EDN*, p. 170, December 16, 1981.
8. Krishna, S. and Kuo, J., Isaura Servin Gaeta "An Analog Technology Integrates Bipolar, CMOS, and High Voltage DMOS Transistors." *IEEE Transactions on Electron Devices*, p. 89, January 1984.

5.9   EXTRINSIC ELECTRONIC DISPLAY ADDRESSING

5.9.1 Fabrication and Application of Thin-Film Transistors to Displays (by L. E. Tannas, Jr.)

The thin-film transistor (TFT) was invented in 1962 by Dr. P. K. Weimer at RCA. It is a unique kind of field-effect transistor in that it is made of a polycrystalline semiconductor and can be constructed on a low-cost substrate such as glass or polymer. As yet, however, there is no product in production using TFTs, because of:

- Successes in MOS FETs
- Technical problems in reproducibility
- Lack of a critical need to support a major development

A critical need now exists in direct-view electronic displays. The successes in MOS have further accentuated the lack of low-cost, low-power, lightweight flat-panel displays that are compatible with the new line of LSI microprocessors, memories, switching power supplies, etc. The best approach is to keep the TFT circuits simple and the dimensions large.

Flat-panel displays have a unique requirement if they must be matrix addressed. Matrix addressing inherently produces "sneak" circuits, which cause at least one-third of the select voltage to be applied across all the nonaddressed pixels. (See Sect. 5.3.5.) Most nonemitting display materials respond to one-third select voltage, such as liquid crystallinity, electrochromism, and electrophoresis. Additionally, some display technologies, such as electroluminescence (EL), require a higher operating voltage

than is readily available with MOS. Also, flat-panel displays can benefit from memory storage at each pixel. The memory allows for a 100% duty factor. Without memory, the duty factor and brightness are inversely proportional to the number of display array rows. The TFT circuit at each pixel eliminates sneak circuits, provides the necessary voltage gain to stimulate the display medium, and facilitates the incorporation of memory elements.

The complete application of TFTs to displays requires four separate layers called stacks:

- The display stack
- The counterelectrode stack including the pixel electrode and a divider capacitor used for high-voltage division
- The TFT stack including all pixel circuit elements (except the counterelectrode stack divider capacitor if used)
- The addressing stack including all the row and column lines with intermediate dielectrics

The following discussion is directed at the application of TFTs to ac thin-film EL. This application contains all the circuit elements that might be used in any TFT application. The major TFT design considerations for a display of vectorgraphic size (240 by 320 lines or more) are given in Table 5-5.

The TFT Stack. Today's transistor is more sophisticated than that developed by Dr. Weimer[1], in that it uses dual gates as shown in Fig. 5-35. Dual gates provide sharper turn-off

Table 5-5   Design Considerations for a Vectorgraphic Flat-Panel Display

TFT advantages	Application to EL	Application to LC and other nonemitters
Switch at each pixel	Not needed, but advantageous	Mandatory for matrix addressing
Memory	Advantage for 100% duty cycle	Used to compensate for slow speed of response
High voltage	Definite advantage. H.V. line driver only option	Not needed at all
Large area	No advantage	Advantage over MOS
Low cost	Must be lower than H.V. line drivers	Must be lower than other extrinsic approaches

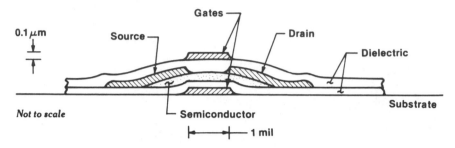

Fig. 5-35. Cross section of a dual-gate TFT.

characteristics and a high β coefficient (geometric factor). The most complete pixel circuit uses two TFTs and a capacitor for display memory, as shown in Fig. 5-36. The second transistor allows the display power to be separate from the addressing logic.

The entire TFT pixel circuit can be fabricated on top of the pixel active area defined by a metal electrode called a counterelectrode, as shown in Fig. 5-37. The TFT pixel circuit can be fabricated in one vacuum pumpdown, using five metal masks and ten additive depositions. Fabrication techniques used at Aerojet Electro Systems[2] are as follows:

- Deposition of materials through chemically etched Kovar metal masks,
- Ball-and-race aligned tooling set for mask and substrate registration,
- Magnetic pulldown for mask clamping to substrate,
- Tool and mask carousel for in-chamber interchanges,
- A three-chimney vapor vacuum-deposition chamber.

The TFT is a thin-film polycrystalline semiconductor device. The most popular and successful devices use CdSe, which is an N-type enhancement mode[3,4,5] semiconductor. Typical properties are as shown in Fig. 5-38. The electrical characteristics of the TFT are accurately governed by the equations[1] given in Table 5-6. Typical TFT performance for high-voltage applications is summarized in Table 5-7.

The β coefficient is at the designer's control. To improve the TFT, the following requirements can be satisfied with changes in the β coefficient and geometry changes:

- High current (I_{ds}) requires thicker dielectric (t).
- Higher transconductance (g_m) requires higher gain (β).
- Lower off current requires dual gates.
- Higher voltage (V_{ds}) requires thicker dielectric (t).
- Lower Miller effect requires reduced gate-to-drain capacitance (caused by gate and drain conductor overlap and channel capacitance and $\epsilon WL/t$).

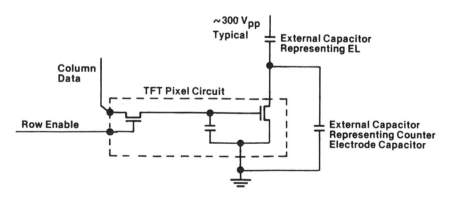

Fig. 5-36. Pixel electrical circuit for driving high-voltage EL.

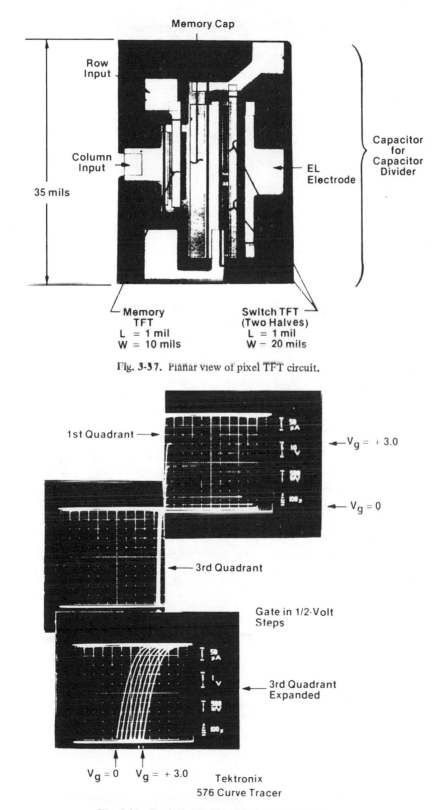

Fig. 5-37. Planar view of pixel TFT circuit.

Fig. 5-38. Typical thin-film transistor parameters.

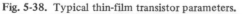

Table 5-6   TFT Theory

Linear region: $I_{ds} = \beta [V_{ds}(V_g - V_0) - \frac{1}{2} V_{ds}^2]$
$g_m = \beta V_{ds}$

Saturated region: $I_{ds} = \frac{1}{2} \beta (V_g - V_0)$
$g_m = \beta (V_g - V_0)$

Geometry term: $\beta = \mu \epsilon \, W/tL$

where

μ = semiconductor mobility
ϵ = dielectric constant
t = dielectric thickness
W = channel width
L = channel length

- Lower on resistance ($R_{ds_{on}}$) requires higher gain (β).

The TFT is a perfectly symmetrical device if the source-to-gate geometry is identical to the drain-to-gate geometry. That is to say, the drain and source can be interchanged. The grounded electrode is the source. I_{ds} is the drain-source current as a function of the source voltage and gate voltage. When the TFT is turned off by making the gate slightly negative (-1 volt), the TFT has the electrical properties of a resistor of approximately 1 to 2 kilohms.

Counterelectrode Stack.   The counterelectrode stack contains a capacitor used in series with the EL capacitance for voltage division. The EL requires approximately 300 volts peak-to-peak. It is desirable to address the display with CMOS, which is limited to approximately 15 volts peak-to-peak. A TFT can be made to drive a load at 300 volts but not with sufficient gain from a 15-volt input signal. One solution for EL display drive is to add a TFT and a capacitor as shown in Fig. 5-36 and use a capacitor divider circuit. This capacitor is built at each pixel in the counterelectrode stack. The second TFT shorts out the divider capacitor to turn on the pixel. When the divider capacitor is shorted out, the full voltage is applied across the EL thin film, which is electrically equivalent to a capacitor.

Electroluminescent (EL) Stack.   The EL stack is first built on the substrate. The material is ZnS with Mn as the primary activator. The stack is optimized for steep brightness-to-applied-voltage performance, which has been achieved as shown in Fig. 5-39. High brightness is not important, since the duty factor is 100%. Brightness control is easily achieved with power frequency control. For frequencies below 5 kHz the brightness is proportional to frequency.

A display with the EL performance shown in Fig. 5-39 is easily made sunlight readable. Since thin-film ZnS is transparent, the use of a black light-absorbing back layer or a reflecting back

Table 5-7   Typical Thin-Film Transistor Parameters

I_{dss} (drain-source current with zero gate voltage) = 10 nA (nanoamps)

g_m (transconductance) = 130 micromhos

$$g_m = \frac{\Delta I_D}{\Delta V_G}$$

$V_{g_{on}}$ (gate voltage to produce 250 μA $I_{D_{SAT}}$) = 3.5 V

$V_{g_{off}}$ (gate voltage to produce 1.0 nA $I_{D_{SAT}}$) = -0.5 V

$R_{ds_{on}}$ (drain-source "on" resistance) = 1-2kΩ

Dual Gate TFT Enhancement Type
Intrinsic N-Type Semiconductor (CdSe)

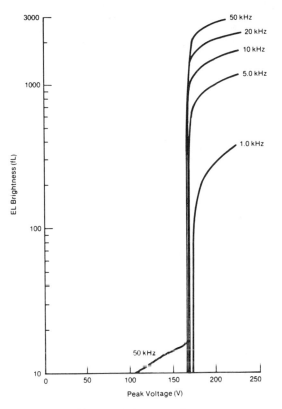

Fig. 5-39. EL brightness performance.

layer with front circular polarizer is applicable to achieve sunlight readability.

Addressing Stack. The addressing stack contains the row and column lines and accommodates the crossovers.

Display. The complete display is shown schematically in Fig. 5-40. The process profile for fabricating the display is outlined in Fig. 5-41. Two basic approaches have been considered for connecting the TFTs to the EL: the monolithic[5] and the sandwich,[6] as summarized in Fig. 5-42. The yield would be improved with the sandwich configuration. The monolithic approach is the most rugged and simplest to make.[7]

The EL is the most efficient light-emitting display except for the cathodoluminescent techniques. The EL panel has a high imaginary power component which cannot be easily saved when it is supplied along with the display logic. In the approach here the logic and display power are supplied from different sources. Therefore, a resonant circuit can be used to save all the imaginary power component.

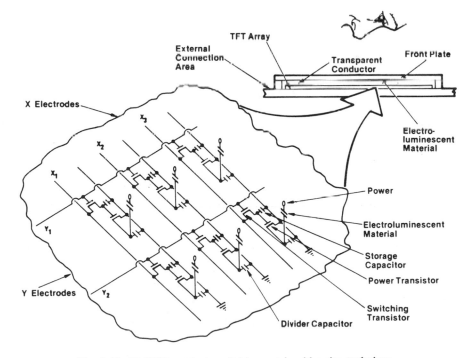

Fig. 5-40. EL/TFT extrinsic switching matrix-addressing technique.

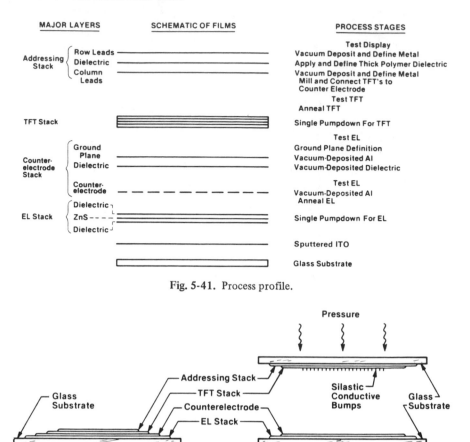

Fig. 5-41. Process profile.

(a) Monolithic   (b) Sandwich

Fig. 5-42. Methods for joining TFTs to display.

REFERENCES FOR SECTION 5.9.1

1. Weimer, P. K. *Handbook of Thin Film Technology*, edited by L. I. Maissel and R. Glang. New York: McGraw-Hill, Chap. 20, 1970.
2. Contract No. DAAB07-77-C-0583 between Aerojet ElectroSystems Company and U.S. Army.
3. Brody, T. P., and Malmberg, P. R. Int. J. Hybr. Microelectron, 2, 29, 1979.
4. Fugate, K. O. "High Display Viewability: Thin-Film EL, Black Layer, and TFT Drive," *IEEE Trans.* ED. 24, No. 7, July, 1977.
5. Kun, Z. K., et. al. "Thin-Film Transistor Switching of Thin-Film Electroluminescent Display Elements." *SID Proceedings*, Vol. 21, No. 2, 1980.
6. Kramer, G. "Thin-Film Transistor Switching Matrix for Flat-Panel Displays." *IEEE Trans.* Ed. 22, 733, 1975.
7. Tannas, Jr., L. E.; Helm, W. J.; and Bassie, D. L. *Integrated Thin Film Transistor Display*. Contract DARK-07-77-C-0583, U.S. Army ERADCOM Report 6341, October 1982.

5.9.2. Extrinsic Matrix Addressing with Silicon Thin-Film Transistor Arrays (by A. I. Lakatos, Webster Research Center, Xerox Corporation).

Introduction. During the last two decades, thin-film transistors (TFTs) and matrix arrays of these devices were explored, developed, and fabricated primarily for large, dot-matrix, flat-panel displays. In electrochromic,[1] electroluminescent,[2] and liquid-crystal[3] displays, TFTs are used as the extrinsic nonlinear switches needed to activate each individual picture element. For these applications, primarily, evaporated, polycrystalline CdSe-type TFTs were used.[3,4,5,6] But in 1980 the first amorphous Si TFTs were built for display application.[7] In the same year laser-recrystallized Si TFTs were also built,[8] and work proceeded on small-grained polycrystalline Si TFTs.[9] The appearance of Si TFTs was significant because these devices have a very wide range of applicability from VLSI to high-voltage switches.[10] Nevertheless, to date, Si TFT arrays were built mainly for use in liquid-crystal flat-panel displays. The purpose here is to present a brief review of the state of the art of the Si TFT arrays built for liquid-crystal flat-panel displays.

The customary use of TFTs in LC displays is illustrated in Fig. 5-43. The source and gate of each transistor are addressed through appropriate electrodes, while the large drain electrode is in contact with the liquid crystal. The drain pad determines the dimensions of the active area of the pixel. Each pixel may also contain a storage capacitor C_s in addition to the TFT. This capacitor is needed when the dielectric relaxation of the combined TFT-LC structure in the off state is shorter than the refresh period. The performance requirements for the TFT were described recently by Castleberry[11] and Luo, et al.[12] The required performance parameters needed for a 500-line display refreshed at TV rates are listed in Table 5-8.

Structures of Si TFT. Three types of Si TFTs are being developed according to the processing of the Si film and the structure of the devices: (1) laser-annealed crystalline, (2) amorphous, and (3) small-grained polycrystalline. Fig. 5-44a illustrates a frequently used structure for laser-annealed crystalline and for small-grained poly-

Table 5-8   TFT Performance Requirements for 500-Line Liquid-Crystal Displays

I_{on} (at $V_G > 0$)	$> 5 \mu A$
I_{off} (at $V_G = 0$)	
with storage capacitor ($C_s = 10$ pf)	< 1 nA
without storage capacitor	< 0.1 nA
Response time, τ_r	$< 3 \mu s$

crystalline transistors, while Fig. 5-44b shows a popular structure for a-Si TFTs. Both structures provide n-channel enhancement transistors. The enhancement mode is particularly desired in order to achieve the <1 nA off currents at $V_G = 0$ as indicated in Table 5-8. In the cases of the laser-annealed and small-grained polycrystalline transistor types, both p- and n-channel devices are possible to build, but the n-channel is usually preferred because of the higher field effect mobility, μ_{fe} for electrons than for holes.[13,14] On the other hand, almost all a-Si-type TFTs built to date have been n-channel, primarily because undoped a-Si:H is preferred and it is usually n-type.[15]

The structure shown in Fig. 5-44a is a thin-film variant of the classical single-crystal bulk MOSFET. In the off condition, the leakage current is determined by the reverse current through the p-n$^+$ junction at the source or at the drain. When the device is turned on with positive

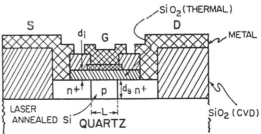

Fig. 5-44a. Cross section of laser-annealed, crystalline-silicon, thin-film transistor.

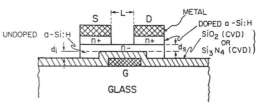

Fig. 5-44b. Cross section of an amorphous silicon thin-film transistor.

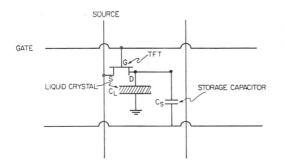

Fig. 5-43. Equivalent circuit for a picture element of a thin-film transistor-aided liquid-crystal display.

gate voltage, a narrow n-type inversion channel is induced at the semiconductor-insulator interface along the source-drain gap. Operation of this device, therefore, requires excellent minority carrier mobility and lifetime, just as in bulk, single-crystal devices. The intentional similarity between the laser-annealed, crystalline Si TFT and the bulk, single-crystal Si MOSFET allows the use of all the silicon n-MOS IC fabrication steps, with the addition of laser annealing.[8] In the case of the small-grained polycrystalline Si device, molecular beam deposition (MBD) technique[16] in addition to the more conventional CVD[17] can also be used with success for the deposition of the Si film. For these devices, the remaining processing steps each may be a low-temperature variant of the usual MOS IC processing.

In a-Si devices, transistor operation depends entirely on the properties of the majority carriers, i.e., electrons in most cases. At zero gate voltage, low off current is achieved because the conductivity of a-Si is low and because the transistor channel is depleted.[7] This depletion is caused primarily by surface or interface states at the semiconductor-insulator interface and by the potential difference between the work function of the gate metal and the a-Si. In the turned-on condition $V_G > 0$, a strong majority carrier accumulation layer is induced in the a-Si

channel. These types of devices require the source-drain contact simply to be "ohmic" or carrier injecting. This can be best achieved with the use of n^+ a-Si films between the undoped a-Si and the electrode metal, which is usually Al. Recently it was shown that direct Al-a-Si contacts are in fact rectifying and that I_{on} at high fields can be increased by an order of magnitude or more if an n^+ a-Si layer is incorporated into the source-drain regions.[18]

Fabrication of Si TFT. For laser-annealed, crystalline Si TFTs, laser annealing and the processing steps associated with it are the only additional steps to customary n-MOS IC fabrication. The major fabrication steps involved are summarized in Table 5-9. Laser fabrication of silicon films on insulator substrates was recently reviewed by Gibbons[19] and by Poate and Brown.[20] These types of devices are built on high-temperature substrates, such as quartz, because fabrication includes thermal oxidation and other high-temperature processing steps. To begin with, high-quality, small-grained polysilicon film is deposited on the amorphous quartz or fused silica substrate using low-pressure, chemical vapor deposition (LPCVD) at 620°C substrate temperature.[21] In order to grow very large grains, or hopefully in the future single-crystal films, the Si film is photolithographically

Table 5-9 TFT Materials and Fabrication Steps

Materials fabrication steps	Laser-annealed crystalline Si	Amorphous Si	Small-grained polycrystalline Si	CdSe
Substrate	Quartz or single-crystal Si	Glass	Glass or quartz	Glass
Semiconductor	LPCVD* at 620°C	Glow discharge at <350°C	LPCVD* at 620°C or molecular beam at <350°C	Thermal Evaporation at <250°C
Gate insulator	Thermal SiO$_2$ at <1000°C	CVD or plasma CVD SiO$_2$ or Si$_3$N$_4$ at <350°C	CVD SiO$_2$ at <350°C	Electron beam Al$_2$O$_3$ at 25°C or anodized Ta$_2$O$_5$ at 25°C
Device patterning	Si IC photolithography			Custom photolithography
Source-drain	Ion implanted n^+ or p^+	Glow discharge deposited n^+	Ion implanted n^+ or p^+	Electron beam evaporated metal

*LPCVD: Liquid-phase chemical vapor deposition.

patterned into 25 μm wide and 75–100 μm long islands. These islands have a narrow neck-type connection to the outer Si film region and, on the other end, they have two chevron-type side channels. This structure was developed to establish a micro-Bridgeman-type crystal growth.[22,23] The crystal is seeded from the narrow neck using a CW ~50 μm diameter laser beam scanned along the length of the island. The entire width of the island is melted and the melt front is propagated down as the laser beam scans over the island. Some of the molten material is allowed to flow out at the end of the island through the side channels. The narrow grooves that define the island are called stress relief grooves[8] and prevent crazing of the Si film. Crazing would take place due to the thermal expansion difference between Si and SiO_2. The key to the growth of single-crystal islands is, however, the control of the temperature profile. Using Ar laser radiation, which is absorbed by Si, Biegelsen, et al.[22] have discovered that the addition of an extra laser light-absorbing layer to the structure and the reduction of radiative heat losses by making the stress relief grooves very narrow will keep the edges of the molten island hot, favoring nucleation at the center of the island. These improvements worked well for Si islands on top of thin SiO_2 or Si_3N_4 films on top of single-crystal Si substrates but not for Si films on top of quartz. This problem was solved by Hawkins, et al[23] who discovered that the use of CO_2 laser radiation (wavelength = 10.6 μm), which is strongly absorbed by the SiO_2 substrate, will create a concave temperature profile across the Si island, i.e., the edges will be warm and the center cool. They also observed (100) cystallographic orientation parallel to the substrate.[24] Laser annealing is usually carried out with 5–15 W lasers scanned at 1–30 cm/s. Laser-annealed Si islands can be ion implanted to form p- or n-type channels and also to form the n^+ and p^+ source-drain regions within the film.[8]

Amorphous Si TFTs are built on low-temperature, glass substrates. The a-Si film is usually a slightly hydrogenated a-Si:H alloy deposited at less than 350°C substrate temperature using glow discharge decomposition of SiH_4.[25] More recently, arc discharge in SiH_4 was also used with success.[26] In addition, sputtered a-Si:F can also be prepared for TFT application.[27] The source-drain contact regions are highly phosphorus-doped, n^+-type a-Si films deposited on top of the undoped n-type a-Si. The SiO_2 or Si_3N_4 gate insulators are put down by low-temperature CVD or by plasma CVD. The devices are patterned using conventional photolithography.

Performance of Si TFT. TFTs are metal-insulator-semiconductor field effect transistors (MISFET). To a first-order approximation, the same equations describe their performance as the ones used for characterizing single-crystal Si MOSFETs. For TFTs, these equations were first derived by Weimer.[28] When the TFT is turned on as a nonlinear switch in the active matrix of a display, it may be operated below saturation, in which case

$$I_{on} = I_D = \mu_{fe}c_i W/L [(V_G - V_0)V_D - \tfrac{1}{2} V_D^2]$$

(Eq. 1)

for $V_D < (V_G - V_0)$ and $V_G > V_0$.

If it is operated in the saturation regime, then

$$I_{on} = I_D = [\mu_{fe}c_i W/(2L)] (V_G - V_0)^2 \quad (Eq. 2)$$

for $V_D > (V_G - V_0)$ and $V_G > V_0$.

In Eqs. 1 and 2, μ_{fe} is the field effect mobility, c_i is the unit area capacitance of the gate insulator, W is the width, L is the channel length of the transistor, and V_0 is the threshold voltage. The turn-on time of the transistor, according to this simple model is

$$\tau = L^2/(\mu_{fe}V_D) \quad \text{(Eq. 3a)}$$

for $V_D < (V_G - V_0)$, $V_G > V_0$, and

$$\tau \sim L^2/\mu_{fe}(V_G - V_0) \quad \text{(Eq. 3b)}$$

for $V_D > (V_G - V_0)$ and $V_G > V_0$.

While Eqs. 1 and 2 describe most reproducible, well-behaved transistors, at least in a limited range of V_G and V_D, Eqs. 3a and 3b are usually taken as a first-order approximation of the actual transient performance. The threshold voltage V_0 and the leakage current I_{off} are functions of the charge density in the semiconductor, charges at the semiconductor-insulator interface, and charges in the gate insulator as well as other factors.[29,30] A reliable control of the semiconductor-interface charges and of trapped charges in the gate insulator at a low $10^{11}/cm^2$ density is a prerequisite for stable, high-performance, thin-film Si TFT technology. It is assumed here that this has been achieved.[13,31] From Eqs. 1 and 2, it can be seen that high I_{on} requires high μ_{fe}, large W, small L, and small V_0. For high-resolution, >50 lines/in. flat-panel displays, the length of one pixel is <0.0200 in., or <508 μm. Leaving space with clearance for the 0.002-in. or 50-μm wide busbars, the maximum W for a TFT will be <400 μm. On the other hand, the use of conventional photolithography over large >5 in. by 5 in. areas will limit L to >5 μm. Then if the TFT is operated in saturation and the gate insulator is 0.3 μm thick SiO_2 with c_i = 1.3×10^{-9} farad/cm^2, V_0 = 1 V, and V_G is a CMOS IC compatible 15 V; in order to obtain I_{on} = 5 μA, the field effect mobility has to be μ_{fe} > 0.5 cm^2/Vs. This can be considered to be the lowest limit on field effect mobility for any type of TFT technology if resolutions higher than 50 lines/in. are contemplated and CMOS IC compatibility is strongly desired. On this basis alone, it can be seen from Table 5-10 that all but a-Si type TFTs are acceptable for large, high-resolution, matrix display application. Researchers in the field are, of course, well aware of this problem with the a-Si TFT, and two different solutions have been reported. Using a novel, hot cathode arc discharge decomposition of SiH_4 for the deposition of a-Si, Matsumura[32] achieved I_D = 3 μA at just 10 V of V_G and 4 V of V_D for a 200 μm wide TFT with a 50-μm channel length. Matsumura calculated the field effect mobility to be of 1.9 cm^2/Vs. At Xerox, Tuan, et al.[33] used a double-gated transistor structure with a 168-μm width and a 10-μm channel length to reach 10 μA at 15 V V_D and 15 V V_G on both gates.

Arrays of Si TFTs.   Since 1980 at a number of research centers, attention has shifted, at least partially, from the fabrication of individual devices to the processing of larger and larger arrays of TFTs. Table 5-10 illustrates this point, but it is not an all-inclusive list of research or development done everywhere. Nevertheless, it represents fairly accurately the state of the art determined on the basis of technical publications. In order to make comparison of the performance of the transistors in the different arrays readily possible, the I_{on}/W and I_{off}/W are shown in Table 5-10.

In 1979-80, the group at the University of Dundee[31,34] was the first to build and apply an a-Si TFT array to drive a small (5 × 7 dot) LC display. Subsequently, a-Si TFT arrays were built for this purpose at Fujitsu[35] and Canon,[36] and these works were published in 1982. Significantly, the a-Si TFT array built at the Canon laboratories is large. It has a 3.8 × 3.8 in. active area and a 240 × 240 element array with a 64/in. density. This represents a factor of 1000 increase in the number of transistors over the number present in the first array built only two years earlier.[31] In addition, the on current per μm of transistor width, I_{on}/W, was also increased by a factor of 8 to 10.

With the rapid emergence of CW laser-induced crystal growth as a feasible process for the fabrication of large-grained (>1 μm) or even single-crystal, thin-film Si on insulator (SOI) devices, laser-annealed, crystalline Si TFTs became a fresh, new switching array possibility for flat-panel LC displays.[29] In 1982, researchers from Mitsubishi published results on a 1.6 × 1.2 in. size array with 160 × 120 elements built on a quartz substrate.[37] This array has 100 TFTs/in. and the average I_{on}/W is 5 $\mu A/\mu m$ at only 10-V gate and source-drain voltage. The paper from Mitsubishi is the only published example for this application of laser-annealed Si TFTs. But papers from Bell Laboratories,[10] Hewlett-Packard,[8] Texas Instruments,[29] and Xerox[38] all mention large-area, flat-panel, extrinsic matrix displays as at least one, if not the primary, application for laser-annealed, crystalline Si TFTs on insulating substrates.

As Table 5-10 indicates, I_{on}/W in laser-annealed Si TFTs is at least a factor of 100 higher than in a-Si TFTs operated at the same V_G and V_D. This can be correlated with and in most cases completely explained by the factor of 100-1000 higher field effect mobility μ_{fe} in the laser-annealed devices. But conversely, I_{off}/W is also generally higher in the laser-

Table 5-10 TFT Arrays Built for LC Displays (Published Works 1980-82)

Origin of work	Year publ.	Ref. no.	TFT type	Active area (in. × in.)	No. of TFTs per array	Channel length $L(\mu m)$	Width $W(\mu m)$	Semicond. thickness (μm)	Gate insul. thickness (μm)	I_{on}/W (A/μm)	I_{off}/W (A/μm)	Mobility μ_{fe}(cm²/Vs)
										TFT Characteristics		
University of Dundee	1980	31, 34	a-Si	?	35 (5 × 7)	40	500	0.5	0.5 (Si$_3$N$_4$)	2×10^{-9}	2×10^{-14}	0.4
Fujitsu	1982	35	a-Si	?	35 (5 × 7)	40	600	0.5	0.3 (Si$_3$N$_4$)	10^{-9}	10^{-13}	0.4
Canon	1982	36	a-Si	3.8 × 3.8	57,600 (240 × 240)	8	100	0.3	0.3 (SiO$_2$)	10^{-8}	5×10^{-3}	0.3
Mitsubishi	1982	37	Laser-annealed crystalline Si	1.6 × 1.2	19,200 (160 × 120)	10	20	0.5–0.7	0.3 (SiO$_2$)	5×10^{-6}	2×10^{-10}	300
CNET	1982	39	a-Si + (L.A. Si for logic)	3.2 × 3.2	100,000 (316 × 316)	50	100	0.75	0.1 (SiO$_2$)	5×10^{-9}	5×10^{-13}	001.00
IBM	1982	17	Small-grained poly-Si	0.2 × 0.2	100 (10 × 10)	10	10	0.75	0.1 (SiO$_2$)	2×10^{-6}	10^{-9}	25

annealed Si TFT. For these reasons, a-Si TFTs are well suited for those applications where >1 μs turn-on time is sufficiently fast and <1 nA off current is desired. Such an application is of course the switch required for the individual picture element in an LC extrinsic matrix display. Laser-annealed devices, on the other hand, have <1-μs response time and are well suited for circuits operated at 10 MHz or higher frequencies. Shift registers for 500-line or larger displays driven at 60 Hz require these types of fast devices. For these reasons, then, a combination of these two Si TFT technologies can be used to create a complete flat-panel LC display, including the necessary logic circuit. At CNET, they chose such an approach and built a very large 3.2 in. $\times$ 3.2 in., 316 $\times$ 316 element a-Si TFT array on a Corning 7059 glass substrate.[39] To create the much faster devices for the shift registers, they laser-annealed the a-Si film at the edge of the substrate.

Finally, Table 5-10 also includes small-grained poly Si TFTs which were built at IBM.[17] These are relatively small arrays, but the authors reported that their performance was very reproducible and stable.

From Table 5-10 a number of interesting observations can be made:

- All Si TFT arrays built by 1983 are smaller than 4 in. $\times$ 4 in.
- The largest working liquid-crystal display built with Si TFTs has 240 $\times$ 240 elements.
- The channel length, L, of Si TFTs built for large-area arrays is in the 8–50-μm range, while the width W varies from 20 to 600 μm. For this application a-Si TFTs are typically an order of magnitude wider than the laser-annealed crystalline and small-grained polycrystalline Si TFTs. The reason for this is of course that μ_{fe} and therefore I_{on}/W are at least two orders of magnitude smaller in a-Si TFTs.

Summary. It is quite clear today that large ($>10^4$ devices) Si TFT arrays can be fabricated that perform adequately as nonlinear switches for flat-panel LC displays.

Since the fabrication processes in most cases are identical or very similar to regular Si IC processing, it is hoped that these arrays will be fabricated at high yields and that they will be operationally stable and easily reproducible. While large-area capability is mentioned in almost all the publications dealing with Si TFT arrays built for flat-panel displays, none of these have actually demonstrated capabilities larger than 3.8 in. $\times$ 3.8 in. for a-Si TFTs and 1.6 in. $\times$ 1.2 in. for laser-annealed crystalline Si TFTs. Similarly, fabrication yield, performance uniformity, and reproducibility still have to be demonstrated on a sufficiently large scale (large area, large number of devices per array, large number of arrays) to establish these important parameters as real, well-controlled parts of the entire technology.

REFERENCES FOR SECTION 5.9.2

1. Barclay, D. G.; Bird, C. L.; and Martin, D. H. *J. of Elect. Materials, 8,* 311, 1979.
2. Luo, F. C., and Hester, W. A. *IEEE Trans. Electron Devices, 27,* 223, 1980.
3. Brody, T. P.; Asars, G. A.; and Dixon, G. D. *IEEE Trans. Electron Devices, 20,* 995, 1973.
4. Anderson, J. C. *Thin Solid Films, 50,* 25, 1978.
5. Wysocki, J.; Poleshuk, M.; and Hudson, R. *Technical Digest IEDM,* 543, 1979.
6. Erskine, J. C., and Cserhati, A. *J. Vac. Sci. Technol, 15,* 1823, 1978.
7. Snell, A. J.; Mackenzie, K. D.; Spear, W. E.; and LeComber, P. G. *Appl. Phys., 24,* 357, 1981.
8. Kamins, T. I. and Pianetta, P. A. *IEEE Electron Device Lett., 1,* 214, 1980.
9. Katayama, Y.; Shiraki, Y.; Kobayashi, K. L. I.; Komatsubara, K. F.; and Hashimoto, N. *Appl. Phys. Lett., 34,* 740, 1979.
10. Leamy, H. J. *Laser and Electron-Beam Interactions with Solids,* edited by B. R. Appleton and G. K. Celler. New York: North-Holland, 1982, p. 459.
11. Castleberry, D. E. 1980 Biennial Display Research Conference 89, 1980.
12. Luo, F. C.; Chen, I.; and Genovese, F. C. *IEEE Trans. Electron Devices, 28,* 740, 1981.
13. Lee, K. F.; Gibbons, J. F.; Saraswat, K. C.; and Kamins, T. I. *Appl. Phys. Lett., 35,* 173, 1979.
14. Gibbons, J. F. and Lee, K. F. *IEEE Electron Devices Lett., 1,* 117, 1980.
15. Spear, W. E.; LeComber, P. G.; and Snell, A. J. *Phil. Mag., B-38,* 303, 1978.
16. Matsui, M.; Shiraki, Y.; Katayama, Y.; Kobayashi, K. L. I.; Shintani, A.; and Maruyama, E. *Appl. Phys. Lett., 37,* 936, 1980.

17. Juliana, A.; Depp, S. W.; Huth, B.; and Sedgwick, T. *SID 1982 Digest*, 38, 1982.

18. Thompson, M. J.; Johnson, N. M.; Moyer, M. D.; and Lujan, R. *IEEE Trans. Electron Devices* 29, 1643, 1982.

19. Gibbons, J. F. *VLSI Design*, 54, March/April 1982.

20. Poate, J. M. and Brown, W. L. *Physics Today*, *35*, 24, 1982.

21. Lam, H. W.; Tasch, A. F.; Holloway, T. C.; Lee, K. F.; and Gibbons, J. F. *IEEE Electron Devices Lett.*, *1*, 99, 1980.

22. Biegelsen, D. K.; Johnson, N. M.; Bartelink, D. J.; and Moyer, M. D. *Laser and Electron-Beam Solid Interactions and Materials Processing*, edited by J. F. Gibbons, L. D. Hess, and T. W. Sigmon. New York: North-Holland, 1982, p. 487.

23. Hawkins, W. G.; Black, J. G.; and Griffiths, C. H. *Appl. Phys. Lett.*, *40*, 319, 1982.

24. Black, J. G.; Hawkins, W. G.; and Griffiths, C. H.; *J. Appl. Phys.* 40, 319, 1982.

25. Street, R. A.; Knights, J. C.; and Biegelsen, D. K. *Phys. Rev. B18*, 1800 1978.

26. Uchida, Y. and Matsumura, M. Denki-Tsushin Gakkai Gijutsu Hokoku SSD80-116 (in Japanese), 1981.

27. Matsumura, H.; Nakagome, Y., and Furukawa, S. *Appl. Phys. Lett.*, *36*, 439, 1982.

28. Weimer, P. K. *Proc. I.R.E.*, *50*, 1462, 1962; and

29. Lam, H. W. see Reference 10, p. 471.

30. Milnes, A. G. *Semiconductor Devices and Integrated Electronics*. New York: Van Nostrand Reinhold, 1980, p. 377.

31. LeComber, P. G.; Spear, W. E.; and Ghaith, A. *Electron Lett.*, *15*, 179, 1979.

32. Matsumura, H. Paper presented at the U.S.-Japan Seminar on Technological Applications of Tetrahedral Amorphous Materials, Palo Alto, CA, July 1982.

33. Tuan, H. C.; Thompson, M. J.; Johnson, N. M.; and Lujan, R. A. *IEEE Electron Devices Lett.* 3, 357, 1982.

34. Snell, A. J.; Mackenzie, K. D.; Spear, W. E.; and LeComber, P. G. *Appl. Phys.*, *24*, 357, 1981.

35. Kawai, S.; Takagi, N.; Kodama, T.; Asama, K.; and Yangisawa, S. *SID 1982 Digest*, 42, 1982.

36. Okubo, Y.; Nakagiri, T.; Osada, Y.; Sugata, M.; Kitahara, N.; and Hatanaka, K. *SID 1982 Digest*, 40, 1982.

37. Nishimura, T.; Akasaka, Y.; Nakata, H.; Ishizu, A.; and Matsumoto, T. *SID 1982 Digest*, 36, 1982.

38. Johnson, N. M.; Biegelsen, D. K.; and Moyer, M. D. see Reference 22, p. 463.

39. LeContellec, M.; Morin, F.; Richard, J.; Coissard, P.; Morel, M.; and Bonnel, M. *SID 1982 Digest*, 44, 1982.

Borkan, H., and Weimer, P. K. *RCA Review*, *24*, 153, 1963.

6

THE CHALLENGE OF THE CATHODE-RAY TUBE

NORMAN H. LEHRER, *Consultant*

6.1 INTRODUCTION

The cathode-ray tube (CRT) is the dominant display device for video and high-resolution applications. It has superior and more than adequate speed (bandwidth) and resolution for the presentation of time-varying pictorial information that is aesthetically satisfying to the human observer. Alphanumerics and graphics are also best displayed on the CRT. A variety of monochrome displays as well as high-quality multicolor presentations can be achieved.

Such applications of the CRT extend across a broad spectrum including the following markets: consumer or entertainment, government (military and nonmilitary), industrial, and medical.

The most familiar CRT display is the home television receiver. In the past thirty years, over 300 million TV sets have been sold to consumers throughout the world.

The same characteristics of the CRT which made it the best choice for use with television have also enabled it to monopolize its newer application to computers. Millions of such computer displays already exist and range in use from passenger reservation displays for airlines to securities quotations for stockbrokers.

The outlook is for further continued growth of the CRT. There is nothing on the horizon at the present time to replace it in applications requiring high speed and high resolution.

A remarkable aspect of the dominance of the high-resolution display market by the CRT is the fact that its precursor was built by William Crookes in about 1879, over 100 year ago. The life cycle of many products is only five to ten years—and here is the CRT, 100 years old with no serious challengers on the horizon. Its durability is even more amazing when one considers the fact that is is one of the few tube types that solid-state devices have failed to replace.

What characteristics of the CRT have permitted it to survive so long? What special or unique advantages does it have that enable it to defy replacement by other technologies?

The CRT has several unique basic attributes which make it superior to other existing technologies for high-resolution and high-speed display applications. These attributes as well as other advantageous characteristics of CRTs are:

1. The speed of response (bandwidth) and resolution of the CRT are capable of satisfying the requirements of the eye for the presentation of high-quality dynamic imagery.
2. The techniques of scanning are not only simple but versatile as well. Scanning can be raster, stroke, or random. Furthermore, they are compatible with resolutions ranging from low to very high.
3. The CRT itself has few parts and requires as little as seven connections. This number increases to eleven if the scanning means is included.
4. The CRT is versatile—one design can be used for a wide number of applications with little or no change in design. The same tube can present alphanumeric, graphic, and pictorial information.

5. The luminous efficiency of the phosphor, tube, and system are all high compared to other approaches.
6. The fabrication of the tube is amenable to mass-production techniques.
7. The CRT can present multicolor displays.
8. The CRT is reliable.
9. The CRT has long life.
10. The CRT display system is relatively inexpensive compared to other display systems.
11. The CRT can present more information per unit time at a lower cost than any other display technology.

The CRT does have several disadvantages. These include the following:

1. The tube depth is comparable to or greater than the diagonal of the display area. This gives the display monitor considerable bulk, which is a disadvantage in certain applications.
2. The cross section of the electron beam has a gaussian current density distribution, thereby degrading the contrast between adjacent resolution elements. This means reduced detail contrast at high brightness.
3. Halation normally limits the number of shades of gray to 10 to 12.
4. The CRT cannot store information at high brightness levels; therefore the display must be refreshed above the critical flicker fusion frequency of the eye.
5. The CRT display is not digital in format. Thus it is not possible to address any one resolution element.

These limitations of the CRT make it vulnerable to new technologies which can outperform it in one or more of the five areas while maintaining comparable performance in the other areas.

It is desirable for anyone whose objective is to replace the CRT, even in a limited way, to have some in-depth understanding of the operation, characteristics, and limitations of the device which he or she is seeking to replace. This chapter has been prepared with that in mind.

6.2 HISTORICAL ORIGINS OF THE CRT

The CRT has its origins in work conducted by several investigators toward the close of the nineteenth century. The experiments conducted by these men were an outgrowth of discoveries made in the fields of electricity and magnetism in the first half of the nineteenth century. The work of Galvani, Volta, Oersted, Ampere, Ohm, Faraday, Lenz, Henry, Gauss, Joule, and Maxwell led to the development of batteries and generators which could provide convenient sources of electricity for experimentation.

In the latter half of the nineteenth century, several investigators were concerned with the phenomena that resulted when a high voltage was applied to an evacuated glass tube containing two electrodes. The luminescence of the resulting gas discharge was observed by these investigators.

John William Hittorf[1] was the first to note that, independent of the glow surrounding the anode, the cathode was a source "of rays of glow which follow rectilinear paths and cause the surface of the glass to fluoresce upon striking it." He found that objects placed between the cathode and the glass walls stop the glow behind the object casting a shadow. He also built a tube with a right-angle bend in it and observed that the rays struck the glass in the leg containing the cathode but did not enter the other leg. Thus, he considered any point of the cathode as the source of a cone of rays which traveled in straight paths.

William Crookes[2] made several key contributions to the evolution of the CRT and is considered to have invented the CRT in 1879. He was able to achieve pressures of 10^{-5} to 10^{-6} torr in the discharge tube. In that pressure range, the luminance of the discharge disappeared and he was better able to observe the cathode rays. He proposed that the cathode rays were actually electrified gas molecules projected from the cathode region with high velocities. The luminance of the glass was "caused by the impact of the molecules on the surface of the glass." The molecules themselves did not generate the light, because the luminance was dependent on the nature of the material they struck.

Crookes demonstrated the mechanical properties of the cathode rays by enclosing a radiometric-type structure in the tube and causing it to turn under bombardment by the rays. He showed that the path of the rays could be markedly affected by an electromagnet, the amount of deflection being proportional to the strength of the magnet. He found that the walls of the glass tube became very hot when the cathode rays were focused on it by means of a magnet. It is of particular interest that Crookes actually built a tube, Fig. 6-1, employing an aperture to provide a narrow spot and a scale to measure the deflection of the beam under the action of the magnetic field. In Fig. 6-1, the negative electrode (A) is a flat aluminum disc with a notch cut in it. The positive electrode (B) is a ring of aluminum. The aperture (C) is formed by an approximate 1-mm-diameter hole in a piece of mica. The deflection amplitude is measured by means of a scale engraved on a piece of German glass (D). Another scale (E) is engraved on a piece of mica to observe the width of the dark space. The path of the rays was observed by means of a screen (F) of mica which was covered with phosphor.

Eugene Goldstein[3] reported in 1886 that, in addition to the first type of rays which came from the front surface of the cathode, there existed a new, second type of rays which originated from the back surface of cathodes where holes had been bored in them. Goldstein called this new type of rays "canal rays," but he was unable to explain either type of rays. The canal rays were later identified to be positive ions. Goldstein was responsible for the use of the term "Kathodoenstrahlen" or cathode ray, which he introduced in 1876 to identify the first type of rays.

In 1895 Jean Perrin[4] published an account of an experiment in which he demonstrated that the cathode rays were negatively charged. He built a tube which incorporated a Faraday cage

tied to an external electroscope. The entrance to the Faraday cage was in line with a hole in the positive anode so that the cathode rays could pass through the hole in the positive anode into the Faraday cage. An electromagnet was used to control the path of the cathode rays so that they could be deflected away from the anode or permitted to enter the hole in the anode and the Faraday cage. By means of this tube, Perrin was able to show that when the cathode rays entered the Faraday cage the electroscope reived a negative charge. Perrin also used a similar structure to demonstrate that positive ions were formed in the tube and were accelerated to strike the cathode. Perrin proposed that "In the neighborhood of the cathode, the electric field is sufficiently intense to break into pieces, or ions, some of the molecules of the residual gas. The negative ions move toward the region where the potential increases, acquire a considerable velocity, and form the cathode-rays; ... The positive ions move in the opposite sense ..."

Joseph John Thomson[5] repeated Perrin's experiments with some slight improvements, confirming that the cathode rays were negatively charged. One objection to the idea that the cathode rays were negatively charged was that no one had been able to deflect cathode rays under electrostatic forces. Hertz had tried to detect this effect by placing two parallel plates inside the tube and causing the rays to travel between them. He did not detect deflection of the rays. Thomson repeated the same experiment, but concluded that the field from the plates was being neutralized by the residual gas ions in the tube. Thomson found that the rays could be deflected by the plates at very high vacuum. The nature of the deflection was consistent with the cathode rays being negatively charged.

Thomson derived an equation which related e/m, where e is the charge on the electron and m, its mass, to the radius of curvature of the path of the rays in a uniform magnetic field, the intensity of the field, and the work done by the beam in delivering a quantity of charge. He also derived an equation which related e/m to the ratio of the electrostatic and magnetic fields necessary to produce the same deflection when they act over the same distance. Using these two equations, he was able to conduct a

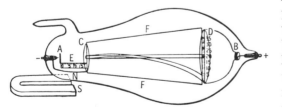

Fig. 6-1. Schematic drawing of Crookes' Tube.

number of experiments which led to a calculation of e/m. The ratio was the largest ever found and, taken with other work, proved that the mass of the cathode rays was quite small compared to ordinary molecules. This marked the discovery of the electron.

In 1896 Ferdinand Braun[6] was apparently the first person to conceive of using the CRT as a display device or indicator of some kind.

Central generating stations had begun to supply power to various communities toward the close of the nineteenth century. The power was both AC and DC. Monitoring the AC output of the generators was a particularly difficult problem at high frequencies. One technique for doing it was to attach a tiny mirror to the moving coil of a galvanometer. The beam of light striking the mirror is deflected back and forth as the voltage varies. The time base is provided by means of a second mirror moving at right angles to the first mirror. This device worked for frequencies of 50 to 60 hertz but was unsuitable for the substantially higher frequencies becoming of interest.

It occurred to Braun that the CRT could be adapted to provide an indicator for the higher-frequency phenomena. He designed and had built a tube which incorporated a cathode, an anode off to one side, and an aperture to provide a narrow beam, very much as Crookes' tube was built. Braun, too, utilized a coil for deflection. But he incorporated a phosphor-coated plate inside and perpendicular to the longitudinal axis of the tube so that the phosphor could be scanned by the electron beam. He first used a rotating magnet, and then a second deflection coil at right angles to the first, for scanning, as illustrated in Fig. 6-2. With such an apparatus, Braun was able to produce an all-electronic device which was a step closer to the modern CRT than Crookes' tube. Of course, Braun's device was crude by today's standards. The cathode was unheated, tens of thousands of volts were required to obtain sufficient emission, the operation was unstable and the beam unfocused, but an important application has been found for this primitive CRT.

As the CRT further evolved, other applications followed.

Paul G. Nipkow of Germany disclosed the first practical television system in 1884. He used a

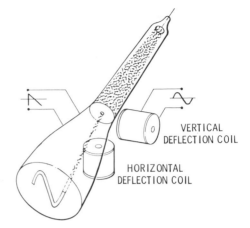

Fig. 6-2. Illustration of Braun's Tube. (From Shiers, G., "Ferdinand Braun and the Cathode-Ray Tube." Copyright © 1974 by Scientific American, Inc. All rights reserved.)

mechanical scanning device both to scan the subject and to create its reproduction. The device was a spinning disc with a spiral pattern of holes punched in it. For pickup, the spinning disc was interposed between an intense light and the subject. Only the light passing through one hole could illuminate the subject. For reproduction, a second spinning disc was placed between the observer and a lamp. The brightness of the lamp was modulated in accordance with the intensity of the light reflected from the subject for any position of the discs, whose motions were synchronized.

The idea of all-electronic television was pursued in the 1920s and culminated in the demonstration by Vladimir K. Zworykin of such a system on November 18, 1929. The pickup was by means of a camera tube, the iconoscope, and reproduction accomplished with a cathode-ray tube called the kinescope. On July 30, 1930, the National Broadcasting Company began operating W2XBS, an experimental television station in New York City. In September 1946 the first postwar TV sets went on sale, marking the beginning of the boom in black-and-white TV.

The development of the present shadow-mask color tube started at the RCA Laboratories in 1949. The initial successful demonstration of the color tube in 1950, although crude by today's standards, led to the eventual overwhelming acceptance of color TV. On October 31, 1953, the first hour-long program in electronic color was broadcast. The initial tube employed a flat aperture mask with the screen deposited on a flat plate mounted inside the tube. In late 1953, under the direction of Norman Fyler at CBS Hytron, a tube was built with a curved mask and the phosphor deposited directly on the curved, inside surface of the faceplate. Such deposition of the phosphor was made possible by use of a newly developed photoresist process. The resulting picture was larger and more appealing than that achieved with the internal flat screen-plate. This curved-mask design became the industry standard for color tube manufacture.

A technological advance of immense significance to cathode-ray tube displays was the development of rare-earth phosphors in 1964.[31,34] The rare-earth red phosphor made possible substantial increases in the brightness of the direct-view color TV tube and provided the basis for the manufacture of practical projection TV displays. Other advances in color tube design incorporated in tube production include the Sony Trinitron[TM] (1968), the black-matrix screen (1969), and the Precision In-line System (1972).

There are many other items of historical interest in the evolution of the CRT, such as its use in early radar systems, the development of projection television, displays combining stored and cursor-generated information, as well as its use in aircraft for vertical-situation, horizontal-situation, and heads-up displays. Also, helmet-mounted displays. The list is seemingly endless.

6.3 BASIC CRT DESIGN AND OPERATION

6.3.1 Monochrome CRTs.
One type of widely used modern CRT is shown schematically in Fig. 6-3. The CRT consists of four essential parts:

1. The Bulb. The bulb, sometimes referred to as the bottle, maintains the high vacuum needed for utilization of the electron beam and also defines the display area. The bulb consists of the clear face or faceplate (upon which is deposited the phosphor viewing screen), the funnel, and the neck. The bulb is normally made of glass, although in some cases the funnel and/or the neck may be made of metal or ceramic. For some applications, a facepanel or implosion panel may be bonded to the face of the tube to provide implosion protection. The facepanel may be of reduced optical transmission and may have antireflective coatings on its external face to improve display contrast, as well as transparent conductive coatings on its internal face to reduce electromagnetic interference.

2. The Electron Gun. The electron gun provides a high-density electron beam whose current can be modulated. In addition, the beam is capable of being focused and deflected. The CRT shown in Fig. 6-3 is electrostatically focused; magnetic focusing can also be used.

3. A Means of Deflection. The deflection means is used to locate the electron beam at the desired point on the viewing screen. In this case, a magnetic yoke or coil is used to provide the deflection. Electrostatic deflection can also be employed.

4. The Viewing Screen. The viewing screen provides the display by luminescing under electron bombardment. The screen consists of a thin layer of phosphor particles coated on the inside of the panel. A very thin film of aluminum is deposited over the phosphor. The aluminum serves to establish the electrical potential at the screen.

The operation of the CRT can be briefly described as follows: The electron beam is scanned over the viewing screen in the desired pattern by means of the deflection yoke when the proper potentials are applied to the CRT. The information to be displayed—that is, the video signals— are applied to the electron-beam current-controlling electrode of the electron gun in synchronization with the deflection signals. Either the grid 1 or the cathode may be used for this purpose. (The grid 1 is the preferred electrode where resolution is critical, because modulation of the cathode varies the beam velocity and thereby the deflection sensitivity of the tube.) By this means, the viewing screen provides a two-dimensional visual display which corresponds to the serial electrical information contained in the

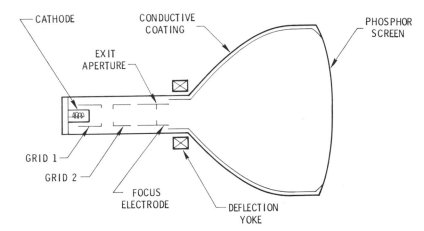

CATHODE

CONDUCTIVE COATING

PHOSPHOR SCREEN

EXIT APERTURE

GRID 1

GRID 2

FOCUS ELECTRODE

DEFLECTION YOKE

Fig. 6-3. Schematic of modern CRT.

electron beam. The display is normally refreshed sixty times per second, both to avoid the sensation of flicker in the eye and to present time-varying information. Lower refresh rates are required by some applications, such as the display of raw radar information or very-high-resolution images containing thousands of scanning lines. In such cases, storage or long persistence must be provided by special CRT phosphors.

The operation of the electron gun can best be understood by dividing the CRT into various electron-optic regions. These regions are described in Section 6.4.

6.3.2 Color CRTs. There are several methods for incorporating color into a CRT:

Mask Techniques. Mask techniques[7] are the most widely used and most successful methods of incorporating color into a CRT. These mask techniques, as illustrated in Figs. 6-4 and 6-5, depend on the principle of color centers or shadowing. A color center is a location in space from which a point source of light casts highlights and shadows. In the case of the aperture-mask color tubes, there are three color centers located about ten inches or more from the faceplate and close

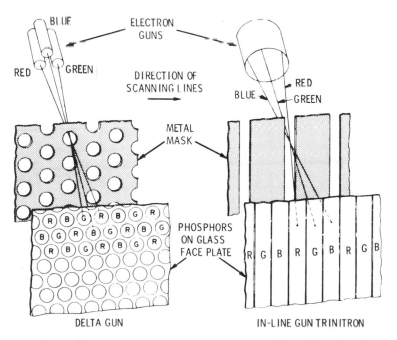

BLUE

ELECTRON GUNS

RED   GREEN

DIRECTION OF SCANNING LINES

BLUE   RED

GREEN

METAL MASK

PHOSPHORS ON GLASS FACE PLATE

DELTA GUN

IN-LINE GUN TRINITRON

Fig. 6-4. Illustrations of aperture-mask and Trinitron color CRTs. (From Herold, ref. 7)

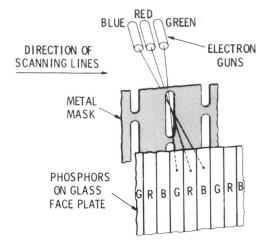

Fig. 6-5. Illustration of slotted-mask color CRT. (From Herold, ref. 7)

to the longitudinal axis of the tube. When a point source of light is placed at one of these positions, it causes a shadow of the aperture mask to fall upon the inside of the faceplate. The illuminated holes projected on the surface are separated from each other by unilluminated areas. By shifting the color center in two 120-degree steps around the longitudinal axis, it is possible to produce three sets of tangent illuminated holes. In the manufacture of the color tube, this principle is used with photoresist techniques to produce faceplates coated with three sets of phosphor dots; one red, one blue, and one green. By making each color center the center of deflection of an electron beam, it is possible to address the red, blue, and green sets of dots separately by means of individual electron guns.

The principle of the in-line gun Trinitron[TM] or grill-mask CRT is essentially the same as that of the aperture-mask CRT. The screen consists of an array of vertical phosphor stripes. The colors of the stripes repeat red, green, blue; red, green, blue, when viewed from the faceplate side of the tube. The electron guns are located in-line instead of in a delta configuration. The Trinitron[TM] has two advantages over the delta-gun aperture-mask tube. First, the vertical resolution is improved because the vertical slots in the grill are entirely free of obstructions. Second, the convergence of the three in-line beams is much simpler than that of the delta gun. A major disadvantage of the Trinitron[TM] is that the mask

cannot be self-supporting and therefore cannot be made spherical in shape. This ultimately limits the size of the tube.

The Precision In-Line system was announced in 1972. The tube is similar to the Trinitron[TM] except that the mask is slotted; that is, it has horizontal supports extending across the vertical slots. These supports make it possible for the mask to be self-supporting, and therefore it can be curved spherically. In addition, the convergence problem is simplified by designing a special deflection yoke which not only deflects the three beams but also converges them.

All three types of color tubes are capable of presenting color television with sufficient brightness and resolution for commercial acceptance. The resolution of the tubes is limited at low brightness levels by the coarseness of the mask, while at high brightness levels it is limited by the spot size of the electron beam. The brightness is limited by the fact that about 80% of the beam current is absorbed by the mask.

Beam Index Tubes. Beam index tubes[8] are similar to grill- or slotted-mask tubes in that they too employ striped color phosphor screens. One or more high-resolution electron guns capable of focusing the beam down to less than the width of one phosphor element are used to assure color purity and selectivity. In place of the mask, however, the screen incorporates an electron-beam-responsive means of signaling when the electron beam has impinged on an indexing region. Color selection is achieved by synchronizing the modulation of the beam in accordance with beam location as derived from the indexing signal.

Vertical stripes are normally employed with the electron beam scanning orthogonally to the stripes. The stripes alternate red, green, blue; red, green, and blue. The synchronizing signal can be generated by providing either a stripe of a material which is a good secondary emitter or a UV phosphor behind one of the phosphor stripes. UV index signal generation stripes are generally preferred because they provide high signal-to-noise ratios and are more reliable.

In operation, Fig. 6-6, the electron beam is scanned at some low level across the phosphor. When the photomultiplier tube senses the indexing signal, the circuits controlling the electron

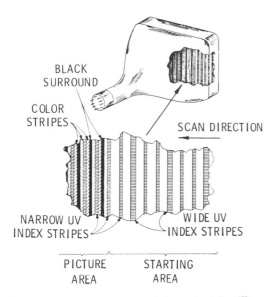

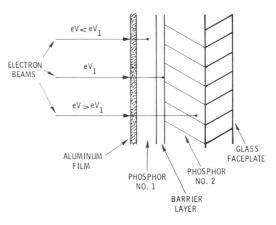

Fig. 6-7. Schematic of penetration phosphor screen.

Fig. 6-6. Illustration of beam index color tube. (From Schwartz, ref. 8)

beam modulate it in synchronization with its position as determined by the index signal.

Conceptually, the beam index tube offers the advantage of lower tube cost, substantial energy savings, high resolution, perfect convergence, and lighter-weight receivers. The operation of the tube is for practical purposes restricted to the raster mode because the indexing signal must be generated by the electron beam as it moves along a straight line. The fact that beam index color TV sets are not yet commercially available indicates that practical problems exist in realizing the advantages of the beam index tube. Some of these problems are associated with the ability to focus the required currents into a sufficiently fine spot to excite only one phosphor stripe whose width is sufficiently narrow to provide the total number of resolution elements needed for a TV display.

Penetration Phosphors. The beam penetration screen[9] is based on the principle that the penetration depth of electrons into solids increases as the beam velocity increases. As the penetration depth of the electron beam increases, so does the depth at which most of the energy is absorbed.

Fig. 6-7 is a schematic of a penetration phosphor screen. The screen consists of a glass substrate on which is deposited a layer of phosphor

No. 2. Phosphor No. 1 is on top of phosphor No. 2 but separated from it by the barrier layer. A thin film of aluminum covers phosphor No. 1.

At beam energies below about 2 kV, the electron beam is virtually completely absorbed in the aluminum and the screen is dark. As the beam energy increases, more and more of the electron-beam energy is absorbed in phosphor No. 1, which is nearest the electron beam. Visible emission from that phosphor diffuses through the barrier layer and phosphor No. 2, appearing as luminescence on the faceplate. At this lower range of beam energies, the barrier layer improves color purity by absorbing the more energetic electrons.

As the beam energy is increased, the electron beam penetrates through the barrier layer into phosphor No. 2, losing more and more of its energy in that phosphor and less and less in phosphor No. 1. As the loss in electron-beam energy progressively shifts from phosphor No. 1 to phosphor No. 2, the color correspondingly changes from that characteristic of phosphor No. 1 to that dominated by the emission from phosphor No. 2. At intermediate beam energies, the colors produced are mixtures of the two emission spectra.

This principle of color penetration can be achieved in several ways with mixed phosphors, superimposed phosphor layers, and onionskin phosphors.

The mixed-phosphor screen employs a mixture of two different color phosphors as the screen. The particles of one phosphor color have a thin, nonluminescent barrier layer coated around them; those of the second phosphor color do not. At low electron-beam energies, the phosphor without the barrier layer is excited while the electrons cannot reach the luminescent regions of the coated phosphor. At high beam energies, both phosphors are excited. This approach has limited color range because the uncoated phosphor is excited over virtually the entire range of electron-beam energies and absorbs the full energy of the electron beam, thereby diluting the color from the coated phosphor.

The superimposed phosphor-layer screen consists of an initial layer of a powder phosphor deposited on the inside of the facepanel. A very thin barrier layer is coated, typically by evaporation, over the initial phosphor layer. A thin layer of fine particles of the second phosphor is deposited over the barrier layer. An aluminum film covers the screen. This approach suffers from several drawbacks. The voltage change required to switch from one color to another is high (10 kV to 15 kV) because of the thickness of the phosphor particles. It is difficult to achieve color uniformity over the entire display area because of nonuniformities in phosphor layer thickness. The luminous efficiency of the screen is low because the optical and electron-beam transparencies of both layers is poor.

Attempts have been made to overcome the shortcomings of the superimposed powder-phosphor screen through the use of transparent thin-film phosphors. Such phosphors are made using evaporation or sputtering techniques from conventional or rare-earth materials. The use of a black backing behind the transparent phosphors permits their application in high-brightness ambients even though the luminous efficiency of the thin-film phosphors is very low. The black backing absorbs the incident illumination, which passes through the transparent phosphors, maintaining contrast in the display which would be washed out in the case of conventional powder-phosphor penetration screens. Such powder-phosphor screens have high reflectances. The superimposed thin-film-phosphor screens have better resolution than their powder-phosphor counterparts because of the continuous nature of the thin film, but their low efficiency limits their use to slow-speed applications.

The onionskin phosphor screen has the appearance of a single-particle layer of a powder phosphor. In reality, however, each particle is comprised of a core phosphor which is surrounded by a thin layer (onionskin) of a second-phosphor. A thin, nonluminescing barrier layer separates the core and onionskin phosphors. The layer of phosphor particles is coated directly on the inside of the faceplate and is covered by an aluminum film. Only the onionskin phosphor is excited by electron beams in the lowest energy range, typical values ranging from 5 kV to 10 kV. The core-phosphor emission predominates at the highest range of electron-beam energies which lie between 15 kV and 20 kV. A noticeable blend of phosphor colors is produced by electron beams at intermediate energies which range from 10 kV to 15 kV.

Several basic limitations characterize all penetration phosphor screens. One problem is the registration of the images generated by the several electron-beam energy levels required for color selection. Second, the luminous efficiency of each phosphor used in the penetration screen is substantially reduced from that obtained in monochrome tubes using those phosphors individually. This loss in efficiency results from the fact that the phosphor closest to the electron beam must be operated at nonoptimal low values of electron-beam energy and that a significant portion of the light emitted from that phosphor is not able to pass through the remote phosphor to reach the faceplate. The effective luminous efficiency of the remote phosphor is reduced because a significant portion of the energy of the electron beam is lost in penetrating the phosphor closest to the electron beam. The color range is limited, in practice, to something less than the two colors characteristic of each phosphor. At low electron-beam energies, the emission is characteristic of the phosphor closest to the electron beam. At high beam energies, however, when the emission from the remote phosphor should predominate, there is always some component from the other phosphor present.

In theory, it is possible to employ three successively superimposed phosphor layers to achieve three primary colors. This approach has not proven to be practical because of the excessive switching voltages required and dilution of the various colors.

The onionskin phosphor, of all the penetration phosphors, has enjoyed the greatest practical success to date. This is probably because it requires the lowest switching voltage. The low luminous efficiency of the phosphor coupled with the requirement for field sequential color selection prevents its application to color TV raster displays. The luminance of tubes employing onionskin phosphors is sufficient at low writing speeds (2,000 to 10,000 in./sec) and 50-hertz refresh rate to make quite practical its application to stroke writers, particularly where better resolution than that of mask color tubes is required and the ambient illumination is reduced.

The onionskin penetration phosphor has been very successfully employed in aircraft night landing simulator displays. The key advantage of the penetration phosphor tube over shadow-mask color tubes in this application where three primary colors are not required is the higher resolution capability of the former.

Current-Sensitive Phosphors. This approach[10,35] has as its basis the fact that the luminance–versus–current density curve varies for different phosphors. The curves fall generally into three classes: linear phosphors, for which the luminance increases linearly with current density; superlinear phosphors, where the luminance deviates in the positive direction; and sublinear phosphors, where the luminance deviates in the negative direction.

Most phosphors exhibit a linear or sublinear behavior in the normal range of current densities. A linear behavior characteristic is usually desired.

Through a combination of a superlinear phosphor of one color with a linear or sublinear phosphor of another color it is possible to fabricate a phosphor screen which changes color with current density. Thus, at low current densities the linear phosphor color would predominate, while at high current densities the color of the superlinear phosphor would become visible.

Current-density-sensitive phosphors do not appear practical at present, based on an evaluation of tubes incorporating such phosphors. The major problems are the limited shift in color with wide changes in current density and the large changes in brightness which accompany a shift in color. The color shift is limited because there is always some color contribution present from both phosphors. Large intensity changes occur between colors because of the low brightness produced by the low currents necessary to produce one color and the fact that an increase in peak current density of about 1000:1 is required to shift to the second color. Chopping the high current density electron beam to reduce its average current density is one method to overcome the brightness disparity problem.

Projection Color TV.   As a result of substantial progress made in the last few years, color TV projectors are now available to the consumer market which are capable of producing very acceptable color television pictures at reasonable prices. The projectors basically employ three different monochrome tubes—one red, one green, and one blue. The appropriate color fields are generated in synchronism on each of the three tubes by means of the electron guns. Both front-throw and rear-throw systems are available. Each tube and its associated lenses are so positioned as to project focused registered images of their respective displays onto a common display screen. The use of refractive optics appears to dominate, but reflective optic systems are also employed. A typical system will utilize three 5.5-in.-diameter tubes to produce a 50-in.-diagonal picture with a highlight brightness of 50 to 100 FL. High-gain projection screens are employed.

6.4   ELECTRON-OPTIC REGIONS OF THE CRT

The detailed operation of the CRT is made clearer by dividing the tube into four electron-optic regions as illustrated in Fig. 6-8. These regions are:

1. The beam-forming region
2. The focusing region

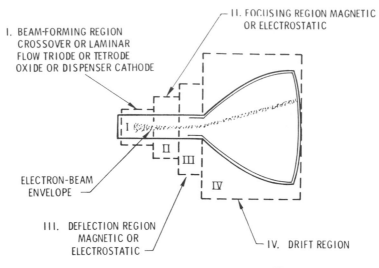

Fig. 6-8. Electron-optic regions of the CRT.

3. The deflection region
4. The drift region

These regions are discussed in detail below.

6.4.1   The Beam-Forming Region

Crossover vs. Laminar Flow.   This region provides the source of the electron beam as well as the means to implement its modulation. Here, the electron beam is produced by the application of an electric field to the surface of the cathode. This electric field is produced by maintaining a positive potential on the accelerating electrode, otherwise known as the grid 2 or first anode, which is located in front of the cathode. The beam intensity is controlled by application of the appropriate negative potentials to the grid 1 electrode, which lies between the cathode and the grid 2. The crossover electron gun[11,12,13] has been the standard electron gun since the development of the CRT. A typical crossover gun CRT is illustrated in Fig. 6-9. With this type of electron gun, the electrons are converged to a crossover just in front of the cathode. This crossover is then imaged on the viewing screen. The function of the crossover is to produce a virtual image of the cathode which is more intense than the cathode itself.

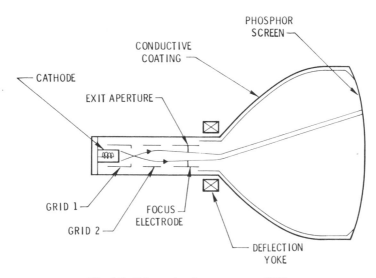

Fig. 6-9. Schematic of crossover gun CRT.

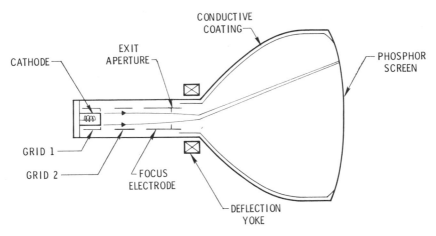

Fig. 6-10. Schematic of direct-replacement laminar flow gun CRT.

In the case of the laminar flow gun, the electrons emitted from the cathode tend to flow in streamline like paths, although many still crossover, until they are converged to a focus at the viewing screen. At least two types of laminar flow guns have been incorporated in commercially available CRTs. One uses an array of apertures maintained at increasingly positive potentials to collimate the beam. The grid 1, which controls the electron beam, operates at positive potentials with respect to the cathode. This type of gun is not directly interchangeable with the crossover gun, and for this and other reasons has not been widely accepted. The second type of laminar flow gun is called the direct-replacement laminar flow gun.[14,15] It has become widely used in military systems. This paper will confine itself, therefore, to the direct-replacement laminar flow gun. Figure 6-10 is a schematic of a direct replacement laminar flow gun CRT.

The crossover gun is designed to achieve an intense source of electron emission by sharply curving the equipotential lines in front of the cathode as shown in Fig. 6-11a. This causes the electrons to abruptly converge to a crossover immediately after leaving the cathode. The electric field is very high at the center of the cathode and falls off rapidly with radial distance. Consequently, the emission density is somewhat conical in shape as shown in Fig. 6-11b. The space-charge repulsion forces increase sharply as the electron beam converges to a crossover.

The direct-replacement laminar flow gun is designed to achieve a uniform, intense source of electron emission. Furthermore, it must do so at the same potentials as the crossover gun for which it is the direct replacement. Both these requirements are satisfied by appropriately shaping the electrodes and adjusting the spacings. In the ideal case, this uniform emission density has the form of a cylinder of emission from a circular cathode. Thus, ideally, the laminar flow gun can provide three times the current from the same cathode area, since the volume of a cylinder

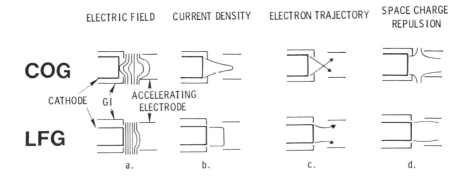

Fig. 6-11. Comparison of crossover and laminar flow beam formation: (a) electric field; (b) current density; (c) electron trajectory; (d) space-charge repulsion.
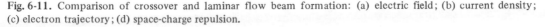

is three times that of a cone. In practice, a factor of between 1.3 and 2 appears realistic, since it is not possible to achieve perfect uniformity of the electric field across the cathode surface. The uniform electric field causes the electrons to move in streamline-like paths. The electron beam may be compressed immediately after leaving cathode. Avoidance of the crossover minimizes space-charge repulsion forces in the beam in the vicinity of the cathode.

Oxide vs. Dispenser Cathode. The oxide cathode,[16] Fig. 6-12, consists of a thin coating, usually a few thousands of an inch thick, of a barium-strontium oxide compound sprayed on the capped end of a small nickel cylinder. The function of the barium-strontium compound is to reduce the work function of the surface so that electron emission may be increased. The dispenser cathode,[17] Fig. 6-12, has a barium compound dispersed throughout a porous tungsten structure. The cathode operates at a higher temperature than that of the oxide cathode, which causes the barium compound within the pores to diffuse to the surface, constantly renewing the emitting region.

Triode vs. Tetrode Structure. In the case of the triode[18] structure, shown in Fig. 6-13, the accelerating electrode is maintained at a poten-

tial of several kilovolts or more; it is frequently tied to the focus or screen potential, thereby eliminating the need for one voltage source. The use of such a high potential on the accelerating electrode requires a large spacing between it and the grid 1 to avoid excessive values of the cutoff voltage. The cutoff voltage is that negative value of the voltage which must be placed on the grid 1 to turn off the electron beam. With such large spacings, the resolution of the tube may be degraded at high currents due to the space charge in the beam depressing the potential in the region between the grid 1 and the accelerating electrode, thereby allowing the beam diameter to blow up in that region.

The tetrode structure, Fig. 6-13, utilizes an additional electrode, the grid 2 electrode, which is operated at a fixed low voltage, usually ranging from 100 to 1000 volts. This low voltage assures a low grid 1 cutoff voltage and also makes the grid 1 cutoff voltage value independent of changes in the focus or screen potentials. Furthermore, the grid 2-to-grid 1 spacing can be small, reducing space-charge effects at high currents in that region.

6.4.2 The Focusing Region. The focusing region can utilize magnetic[19] or electrostatic[19] focusing design alternatives, as shown in Fig.

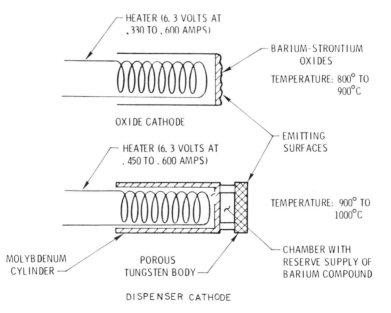

Fig. 6-12. Comparison of oxide and dispenser cathode structures.

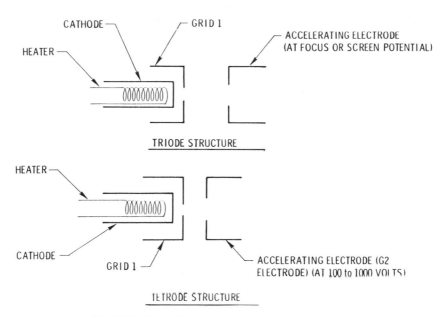

Fig. 6-13. Comparison of triode and tetrode structures.

6-14. A magnetic field, essentially parallel to the longitudinal axis of the tube, is created by means of a coil around the neck of the tube. The divergent electron beam enters the magnetic field and the resulting forces impart a focusing action on the beam. Magnetic focusing provides the highest resolution, probably because the lenses have fewer aberrations than the electrostatic ones.

An electrostatic lens, appropriately shaped by electrode design and with the proper applied

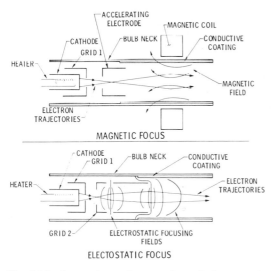

Fig. 6-14. Comparison of magnetic and electrostatic focusing.

voltages, converges the divergent electron beam to a focus at the screen. There are two types of electrostatic lenses: the bipotential focusing lens and the unipotential focusing lens.

The bipotential-focus design,[19] shown in Fig. 6-15, is also known as the high-voltage-focus gun because the focus potential is 15% to 25% of the viewing-screen potential. The lens is called bipotential because one element of the lens is formed between the focus electrode and the G2 electrode, which is at a lower potential, and the second element is formed between the focus and an electrode at viewing-screen potential. The resolution attained with this type of gun cannot match that provided with magnetic focus, but it typically is 30% to 50% better than that which can be achieved with a unipotential-focus gun. The focus electrode may draw current, depending on the size of the aperture used to trim the beam. The smaller the aperture, the larger the focus electrode current and, over a wide range, the smaller the spot size. The resolution is quite sensitive to changes in focus voltage.

The unipotential-focus gun design,[20] shown in Fig. 6-16, is also known as the low-voltage-focus gun or the einzel gun. It is called the unipotential gun because the electrodes on both sides of the focus electrode are the same potential, which is higher than that on the focus

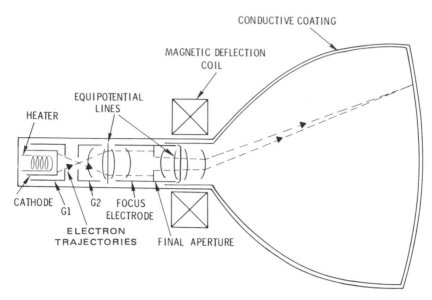

Fig. 6-15. Schematic of bipotential gun CRT.

electrode. The focus voltage is usually 0 to 500 volts and frequently uses the same supply as the grid 2 electrode. Note that the electron lens here is different from that of the bipotential gun because the equipotential lines from the focus electrode intrude between those provided by the unipotential electrodes. Consequently, this lens has severe aberrations which degrade the resolution. Aperturing the beam improves the resolution, but at the expense of grid drive. The advantages of the unipotential gun are that the focus electrode draws negligible current and

the resolution is insensitive to moderate changes in the focus voltage. Electron guns of comparable length and diameter, operating at the same viewing-screen potential, are assumed in comparing their performance.

6.4.3 The Deflection Region. Either magnetic or electrostatic fields can be used to deflect the electron beam. Magnetic deflection,[21] illustrated in Fig. 6-17, employs two pairs of coils, the coils in each pair being located on opposite sides of the neck. The magnetic fields

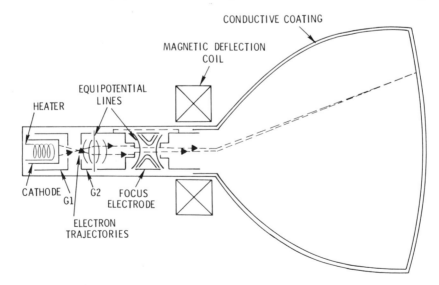

Fig. 6-16. Schematic of unipotential gun CRT.

TERMS:

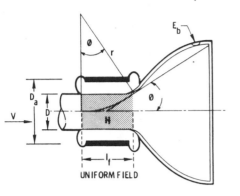

H	=	uniform magnetic field $\cdot$ ni/D_a
n	=	Number of Turns of One Pair of Coils
i	=	Deflection Current
D_a	=	Inside Diameter of the Yoke Core
v	=	Velocity of the Electron $\cdot$ $2E_b e/m$
E_b	=	Beam Acceleration in Kilovolts (Anode Voltage)
e	=	Unit Charge of an Electron
e/m	=	Charge to Mass Ratio of Electron or Ion to be Deflected
l	=	Length of Field H
$\emptyset$	=	Deflection Angle
r	=	Radius of Curved Electron Path Within Field H
D_n	=	Inside Diameter of Yoke Coils
L	=	Yoke Inductance

$$\text{Sin }\theta \cdot i\left(\frac{L}{E_b}\right)^{1/2}\left(\ell_f\frac{1}{D_a D_n}\frac{e}{2m}\right)^{1/2}\times \text{ constant}$$

*Constant $= 1.26\times10^{-8}$ henry with dimensions in centimeters.

Fig. 6-17. Magnetic deflection. (From Syntronic Instruments, Inc., Addison, Illinois, Deflection Yoke Catalog.)

produced by each pair of coils are at right angles to each other to provide horizontal and vertical deflection of the electron beam. By adjustment of the current in the coils, the force acting on the electron beam can be controlled to deflect the beam to any desired point on the screen. The deflection amplitude is directly proportional to the coil current and inversely proportional to the square root of the viewing-screen potential.

Electrostatic deflection,[22] illustrated in Fig. 6-18, utilizes two sets of metal plates set at right angles to each other to provide horizontal and vertical deflection when a difference in potential is applied between the two plates in each pair. The deflection amplitude is inversely proportional to the viewing-screen potential.

A summary of some of the characteristics of various combinations of the focus and deflection methods is shown in Table 6.1. The vertical resolution typically provided by these combinations is expressed in TV lines at .5m where m stands for modulation as described in Section 6.7.1. Electrostatic deflection provides the highest deflection speeds available, but only moderate resolution. It is normally used in combination with electrostatic focus. Magnetic deflection provides slower deflection speeds and exhibits moderate resolution when used with electrostatic focus; this combination is the most widely used. Magnetically deflected, magnetically focused tubes provide the highest resolution.

6.4.4 The Drift Region. This is a field-free region at viewing-screen potential in which the electrons continue in their convergent paths to be brought to a focus at a viewing screen without being subjected to further externally imposed electric and magnetic fields. The beam does tend to spread due to internal space-charge repulsion forces and the radial component of the thermal energy distribution. The viewing system, comprised of the facepanel and phosphor screen, lies within the drift region and is discussed in Section 6.6.

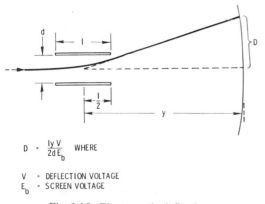

$$D = \frac{l y \, V}{2d\, E_b} \quad \text{WHERE}$$

V = DEFLECTION VOLTAGE
E_b = SCREEN VOLTAGE

Fig. 6-18. Electrostatic deflection.

Table 6-1.   Focus-Deflection Combinations

Deflection	Focus	Characteristics	Application	Vertical Resolution TV lines @ 0.5 m
Electrostatic	Electrostatic	High speed, moderate resolution	Scope tubes	450 to 500
Magnetic	Electrostatic	Moderate to high resolution	TV, computer terminal tubes widely used	700 to 900
Magnetic	Magnetic	Highest resolution	Projection recording tubes	1600 to 2000
Electrostatic	Magnetic	—	Not generally employed	

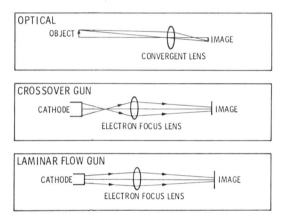

Fig. 6-19. Optical analog of electron gun.

6.5   LIMITATIONS ON ELECTRON-GUN PERFORMANCE

6.5.1 Optical Analog of the Electron Gun. An optical analog of an electron gun is shown in Fig. 6-19. The function of the electron gun and its focusing system is to provide a dense, sharp, electron-beam image on the phosphor screen. This function is similar to that of an optical system that produces an intense image of a bright source. The source serves as the object for a lens which focuses its image into a small spot. In the optical system, the image brightness is dependent upon source intensity and distribution. In this simple comparison, the spot size produced by the electron gun is also dependent upon source intensity (cathode-current density) and distribution. The electrons are charged, however, and have an energy distribution, while the photons are uncharged (neutral). The practical limits on electron-gun performance are magnification, cathode loading, and lens aberrations. The fundamental limits

on electron-gun performance include the thermal energy distribution and space-charge effects.

6.5.2 Magnification. The effect of magnification on spot size is shown in Fig. 6-20. As mentioned above, there is an analogy between electron optics and light optics. For light optics, when the object and image lie in media with different indices of refraction, the lateral magnification, M, of an object, y, is given by

$$M(\text{optical}) = \frac{y'}{y} = \frac{S'n}{Sn'}   \text{(Eq. 6-1)}$$

where S represents the object distance, S' the image distance, and n is the index of refraction of the media. It is also true for a CRT that the image and object lie in regions with different indices of refraction, since the potential is analogous to the index of refraction. For electron optics, the lateral magnification is given by

$$M(\text{electron-optical}) = \frac{y'}{y} = \frac{S'}{S}\left(\frac{V_e}{V_s}\right)^{1/2}$$

$$\text{(Eq. 6-2)}$$

where

V_e = potential in region of cathode
V_s = screen potential

The lower the value of magnification, the smaller the spot size, within limits. Equation 6-2 indicates that there are two simple ways to improve resolution by decreasing the magnification. One way is to increase the screen potential V_s. The second way is to lengthen the tube to permit increasing the object distance of the electron gun.

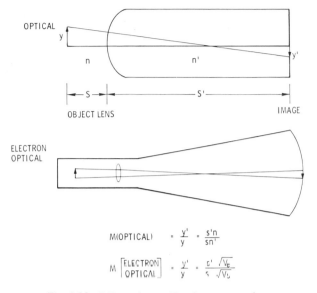

$$M(\text{OPTICAL}) = \frac{y'}{y} = \frac{s'n}{sn'}$$

$$M\begin{bmatrix}\text{ELECTRON} \\ \text{OPTICAL}\end{bmatrix} = \frac{y'}{y} = \frac{s'}{s}\frac{\sqrt{V_e}}{\sqrt{V_o}}$$

Fig. 6-20. Effect of magnification on spot size.

6.5.3 Cathode Loading.

David Langmuir[23] derived an equation for the peak current density obtainable in the electron-beam image by using a model based on an optical analogy. He derived the expression

$$J = \frac{J_0}{M^2}\left[1 - (1 - Z)\,\epsilon\,\exp - \left(\frac{Ve}{kT}\right)\left(\frac{Z}{Z - 1}\right)\right]$$

(Eq. 6-3)

where:

J = current density at focused spot (image)

J_0 = current density at cathode (cathode loading)

$Z = M^2 \sin^2 B$

M = the linear magnification

B = half angle of cone of convergence to focused spot

V = potential of screen with respect to cathode

k = Maxwell-Boltzman constant

T = absolute temperature of cathode in degrees Kelvin

e = absolute value of charge on the electron

Langmuir made the following assumptions: the electrons leave the cathode with a Maxwellian velocity distribution, the focusing system obeys the Abbe sine law, the focusing system is free from aberrations, and space charge is neglected.

It is of particular interest to determine the limiting values of J as the magnification of the electron-optical system changes. The magnification is easily varied by increasing or decreasing the length of the electron gun to respectively decrease or increase the magnification.

For M large (comparable to unity)

$$J = \frac{J_0}{M^2}$$

For M small (much less than unity)

$$J \cong J_0\left(\frac{Ve + 1}{kT}\right)\sin^2 B$$

For most CRTs, the magnification M is between 0.5 and 1. In either case, however, the current density in the focused spot T is directly proportional to J_0, the current density at the cathode. This means that to achieve high resolution at the screen—that is, high current densities in the focused spot—high cathode-emission current densities are required. One should not conclude, however, that continued increase in the emission-current density will result in higher and higher current densities in the image. As the emission-current density continues to increase, space charge will limit the value of the peak current density at the screen.

6.5.4 Lens Aberrations.

Lens aberrations of one type or another constitute a major limiting factor in the resolution of the modern CRT. Even if the electron-optical system were perfectly fabricated, spherical lens aberrations would still be present. Figure 6-21 illustrates how a lens contributes spherical aberration, which is the inability of the entire lens to focus an image at the same point. The larger the radial fraction of the lens used, the worse the error, as shown in Fig. 6-21.

6.5.5 Thermal Effects. Langmuir's equation takes into account the effect that the thermal velocity distribution has on the resolution of the electron beam at the screen. Referring to Eq. 6-3, it has already been pointed out that for values of the magnification M close to unity, which is where most CRTs operate,

$$J = \frac{J_0}{M^2}$$

Thus, for most CRTs the electron-beam energy distribution does not limit the resolution. One should not conclude from this that the temperature of the cathode is not critical to tube resolution. The cathode emission-current density J_0 is also a function of temperature increasing at higher temperatures. Thus, over some restricted temperature range it would be anticipated that for those tubes where M is close to

unity, higher temperatures will provide better resolution because of increased emission current densities.

For M small, much less than unity,

$$J = J_0 \left(\frac{Ve + 1}{kT} \right) \sin^2 B$$

This case is appropriate for higher-resolution tubes where the magnification is about 0.05 and the electron beam is highly apertured. If the cathode loading were held constant and the temperature of the cathode increased, the resolution would be decreased due to the thermal energy distribution of the electrons. In actual practice, however, the cathode loading increases superlinearly with temperature, and therefore the increase in the thermal energy distribution with temperature is more than offset by the increase in the cathode loading.

6.5.6 Space Charge. Space-charge effects limit resolution because the mutual repulsion of electrons imposes limits on their packing density. These space-charge effects act to nullify the focusing forces exerted on the electron beam.

In the crossover gun, space-charge effects are encountered in both the crossover and drift regions. In the laminar flow gun, space-charge effects are reduced near the cathode because the number of electrons crossing over is reduced, but are of concern in the drift region.

Calculations[24] indicate that at the high-screen voltages used today, space-charge effects in the drift region do not generally limit resolution which is more likely to be limited by such effects in the grid 1-grid 2 region.

6.5.7 Final Spot Size. The final spot size achieved in a CRT may be considered to consist of the unaberrated spot size increased by contributions from the various aberrations. It is not possible to directly measure the component line width associated with any particular aberration although it may be possible to calculate it. The final spot size is given by the following equation:

$$d_s = (d_m^2 + \Sigma_A \, d_a^2)^{1/2}$$

where

d_s = diameter of spot at screen
d_m = diameter of unaberrated spot

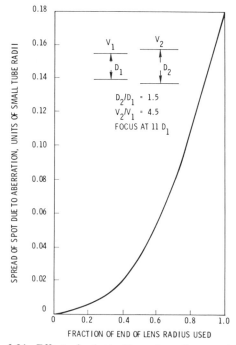

Fig. 6-21. Effect of spherical lens aberration on minimum spot size. (From Spangenberg, Karl R. *Vacuum Tubes.* New York. McGraw Hill Book Company, Inc., 1948.)

d_a = component line width contribution from each aberration

Σ_A = sum over all aberrations

6.6 THE VIEWING SYSTEM

6.6.1 Cathodoluminescence.
The phenomenon of cathodoluminescence[25] is utilized in the CRT to produce the visible image. Electrons ranging in energy from several volts to over one million electron volts produce cathodoluminescence.

CRTs utilize both the fluorescent radiation—that is, that emitted during the excitation process—and the phosphorescent radiation which is emitted after excitation has ceased. Luminescence that persists more than 10^{-8} seconds after excitation has ceased is termed phosphorescence. This is based on the experimental observation that the natural lifetime of an excited state is 10^{-8} to 10^{-9} seconds.

The efficiency of any luminescent process may be calculated by comparing the ratio of the luminous-energy output to the electron-beam-energy input. Only the useful luminous output is considered for meaningful application to CRTs. Useful in this case means only the luminous flux emitted from the front surface of the phosphor screen viewed by the observer, since radiation out the back surface would be lost to the observer. Most materials exhibit cathodoluminescent efficiencies of less than 1% and are not particularly useful from a practical standpoint. The term "phosphor" is used to identify that class of materials that exhibit superior luminous efficiencies, which usually range between 1% and 20%.

It has been experimentally determined that good phosphors are usually crystalline materials. They are exceedingly sensitive to impurities and crystal structure. These phosphors are generally characterized by a host crystal to which is added one percent or less beneficial impurities called activators. The important groups of cathodoluminescent materials are:

1. The rare-earth compounds such as gadolinium oxysulfide with terbium activation. These compounds are the most recent additions to the phosphor family and are of great practical importance. They are capable of high brightness and are frequently characterized by line emission spectra.
2. The silicate phosphors such as zinc ortho-silicate, manganese-activated. This is one of the earliest materials used, but it is still unmatched for performance in many areas. The zinc orthosilicate phosphor, manganese-activated, know as P-1, can achieve the highest luminance levels of any phosphor, for example.
3. The sulfides of zinc and cadmium activated with manganese, silver, gold, or copper. These phosphors are capable of providing superior color choice and higher efficiencies at reduced brightness levels.
4. The oxide phosphors such as self-activated zinc oxide and aluminum oxide activated with transition metals.
5. The alkali halides activated with thallium or other heavy metals.
6. Organic crystals, such as anthracene activated with naphtacene. Scintillation counters are one application of these materials.

The organic materials cannot be considered for application to CRTs because they outgas in vacuum and poison the cathode. Thus, the phosphors used in cathode-ray tubes are inorganic crystalline materials with various activators used in the host material.

Some of the most efficient phosphors are wide-band gap or insulating materials which have been doped with impurities. The unactivated materials have little or no luminescence.

As a result of various experimental observations, a theoretical explanation of catholuminescence has been developed based on the electron energy states in an instructor.

Figure 6-22a is a simplified picture of the energy levels of a single atom, while Fig. 6-22b is the same for an insulator. In Fig. 6-22a the curved lines represent the potential energy difference as a function of distance from the atom in free space. In Fig. 6-22b the horizontal axis represents distance along a crystal, the various nuclei are represented by positive signs, and the curved lines represent the potential. The idealized energy levels are shown for the single atom and the crystal insulator.

In the case of the insulator, there are two classes of levels: filled and unfilled. There can be no electron conductivity unless by some mechanism electrons are given sufficient energy to excite them from the filled band into the con-

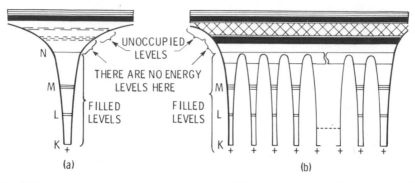

Fig. 6-22. Energy levels for (a) single atom, and (b) insulator. (From Soller et al., ref. 21)

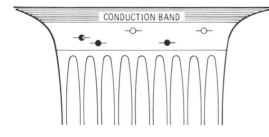

Fig. 6-23. Energy levels for an insulator with impurities. (From Soller ėt al., ref. 21)

duction band. Studies on efficient phosphors have indicated that the function of the impurities is to introduce various types of energy levels into the forbidden band. Figure 6-23 shows the possible states introduced by impurities.

Metallic impurities are thought to introduce filled electronic states just above the filled band. The second type of impurity level is due to an excess of the nonmetallic element introducing unfilled states just below the conduction band. A third type of impurity level assumes a partially filled state perhaps in the middle of the forbidden band.

Electrons in the filled band are excited into the conduction band when the crystal is bombarded with high-energy electrons. Under such electron bombardment and for a short period thereafter, vacancies will exist in the filled band and excess electrons in the conduction band. For electrons to return to the filled band, recombination must occur. This recombination is not particularly efficient unless electron and hole traps are present.

Vacancies (holes) in the filled band are initially mobile. These vacancies are attracted to the filled impurity states that are just above the filled band and are trapped at the impurity sites. Then an electron from the conduction band can refill this hole by dropping from the conduction band into the trapped hole. This loss of energy is frequently accompanied by the emission of a photon or quantum of light. If the unoccupied states lie just below the conduction band, electrons from the conduction band may be trapped in those states. At room temperature, these electrons may then be re-excited into the conduction band to recombine with holes in the filled band or in partially filled states in the forbidden band.

The impurity levels associated with the recombination light-emission process are not all equally spaced with respect to the lowest level of the conduction band, and therefore the emitted radiation will not be confined to a single wavelength but, extend over a considerable range of wavelengths depending on the distribution of the recombination centers.

The difference in energy between the highest level of the filled band and the lowest level of the conduction band normally ranges from 2.5 to 4 electron-volts. This determines the wavelength of the emitted radiation, since the product of Planck's constant and the frequency of the emitted radiation is equal to the loss in energy in going from the higher to the lower state. Thus the nature of these transitions determines the emission spectra obtained from any phosphor. Broad emission spectra peaks will prevail when the states between which the transitions occur are spread out. Confinement of these to a reduced range of values will favor narrower spectral peaks as well as line spectra.

It should not be inferred from the above that the excitation-emission process is completely understood. The above description is only one possible scenario based on studies performed until the early 1980s.

6.6.2 Phosphors. Most phosphors[25,26] used in CRTs are registered with the Electronic Industries Association. At the time of registration, information is supplied by the manufacturer about each tube and phosphor and a type number is assigned to each. For many years, the tube type number has appeared followed by the letter "P" with an identifying number which tells what phosphor it is. This phosphor identification system was revised in 1982 to be replaced by the Worldwide Phosphor Type Designation System (WTDS).[32] In place of the "P" number, one or two letters will be used. The "first letter" will normally indicate the fluorescent color which is seen during phosphor excitation. This will not be the case for long persistence phosphors in which case it will denote the phosphorescent color seen after excitation has ceased. The "second letter" is optional and pertains to other screen properties. Complete information on both older phosphors as well as the newer ones is included in the EIA publication TEP 116, "Optical Characteristics of Cathode Ray Tube Screens" which can be obtained from the Electronic Industries Association, 2001 Eye Street, Washington, D.C. 20006.

Table 6.2 is a list of commonly used phosphors. The older "P" numbers as well as the newer WTDS designations are included. Under the column "Color," "Fl" stands for fluorescence and "Ph" stands for phosphorescence.

P-4 is the most widely used black-and-white phosphor for television tubes. Although efficient at low current densities, it does tend to saturate at higher currents. Recently developed rare-earth phosphor P-45 is a better choice for higher brightness displays, although it is less efficient at low current densities.

P-1 is the best phosphor for achieving high brightness at high current densities except, perhaps, at low writing speeds. It is outstanding for use in projection CRTs. It also has long life.

P-20 and P-31 provide efficiency at low current densities but saturate rapidly. The persistence of P-31 is somewhat longer than that of P-20.

P-43 and P-44 are not as efficient as P-1 at

Table 6-2.   List of Selected Phosphors

New WTDS Designation	Old P No.	Composition	Peak wavelength, nanometers	Decay time to 10% level	Color Fl	Color Ph	Applications
GJ	P1	$Zn_2SiO_4:Mn$	525	24 ms	YG	YG	High brightness projection
GL	P2	$ZnS.Cu$	543	35 to 100 µsec	YG	YG	Scope tubes
WW	P4	$ZnS:Ag +$ $ZnS\ CdS:Ag$	440 565	25 µsec 60 µsec	W	W	Black and white
BJ	P5	$CaWO_4$	430	25 µsec	B	B	Photo recording
GM	P7	$ZnS:Ag +$ $ZnS\text{-}CdS:Ag$	440 560	40 to 60 µsec 6.4 sec	B	YG	Long persistence radar, scopes
BE	P11	$ZnS:Ag$	460	25 to 80 µsec	B	B	Photo recording
AA	P16	$Ca_2MgS_2O_7:Ce$	385	.1 µsec	UV	UV	Flying spot scanners, photo recording
KA	P20	$ZnS:CdS:Ag$	520–560	.05 to 2 msec	Y to YG	Y to YG	High efficiency, P-4 yellow component
X	P22B	$ZnS:Ag$	440	25 µsec	B	B	Color TV
X	P22G	$ZnS:CdS:Ag$	530	60 µsec	YG	YG	Color TV
X	P22R	$Y_2O_2S:Eu$	627	.9 ms	R	R	Color TV
GH	P31	$ZnS:Cu$	522	40 µsec	G	G	High efficiency, scopes
GR	P39	$Zn_2SiO_4:Mn:As$	525	150 ms	YB	YG	Long persistence, low frame rate displays
GY	P43	$Gd_2O_2S:Tb$	544	1 ms	G	G	High brightness, spectral filter displays
GX	P44	$La_2O_2S:Tb$	540–545	1 ms	G	G	High brightness, spectral filter displays
WB	P45	$Y_2O_2S:Tb$	5450	2 ms	W	W	High brightness
VA	P49	$YVO_4:Eu +$ $Zn_2SiO_4:Mn$	615 to 619 525	9 ms 24 ms	OR YG	OR YG	Penetration color, simulators
KJ	P53	$YAGaG:Tb$	543	7 ms	YG	YG	High brightness at low speeds

high current densities, but both exhibit narrow spectral peaks lending themselves to use with narrow-band spectral filters as well as holographic lenses.

P-7 is a double-layer, cascade phosphor. The double-layer refers to the fact that the screen consists of two phosphor components. Initially, the first component is deposited on the inner surface of the faceplate, and then the second component is deposited, in turn, over the first, forming a double-layer screen. When the electron beam strikes the second component, it emits a bright flash, with some of the radiant energy being absorbed by the first component. This absorbed energy causes the first component to reemit luminous radiation over an extended time period at low brightness levels. The word "cascade" refers to this excitation-absorption-reemission process. The duration and intensity of information displayed on such a cascade screen is usually superior to that which can be obtained with single-component screens. For the P-7 phosphor, the bright flash is blue and the persistent image yellow-green, which is of substantial help when the display is used for detection. The P-7 phosphor is widely used for radar and other long-persistence displays.

P-39 is a single-component long-persistence phosphor frequently used for reduced frame rate systems.

P-49 is an onion-skin, two-color penetration phosphor.

P53 is a recent addition to the list of registered phosphors. It has been reported[33] that at writing speeds of 1000 in/s P53 has twice the brightness of P1 achieving a peak line luminance of about 50,000 fL.

The letter "X", corresponding to the P-22 phosphor, denotes the set of three different color phosphors used in CRTs for consumer color television sets. P22B, P22G, and P22R represent a set of components of "X".

Phosphor luminous efficiency increases with increasing beam energy over a wide range. Depending on phosphor thickness, the luminous efficiency curve starts to saturate at about 15 kV and arrives close to the saturation value at 20 to 22 kV. Thinner films shift this curve to lower values of voltage and luminous efficiency, while thicker films tend to raise those values. Thus, when the maximum luminous efficiency of a phosphor is desired, the CRT should be op-

erated at the highest practical screen potential, certainly at 20 kV or above.

Current density also has a marked effect on the luminous efficiency of the phosphor. The current density referred to is the average current density in the focused spot. For most phosphors, the luminous efficiency is constant below 20 to 30 $\mu A/cm^2$. At current densities beyond that range, the luminous efficiency may decrease significantly.

6.7   CRT RESOLUTION AND CONTRAST

6.7.1 Defining CRT Resolution.
Two factors determine the resolution of a CRT. One is the current-density distribution in the electron beam when it strikes the phosphor. The second is the optical properties of the phosphor itself. The particle size and screen thickness of conventional powder phosphors limit the resolution capability of the screens to less than that which can be achieved with the electron gun at lower currents as a result of optical scattering in the phosphor. Phosphor resolution is improved for powder phosphors by reducing the screen thickness and particle size (using techniques such as ballmilling or cataphoretic deposition) or employing thin-film transparent phosphors, but this improvement is achieved at the expense of phosphor luminous efficiency.

Electron-beam and powder phosphor technology is such today that an .0008-in. line width can be achieved in a 1-in.-diameter CRT. Line widths ranging from .002-in. in a 5-in.-diameter CRT to .007-in. in a 25-in.-diameter CRT are possible.

Defining and measuring the resolution of a CRT is a complex task. CRT resolution, according to the Standards and Definitions Committee of the Society for Information Display, is a "measure of the ability to delineate picture detail; also, the smallest discernible and measureable detail in a visual presentation. . . . Resolution may be stated in terms of modulation transfer function, spot diameter, line width, raster lines or television lines." There is no single number by which it is possible to characterize CRT performance. No matter which term is selected, a meaningful and reproducible statement of the resolution must include both the operating parameters of the tube and the specifications of the display to avoid the ambiguities that can easily occur because of the large number of trade-offs possible in a CRT. Examples of some of the parameters that should be specified are screen potential, focus voltage, and focus electrode current for electrostatic guns, as well as grid No. 1 cutoff voltage and grid drive.

The display specifications depend upon the nature of the presentation; selection of a suitable phosphor with regard to color, persistence, and luminous efficiency is an obvious first step. For bistable displays where the "on" brightness level is fixed at a single value, the resolution can be specified for that condition. Whenever the "on" brightness level may vary or shades of gray may be shown, it may be necessary in critical applications to specify the resolution at both high and low brightness levels, since optimum focusing at one brightness level may cause defocusing at the other brightness level. Consideration must be given to establishing resolution requirements for uniformity over the display area, both horizontally and vertically. Specialized measurement techniques may be required if the display is comprised of dots produced by short pulses of the electron beam.

The resolution of a stroke-written display must be established at a value of electron-beam writing speed and repetition rate. The maximum line width, or maximum and minimum if required, can then be specified for a particular value of peak line brightness. The parameters for a raster display should include the size of the raster, the frame repetition rate, the number of fields per frame, the number of active lines per field, how the fields are interlaced, and the horizontal writing time per line. The resolution specification for raster displays is more complex than for stroke-written displays because the resolution elements or pixels are more densely packed. The acuity of the eye (that is, the ability of the eye to resolve detail) depends on other factors, such as image contrast and brightness. The higher the luminance and contrast, beyond some fixed number, the greater the resolution of the eye. This is particularly important with respect to CRTs, and especially for higher resolution raster displays, because as the luminance of a CRT increases, the contrast between adjacent resolution elements decreases. (This decrease in contrast primarily results from the fact that the profile of the electron beam is not square but Gaussianlike.)

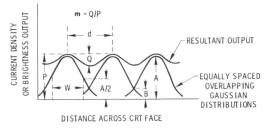

Fig. 6-24. Luminance distribution on CRT face.

The key question is: In order to achieve a specified raster luminance, resolution, and contrast, what line width is required? Assuming the current density in the scanning beam to be Gaussianlike, it is possible to calculate the resulting contrast between adjacent resolution elements as a function of the width of the line, W, at one-half its peak luminance value, and the separation of the lines, d. For simplicity, the contrast between adjacent resolution elements can also be expressed as the modulation. This is the ratio of the minimum luminance between overlapping Gaussian distributions to the maximum luminance at the peak of the Gaussian distributions that represent the raster lines.

Figure 6-24 portrays the idealized Gaussian luminance distribution produced by the electron beam on the face of the CRT. The ordinate represents the luminous output of the phosphor, the abscissa the distance across the raster lines on the screen. In reality, the current-density distribution in the beam may not be Gaussian. The halo that is present around the pixel in a CRT which employs a particulate screen will also contribute to a non-Gaussian luminous distribution. The Gaussian model is good to a first approximation, however, for analyzing resolution and contrast.

The trace lines of the CRT are represented by equally spaced overlapping Gaussian distributions. The resultant output is shown by the envelope of the curves. The amplitude of each Gaussian is A, while B is the value of each Gaussian at the crossover point with the adjacent Gaussians. The modulation, m, is equal to Q/P.

The important factor to determine is how the modulation m changes as the ratio d/W increases. Figure 6-25 is a graph which relates the percent modulation to the relative line separation.

Note that there are two curves shown in Fig. 6-25. The upper curve indicates the modulation

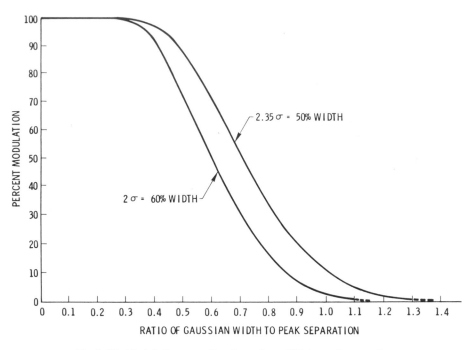

Fig. 6-25. Modulation vs. ratio of gaussian width to peak separation.

as a function of the line width measured at the 50% intensity or 2.35 sigma points. The lower curve indicates the modulation when the line width is defined at the 60% intensity or 2.00 sigma points. Two different techniques are used to obtain the line widths at these points.

Measurements of the line width at the 50% intensity points are obtained by means of a slit analyzer. With this technique a line is displayed on the CRT under the desired conditions. The slit analyzer is then used to scan the line tracing out the intensity profile of the line on a scope or x-y recorder. Measurements made by this technique not only give a complete profile of the line but also are accurate, repeatable, and objective. The width at the 50% intensity points can be measured from the calibrated trace of the beam profile on the x-y recorder.

The shrinking raster technique is frequently used for measuring the line width at the 60% intensity points. This method is used where the equipment is inadequate to perform measurements by more precise techniques. Implementation of the shrinking raster method starts with an unmodulated raster adjusted for a selected luminance level displayed on the tube. The number of raster lines is known. The vertical raster size is then progressively reduced until

the lines appear to be just merged. This occurs when the 60% intensity points on each line have intersected the same point on each of the neighboring lines. The height of the raster is measured in this merged raster condition and the line width calculated by dividing it by the number of lines. This is not considered a highly reliable measurement method because of the subjective nature of the measurements.

The use of the curves in Fig. 6-25 is straightforward. Assume that one desires to display 500 raster lines in a 10-in.-high raster with 50% modulation between the lines. The distance between the peaks is 10/500 in. or .020 in. The ratio of the 2.35 sigma width to peak separation must be 0.7 for a modulation of 50%. Therefore, the 2.35 sigma points are 0.7 × .020 in. or .014 in. apart. Thus the line width at the required luminance cannot exceed .014 in. if the modulation is to be at least 50%.

6.7.2 Contrast and Gray Scale in the CRT. Contrast describes a state of difference expressed in terms of ratios. The contrast ratio of a CRT expresses the ratio of the luminances of bright and dim regions of the display. Two types of contrast ratios[27] are associated with the CRT. One is the detail contrast ratio and the other is the range or large-area contrast ratio. The detail contrast ratio is the ratio of the luminance of excited, bright areas to that of unexcited, dim areas which are closely surrounded by the excited areas. The range (large area) contrast ratio is the ratio of the luminance of the brightest area of the display to that of the dimmest area. The first ratio is more important than the second because it is the detail contrast ratio which ultimately determines the ability to display shades of gray in a pictorial presentation. The fundamental question is: What is the maximum contrast ratio with which adjacent resolution elements can be displayed?

Halation plays a major role in limiting the detail contrast ratio. Figure 6-25 illustrates how halation occurs in a CRT. A small region of the phosphor is excited by the electron beam. Light rays which enter the glass at normal or nearly normal angles will be substantially transmitted through the facepanel. As the angle of incidence on the front surface of the facepanel is increased, the transmitted ray will be refracted away from the normal and a small percent will be reflected. When the angle of incidence is equal to or greater than the critical angle, however, the entire amount of light will be totally internally reflected and returned to the inner surface of the CRT facepanel. The internally reflected rays strike the glass-phosphor interface at some distance from the place where they were emitted. Upon striking the glass-phosphor interface, most of the rays are scattered with a Lambertian distribution due to the granular nature of the phosphor powder. Some of these rays are again totally internally reflected and then give rise to successive internal reflections. Thus, the bright center spot is usually surrounded by a series of concentric circles of diminishing brightness.

The net effect of this halation is to reduce the detail contrast ratio to a value between 10 and 15. The maximum value of the range contrast ratio is typically 50:1.

Filtering techniques are one effective way to increase the contrast ratios. Filters, which are bonded to the faceplate, increase the contrast ratios because the oblique rays which produce halation must travel through the filter before reaching the front surface/air interface that causes total internal reflection. These rays, therefore, undergo greater filtering than the normal rays produced by the image. A detail contrast ratio of 25:1 and a range contrast ratio of 100:1 can be achieved using neutral density filters of 20% transmission under low ambient conditions.

The high reflectance (80%) of the phosphor also reduces the available display contrast as the ambient illumination increases because the reflected incident illumination increases the background brightness level. The use of neutral density filters, either incorporated into the facepanel or bonded to it, can overcome this problem, but at the expense of light output. These filters increase the contrast by filtering the incident radiation twice, while the tube output is filtered only once before emerging from the display. Thus, the use of filters aids contrast by both reducing the effects of halation under all ambi-

SPREAD OF LIGHT FROM INTERNAL REFLECTIONS

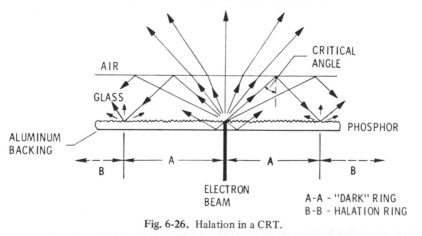

Fig. 6-26. Halation in a CRT.

ent conditions and reducing high reflectance at moderate to bright ambients.

Gray scale refers to the number of shades of gray which a CRT can display. Several factors determine the gray scale a display can achieve. These factors include the definition of the luminance change associated with each gray step, the dynamic range, the contrast ratio inherent in the display device, ambient lighting, and filtering. One step of the gray scale is frequently defined as a square root of two factor change in luminance.

The inherent gray-scale capability is limited by the detail contrast ratio. As discussed above, without filtering, a CRT is capable of a 15:1 contrast ratio. This is equivalent to about nine shades of gray. With a 20% filter, the contrast ratio increases to 25:1, which can produce about ten shades of gray. The above numbers assume a zero ambient condition. For a 10-fc ambient, assuming a 100-fL maximum highlight luminance, the contrast ratio is reduced to about 13:1 or eight shades of gray.

In the past several years, CRTs have been applied to airborne high-brightness environments[28] which would have previously washed out the contrast in the display. These applications in which several shades of gray are visible in a 10,000 fL ambient have been made possible by significant advances in CRT and related technology. These advances include the use of narrow-band-emitting P-43 rare-earth phosphors in conjunction with spectral filtering techniques.

6.8 THE LIFE OF THE CRT

The mechanisms of CRT failure are well understood and quantified. The performance of tubes generally degrades with operating time, or more specifically, in proportion to the cumulative beam current deposited on the phosphor. The tubes are, therefore, said to undergo an aging process. A progressively dimmer picture is a manifestation of the aging process. The rapidity of the aging process depends upon how the tube is operated. Tubes operated at low luminance levels are going to age much more slowly than those operated at high levels. Given a particular set of operating parameters—that is, specific

luminance levels and operating time at each level—it is possible to predict the life of the CRT.

One definition of CRT life is the operating time required for the luminance of the tube at a specific resolution to be reduced to 50% of its initial value. Part of the degradation in tube luminance with time can be compensated for by increasing the input signal level, but ultimately that leads to a loss of resolution. Of course, the useful operating life of the tube can be substantially greater than the time required to degrade to 50% luminance if the tube is used in a reduced ambient where a dim picture can be seen.

Sources of catastrophic failure include loss of vacuum and inter-element shorts. These are very rare today because of the maturity of the CRT manufacturing process.

The two major failures which contribute to the aging process in CRTs are loss of cathode emission and a reduction in phosphor efficiency. These are described below.

6.8.1 Cathode Life. The life of the cathode is markedly affected by cathode loading—that is, the current drawn per unit area from the cathode at maximum grid drive. The designer of the cathode-ray tube, by adjustment of the electron-gun geometry, selects the minimum value of cathode loading based on the resolution requirements of the CRT. Langmuir's equation shows that the current density in the image is directly proportional to the emission current density drawn from the cathode. Thus, the current density in the focused spot is directly proportional to the cathode loading, within limits. High-resolution CRTs require high values of cathode loading. The current densities available from oxide cathodes have been measured,[16] as shown in Fig. 6-27, and the maximum values found range from 0.1 A/cm^2 at 700°K to 100 A/cm^2 at 1200°K. Practical values of oxide cathode current densities for application to CRTs are limited by two factors: At temperatures below 1000°K, the cathodes are "poisoned" (degraded) in several hundred hours. At temperatures above 1150°K, the cathode material may sublime in several hundred hours. Thus, CRT oxide cathodes are operated between approximately 1000°K and 1130°K.

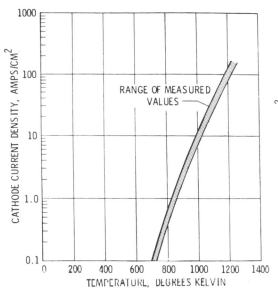

Fig. 6-27. Current densities from oxide cathodes.

DISPENSER CATHODE CURRENT DENSITIES

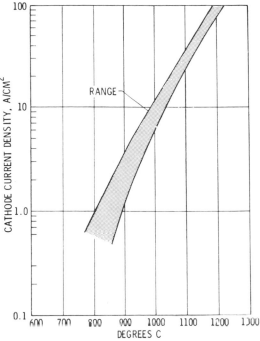

Fig. 6-28. Current densities from dispenser cathodes.

The maximum average emission current density which can be drawn from an oxide cathode is therefore limited by the anticipated life of the cathode. The higher the current density, the faster the poisoning of the cathode. In practice, 0.3 A/cm² is considered the optimum maximum average loading; it provides CRTs with about a 10,000-hour life.

Emission current densities from dispenser cathodes[17] have been measured and found to be similar to the values obtained with the oxide cathode, as shown in Fig. 6-28. The optimum operating temperature for the dispenser cathode is about 100°C higher than for the oxide cathode. At that temperature it is possible to draw in excess of 1 A/cm² without degrading the cathode.

The life of the dispenser cathode[17] is limited by the evaporation of the barium compound from the pores of the tungsten. Depending upon the particular dispenser cathode, lifetimes substantially in excess of 20,000 hours appear practical.

The necessity for the use of crossover guns in CRTs arose from the limitations on beam current density (resolution) imposed by the available emission current density from the cathode. Formation of the crossover increases the beam current density (resolution) which now becomes limited by space-charge repulsion in the cross-

over rather by the emission current density from the cathode.

6.8.2 Phosphor Life.

The luminance of a phosphor decreases under continued electron bombardment. Figure 6-29 indicates the relationship observed between the luminance of the phosphor and the time it has been exposed to a constant bombarding current density. The phosphor luminance (luminous efficiency) is found to decrease with the number of coulombs deposited.

The relationship between I, the aged intensity of the phosphor, I_0, the initial phosphor intensity, and N, the number of electrons deposited per cm², is given by Pfahnl's Law[29]:

$$I = \frac{I_0}{(1 + CN)}$$

In the above equation, C is the burn parameter (rate of degradation) characteristic of each phosphor. Its reciprocal is the number of electrons per square centimeter needed to reduce the phosphor luminance to one-half its initial value. Note that the smaller the value of C, the

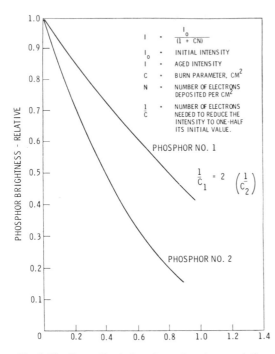

$$I = \frac{I_0}{(1 + CN)}$$

I_0 = INITIAL INTENSITY

I = AGED INTENSITY

C = BURN PARAMETER, CM^2

N = NUMBER OF ELECTRONS DEPOSITED PER CM^2

$\frac{1}{C}$ = NUMBER OF ELECTRONS NEEDED TO REDUCE THE INTENSITY TO ONE-HALF ITS INITIAL VALUE.

PHOSPHOR NO. 1

$$\frac{1}{C_1} = 2 \left(\frac{1}{C_2} \right)$$

PHOSPHOR NO. 2

Fig. 6-29. Generalized phosphor aging characteristics.

Table 6-3. Aging Constants and Luminous Efficiencies of Selected Phosphors

Phosphor	Efficiency (lumens/watt)	Life to 50% of initial brightness (C/CM^2)
P1	30	> 100
P4	30(color depend.)	25
P7	20	25
P20	40–70	25
P22(G)	40–65	25
P31	35–45	25
P39	20	> 25
P40	20	25
P43	40	> 50
P44	35	> 50
P45	20	> 50

larger the number of electrons required to reduce the efficiency of the phosphor by a given percentage.

Figure 6-30 indicates the aging curves for several phosphors. It can be seen from these

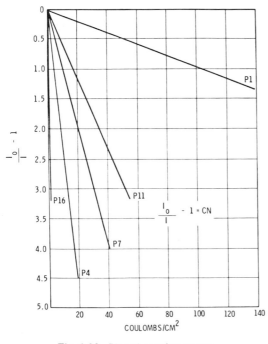

$$\frac{I_0}{I} - 1 = CN$$

Fig. 6-30. Phosphor aging curves.

curves that P-1 phosphor has the greatest resistance to phosphor aging. Table 6-3 is a table of luminous efficiency and aging constants of several phosphors.[30] Note that efficiencies are given for lower values of the current density. It should not be concluded that at high current density levels the values of luminous efficiency are directly related to the values at the lower current densities, since the saturation of light output with increasing current density varies from phosphor to phosphor. The values of luminous efficiency given for P-31 are lower than the 70 lumens/watt reported elsewhere.

Table 6-4 presents an indication of the effect of the tube size on phosphor life. Three widely used tube sizes have been selected. The 21-in. size is typical of many computer terminal and television displays. The 5-in. X 5-in.-square format is frequently used in military airborne high-brightness multipurpose displays. The 9-in. diagonal tube represents an intermediate size. The display area of each tube is given in both square inches and square centimeters.

The operating time to 50% intensity, assuming a screen current of 200 microamps and a 100% duty cycle raster, has been calculated for both the P-1 and P-4 phosphors. The luminance area factor indicates the ratio of the various display sizes, and therefore the ratio of the various display luminances, given a constant screen current.

Note that the P-1 phosphor is capable of 18,000 hours operation before the luminance is reduced by a factor of two for the 21-in. display.

Table 6-4.   Effect of Tube Size on Phosphor Life

Tube size	Useful screen size (in.)	Area in.^{2}	cm^2	Time to 50% intensity @ 200 μA current (hours) P1	P4	Luminance area factor
21" diagonal	12 × 16	192	1,239	18,000	4,500	1
9" diagonal	5.4 × 7.2	39	252	3,656	914	4.9
5" × 5" square	4 × 4	16	103	1,496	374	12.02

For the 5-in. × 5-in. display, the time is reduced to about 1,500 hours. It should be noted, however, that the small display operates at 12 times the luminance of the larger display. Similar numbers are presented for the P-4 phosphor.

6.9   APPLICATIONS AND TYPES OF CRTS

6.9.1 CRT Applications. The major markets for CRTs can be divided into four broad areas: consumer, industrial, government, and medical. Government applications can be further divided into military and Federal Aeronautics Administration (FAA) or air-traffic-control displays. The basis for differentiation is that each of these markets generally requires specialized CRTs designed for a specific application. This is not to say that the same CRT cannot be used in more than one of the above markets.

The following is not a complete listing of CRT applications, but rather an attempt to provide the reader with an indication of the large number and types of CRT applications. The development of the computer with its requirement for high-speed readout has significantly expanded the number of CRT displays and continues to do so. The list may be redundant in that in some cases both a class of displays may be shown as well as a specific example.

Consumer Applications of CRTs:

Television, black-and-white
Television, color
Cable TV
Video games
Home computer terminal
Video security systems
Video tape players

Large-screen TV projection display; black-and-white, color
Picturephone

Industrial Applications of CRTs:

Airline passenger reservation monitors
Airline flight status monitors
Message consoles
Graphic consoles
Interactive consoles
Word-processing monitors
Text-editing monitors
Computer output microfilm or microfiche
Character generation
Video surveillance for crime prevention
Video surveillance for traffic control
Electronic typesetting
Stock quotations
Computer-aided design
Weather radar monitor
Closed-circuit television; black-and-white, color
Television; black-and-white, color
Projection television, color
Teleconferencing
Motion picture production
Hairstyling displays
Airport X-ray baggage inspection monitor
Flying spot scanners
Picturephone
Utility power distribution system status monitor
Flight simulators

Government Applications of CRTs:

Military, Airborne:

Vertical-situation display (computer-generated)
Horizontal-situation display

Sensor displays
  Radar (ground map, air-to-air, moving-target indicator)
  Electro-optical (missile TV, low-light-level TV, FLIR)
  Line scanners (IR, side-looking radar, moving-target indicator side-looking radar)
Computer-generated information display
  Flight control
  Fire control
  Instrument information
Multipurpose displays
High-brightness displays
Projection displays
Helmet-mounted displays
Antisubmarine warfare displays
Electronic countermeasures displays
  Activity displays
  Direction-finding displays

Military, Ship- and Ground-Based:

Radar displays (2D and 3D)
FLIR displays
Command and control displays
Projection
Electronic countermeasures displays

Federal Aeronautics Administration:

Airport surveillance radar
High-brightness tower radar displays

Medical Applications of CRTs:

Information systems
Photorecording for radioisotope scans
Photorecording for brain scanners
Heartbeat monitors

6.9.2 Types of CRTs. Two broad classes of CRTs have evolved to satisfy the applications listed above. Entertainment or television tubes have been developed to satisfy the consumer market for black-and-white and color TV. These tubes are characterized by moderate performance and low cost. Low cost is possible because the market for these tubes is so vast that it has justified the capital investment needed to secure the economies of mass production. The second class of tubes are called special-purpose CRTs. These tubes enjoy a limited marketplace

and are frequently custom-designed to fit a specific requirement. Special-purpose CRTs usually exhibit high or unusual performance and frequently cost five to ten times as much as entertainment tubes.

The reader is referred to the *Proceedings of the Society for Information Display 24*, No. 4: (Fourth Quarter, 1983) for a description of the standardization and identification of various types of CRTs. Information on Industrial CRT test methods and bulb criteria are also included in that issue. Such information is also directly available from the Electronic Industries Association whose address is given in Section 6.6.2.

The summary below of the various classes of tubes indicates their general characteristics and applications.

1. *Classification: Entertainment or Television Tubes.*
   Characteristics: Available in color and black-and-white. Low cost. Moderate luminance and resolution. Minimal overall length. Deflection angle: 90 to 114 degrees. Size: 5 in. to 25 in. Magnetic deflection and electrostatic focus. Monochrome: Low-voltage focus. Color: High-voltage focus.
   Applications: Consumer black-and-white and color television. Computer terminals.
2. *Classification: High-Resolution Information-Display Tubes.*
   Characteristics: Employ a variety of phosphor screens. High resolution at moderate brightness. Extended length. Deflection angle: 70, 90, and 114 degrees. Size: 3 in. to 25 in. Magnetic deflection and electrostatic focus. High-voltage focus (sometimes using unipotential electron guns).
   Applications: High-resolution television monitors. High-resolution computer terminals. Air-traffic-control displays. Moderate-brightness airborne military displays. Moderate-brightness military ship- and ground-based displays. CAD/CAM and motion picture displays.
3. *Classification: X-Y CRTs.*
   Characteristics: Monochrome screen. Moderate resolution at moderate brightness.

High-speed random-scan capability. Long. Narrow deflection angle. Size: 3 in. to 16 in. Electrostatic focus and electrostatic deflection. High-voltage focus or apertured low-voltage focus.

Applications: Instruments using X-Y display formats. Oscilloscopes. Electronic countermeasures displays: spectrum analyzers and direction-finding. High-speed transients. Medical waveform monitors. Auto ignition analyzers.

4. *Classification: Photorecording CRTs.*
Characteristics: Monochrome screen. High-resolution requirements imposed on gun and phosphor screen. Geometric accuracy. Long. Narrow deflection angle. Size: 3 in. to 10 in. Magnetic focus and magnetic deflection.

Applications: Photorecording processes in which a lens is used to image the display on a photosensitive medium. Computer output microfilm. Medical electronics: nuclear and X-ray scans. Graphic arts.

5. *Classification: Flying Spot Scanners.*
Characteristics: Very short persistence phosphor. Monochrome screens. High resolution. Long. Narrow deflection angle. Size: 3 in. to 5 in.

Applications: Flying spot scanner equipment. Microfilm or microfiche reading. Conversion of color film to video display.

6. *Classification: Projection CRTs.*
Characteristics: High brightness and resolution. Various monochrome phosphors. Long. Size: 3 in. to 7 in. Narrow deflection angles. Magnetic deflection. Magnetic or high-voltage electrostatic focus.

Applications: Commercial monochrome and color large-screen projection TV. Military monochrome and color command and control large-screen projection displays. Flight simulation displays. Military airborne head-up displays.

7. *Classification: High-Brightness CRTs.*
Characteristics: High brightness and resolution. Usually P-43 phosphor or P-1 phosphor. Long. Size: 5 in. to 9 in. Deflection angle: 45 to 90 degrees. Magnetic deflection and high-voltage electrostatic focus.

Applications: Airborne multipurpose displays. Air-traffic-control bright tower displays.

8. *Classification: Long-Persistence CRTs.*
Characteristics: Long-persistence phosphors P-7, P-38, P-39. Moderate resolution and low stored brightness. Minimal length increasing with resolution. Size: 3 in. to 25 in. Deflection angle: various. Magnetic or electrostatic deflection. Magnetic or electrostatic focus.

Applications: Ground, naval, and airborne radar systems used in reduced ambients. Sonar, Foward-Looking IR, and other long-frame-time sensor systems. Slow-scan TV systems. Medical systems. Obsoleted in many applications by the use of solid-state memories.

9. *Classification: High-Resolution Color CRTs.*
Characteristics:
(a) High-resolution mask tubes (large screen): High-resolution for color. Three primary colors. Moderate to low brightness. Size: 14 in. and 19 in. Minimal length. Deflection angle: 90 degrees. Eighty to 100 resolution elements per inch.

(b) High-resolution mask tubes (small screen): Three primary colors. Resolution: Eighty to 100 elements per inch. Size: 7 in., 8 in. Deflection angle: about 50 degrees. Moderate to low brightness.

(c) High resolution color penetration tubes: Moderate resolution and low brightness. Two primary colors. Minimal length. Deflection angle: various.

Applications: High-resolution mask tubes (large screen): Any high-resolution multicolor display where low to moderate brightness is acceptable. Suitable for a broad class of applications including high-resolution television and high-resolution computer displays.

(b) High-resolution mask tubes (small screen): Suitable for use as a small multicolor display with TV-type resolution at low to moderate brightness levels. When used with appropriate filter can be used in airborne applications at high ambients for bistable (on-off) displays.

(c) High-resolution color penetration tubes: Limited to two-color systems where

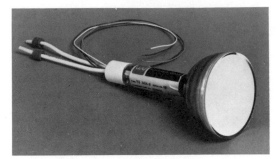

Fig. 6-31. Photograph of 3-in. round laminar flow gun CRT.

higher resolution is required than that provided by mask tubes. Slower speed makes the tube more suitable for use with stroke writers than raster generators. Flight simulators are one application. Recent advances in the resolution of mask tubes may have obsoleted this tube.

6.9.3 Examples of High-Performance CRTs.

Three-Inch Round CRT. This tube, as shown in Fig. 6-31, employs a laminar flow gun with a dispenser cathode. It is employed as a projection CRT in airborne military heads-up displays. This CRT is capable of achieving a peak line luminance of 20,000 fL at a writing speed of 5,000 in./sec and a 60-hertz refresh rate utilizing P-1 phosphor. The screen potential is 18 kV and the spot size .008 in.

Four-Inch Square CRT. Figure 6-32 is a picture of a tube which uses a laminar flow gun with dispenser cathode. The application is a multisensor display in an airborne high-bright-

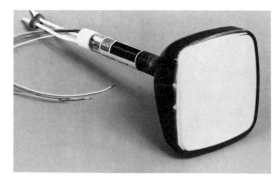

Fig. 6-32. Photograph of 4-in. square laminar flow gun CRT.

ness environment. The tube is capable of a maximum TV raster luminance of 7,000 to 10,000 fL with P-43 phosphor. The screen potential is 20 kV and the spot size about .006 in. with dynamic focus. This CRT is normally used with a matched P-43 spectral filter to enhance contrast in a high-brightness ambient.

6.10 DRIVING THE CRT

6.10.1 CRT Circuits. A typical circuit for driving an electrostatically focused and magnetically deflected CRT is shown in Fig. 6-33.

The input is provided in the form of video information and timing signals. In the case of composite video, all the information is combined in one channel. The supplies required to produce a display on the CRT may be divided into four categories: High-voltage, low-voltage, deflection generation and amplification, and video and blanking supplies. This does not include the power to these supplies. The high-voltage supply provides the voltage to the viewing screen. Note that it is shown in series with a protective resistor. Sometimes this resistor is incorporated in the supply itself. The purpose of the resistor is to prevent damage to the tube and its associated circuits in the event of a high-voltage arc. The capacitance of the viewing screen may be a few hundred picofarads, and in the event of an arc the resistor limits the current which can be drawn from the supply and therefore protects both the tube and the supply. The regulation of this supply is dependent on display resolution requirements since fluctuations in the screen voltage decrease the resolution of the tube.

The focus supply may also be a high-voltage supply if a bipotential or a high-voltage unipotential gun is employed. Sometimes this supply is provided by a bleeder off the high-voltage supply. The danger in using a bleeder supply is that if the focus electrode draws excessive current it may decrease the value of the bleeder focus voltage. In high-resolution applications, a dynamic focus correction is provided to correct for the defocusing of the beam as it moves away from the center of the tube. The dynamic focus voltage usually has a parabolic wave form. The G2 supply provides a regulated low voltage between 300 and 1000 volts. The G2 does not draw significant current.

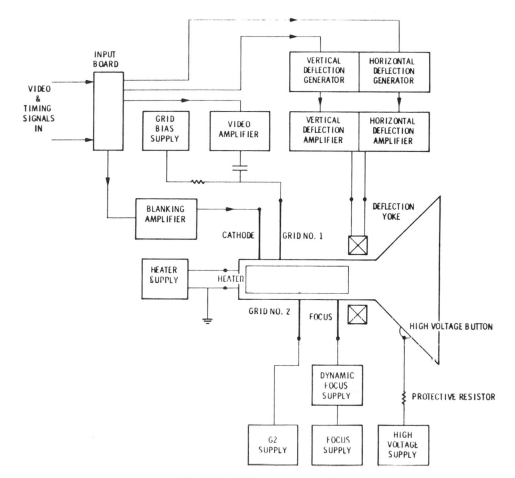

Fig. 6-33. CRT circuit schematic.

In some tubes using unipotential guns, the tube is designed so that both the focus and G2 voltages are the same. In such a case, the G2 supply may supply voltage to both the G2 and the focus electrode, in which case the separate focus supply would be eliminated. The heater supply, a low-voltage supply, is frequently a floating supply which can have one side grounded or be operated near cathode potential. This is very important because the heater-cathode insulation cannot stand more than about 200 volts. If the heater is permitted to float or be biased so that the heater-cathode potential exceeds 200 volts, the insulation may break down and short the cathode to the heater, in some cases making the tube inoperative. The grid bias supply provides a negative bias beyond cutoff through a decoupling resistor to the grid. The supply is usually adjustable between -40 and -60 volts.

The video signals are fed to the video amplifier, and the amplified signals are then coupled into grid No. 1. The horizontal and vertical timing signals are fed to vertical and horizontal deflection generators whose outputs are magnified by their respective amplifiers. The horizontal and vertical windings in the yoke are driven by the output of the deflection amplifiers.

Note that the video signal is applied to the grid No. 1, and the blanking signals are applied to the cathode. Sometimes the video is fed to the cathode and the blanking signals are applied to the grid. The preferred mode is to apply the video signals to the grid, since applying them to the cathode will affect the deflection sensitivity of the yoke and consequently degrade the resolution.

6.10.2 CRT Grid Drive Characteristics. A knowledge of grid drive characteristics[12] is of

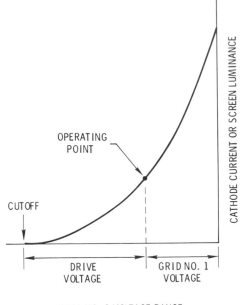

Fig. 6-34. CRT grid drive characteristic.

the screen. The equation for the cathode current as a function of grid drive is

$$I_c = KV_d 3.5 \, V_c^{-2.0} \, \mu A$$

where:

I_c = cathode current
K = the modulation constant
V_d = drive voltage which is the positive operating voltage of grid 1 relative to its cutoff value
V_c = cutoff voltage

In reality the exponents of V_d and V_c vary around the values given above. Figure 6-35 is a plot of the cathode current versus grid drive for several values of grid cutoff voltage. Note that the uppermost curve indicates the value of cathode current when the drive equals the cutoff voltage, or when the cathode and grid are at the same potential.

From observation of Fig. 6-35 it can be seen that for any given application the greater the required cathode current, the more negative the cutoff voltage of the grid 1 must be. For a specific tube, the grid 1 cutoff voltage can normally be controlled by varying the grid 2 voltage; the

vital importance in designing the circuit to operate any CRT. Figure 6-34 is a typical grid drive characteristic. Such a curve indicates a relationship between the cutoff voltage, the drive voltage, and the current or luminance of

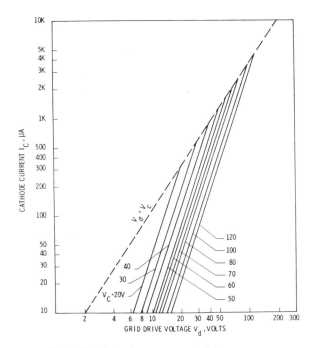

Fig. 6-35. Cathode current modulation curves.

more positive the G2 voltage, the more negative the required grid 1 cutoff voltage.

The value of modulation constant, 3.5, assumed in Fig. 6-35, is higher than is normally encountered with the crossover gun using an oxide cathode. More realistic values of K probably range from 60% to 70% of the 3.5 used to calculate the curves in Fig. 6-35. Values of K for the laminar flow gun with dispenser cathode probably range from 100% to as much as 150% of the 3.5 value assumed for K.

6.11 OVERVIEW OF CRT PERFORMANCE

How does one quantify the challenge of the CRT? It would be ideal if one single number, a figure of merit, could be arrived at which completely characterized the performance of the CRT. Such a number could then be used to compare the performance of the CRT with the figures of merit of other display devices. Unfortunately, this is not the case. The CRT, although of simple basic design, is available in numerous versions and is capable of so many design tradeoffs that no one single number can completely characterize any one CRT.

The data presented in Table 6-5 is an attempt to quantify the challenge of the CRT in terms of a number of key performance parameters. The table lists the values of a number of key performance parameters available through CRT technology. No one tube can incorporate all the maximum values presented. It is possible, however, to choose a few parameters and design and build a CRT that would meet those selected requirements at the expense of some others. Innumerable CRT designs are possible by trading off one or more parameters against other sets. A brief discussion of each item in the table is presented below.

Resolution. The common monochrome television or entertainment tube is capable of reproducing the required 500 by 500 pixels for commercial television displays. Refinements in the phosphor and in the electron gun make possible displays with 5,000 by 5,000 pixels or

Table 6-5 Summary of CRT Characteristics

Parameter	Performance	Units
1. Resolution		
Monochrome	500 by 500 to 5,000 by 5,000	Pixels
Color	500 by 500 to 1,000 by 1,000	Pixels
2. Luminance		
Monochrome	100 to 10,000	fL
Color	50 to 100	fL
3. Contrast ratio, detail	10 to 25	
4. Contrast ratio, range	70 to 100	
5. Shades of gray (0 ambient)	10 to 12	
6. Shades of gray (10^4 fL ambient)	6 to 8	
7. Phosphor luminous efficiency	20 to 60	Lumens/watt
8. Tube luminous efficiency	16 to 45	Lumens/watt
9. System luminous efficiency	6 to 15	Lumens/watt
10. Depth/diagonal ratio	1:1 to 4:1	
11. Size (direct view)		
Monochrome	1 to 25	Inches
Color	12 to 25	Inches
12. Modulation	0.1 (at maximum resolution)	
13. Storage	Not controllable	
14. Duty cycle	30	Frames/sec
15. Spot shape	Gaussian	
16. Life	1,000 to 10,000	Hours
17. Cost per resolution element	.04	Cents
18. Bandwidth		
Video	0 to 500	Megahertz
Deflection	0 to 10	Megahertz

more. Such high-resolution displays may require the use of long-persistence phosphors and reduced frame rates to conserve the bandwidth of the video and deflection systems.

Color tubes of both the mask and penetration type can achieve the 500-by-500-element resolution needed for commercial color TV. Recent advances in color TV tube fabrication, which include increased hole density in the mask and the improved performance of the electron gun, have made possible tubes with 1,000-by-1000-pixel resolution.

Luminance. Monochrome tubes for use in home TV and computer terminals can easily provide 100 or more fL of brightness. Special tubes built for use in military aircraft are capable of achieving 10,000 fL or more. Projection CRTs are capable of 20,000 fL or more.

Color TV tubes for use in home television receivers can achieve a brightness of 100 fL. Higher-resolution tubes for specialized applications exhibit reduced brightness because of a reduction in the transmission of the aperture mask. The brightness in such a case may be as low as 10 to 20 fL.

Contrast Ratio, Detail. Halation limits the detail contrast ratio to about 10 in a CRT with a clear faceplate in low ambients. Through the use of light-absorbing filters, and at the expense of reduced brightness, it is possible to increase the detail contrast ratio to 25:1.

Contrast Ratio, Range. The range of large-area contrast ratio is usually limited by the ambient light rather than by the CRT itself. Values of 70 to 100 are possible in zero ambient.

Shades of Gray (0 Ambient). The number of shades of gray is predicated on the detail contrast range of the tube. In low ambients, 10 to 12 shades of gray are possible. In this case, each shade of gray is based on a $2^{1/2}$ change in brightness. Thus 10 shades of gray (the background is the first shade) requires a $(2^{1/2})^9$ or 22.4:1 brightness or contrast ratio.

Shades of Gray (High Ambients). High ambients, in the order of 10^4 fL, are frequently encountered in airborne displays. In such a situation, a normal CRT display is washed out due to the reflectance of the phosphor. Through the use of spectral filtering techniques in combination with a high-brightness P-43 phosphor CRT, it is possible to see 6 to 8 shades of gray in such an environment.

Phosphor Luminous Efficiency. The luminous efficiencies of several phosphors were shown in Table 6.3. While efficiency varies with phosphors and operating conditions, efficiencies as high as 70 lumens/watt are realizable in practice. These are practical values of luminous efficiency since the only luminous flux taken into account is that which radiates from the front surface of the screen and can be seen by an observer.

Tube Luminous Efficiency. When calculating the overall luminous efficiency of the tube, energy losses in the filament and in the electron beam must be considered in addition to the conversion efficiency of the phosphor. A typical filament requires about 2 to 4 watts of power none of which is converted into useful light. In high-resolution tubes, anywhere from 10% to 70% of the beam current may be intercepted by the electrodes of the electron gun.

System Luminous Efficiency. This value of luminous efficiency takes into account all the power to drive the display. This includes not only the high- and low-voltage supplies necessary to drive the tube, but the video and deflection channels as well. The high value of 15 lumens/watt is for monochrome displays, while the low value of 6 lumens/watt is for color TV displays.

Depth/Diagonal Ratio. Television or entertainment tubes usually have the widest deflection angle and least depth; their ratio of depth to diagonal is about unity. As tube performance improves, the ratio increases so that for some 5 in. and 7 in. tubes the ratio is almost 4:1.

Size. Monochrome: Standard entertainment tube types are available in sizes ranging from 5 in. to 25 in. with deflection angles of 70, 90, 110, and 114 degrees. Custom bulb types have been made as small as 1 in. and as large as 36 in. Bulbs are available with flat and curved facepanels. *Color:* Normal color TV bulbs range in size from about 10 to 25 in. Some 5-in. color CRTs have also been fabricated.

Modulation. The gaussianlike shape of the electron beam plus the halation of the viewing screen severely limit the modulation and therefore the detail contrast ratio attainable with a CRT. If the separation between active lines is equal to the line width (measured at the 50% intensity points), then the maximum achievable modulation is about 10% in an unexcited resolution element surrounded by two fully excited ones. This limits the detail contrast ratio to about 1.1, or not even one shade of gray with no ambient lighting.

Storage. Cathode-ray tubes depend upon the phosphorescent properties of the phosphor to achieve storage. The brightness of such phosphorescent emission is very low, perhaps 1/1000 of the fluorescent emission obtained during excitation. Therefore such storage can only be observed in reduced ambients. In addition, the persistence is a characteristic of the phosphor and cannot be controlled.

Duty Cycle. The image on the phosphor must be continuously regenerated because the phosphor cannot controllably store information at high brightness levels. Studies have shown that to minimize flicker the CRT display must be refreshed at a frequency of about 60 Hz depending upon the phosphor. A 30-Hz frame rate is satisfactory where there is a 2:1 interlace with a 60-Hz field rate.

Spot Shape. The nature of the electron optics and the halation at the facepanel is such as to produce a gaussianlike intensity distribution in the luminous spot at the viewing screen. This gaussianlike distribution imposes fundamental limits on the resolution and contrast that can be obtained with a CRT.

Life. High-brightness tubes for airborne applications can achieve 1,000 hours life using oxide cathodes. The use of dispenser cathodes can increase the life to 5,000 or more hours as limited by the phosphor. Entertainment monochrome tubes using oxide cathodes are capable of running in excess of 10,000 hours. Color TV tubes can run 8,000 to 10,000 hours using oxide cathodes. Further improvements in the life of color TV tubes which could be obtained with dispenser cathodes normally do not justify the substantial increase in CRT cost at the present time when compared to tube replacement cost.

Cost Per Resolution Element. Small monochrome television sets and computer terminal display monitors are available for under $100. This equates to a cost of .04 cents per resolution element. The price for color is about four times that number.

Bandwidth. The video bandwidth of conventional CRT gun structures is limited to an approximate value of 50 MHz by the input capacitance of the gun. This value can be increased by a factor of 10 through the use of special designs such as coaxial inputs.

The maximum deflection frequency of conventional magnetic tubes is limited by the yoke inductance to 100 kHz for raster scanning and 1 to 3 MHz for stroke writing. The use of conventional electrostatic deflection structures extends this value to about 10 MHz as limited by the transit time of the electron beam through the deflection plates.

REFERENCES

1. Hittorf, J. W. *Veber die Electricitatsleitung der Gase*, Annalen der Physik und Chemie', Vol. 136, 1869. Translation from W. F. Magie, *A Source Book in Physics*, New York and London: McGraw-Hill Book Co., 1935.
2. Crookes W. *On the Illumination of Lines of Electrical Pressure, and the Trajetory of Molecules*, Philosophical Translations, Part I, 1879. Also, W. F. Magie, *ibid.*
3. Goldstein, E. *Ueber eine noch nicht untersuchte Strahlungsform an der Kathode Inducirter Entlandungen*, Sitzungsberiche der Koniglichen Akademie der Wissenschaftern zu Berline, July 29, 1886. Translation from W. F. Magie, *ibid.*
4. Perrin, J. "Nouvelles Propriétés des rays Cathodiques," *Comptes Rendus*, Vol. 121: 1130 (1895). Translation from W. F. Magie, *ibid.*
5. Thomson, J. J. "Cathode-Rays," *Philosophical Magazines*, Vol 44, Series 5: 293 (1897). Also, W. F. Magie, *ibid.*
6. Shiers, G. "Ferdinand Braun and the Cathode-Ray Tube," *Scientific American*, Vol. 230, No. 3: 92–101 (March 1974).
7. Herold, E. W. "History and Development of the

Color Picture Tube. *Proceedings SID 15*, No. 4: 141-9 (Fourth Quarter, 1974).

8. Schwartz, J. W. "Beam Index Tube Technology." *Proceedings SID 20*, No. 2: 45-53 (Second Quarter, 1979).

9. Galves, J. P. "Multicolor and Multipersistence Penetration Screens." *Proceedings SID 20*, No. 2: 95-103 (Second Quarter, 1979).

10. Sisneros, Faeth, David, and Hillborn. "Current-Sensitive, Single-Gun Color CRT." *Information Display 7*, No. 4: 32-7 (April, 1970).

11. Moss, H. "The Electron Gun of the Cathode-Ray Tube, Part I." *Journal of the British Institution of Radio Engineers 5:* 10-22 (January–February 1945).

12. Moss, H. "The Electron Gun of the Cathode-Ray Tube, Part II." *Journal of the British Institution of Radio Engineers* 6: 99-128 (June 1946).

13. Moss, H. "Engineering Methods in the Design of the Cathode-Ray Tube." *Journal of the British Institution of Radio Engineers* 5: 204-223 (October–December 1945).

14. Silzars, A., and Bates, D. J. U.S. Patent 3,740,607, Laminar Flow Electron Gun and Method (June 19, 1973).

15. Lehrer, N. H. Application of the Laminar Flow Gun to the Cathode Ray Tube. *SID Journal:* 7-11 (March-April, 1974).

16. Wagener, Phillip S. *The Oxide Coated Cathode*, Vol. I and II. London: Chapman and Hall Ltd., 1951.

17. Zalm, P., and van Stratum, A. J. A. "Osmium Dispenser Cathodes," *Phillips Technical Review 27*, No. 3/4: 69-75 (1966).

18. Parr, G., and Davie, O. H. (eds). *The Cathode Ray Tube and Its Applications*. 23-6. London: Chapman and Hall Ltd., 1959.

19. Maloff, I. G., and Epstein, E. W. (eds). *Electron Optics in Television*. 100-123, 163-5. New York and London: McGraw-Hill Book Co., 1938.

20. Cosslett, V. E. *Introduction of Electron Optics*. 68-71, 128-9, 142-162. Oxford: Clarendon Press, 1946.

21. Soller, Theodore, Starr, Merle A., and Valley, George E., Jr. (eds). *Cathode Ray Tube Displays*. Vol. 22 Radiation Lab Series, Lexington, Mass.: Boston Technical Publishers, Inc., 1964.

22. Millman, J., and Seely, S. *Electronics*. 63-86. New York and London: McGraw-Hill Book Co., 1941.

23. Langmuir, D. B. "Theoretical Limitations of Cathode Ray Tubes." *Proceedings I.R.E.* 25: 977 (August 1937).

24. Moss, H. "On the Limit Theory of Circular Electron Beams." *Proceedings of the Fourth Symposium on Electron Beam Technology*, Sponsored by Alloyd Electronics Corp., Cambridge, Mass. (March 1962).

25. Leverenz, Humbolt W. *An Introduction to the Luminescence of Solids*. New York: John Wiley & Sons, Inc., 1950.

26. Larach, S., and Hardy, A. E. (eds). "Some Aspects of Cathodoluminescent Phosphors and Screens." *Proceedings of SID 16*, No. 1: 20-29 (1975).

27. Law, R. R. "Contrast in Kinescopes." *Proceedings of I.R.E.* 27: 511-24 (August 1939).

28. Carroll, G. "Contrast Enhancement of CRT Displays." *Digest of Technical Papers of the 1977 International Symposium of the SID:* 118-9 (1977).

29. Pfahnl, A. "Aging of Electronic Phosphors in Cathode-Ray Tubes." *Advances in Electron Tube Techniques:* 201-8 (September 1960).

30. Seats, P. "Fundamentals of Cathode-Ray Tubes." *Digest of Technical Papers of the 1976 International Symposium of the SID:* 172-3 (1976).

31. Wickersheim, K. A., and Lefever, R. A. "Luminescent Behavior of the Rare-Earths in Yttrium Oxide and Related Hosts." *Journal of the Electrochemical Society* Vol. 3, No. 1: 47-51 (January 1964).

32. Keller, P. "Recent Phosphor Screen Registrations and the Worldwide Phosphor Type Designation System." *Proceedings SID 24*, No. 4: 323-8 (Fourth Quarter, 1983).

33. Critchley, B. R. and Lunt, J. "Garnet Phosphor for Heads-Up Display." *Digest of Technical Papers of the 1983 International Symposium of the SID*: 122-3 (1983).

34. Levine, A. and Palilla, F. C. *Applied Physics Letters* 5, No. 6: 118 (1964).

35. Takeuchi, O., Kusama, H., Kambayashi, K., and Yukawa, T. "An Improved Current-Sensitive CRT Display." *Proceedings of the 3rd International Display Research Conference:* 140-2 (1983).

7

FLAT CATHODE-RAY TUBE DISPLAYS

WALTER F. GOEDE, *Northrop Electronics*

7.1 INTRODUCTION

In this chapter, the motivation for flat CRTs, the history of their development, and a general review of the principles of operation of the component parts will be presented. In addition, a review of the technical achievements of devices built to date will be given.

In the motivation section, some fundamental advantages of flat CRTs as compared to other technologies will be presented. In particular, several potential, new market areas will be discussed as possible application areas for a flat CRT. Some existing market areas which are already being served by flat displays will also be outlined. Next, some basic history of the development of various flat CRTs will be given. Major developments will be highlighted along with a brief description of the current status of each particular device.

In the technical description section, flat CRTs are described in terms of their fundamental parts. The intent of this section is to give the reader a basic understanding of the functions and requirements placed on each of the major component parts of all flat CRTs. Detailed descriptions are first given of the various cathode structures which have been applied to flat CRT devices. Next, the beam positioning and modulation fundamentals are described. Following this is a discussion of various techniques which have been used to enhance the displays' luminance. Next, a very brief discussion is given of phosphor screens and their impact on overall display capability. Finally, a discussion of the vacuum envelope and required processing is presented.

In the last section, a review of the performance achievements of various companies which have built flat CRTs is given. This includes a review of over twenty different devices. Performance characteristics and capabilities are compared and summarized.

For reference, Fig. 7-1 presents a family tree history for the flat CRT types, identifying the major categories of CRT types and most of the companies which have participated in this field. Over ten of the companies listed still have active programs as of mid 1984.

7.2 MOTIVATION AND GOALS

As discussed in Chap. 1, the dream of achieving a flat-panel display has been around for many years. These dreams have been spurred by both the motivation to achieve a flat-format configuration for packaging considerations and the desire to achieve either improved performance or performance capacities not possible via conventional CRTs. As shown in Chap. 6, the conventional CRT has made it very difficult for the flat-panel displays to achieve this second objective, as it has been continually improved and extended into new areas.

Nevertheless, there are a number of reasons why research has continued on flat-panel displays. While most researchers have concentrated on solid-state versions, a number of companies have tried to achieve flat-panel CRTs. The potential advantages of flat CRTs over other flat-panel approaches include (1) established and efficient color materials and techniques, (2) good aesthetics, (3) high luminous efficiency

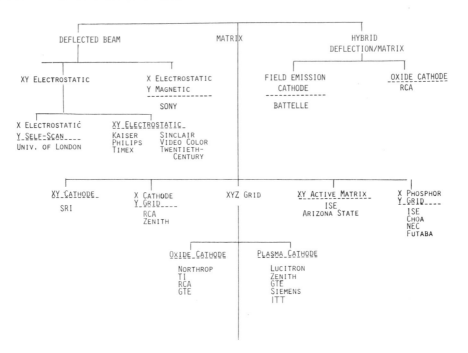

Fig. 7-1. Family tree of flat CRT developments.

materials (phosphors), (4) high luminance, (5) good multiplexibility, (6) ease of achieving gray scale, (7) simplified electronics, (8) format and size flexibility, (9) use of established and well-understood technologies, and (10) fast response. It is these characteristics which make many researchers believe that the flat CRT is the only known panel device potentially capable of establishing itself in the home entertainment marketplace. However, even the more optimistic researchers do not believe that these flat CRTs are capable of competing head-on with the conventional CRT. Rather, the prevalent strategy for those currently trying to develop flat CRTs is to concentrate on areas where the CRT is weak, i.e., in applications where the need is for either a very small display (<5-in. diagonal) or a very large display (>30-in. diagonal). Desired characteristics for these two possible applications are detailed in the following paragraphs.

Many market analysts believe that a sizable market will develop for a personal portable consumer television. In particular, several Japanese companies appear to be firmly convinced that such a device could be very popular. The characteristics of importance for a consumer portable TV are: (1) ultra low power, (2) low cost, (3) very small size and depth, (4) view-ability in outdoor environments (requires high contrast and good luminance), and (5) good aesthetics.

Although many believe that liquid crystals have the best approach for a monochrome version of this small-size display (because of their extremely low power dissipation), the possibilities for achieving a low-cost, natural, multicolored display with liquid crystals are low. In addition, LCDs have switching-speed and viewing-angle problems which may limit user acceptance. Thus, the techniques already well established using tridot phosphors for color in conventional CRTs make it conceivable that a small flat-panel CRT could be the only technology capable of satisfying the long-term consumer desire for a natural multicolor small flat television. Recent developments in the small-size monochrome flat CRT area also indicate that this device may also be the best short-term solution to the desire for a portable TV as well.

In the large physical size end of the market, a substantial market exists for large-screen display devices. Currently these markets are being served by projection CRT displays. However, projection displays have had only limited success due to their high cost, cumbersome packaging requirements, and limited viewing angle. A flat-panel device which could literally hang on the wall and not require space for either front or rear projection could offer significant advantages. In addition to the desirable characteristics listed above, the flat CRT appears to be one of the few technologies capable of being made in large enough sizes to satisfy this application. For this reason, several of the large television companies are working on flat CRT devices.

It is these two applications in particular which have renewed interest in the flat-panel cathode-ray tube. However, it should be emphasized that a number of other application areas exist in between these two extremes. For example, the application to computer graphics and alphanumeric readout type devices should certainly not be overlooked. In fact, the vacuum fluorescent display (a form of flat CRT) has already established a sizable market in areas requiring small to moderate numbers of digits and alphanumerics. Since the late 1970s, production quantities have been measured in the tens of millions of tubes per year. Through this high volume, costs have been reduced to the point where this device has the lowest cost per character in many applications.

If any flat-panel display is to be truly successful, it must be capable of low cost, which in turn can probably only be achieved through a combination of a simple structure and high-volume manufacture. The best chance of achiev-ing the high-volume requirement is through the consumer marketplace. If successful in penetrating the consumer market in either the very small or very large size displays described above, the same technology will also be able to attack the intermediate size marketplaces being pursued by other flat-panel display devices. If this is the case, the low cost achieved from the previously described high-volume markets, should give the flat CRT an overall advantage for many additional application areas.

To set the stage for the remainder of this chapter, a set of application goals will be identified. The three application areas include: a small portable personalized consumer TV, a medium-size vectorgraphic display, and a large-screen flat-panel display in sizes able to compete with existing projection CRT displays. Table 7.1 identifies some important characteristics a display must have to successfully meet the requirements of these three application areas. Throughout this chapter, the goals identified in this table should be kept in mind to see how the various approaches comply with these needs.

All parameters in this table mean total display-system performances as opposed to just that of the display device. For example, the characteristic "Power" is meant to include all power dissipated in the entire display subsystem including the power dissipated in such items as

Table 7-1. Desired Performance Characteristics for Three Potential Applications

Characteristics	Small Portable TV	Medium Size Vectorgraphic	Large-Screen Video
Size (diagonal)	2" to 5"	15" to 19"	36" to 50"
Matrix size	240 × 320 min	512 × 512 min	512 × 640 to 1024 × 1280
Resolution	~100 lpi	50 to 100 lpi	15 to 40 lpi
Total power[1]	<2 watts	100–300 watts	<500 watts
Color	Mono acceptable Full color desired	Mono (acceptable) Multicolor desired[3]	Full color mandatory
Viewing environment	Suitable for outdoors (1000–5000 fc)	Well-lit office (~100 fc)	Home and industry (>50 fc)
Viewing angle	±20° min[4]	Lambertian	Lambertian
Cost[2]	<$200	$300–$2,000	$800–$2,500

[1] Represents total display system power (display, drivers, power supply, etc.).
[2] Total receiver or terminal cost including power supplies, drivers, etc. (1983 dollars).
[3] Many applications for mono displays exist; however, if available for modest cost, a technology with color capability should dominate (6 to 8 colors acceptable).
[4] For small personal displays, a small viewing angle could be tolerated, but a display capable of larger angles would be preferred in the marketplace.

the cathode, power supplies, deflection circuits, drivers, etc. As can be seen, each application area has its own unique set of characteristics. However, note that color is a desirable characteristic in all cases and is considered mandatory for the very large screen display.

7.3 HISTORY

In this section, a chronological review of the development of flat-panel CRTs will be given. The patent literature abounds with concepts for flat CRT's. However, in this section, only those developments in which working devices were actually fabricated and results reported in the open literature will be covered. Operational and performance details will be given in a later section.

Based on patent application dates, the first significant flat CRT was proposed in September 1952[1-3] by a professor at the University of London—Professor D. Gabor. This device, hereafter described as the Gabor Tube, was an extremely interesting and complex device. Essentially all work was carried out at the university, and development continued until approximately 1969.

Another somewhat similar device was independently conceived at nearly the same time by R. W. Aiken while at the University of California Radiation Lab, in the spring of 1951. After making a few small externally pumped models to prove feasibility, the patent rights were sold to Kaiser Industries. Aiken joined Kaiser, where he continued to work on this device for both Navy and commercial color television applications.[4-7g] The Aiken patent was applied for in May 1953. Although both the Gabor Tube and the Aiken Tube used a conventional electron gun to launch an electron beam parallel to the phosphor, and both used electrostatic deflection to guide the beam, the devices were otherwise quite different. Work has continued on versions of the Aiken Tube on and off through the years, and one such device is currently being used in a miniature portable television marketed by Sinclair Research and Timex of Scotland.

In the late 1950s through early 1960s, Philips Mullard developed another unusual flat CRT dubbed the Banana Tube.[8-13] As was the case with the Gabor and Aiken tubes, much of the motivation for this work was the desire to market a color television display. During this time period, most experts believed that the now dominant shadow-mask technique for achieving a multicolor display would never work. Although working models were demonstrated, the device could not compete with the shadow-mask CRT.

About the same time that the ac plasma display was being invented at the University of Illinois (1964), another flat-panel CRT display program was being initiated at Northrop Electronics.[14-22] This display, called the Digisplay, conceived by L. Jeffries and D. Hultberg, differed from earlier devices in that it utilized an area cathode and a series of switching grids to sequentially control which area of the phosphor was excited. The Northrop program was active from about 1962 to 1973. Its goals were to achieve rugged, high-brightness alphanumeric and video displays for military use. In 1973, the patent rights and equipment were sold to Texas Instruments. At TI, the major effort was directed at trying to adapt the Northrop device to serve as a direct competitor to the home television CRT.[23-24] The work was continued until approximately 1979 when the effort was abandoned, as cost could not be reduced sufficiently to compete with a conventional color TV receiver.

In 1967, a Japanese company named ISE introduced the first commercially successful flat CRT. ISE called its device a vacuum fluorescent display (VFD).[25-36] This device has been quite successful in applications requiring small numbers of digits and alphanumerics. It uses a multifilament area cathode similar to that developed at Northrop and uses a low-voltage (50 to 100 volts) ZnO phosphor (instead of the more conventional television phosphors which operate at many kilovolts). This technology is one of the most successful flat-panel ventures and has a good manufacturing base which may allow it to expand to larger sizes.

Stanford Research Institute (SRI) had still another concept for a flat CRT.[37-42] The novel part of this approach was the use of an array of field emitters as the electron input. This work was initiated in the mid 1960s, but was never pursued very actively. The field emission cathode work is still active for use in special applications, but all work was discontinued on the CRT version in the early 1970s. Another display using field emitters was under development at Battelle in Switzerland in the mid to late 1970s.[43,44] In this device the field emitter matrix was addressed by two electron guns.

Another university-conceived flat-panel CRT was reported by workers at Arizona State University.[45] This work, headed by Professor M. Sirkis, used an area-emitting hot filament cathode similar to that used in the early Northrop Digisplay. The unique part of this device was its implementation of an active matrix-addressing scheme to give a frame storage feature. The active matrix consisted of a thin-film transistor (TFT) and storage capacitor at each pixel of the matrix display. Work began in approximately 1970 and continued through 1977. In 1976 this work was transferred from the university to Sperry Phoenix who sponsored the early work. All work was discontinued in 1977.

Several television companies quite naturally have also been pursuing flat CRT concepts. Due to the highly competitive nature of their business, there have, however, been few public disclosures. GTE Sylvania worked in partnership with Northrop for a short time on the Digisplay, but discontinued work when a corporate decision was made in the early 1970s to withdraw from the computer and computer terminal business.[22] Meanwhile, Zenith and RCA were independently developing their own flat CRT concepts. Unlike Sylvania's, both these efforts were aimed at the home television marketplace for large diagonal displays.

The Zenith display was first reported at the 1976 SID Symposium (the first patent was issued in 1974).[46] The unique features of this display were its use of a cold ion feedback cathode and a series of electron multipliers (dynodes). Unlike previous approaches which operated in high vacuum ($<10^{-6}$ torr*), this device operated in the 10^{-3} torr pressure range. The background gas was primarily helium. In addition, the structure was internally supported,

so that devices could be made quite large without requiring thick faceplates to withstand atmospheric pressure. The current status of this device at Zenith is unknown but it is believed to have been discontinued. Some rumors indicate that Zenith is still pursuing a different type of flat CRT.

Another group at Zenith developed a hybrid plasma/CRT which used a plasma discharge as the electron source for a flat CRT somewhat similar to the Northrop device. This device was also discontinued. However, the workers on this device left Zenith and formed their own company called Lucitron.[47,48] Siemens has a similar hybrid plasma/CRT under development.[49-51] The historically significant part of the hybrid plasma/CRT work is the unique method of extracting electrons from a low-pressure (approximately one torr) gas discharge. As techniques have been developed for self-scanning a gas discharge (see Chap. 10 on the Burroughs Self Scan device), this device offers the potential of greatly reducing the drive electronics as compared with most other flat-panel devices.

Two years after the disclosure by Zenith of their flat CRT, RCA reported on a very similar structure at the 1978 SID Symposium.[52,53] This device apparently had been under development since 1972, and the first patent had been issued in 1977. This device also used an ion feedback cold cathode and multiplier dynodes and had a self-supporting structure. Although it too was discontinued, RCA has continued working on similar structures.[54-57] These use a multiplicity of electron guns and electron-beam guides to steer the electron beams to the appropriate place on the phosphor.

More recently, flat CRT concepts have been described by Philips and Sony. The Philips approach uses a channel plate type multiplier for enhancing the brightness in a structure which revives many concepts originally put forth in the Gabor and Aiken displays.[72,72a,73] Sony introduced its flat CRT hand-held TV version to the marketplace in mid 1982, with almost no advance publishing of technical details.[64,65] This device is also an Aiken derivative, and has many similarities to the Sinclair/Timex version mentioned earlier. It uses magnetic focusing in one axis.

---

*A torr is a unit used to measure the degree of vacuum. One torr is defined as 1/760 of a standard atmosphere and is approximately equal to 1 mm of mercury. One torr equals approximately 133.3 Pascals.

Table 7-2.   Flat CRT History

Gabor Tube 1952–1969	Field Emission 1970–1980
University of London	SRI
Aiken Tube 1953–present	Battelle
Kaiser	Thin-film transistors 1970–1975
Video Color	Arizona State
Twentieth Century	Ion feedback 1970–1978
Sinclair	RCA
Sony	Zenith
Banana Tube 1955–1963	Hybrid plasma 1976–present
Philips	Lucitron
Digisplay 1963–1979	Siemens
Northrop	AEG–Telefunken
Texas Instruments	GTE
GTE	Zenith
Vacuum fluorescence 1966–present	Electron guides 1977–present
ISE	RCA
Futaba	Channel electron multiplier
CHOA	1978–present
NEC	Philips
Hangzhou Univ. (China)	

As can be seen by the preceding discussion, a large number of configurations for achieving a flat CRT have been investigated and a number are still under development. Table 7-2 summarizes the history of most of the significant flat CRT display developments, and the companies which have participated. Although most of these displays are still not a commercial reality, the hope for achieving a large-screen flat television keeps researchers interested. Compared to other flat panels, the flat CRTs appear to have better potential for achieving full color, good gray scale, and large size.

7.4   FUNCTIONAL AND TECHNICAL DESCRIPTIONS

This section will present an overview of the operating principles common to all flat CRTs. The intent of this section will be to show how the various components interact. Sections 7.5 to 7.9 will identify the desirable operating characteristics of each portion of the device. Descriptions of how these components have been used in actual devices (and the performance characteristics achieved) will be given in Section 7.10.

Rather than discuss the principles of operation of each device independently, it will be instructive to study the characteristics that are common to all flat CRT's. In so doing, a better appreciation of fundamental characteristics can be obtained, and ideas for future research will become more apparent. To further this goal, we will subdivide the flat CRT into its fundamental sections. These include the cathode, the section that provides beam positioning and modulation, the brightness enhancement section, the phosphor, and, finally, the vacuum envelope.

In this book, the term flat CRT is reserved for devices which use an electron beam(s) hitting a phosphor as the method of light generation. Thus, all such devices must have a method for generating an electron beam(s). This portion of the device is called the cathode. Next, a technique for getting the beam(s) to the right part of the phosphor target at the right time and intensity level must be employed. This section of the device will be called the "beam positioner and modulator." In some cases, positioning and modulation are performed separately, or in other cases one of these functions is combined in the cathode section. If a separate focusing function is required, it is also usually located in this section.

In many of the flat CRTs, it was found that the cathode was incapable of supplying enough beam current for suitable luminance. To overcome this problem, many devices have resorted to incorporating a brightness-enhancement technique. As this is an important function, this subject will also be described separately.

Table 7-3. Desired Cathode Properties

Small-area uniformity[1]	±10% max
Large-area uniformity	±30%
Current output	100 μA for single-element cathode, 4 to 10 mA/cm^2 for area cathodes (line-at-a-time addressing)
Low power	<5 watts for single-element, <0.1 watt/cm^2 for area cathode
Life	>10,000 hr
Stability	Materials should be air stable and capable of withstanding air and vacuum bake of >450°C
Operating environment	Should work in moderate vacuum (<10^{-6} torr)
Ruggedization	Should be capable of withstanding harsh tube environments such as high shock and vibration
Size	Should be capable of being used in large variety of sizes and be less than 1 cm thick
Cost	Should be capable of mass manufacturing to achieve low cost
Energy distribution	Electrons emitted should all be of same kinetic energy and be emitted essentially perpendicular to the surface

[1] Depends on the spatial frequency and length of the defect. Can be ±1% or lower if pattern is repetitive. If pattern is of a long line type defect which occurs periodically, the eye is amazingly sensitive. Concepts using modular type construction are particularly plagued by this problem.

The light emission takes place at the cathodoluminescent phosphor screen. As the materials and characteristics of the phosphor have been described in detail in Chap. 6, only brief mention will be given in this chapter. The last major element of the flat CRT is the vacuum envelope which houses all the components. Although the functions of this component are seemingly obvious, in practice the achievement of a suitable enclosure has been elusive. In the following sections, each of these topics will be described in more detail.

7.5 CATHODES FOR THE FLAT CRTS

Depending on the specific display device configuration, the cathode can range from the simplest part of the overall structure to the most complex. The requirements which the cathode must satisfy are generally quite severe. Table 7-3 lists some of the characteristics an ideal cathode should possess. Various approaches which have been attempted, as well as the companies which performed most of the work in each area, are shown in Table 7-4. Each

Table 7-4. Cathode Types and Users

Cathode Type	Companies Who Investigated
1. Single-point cathode electron guns	Battelle, Kaiser, Sinclair, University of London, Video Color, Philips, Sony, Timex
2. Individual cathode per resolution element	Battelle, Stanford Research
3. Large-area cathodes	
A. Thermionic oxide coated	
Large area, low temp.	Northrop, Stanford Research
Screen mesh	Northrop
Button matrix	Northrop
Multifilament	Arizona State, CHOA, ISE, NEC, Northrop, Futaba, Texas Instruments
B. Ion feedback	RCA, Zenith
C. Photocathodes	RCA
D. Gas discharge	Lucitron, Northrop, Siemens, Telefunken, ITT
E. Radioactive	Northrop

of these approaches will be described in detail in the remainder of this section. How various companies applied these cathodes in a functional device will be described in Section 7.10.

It should be noted that many of the cathodes to be described proved unsuccessful in practice. However, they are included here for completeness and, hopefully, to illustrate potential areas of difficulty in future research.

7.5.1 Single-Point Cathode.

In this type of cathode a single electron beam is deflected to the appropriate place on the phosphor. In a flat CRT configuration, an electron gun similar to that used in a conventional CRT is repositioned so that the beam is launched parallel to the phosphor screen and then is eventually redirected so that the beam hits the phosphor at as close to a 90-degree angle as possible (in practice the angle is often much less). In general, the electron gun is nearly identical to those described in Chap. 6 and must satisfy essentially the same requirements in terms of spot size and beam current. The emitter is a thermionic oxide–coated cathode.[58] As in a conventional shadow-mask color CRT, three independently controlled beams are usually needed for color applications. This cathode type has been relatively easy to incorporate into a flat CRT structure as it has already been well developed for conventional CRTs.

In the absence of a brightness-enhancement technique (described in Section 7.7), the single cathode has to supply enough beam current to ensure adequate luminance. Equation 7.1 describes the relationship between beam current I, desired area averaged luminance L, the accelerating voltage V, the screen area A, the phosphor proportionality constant K, and the phosphor exponent η.

$$I = \frac{LA}{KV^\eta} \qquad \text{(Eq. 7.1)}$$

As can be seen, the required current is a direct function of the desired luminance and screen area. For large devices, the demand on beam current rises rapidly. (Note that both K and η vary depending on operating conditions and fabrication considerations). For moderate screen

sizes and luminance values ($\sim$100 fL and 25-cm diagonals), beam currents of 10 μA to 1 mA are typical if reasonably efficient phosphors are used. As this type of cathode is identical to that used in a conventional CRT as discussed in Chap. 6, no further coverage will be given here.

7.5.2 Large-Area Thermionic-Oxide-Coated Cathodes.

A number of different approaches have been attempted to achieve practical large-area flat cathodes. They will be discussed in terms of principles of operation, advantages, disadvantages, and actual accomplishments.

Equation 7.2 can be used to predict the required cathode current density for any large-area cathode. In μa/hole,

$$I = \frac{LAN}{KM} \qquad \text{(Eq. 7.2)}$$

where

  L = desired spot luminance in fL
  A = area of spot in cm^2
  N = total number of elements
  K = phosphor efficiency in fL/μA/cm^2 at the correct voltage and current density
  M = number of multiple beams

For example, suppose one desires to achieve 50fL average spot luminance in a 512 $\times$ 512-element display using a P-20 phosphor (K = 420 fL/μA/cm^2 at 10 kV). In this display, the spot size is to be 250 μm (0.01 in.) and 32 multiple beams are to be used. Under these conditions, the required output current at the phosphor would have to be 0.5 μA/hole. This corresponds to a current density of approximately 1 mA/cm^2. Note that the phosphor efficiency value must be obtained at the correct anode voltage and current density.

Large-Area–Low-Temperature Cathode. Several groups have tried to develop a low-temperature–large-area cathode capable of supplying a continuous emission surface. A typical structure is illustrated in Fig. 7-2. Here a modified triple carbonate (Ba, Ca, and St) coating material is applied to a substrate with a metallic electrode.[58] The return conductor, of course, is required for a conduction path return for elec-

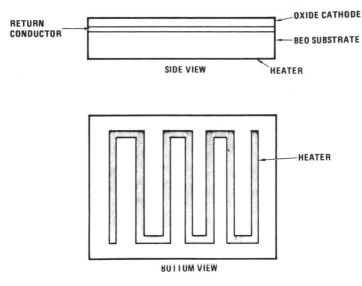

RETURN CONDUCTOR

OXIDE CATHODE

BEO SUBSTRATE

SIDE VIEW   HEATER

HEATER

BOTTOM VIEW

Fig. 7-2. Large-area–low-temperature cathode.

trons emitted from the coating material. On the reverse side of the dielectric substrate, a serpentine heating element is applied to raise the substrate to the appropriate temperature. In devices tried, a typical substrate material was BeO because of its good thermal conductivity.

The use of this indirectly heated structure had several potential advantages. First, being indirectly heated, no voltage drop occurs across the cathode. This would result in a uniform energy distribution in emitted electrons across the cathode. In directly heated large-area cathodes, a voltage drop across the cathode can result in a nonuniform display, as electrons emitted at different energies may create modulation difficulties. Second, a large-area cathode could be operated at a very low cathode-loading density. Two to three orders of magnitude less density would be required for a line-at-a-time-addressed display. As a result of the low cathode loading, long life should be possible. In addition, small defects in the emission surface, caused either in manufacture or by damage from ion bombardment, should not be noticeable. Also, the high emission currents available would eliminate the necessity for a brightness-enhancement technique elsewhere in the structure. Finally, the near-planar emission generally simplifies addressing and modulation. Such a cathode is conceptually simple and should be relatively inexpensive.

Unfortunately, one very serious problem has prevented this technique from becoming practical. That is the extremely high power dissipation required to heat the cathode to a useful temperature. Test performed at one laboratory indicated that the power input required to reach operating temperature (approximately $500^\circ C$) was about 1.5 watts/cm^2. This heat load to the tube is much too high. In addition, emission uniformity was poor. These problems resulted in all known work being discontinued on this approach.

This type of approach in general can be eliminated from contention by considering basic fundamentals of thermal emission. Typically, in the absence of space-charge limiting, emission increases as a large power of temperature ($T^{n\ \text{where}\ n>3}$), whereas the power required to heat a given area to a temperature (T) goes up approximately with the area to be heated (this of course depends on many other thermal factors). Thus, to achieve a given output current at the lowest power, it is obvious that it is better to heat a small area to a high temperature than it is to heat a large area to a low temperature.

Screen Mesh Cathode. To reduce the amount of area to be heated from that in the previous approach, but still provide electron input over a large area, several groups have experimented with both woven and electroformed wire

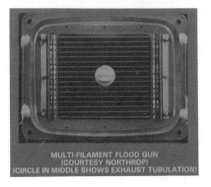

MULTI-FILAMENT FLOOD GUN
(COURTESY NORTHROP)
(CIRCLE IN MIDDLE SHOWS EXHAUST TUBULATION)

Fig. 7-3. Multi-filament flood gun. (Circle in middle shows exhaust tabulation.) (Courtesy of Northrop)

screens. Typically, a nickel screen mesh of from one to sixteen wires per millimeter was coated with a standard oxide-cathode mixture. A separate busbar was attached to two opposite ends of the screen with tension applied to maintain flatness. In operation, a voltage would be placed across these two opposite ends, and heating current would be passed through the wires connecting the two buses. No heating current would flow through the wires running parallel to the busbars, since they remain at equal voltage levels. These cross-wires serve primarily as mechanical support. The number of wires per unit length (mesh size) can be optimized for uniformity. This approach appears to meet most of the desired characteristics described earlier except for power and uniformity. In cathodes of this type fabricated at Northrop, it was found that achieving uniform emission was quite difficult. In addition, the mesh tended to stretch and deform as it was heated. As is the case with nearly all oxide-coated cathodes, during manufacture a high-temperature-forming step is required to activate the cathode. This requires that the entire cathode be heated, in vacuum, to approximately 1000 to 1100°C. At this temperature, power input for a 1 wire/mm electroformed nickel mesh was about 0.8 watts/cm². Operating power (~800°C) required about 0.3 watts/cm².

Button Matrix. Another approach consisted of mounting an array of very small thermionic cathodes on a frame which served both as mechanical support and electrical interconnection.

When current was passed through the frame, the small cathodes were heated and caused to emit electrons. A series of apertures and defocusing screens were then used to attempt to spread the emission uniformly over the display active area. A small matrix of cathodes of this type was fabricated and tested at Northrop. Ample emission was obtained, but uniformity was very poor. Several tests were made to improve the uniformity, but only small improvements resulted. The electrons from each cathode could be made to spread uniformly only over a circular area of approximately one centimeter diameter.

Multifilament Flood Gun. To date, the most successful of all area-type approaches described in this chapter used a number of small-diameter wires for the cathode. Figure 7-3 pictures one of the cathodes used by Northrop. In these approaches, the cathode wires are made from tungsten wire—usually with a small percentage (~3%) of rhenium added to improve shock resistance. Wire size is selected to be as small as feasible to minimize power. Typically, wire sizes of approximately 16 μm (.0006 in) are used. The wire is then electrophoretically coated with a standard triple carbonate mixture. Coated-wire diameter is typically 50 to 75 μm. The wires are then strung from one end of the display to the other and operated under spring tension. The tension is required to maintain the planar structure during heating. Figure 7-4 shows a side view of the filament structure and typical spacings and voltages found to give uniform distribution in an early Northrop device. The power required was relatively low and the uniformity acceptable. However, approximately only 10% of the current emitted by the filaments was capable of being utilized. Typical usable-current output was approximately 0.1 to 0.5 mA/cm².

Substantial improvements in performance were made by Texas Instruments to this structure. They reported achieving up to 5 mA/cm² input to the device. In a color video version of this device, they achieved a 4 mA/cm² output from a structure having 40 wires to cover a 15 × 20-cm active area device. Cathodes were on 0.5 cm centers, and total cathode power was

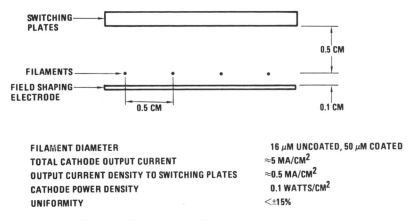

FILAMENT DIAMETER	16 μM UNCOATED, 50 μM COATED
TOTAL CATHODE OUTPUT CURRENT	$\approx$5 MA/CM^2
OUTPUT CURRENT DENSITY TO SWITCHING PLATES	$\approx$0.5 MA/CM^2
CATHODE POWER DENSITY	0.1 WATTS/CM^2
UNIFORMITY	$<\pm$15%

Fig. 7-4. Northrop multifilament cathode—side view.

16 watts (0.05 watts/cm^2). The structure used by Texas Instruments is shown in Fig 7-5.[24] As can be seen, the primary difference relative to the Northrop cathode was the field-shaping electrodes added between and slightly behind the emitting filaments. In addition, workers at Texas Instruments developed an elaborate and accurate computer program to model structure changes, allowing quick evaluation of geometry changes. The program was based on performing an exact analysis based on Maxwell's equations and included the effects of space charge.

Despite initial impressions, the multifilament cathode was both rugged and had excellent life characteristics. In one application, the Northrop device was to be used as a gun sight in a military fighter.[21] For accuracy, the display had to be essentially hard-mounted to the same bulk-head as a 30-mm Gatling gun cannon. The shock and vibration environment was extremely severe. By proper tensioning of the 3.5-in. long filaments, supported only at the ends, this display was able to operate under the required conditions. This was due to the low mass of the wire filaments and the relatively small importance of slight changes in cathode position due to vibration.

Life tests of cathodes of this type at Texas Instruments have shown no degradation after over 3,000 hours of full up cycled operation.[24] Similar life tests at Northrop verified over 10,000 hours of operation including thermal cycling. Life tests at ISE on similar cathodes used in vacuum fluorescent displays (VFD) have shown even longer operation with no degradation. In the VFD, it is not unusual to see cath-

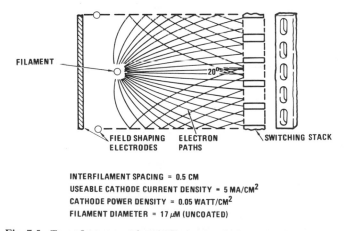

INTERFILAMENT SPACING = 0.5 CM
USEABLE CATHODE CURRENT DENSITY = 5 MA/CM^2
CATHODE POWER DENSITY = 0.05 WATT/CM^2
FILAMENT DIAMETER = 17 μM (UNCOATED)

Fig. 7-5. Texas Instruments' multifilament cathode. (After Scott, 1978)

ode filaments strung over 12 inches and still be used in rugged environments.

7.5.3 Large-Area Nonthermionic Cathodes.

Individual Cathode Per Resolution Element. In this approach, each resolution element of the display is supplied current from its own individual cathode. In some concepts, the cathodes are modulated and switched directly, while in others this function occurs in other elements. The primary requirements for a cathode of this type are:

- The individual elements must be able to be batch-fabricated to achieve low cost.
- The individual elements must have uniform characteristics including current output, aging, addressing, and spot size.
- If the cathode elements are modulated directly, they must have fast and uniform modulation characteristics as well as a large dynamic range.
- The elements must have sufficient current output to achieve the desired luminance.

To date, only one known emitter of this type has been evaluated. This has employed the use of field emission from a sharp point. As the name implies, a field emission cathode is one in which current is generated, usually from a sharp point, by a high electric field. In 1928, Fowler and Nordheim[42] showed that field emission followed the equation

$$J = \frac{A F^2}{t^2(y)\phi} \exp \frac{-B v(y)\phi^{3/2}}{F} \qquad \text{(Eq. 7.3)}$$

As discussed by Spindt,[37] J is the emission current density in A/cm^2, A and B are constants, F is the field at the tip, ϕ is the work function in eV, and $v(y)$ and $t^2(y)$ are slowly varying functions of y where

$$y = \frac{3.79 \times 10^{-4} F^{1/2}}{\phi} \qquad \text{(Eq. 7.4)}$$

Both $v(y)$ and $t(y)$ are tabulated in the literature.[38] The field at the tip is

$$F = cV \text{ volts/cm} \qquad \text{(Eq. 7.5)}$$

where V is the voltage applied to the diode structure and

$$C = f(r, R, \theta)\text{cm}^{-1} \qquad \text{(Eq. 7.6)}$$

The relationship between the field at the tip, the tip radius (r), the anode-to-tip spacing (R), and the emitter cone half angle (θ) is very complex. However, for the purpose of this discussion it is sufficient to know that as r, R, and θ decreases, C increases. Thus, to reduce the required voltage (a desired feature), the cathode-to-anode gap should be made as small as possible.

Figure 7-6 shows a conventional field emitter. In this structure a short segment of fine wire which has had a sharp point electrolytically etched at one end, has been mounted on a hairpin filament for support. After cleaning in vacuum by heating to incandescence, cold electron emission can be obtained by applying a high voltage to the anode, which in this case is about 1 cm from the tip. The electric field required for emission is approximately 10^7 V/cm. Thus, even with the local field enhancement resulting because of the sharpness of tip, anode potentials on the order of kilovolts are required for field emission if the anode is about 1 cm from the tip. This is obviously too high for use in a display device. Such a structure also tends to have short life because of ion bombardment, as gas particles ionized by the process tend to be directed to the tip. The ion bombardment tends to reshape the tip and reduce life.

In the late 1960s researchers at Stanford Research Institute (SRI) developed a thin-film structure which offered solutions to many of the problems of previous concepts and could be made in a moderate-size matrix. Figure 7-7 shows a schematic of this type of cathode.[37-41] These arrays have been developed to support a number of potential applications including use in a flat CRT, and as cathodes for high-power electron tubes.

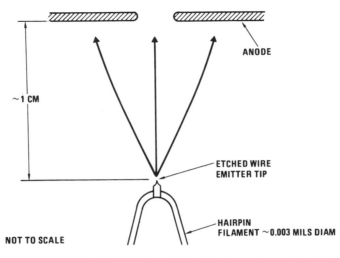

~1 CM

NOT TO SCALE

ANODE

ETCHED WIRE
EMITTER TIP

HAIRPIN
FILAMENT ~0.003 MILS DIAM

Fig. 7-6. Conventional field emitter and anode. (After Spindt, 1979)

These results have shown that field emission arrays can be made which operate below 100 volts anode potential and can have emission currents per cathode of over 10 mA. In fact, for special applications, currents of over 100 mA at about 200 V have been tested. Because of the low-voltage operation, sputtering caused by ion bombardment has also been greatly reduced, and tests have shown that at 20 μA per cathode, over 40,000 hours of operation is possible. Thus, based on the current requirements established earlier, it appears that such a cathode would be capable of operating with good life and high brightness in a flat CRT without requiring brightness enhancement.

Unfortunately, other problems have prevented such a structure from being implemented. These include the need to operate in ultra-high vacuum (10^{-9} torr), nonuniformity from cathode to cathode, inability to be made uniformly over large areas, and high cost. How-

ever, research is continuing on this type of cathode, primarily for use in special-purpose CRTs and high-energy microwave tubes.

Ion Feedback Cathode. A relatively recent new approach to a cold area cathode was reported in 1976 by Zenith and in 1978 by RCA.[46,52] The important features of this structure are shown in Fig. 7-8.[53] In this case, the cathode is typically just a plain piece of sheet metal.

Emission occurs when the cathode is struck by a high-energy ion, which creates secondary electrons in accordance with the impact energy of the ion and the secondary emission coefficient of the cathode material. The electric fields in the device are devised such that the secondary electrons leaving the cathode are accelerated to the first dynode, where more secondaries are produced, and so on. As the function of the dynodes will be discussed more fully in the brightness-enhancement section, for now it is

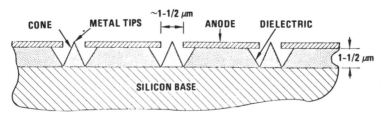

CONE   METAL TIPS   ~1-1/2 μm   ANODE   DIELECTRIC

SILICON BASE

1-1/2 μm

SCHEMATIC OF A THIN-FILM EMISSION CATHODE ARRAY

Fig. 7-7. Spindt-type field emission cathode. (After Spindt, 1979)

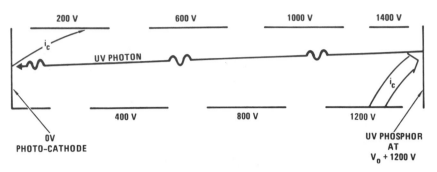

Fig. 7-8. RCA ion feedback cathode. (After Endriz, 1979)

sufficient to note that the electron current builds up from stage to stage.

The entire device was operated in relatively poor vacuum in the range of 10^{-3} to 10^{-4} torr. The background gas usually consisted of helium, neon, argon, or xenon (or a mixture of those gases). The electrons heading toward the phosphor had a finite probability of ionizing some of these gas atoms. The probability dP for an electron to ionize a gas atom over a path length dx is

$$dP = p\, \epsilon\, dx, \qquad \text{(Eq. 7.7)}$$

where p is the pressure and ϵ is the ionization function of the gas species (varies as a function of electron energy). For the gases of interest, ϵ reaches a maximum between 60 and 200 eV; thus, a high probability of ionization exists (for further details, see Reference 52). Of the ions formed, it was found that about half were accelerated back to strike the cathode, which in turn created more secondaries. The remaining ions were collected on the dynodes. As can be seen, once initiated, a feedback process is developed which will cause the electron current to increase until some saturation effect occurs. Typically, from initiation to saturation, the entire process occurs in well under a microsecond so that the effect can be used in devices where fast response is required.

The net result is a cathode structure which draws power only when activated for addressing a particular element on the phosphor. As the reader has probably noted, the mechanism, as described thus far, works fine once started, but no turn-on mechanics has yet been described. Initial turn-on is dependent on some stray event, such as a cosmic ray ionizing a gas atom or a random field emission from a sharp point. However, once started, metastable atoms created in the normal operation of the device usually persist in the region long enough to give reliable start-up between refresh cycles. A cathode is switched on by changing the first dynode potential from 0 to 200 volts.

The main problems of this technique were a result of its relatively complex structure and necessity for operation in a background pressure of a noble gas. Maintaining the proper pressure and mixture of such gases is often difficult. Sputtering of cathode material onto the dynodes may also be a problem. As uniformity and current output are adequate, it appears that the prime problems of this technique relate to cost. Work appears to have been discontinued on this cathode technique at both Zenith and RCA.

Photocathodes. Photocathodes are materials which emit electrons when struck with light of the proper wavelength. Two separate and distinct problems are associated with the development of a photocathode for use in flat CRTs. The first is related to the fabrication and operational characteristics of the photocathode itself, and the second is that of the characteristics of possible light sources which power the photocathode. Since it is highly desirable for the cathode to be air stable, virtually all of the most efficient photocathodes are eliminated from consideration. The type of light source employed depends on the emission characteristics of the photocathode. The light source can either be external or internal to the tube (the light from a phosphor can also serve as the stimulus to the photocathode).

Currently, the only cathodes that can be opened to the air and which have useful quantum efficiency are the ultraviolet types, having sensitivity from about 1050 Å to 1450 Å, peaking around 1200 Å. The most efficient external light sources for this region are hydrogen, deuterium, or nitrogen arcs. These require approximately 50 watts to operate, which is well above the desired input-power level described in Table 7.3. For these reasons, no known flat CRTs have been fabricated using photocathodes with external light sources. No further discussion of this type of photocathode will be given.

A more interesting use of a photocathode as an electron input device was proposed by Endriz of RCA and Schwartz of Zenith.[46,53] In this device, the light source was the output phosphor itself. The device consisted of using the UV light returned from the phosphor (the phosphor was a mixture of both visible and UV-emitting materials) to the cathode as a feedback loop and resulted in a cathode which required power only when addressed. As shown in Fig. 7.9, UV photons returned from the phosphor created an input electron current for the multiplier dynodes.

As can be seen from the figure, this approach is quite similar to that described in the ion feedback section. The primary difference is that instead of gas ions striking a metal plate to generate the electron input, here photons strike a photocathode. The photocathode materials reported include barium metal and cesium antimonide, with the barium type the preferred choice. The phosphor used was cerium-doped lanthanum phosphate ($LaPO_4$:Ce). The purpose of the dynode structure shown in the figure will be described in more detail in the brightness-enhancement section (7.7).

The advantage of this structure over the ion feedback structure resulted from eliminating the need for operating the entire device in poor vacuum. In addition, long decay time phosphors can be used, which ensures good start-up properties when the device is operated in a television scan format. Obtaining reliable start-up from metastable atoms was a problem in the ion feedback version. Problems of ion sputtering are also overcome. Unfortunately, photocathodes are typically very hard to manufacture and control in large-area devices.

Gas-Discharge Cathodes. Another possibility for a flat CRT cathode is to use the electrons or ions formed in a low-pressure gas discharge to excite a phosphor screen. Early reports of such a possibility highlighted some of the potential advantages and disadvantages.[15,59] The most obvious disadvantage is the need to operate the device in the 10^{-1} to 1 torr range. This relatively "dirty vacuum" environment could lead to potential life or breakdown problems. Assuming that this problem can be overcome, the next problem involves how to extract the electrons (or ions) in sufficient quantities from the discharge. This had been accomplished as early as 1955.[60]

Despite the apparent problems, several structures using such devices have been fabricated. Both Lucitron and Siemens have reported on this type of cathode.[47-51,77] However, few details have been published on this subject. Design considerations for this specific type of cathode are covered in Chap. 10. One of the interesting features of the cathode is its relatively high current output. Currents in the several-milliamp-per-element range are possible, which

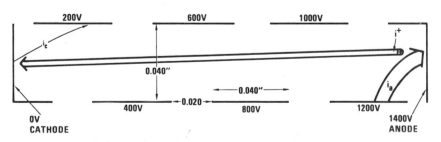

Fig. 7-9. RCA flat CRT using photocathode. (After Catanese, 1979)

TABLE 7-5.   Comparison of Cathode Approaches

Technique	Uniformity	Power Dissipation	Technology	Cathode Cost	Thickness	Lifetime	Compatible With Flat CRTs
Electron gun(s)	E	~3 Watts	Well developed	Very low	Long	10000[1]	Yes
Coated screen mesh	F	3 Watts/cm^2	Should be simple	Low	<1 cm	>5000[2]	Yes
Multifilament flood gun	G to E	.05 Watts/cm^2	Developed up to 10" X 10"	Low-mod	<1 cm	>5000[2]	Yes
Low-temperature coated substrate	F	1.5 Watts/cm^2	ND	Low	<1 cm	U	High thermal load
Malter emission	U[3]	N	ND	Med-high	<1 cm	U	U
SRI field emission	U[3]	Low	ND	High	<1 cm	U	Yes[5]
Ion feedback	G to E[10]	Low	ND	Mod-high	1 cm[15]	U	Yes
Photocathode feedback	G to E[10]	Low	ND	High	1 cm[15]	U	Yes
Plasma cathodes	G to E[10]	.15 Watts/cm^2	R	Low-mod	<3 cm	U	U
Radioactive Sources	G to E	N	ND	Med-high	<1 cm	Good	U

1. Based on typical oxide-cathode life characteristics
2. Estimated
3. Early results show uniformity is inadequate
4. Normal vacuum-tube operation, $\sim 10^6$ torr
5. Requires special processing for ultra-low vacuum, $\sim 10^9$ torr
6. Requires special atmosphere (~ 1 torr)
7. Requires output multiplier
8. Desirable during development
9. Depends on spot size and structure
10. Uniformity achieved via feedback
11. As of 1984 (A = active, D = discontinued)
12. Operates at 10^3 to 10^9 torr

would provide a very high output brightness. The Lucitron device is particularly interesting as the features of self-scanning (similar to that used in the Burroughs dc Self-Scan^{TM}) and xyz beam positioning (as used in the Northrop Digisplay) are employed in addressing the display. This results in a great reduction in the number of driving circuits required.

Radioactive Sources. Several devices have tested the use of low-output radioactive materials as an electron input for flat CRTs. In one such approach, a titanium/copper foil that had been exposed to radioactive tritium (H_2^3) gas was used. The tritium reacted with the titanium to form titanium tritide. Since tritium is a relatively pure beta-emitter with a 12-year half-life, the tritium cathode had the advantage of being a compact, long-life, zero-power, area electron source with good uniformity. Unfortunately, however, it had several serious short-comings, including (1) maximum electron output current density of only 10^{-10} A/cm^2; (2) helium gas was produced by the tritium decay and released into the vacuum; and (3) concern over safety of the radioactive source. The gas release required continuous pumping of the display device to maintain a good vacuum. A device

Air Stable	Vacuum Bake Possible	Re-exposure To Air Possible	Vacuum Required	Ruggedization Possible	Output Current	(11) Status	Largest Demo'd Active Area
Yes	Yes	Yes	4	Yes	High[9]	A	36″ Diag
Yes	Yes	Yes	4	Yes	~10 ma/cm^2	D	3″ × 3″
Yes	Yes	Yes	4	Yes	~5 ma/cm^2	A	To 10″ × 10″
Yes	Yes	U	4	Yes	Low	D	2″ × 2″
Yes	U	U	5	Yes	High	D	TC[14]
Yes	Yes	Yes	5	Yes	~20 to 100 μA/TIP	D	.5″ × .5″
Yes	Yes	U	12	Yes	Low [7,15]	D	1″ × 1″
ND	U	ND	5	Yes	Low [7,15]	D	TC [14]
Yes	Yes	Yes	6	Yes	1A/cm^2	A	6″ × 8″
Yes	250°C max	Yes	13	Yes	Very low[7]	D	1″ × 1″

13. Gas emission requires that tube has pump
14. TC = only small test cells fabricated
15. Excludes multiplier dynodes which add considerable thickness)

 U = Unknown
NA = Not available
 E = Excellent
 G = Good
 F = Fair
 N = Negligible
ND = Needs development
 R = In early research

using such a low current output would require a brightness multiplier such as those described in Section 7.7. Because of the fundamental nature of these problems, work on this cathode approach has been discontinued.

7.5.4 Cathode Summary. The preceding sections have attempted to highlight the principles of operation of a variety of cathodes which have been used in flat CRTs. Table 7-5 gives a comparison of these cathodes based on the desired performance criteria established earlier. From this chart, the following observations can be made:

- For flat CRTs of the beam-deflection type, the standard electron gun is quite satisfactory. Single or multiple (hundreds) electron guns may be used.
- For cathodes requiring a large-area electron input source, several candidates appear capable of satisfying most of the important requirements. These are the multifilament flood gun, the ion and photocathode feedback type, and the plasma cathode. However, except for the first of these, little data is available for comparison and the true status and capabilities of these other cathodes is presently unknown.

7.6 BEAM POSITIONING AND MODULATION TECHNIQUES

The most innovative and important part of most flat CRTs is the beam positioning element. In most cases, beam modulation is also included in this portion of the device. Where possible, these two functions will be discussed together. Table 7-6 identifies the most common techniques used and the companies which experimented with each technique. Table 7.7 identifies the characteristics that an idealized flat CRT should have in terms of beam positioning and modulation characteristics. As the desired characteristics for the techniques using beam deflection (such as in the Aiken or Gabor tubes described later) are significantly different from those approaches which have a separate beam for each resolution element (matrix addressed), the requirements are discussed separately. In the following sections each technique in Table 7-6 will be described and measured against capabilities identified in Table 7-7. As many variations within each technique have been tried, only the

Table 7-6. Beam-Positioning Techniques and Users

Deflected Beam	Primary Companies
XY electrostatic	Kaiser, Philips, Sinclair, Timex, Video Color
X electrostatic, Y, electrostatic, self-scan	Univ. of London
X electrostatic, Y magnetic	Sony
Independent XY	Battelle
XY electrostatic in beam guides	RCA
Matrix Addressed	
XY discrete cathode	SRI
X cathode, Y control grid	RCA, Zenith
XYZ control grids	Northrop, GTE, RCA, TI, AEG-Telefunken, CHOA
X phosphor, Y control grid	ISE, Futaba, CHOA, NEC
XY in single control grid with active matrix	Arizona State, ISE, Sperry
XY self-scan and XYZ control grids	Lucitron, Zenith, GTE, Siemens, AEG-Telefunken

Table 7-7. Desired Beam-Positioning and Modulation Characteristics

Matrix addressed approaches

1. The beam positioning or modulation technique should not introduce nonuniformities into the output image, and ideally should reduce any nonuniformity that may be cathode related.
2. The beam spot size, switching and modulation characteristics should be the same throughout the display.
3. The technique, as a minimum, must be capable of displaying a video-picture at TV rates.
4. The number of drivers should be minimized.
5. The voltage and the impedance to be driven should be compatible with low-cost integrated drivers.
6. The modulation characteristics of the individual cells must be closely matched and be capable of wide dynamic range.
7. The technique should be adaptable to meeting different size and performance requirements.
8. The technique should be amenable to mass production.

Deflected beam approaches

1. Low power deflection circuits.
2. Minimum number of circuits and different power supply voltages.
3. Predictable and repeatable positional accuracy.
4. If distortions are inherent in the deflection process, they should be easily correctable to high precision.
5. Resolution, luminance, contrast, and modulation characteristics should be independent of screen location.
6. The beam should be controllable over a large dynamic range with a smooth and reproducible characteristic.
7. Modulation voltages should be small (5 to 30 volts) and driving capacitance should be low to ensure high bandwidth and low power.
8. The beam should be capable of high speed deflection, at least up to television speeds (5 MHz min).
9. The technique should be amenable to mass production.

most significant or interesting of each approach will be described.

7.6.1 Deflected-Beam Approaches

XY Electrostatic Deflection Grids. This is perhaps the easiest to describe of all the flat CRT approaches. In this technique, modulation and deflection occur in the electron gun exactly as in a conventional electrostatic CRT gun. Deflection of the beam is electrostatic in both X and Y axes. Figure 7-10 illustrates the funda-

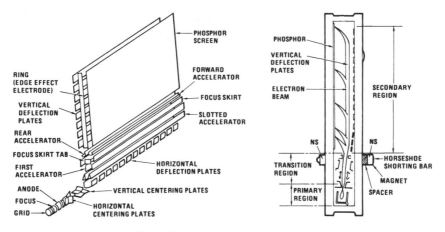

Fig. 7-10. Aiken Tube construction.

mentals of this approach as used in the Aiken Tube.[4-7] The electron gun is off to the side and launches the beam parallel to the display screen. In the configuration shown, the beam travels horizontally along the bottom of the tube through a series of horizontal deflection plates. If the plates were all held at gun potential, the beam would travel all the way to the right edge as it drifts through the field free region. However, if the voltage on one of the U-shaped plates is decreased, the beam will be deflected upward. By controlling the time and magnitude of the voltages in the proper order, the beam can be deflected upwards at any point. In this portion of the tube, deflection voltages are in the order of 500 to 1000 volts. Approximately 10 horizontal deflection elements are required to address any element in a horizontal line (approximately 600 positions per line).

Once deflected upwards, the beam first traverses a transition region which serves to isolate the low-voltage horizontal deflection section from the high-voltage vertical deflection section. The beam then enters the vertical deflection region, which consists of a set of vertical deflection plates on one side (these may be transparent electrodes if the tube is viewed from this side) and the phosphor screen on the other.

For no deflection, both the phosphor and the deflection electrode elements are at the same potential and the beam continues upward through the field free region. However, if the voltage on one of the electrodes is lowered, the beam is directed towards the phosphor. As the beam approaches the phosphor, a natural focusing action also occurs.[4] Note that as shown in Fig. 7-10 the beam hits the phosphor at less than 90-degree incidence. Once again, by varying both the timing and voltage applied to the vertical deflection plates, the beam can be caused to hit any point along the upward trajectory. As in the horizontal section, approximately only 10 plates are required. However, the voltage swing is much higher, on the order of 85% of the anode voltage.[5] Switching these high voltages is an obvious problem.

Conceptually this structure is relatively simple, but several problems exist in achieving performance competitive with conventional CRTs. The achievement of a high current yet small uniform spot size throughout the tube is one (especially considering the long drift regions), while handling the high-voltage switching is another. A good discussion of the techniques and theory of the deflection and focusing mechanisms can be found in Reference 4. Through the years, a number of variations on this technique have been developed in an attempt to overcome these difficulties. One such device, which has been called the Sinclair Tube,[61-63] is shown in Fig. 7-11. The two sets of deflection plates control the point at which the beam strikes the phosphor. This technique eliminates the high-voltage switching encountered in the Aiken Tube. As with all the devices covered in this section, performance results and further details, will be given in Section 7.10.

X Electrostatic, Y Magnetic. Another variation on the Aiken Tube was first announced by

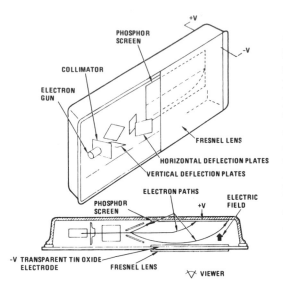

Fig. 7-11. Sinclair flat CRT. (Courtesy of Sinclair)

Sony in 1982.[64,65] This device is very similar to the Sinclair structure except that magnetic deflection is used in the vertical direction. Figure 7-12 illustrates the construction principles used in this device, which was introduced for sale in 1982 for just over $200. An improved version was introduced in 1983 for under $200.

X Electrostatic Deflection, Y Self-Scan. A device somewhat similar to the Aiken Tube was the Gabor Tube, first described in the early

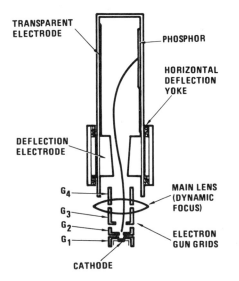

Fig. 7-12. Sony flat CRT. (Courtesy of Sony)

1950s.[1-3] As in the Aiken Tube, a somewhat conventional electron gun was used for beam formation, deflection and modulation. A cutaway view of the display structure is shown in Fig. 7-13. The gun is mounted behind the display surface and the beam is launched downward parallel to the phosphor screen. X-axis deflection is accomplished with the somewhat conventional electrostatic line deflection plates. These plates deflect the beam back and forth parallel to the base plate with an angular excursion of about ± 15 degrees. The beam then enters a rather complicated and novel element called the reversing lens. As a complete description of this element is beyond the scope of this chapter, suffice it to say that the purpose of this element is to reverse the beam back on itself by 180 degrees, shift the beam to the viewer side of the magnetic screen divider, while at the same time increasing the divergence angle by a factor of approximately four (to ± 60 degrees). With the capability of this large divergence, the beam can scan across the entire width

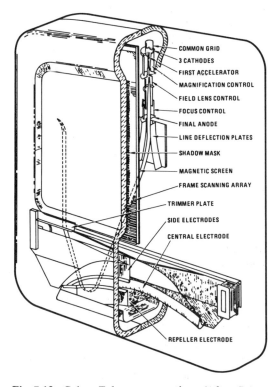

Fig. 7-13. Gabor Tube—cross section. (After Gabor, 1958)

of the picture screen just a few inches above the bottom of the tube. Thus, the bottom few inches of the envelope structure does not have phosphor applied. From this elementary description, it can be seen that the X deflection, consisting of the line deflection plates and the reversing lens, makes it possible to scan across the entire line.

The Y deflection technique is particularly interesting as no external connections to any internal element are required to step the beam from line to line. As the scanning action is internal to the tube, this technique will be referred to as self-scanning, a term also used in various plasma displays (see Chap. 10). The details of this scanning array are shown in Fig. 7-14. The purpose of this element is to deflect the beam towards the screen at the appropriate time. As in the Aiken Tube, this action also tends to focus the beam. A further function of this element is to automatically step the beam position down a line at a time after each horizontal line scan, and also to retrace the beam to the top of the display and reset the potentials at the end of the field scan.

The array consists of approximately 100 conductors on an insulating substrate. The conductors are not connected directly to any voltage source, but are voltage-controlled by the electron beam via secondary emission (see Section 7.7.2 for further discussion of secondary-emission fundamentals). Referring to Fig. 7-14, assume that below the four conductors shown in the middle of the figure, the conductors are charged to the phosphor screen potential. Above the four conductors, the conductors are charged substantially negative, say to one-fourth screen potential. The four conductors serve as a transition region between the two voltage levels. As the beam moves upwards from the bottom, it is in a zero field region until it reaches the transition region, where it is deflected to the screen. Note that this is similar to the action of the vertical deflection plates in the Aiken Tube.

The second function of this array is to automatically perform the vertical scanning. This is accomplished in the U-shaped loops at either side of the active area. Although the entire process is too complex to be completely described here, it is sufficient to note that during the active line scan, the voltages on the conductors remain unchanged. However, when the beam enters the loop shown in the right-hand side of Fig. 7-14, the electron beam strikes on the conductors rather than the phosphor screen. This occurs during the time allotted to line fly-back.

The conductor pattern is deposited such that an offset exists on the conductors around the loop. Thus, as the electron beam is deflected horizontally into this loop, it begins to bombard conductors which are physically below the hor-

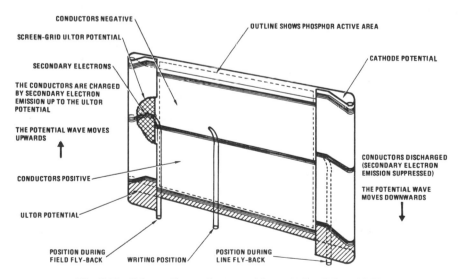

Fig. 7-14. Gabor self-scanning control layer. (After Gabor, 1957)

izontal line just scanned. If the amount of conductor offset is properly designed, the electron beam will now bombard one or more of the conductors which were part of the transition region described above. The electrons accumulate on these conductors (secondary emission ratio less than 1) and discharge the conductors to a lower potential. Now, after fly-back, when the beam begins at the left-hand side of the screen, at the start of the next horizontal line sequence, it will find that the transition region has moved down an amount equivalent to one line position. It is thus possible to change the charge pattern so that after each successive active scan period the scan is automatically lowered by one line.

At the end of the first field, the beam has discharged the entire conductor array. The array is recharged by moving the beam into the loop on the left-hand side of the structure during vertical retrace. This loop differs from the right-hand loop in that it has a slightly larger offset and has a screen collector grid which is maintained at phosphor potential. The grid acts as a collector of secondary electrons. When the beam enters the bottom of the left-hand loop, it bombards the conductor pattern with an energy level such that the secondary emission ratio is greater than one. The excess electrons are collected by the grid, thereby causing the conductor elements it is hitting to recharge to screen potential. As this occurs, the beam is caused

to self-scan upwards. By the end of vertical retrace, the beam has moved back up to the upper left-hand corner of the display, and has completely recharged all the conductors back to screen potential. For further details of this extremely interesting device, the reader is referred to References 1 and 2. Being one of the first of the flat panel display concepts, development was hampered by lack of suitable materials and drivers. Some of the Gabor Tube principles are now being revived in devices such as the Philips Channel Electro Multiplier tube. See Section 7.10.4 for details.[72,73,77]

Independent Beam Addressing of XY Emitter. In this approach, two independent conventional electron guns are used to address the rows and columns of an array of individually controllable electron emitters. This approach, developed by Battelle in Switzerland, uses two scanned electron beams to address an XY matrix of field emission cathodes.[43,44] Figure 7-15 illustrates the basic principles of this device. The electron beams are used to define a specific row and column of the field emitter matrix. Current can only flow through the emitter defined by the intersection of the selected row and column, as the current from the electron gun provides the return path for emitted electrons. The advantage of this device is that true matrix positional accuracy is achieved while retaining the simplicity of electron-beam addressing. Few

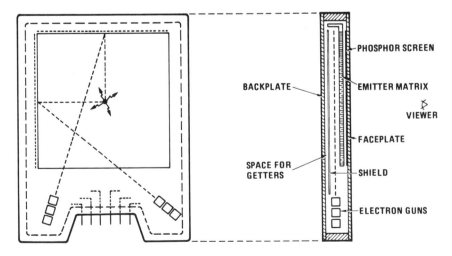

PHOSPHOR SCREEN

BACKPLATE

EMITTER MATRIX

VIEWER

FACEPLATE

SPACE FOR
GETTERS

SHIELD

ELECTRON GUNS

Fig. 7-15. Battelle flat CRT. (Courtesy of Battelle)

details are available on this device, and it is believed that work has been discontinued due to lack of funding.

Electron-Beam Guides with XY Electrostatic Deflection. A new twist combining electrostatic deflection with an electron-gun type cathode has been reported by RCA. This device, first reported at the 1980 SID Symposium, is unusual in that it combines the principles of electron deflection (as in the Aiken Tube) with some of the matrix-addressing principles of an X cathode control–Y control grid approach described in a later section. Also, to enable fabrication in very large sizes, it has a self-supporting structure as shown in Fig. 7-16. The cathode structure consists of a large number of electron guns (120 guns would be used in a 127-cm-diagonal 640-pixel-wide color structure). The electron guns are located at the bottom of the display and launch their beams upward parallel to the phosphor screen. The multiple electron beams are guided to the proper line and deflected towards the screen by a structure called an electron-beam guide.

Two of the various types of guides tested by RCA are shown in Fig. 7-17. In the slalom guide adopted for a flat cathodoluminescent display (Fig. 7-17a), the actions of periodic electrostatic focusing restricts the electron beams to a sinusoidal path over and under the slalom wires. At the desired line, a negative voltage pulse is applied to the extract electrode, which forces the beams toward the phosphor screen.

Since there are 640 pixels per horizontal line (three colors per pixel) and only 120 electron beams, horizontal deflection is required to address the 16 pixels within each module. This occurs in the 7.6 cm-region between the beam guide and the screen by applying a varying voltage to the horizontal scan electrodes on the support vanes shown in Fig. 7-16. An alternative beam guide structure is shown in Fig. 7-17b. This type has been used in the most recent devices. As of late 1983, work was still active on this device, and thus far RCA has tested single modules 2.5 cm wide $\times$ 33 cm long and a multiple module display 12.7 cm wide $\times$ 25.4 cm long (five 2.5 cm $\times$ 25 cm modules). The five-module version has been demonstrated in both monochrome and color.

X Electrostatic Deflection, Y Mechanical. This complex system developed by the Philips Mullared Research Labs in 1955 was aimed at reducing the cost of achieving a full-color display relative to the competing shadow mask color cathode-ray tube.[8-13] The expected cost savings in the tube would be used to offset the increased costs in the receiver electronics and added mechanical and optical components. Unlike all the other devices described in this chapter, which generate a full fixed raster, this tube only generates a three-line-wide image (one line for each color). The line is then optically projected to create a virtual image in conjunction with a stationary hyperbolic mirror. Figure 7-18 shows the long cylindrical Banana Tube and

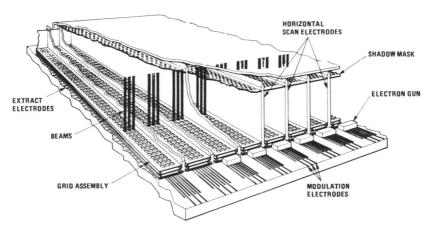

Fig. 7-16. RCA modular flat display. (Courtesy of RCA)

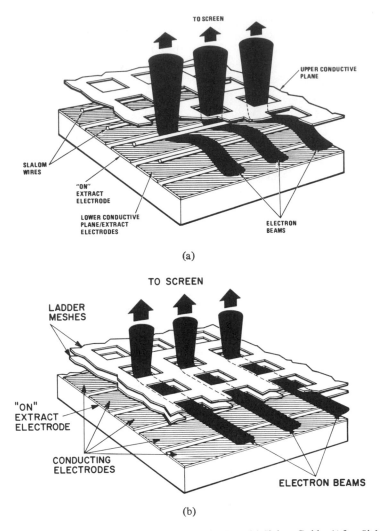

Fig. 7-17. RCA electron beam guide structures. (Courtesy of RCA). (a) Slalom Guide. (After Siekanowicz, 1980). (b) Ladder Guide.

how it was used in the opto-mechanical projection system. Note that to the observer the image appears to be located behind the hyperbolic mirror (virtual image).

7.6.2 Matrix-Addressing Approaches. In this section we will discuss techniques which have been applied to perform the beam positioning (or selection) and modulation in flat CRTs which do not use beam-deflection techniques. In most flat-panel display approaches (liquid crystals, plasma, etc.), the display material is directly controlled by application of the appropriate control signal (voltage, current, electric field, or magnetic field). For example, in an electroluminescent display, a voltage applied across the material causes the material to emit light. This is usually not the case in flat CRT devices. Rather, a control layer(s) is used to selectively control the passage of electrons generated in a separate cathode structure. The electrons which are allowed to pass in turn excite a cathodoluminescent phosphor screen. This control layer functions much like a grid in a triode vacuum tube.

As can be seen by the above discussion, it is generally not possible (nor necessarily desirable) to control all the elements of a matrix

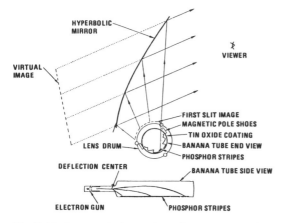

HYPERBOLIC MIRROR

VIRTUAL IMAGE

VIEWER

FIRST SLIT IMAGE
MAGNETIC POLE SHOES
TIN OXIDE COATING
LENS DRUM
BANANA TUBE END VIEW
PHOSPHOR STRIPES
DEFLECTION CENTER
BANANA TUBE SIDE VIEW
ELECTRON GUN
PHOSPHOR STRIPES

Fig. 7-18. Banana Tube and optics system. (After Schagen, 1961)

device with a single control layer. Several alternative methods have been employed to achieve matrix addressing in flat CRTs. These include:

- Direct XY matrix addressing of a discrete element cathode (an individual cathode emitter per resolution element)
- X-axis control of the cathode and a Y axis control grid
- X and Y axis control in separate control grids (includes XYZ or 3-axis addressing)
- X-axis control of the phosphor and a Y-axis control grid
- X and Y axis control in a single control grid via incorporation of an active control element (e.g., thin-film or silicon transistor)

Each of these approaches will be described in the following subsections.

XY Matrix Addressing of a Discrete Element Cathode. As described in the cathode section, one approach to achieving a flat CRT is to have a matrix of individually controllable cathodes. In this approach, direct XY matrix-addressing theory, as described in Chap. 5, can be applied. Line-at-a-time addressing (LAT) is usually employed to maximize brightness and reduce switching speed.

In one such device, Stanford Research Institute (SRI) proposed using an array of field emission cathodes which were controlled by an XY electrode pattern. To improve the unifor-

mity between pixels, each pixel derived its current input from a number of individual field emitters. This allowed differences between individual emitters to be averaged out. As the field emitters are capable of being made in very high resolutions (40 elements/mm), a high-resolution device using multiple emitting elements per pixel could still be realized. The device was constructed such that the X or horizontal-axis electrodes were connected to the bases of a row of pyramid field emitters. The Y or vertical-axis electrodes were then used to apply the field between the tips of the pyramids (as shown earlier in Fig. 7-7) and the gate control anode. Modulation was simultaneously performed by the Y-axis electrodes.

X-Axis Cathode Control, Y-Axis Control Grid. This technique typically utilizes an area-type cathode which is capable of being segmented into individual rows. A grid or control plate which is segmented into individual columns is then used to simultaneously control and/or modulate the output of the selected cathode line. Figure 7-19 illustrates this relatively straightforward concept. The segmented grid or control plate is typically a perforated dielectric material to which a metalized conductor pattern is applied. Figure 7-20 shows a cutaway view of such a grid. As with all materials used in a flat CRT, it is important that the materials used to fabricate the control plate be compatible with vacuum-type structures. This requires that the materials must have good high-vacuum properties (low outgassing characteristics), must be able to withstand high-temperature processing, and must be capable of being manufactured at low cost. Such demands drastically reduce the number of candidate materials and technologies which might be applied to the fabrication of such parts.

Although this technique seems extremely simple and obvious, it should be noted that in practice only a few devices have attempted to use this approach.[46,52,53] This is due to the fact that (as shown in the discussion of cathode techniques) only a very few types of area cathodes have been developed which are capable of being segmented such that they may be

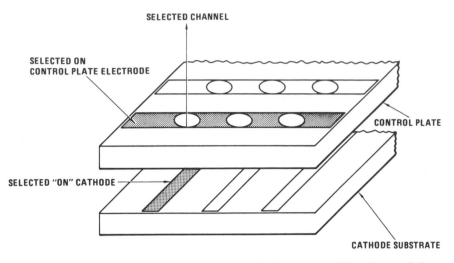

Fig. 7-19. Beam-positioning scheme using X-axis cathode control and Y-axis control plate.

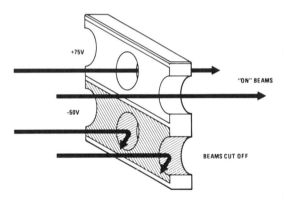

Fig. 7-20. Switching action in typical control plate.

addressed in a line at-a-time mode. Of those cathodes discussed previously, only the field emission, ion feedback, and photocathode feedback "cold" cathode structures were capable of this feature. As of mid 1984, no current research was reported in this area.

X and Y Axis Control in Separate Control Grids. An obvious next step from the approach described in the previous section is to use two control grids, one for each control axis. This allows one to use a nonsegmented area cathode. As segmented area cathodes are typically difficult to fabricate, this approach offers a potential improvement over the previously discussed technique. In this two-control-grid configuration, one grid would select a line to be addressed. The output of this first grid is thus a single line

of electron beams. This serves as the input to the second control grid. This second grid then modulates each of the electron beams. The output of this plate is then allowed to pass on to the phosphor screen. Figure 7-20 illustrates the passage or cutoff of electrons entering one such control plate.

To use such a structure to generate a TV-type picture of say 500 × 500 pixels would require an area-type cathode and two control grids each having a 500 × 500 array of holes. Each grid (or plate) would have an electrode pattern consisting of either 500 rows or columns. Each electrode conductor would have a separate lead attached and would be driven by a separate driver. This would necessitate making electrical contact, through the vacuum tube, to 1000 individual conducters. In addition, the two grids have to be mounted such that the arrays of holes remain in perfect alignment throughout tube fabrication and operation. Primary problems are the large number of connections and drivers required. No structures utilizing two separate control grids have been reported.

XYZ Grid Addressing. One particularly novel approach to overcome some of the problems in the two-control-grid XY approach was developed by Northrop in the early 1960s.[14-21] As this technique has been extensively investigated by a number of laboratories, and has been ap-

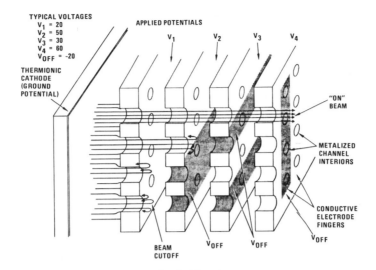

Fig. 7-21. Cross section of switching plates showing beam cutoff.

plied to other non-CRT flat panels, it will be described in more detail than most of the other techniques. This modified approach has often been referred to as an XYZ matrix addressing approach. Of the matrix-addressed flat CRTs discussed in this chapter, this was the first to be conceived and implemented. In its earliest configuration it consisted of a large number of binary control grids (called plates in this section). In its initial configuration, the control

plates performed a dual function: beam control and brightness enhancement. The discussion of the brightness-enhancement version will be reserved for a later section as it is not necessary for understanding of the operation.

Figure 7-21 is a cross-sectional view of several plates, each performing a binary subdivision of the electron input.[14] Figure 7-22, which further describes the principles of operation, is an exploded view of a simple 8-element-by-8-element

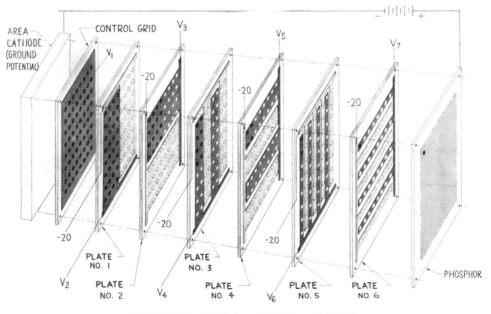

Fig. 7-22. Exploded view of Northrop flat CRT.

display. In this configuration, each plate requires only two electrical connections, and it is the simplest mode of operation to describe. However, in practice, multilead electroding patterns are used to combine the function of several plates. The multilead concept will be described later. The principle components are:

1. An *area cathode* of one of the types described earlier (usually a multifilament type).
2. A series of thin, electroded *switching plates* each containing an array of channels through which the electrons from the area cathode flow on their way to the phosphor display screen. These plates not only confine the electron flow to physically defined channels, but also cause splitting of the incident electron flood by stopping the electron flow in a selected group of channels. Digital addressing signals are applied to the switching plates. In turn, each plate allows only a fraction (one-half in this example) of the number of beams to exit the plate as that entered. As will be seen, other decoding factors have been shown to be more practical.
2. A video *modulation plate* or *grid*, to modulate the electron-beam current with the video signal.
4. *Dielectric spacers* (not shown) located between the plates, to increase the mechanical rigidity of the structure. When finally assembled, the layers of switching, modulating, and spacing plates were frit sealed together to form a rugged structure.
5. For a *target*, either a monochrome or multicolor phosphor may be used. The scanning electron beam writes its information on the phosphor target where it can be viewed through the faceplate on which the phosphor is deposited.

Following through the operation of the device shown in Fig. 7-22, electrons travel from the area cathode through a series of apertured plates (switching plates and modulation plate) to the phosphor target. The switching plates are constructed from glass substrates and contain an array of channels through which the

electrons flow. The surfaces of these plates are coated with conductive electrode patterns which connect groups of channels according to a predetermined coding scheme. These plates act collectively to cut off electron flow so that single or multiple electron beams emerge from the last plate at discrete positions determined by the digital addressing signals and the actual electrode coding utilized. The scanning beam, or beams, write the information on a phosphor target where it can be viewed through the faceplate.

The electron passage and cutoff processes that occur in the switching plates are identical to those illustrated in Figs. 7-20 and 7-21. Referring to the four opened channels in Fig. 7-21, electrons emitted from the area cathode see an accelerating potential V_1 where V_1 is a positive voltage relative to the grounded cathode (typically V_1 is 20–30 volts). Some electrons collide with the surface; others enter the holes of the first switching plate. Electrons leaving the top two channels, however, see a retarding potential (approximately – 20 V) on the second switching plate. Because this voltage is lower than the initial voltages of the most energetic electrons, the forward motion of the electron beam is reversed and beam cutoff occurs. Since all electrons within the channel originated at the cathode, it is necessary only to apply a channel voltage slightly more negative than the cathode voltage to achieve cutoff (the negative voltage is required to account for the Maxwellian velocity distribution of the emitted electrons and to prevent pull-through from subsequent plates).

Of the two electron beams leaving the second plate, the lower one is cutoff by a retarding voltage applied to the corresponding channel in the third switching plate. The top beam sees a positive voltage of magnitude V_3. Typically, V_3 is smaller than V_2. It should be noted that each plate has a lens effect on the electron beam. "On" regions of plates have alternating voltages (high, low, high, etc.), but always positive relative to the cathode. Only the top beam passes through all four plates unaffected.

These operational principles can now be applied to the total number of switching plates required for the 8 × 8-dot matrix device shown in the exploded view Fig. 7-22. This figure illustrates the means by which a single electron beam is obtained from the last plate, at a discrete position determined by the digital addressing signals. As previously mentioned, the electrode pattern follows a binary coding scheme which electrically divides each plate in half. Potentials are applied to the electrodes such that half the channels have an electron-accelerating potential and half have an electron-retarding potential. In the first switching plate of Fig. 7-22, the 32 channels in the left half of the plate have a positive potential applied to them so the electrons come through; the 32 channels in the right half have a negative potential which causes the electron trajectories to reverse and thus produce beam cutoff. The electrode pattern on the second switching plate is identical to that of the first except for a 90-degree rotation. Electrons coming out of the 32 channels in the left half of the first plate enter the corresponding channels in the second plate. Since the channels in the lower half of this plate are biased off, electrons emerge only from the 16 channels in the upper left quadrant and proceed to the third plate. In the third plate, the eight channels in the right half of the quadrant are biased off so that the electron flow is bisected down to eight elements "on." As can be seen, this selective splitting process continues in the following plates until only a single beam emerges from the final switching plate—in this case, in the upper left corner. The position of this beam

can be changed simply by reversing the polarity of the potential on one or more plates with an electronic switch.

The display shown in Fig. 7-22 for reasons of simplicity has binary half-splitting switching plates. However, it was shown in actual practice that other decoding schemes are more efficient than the simple two-leads-per-plate technique shown. These decoding schemes involve the "combining" of two or more decoding patterns on one plate. Referring to Fig. 7-23, we can see how one might apply this technique.

From this simplified figure showing the switching patterns for an 8 X 8-element display, one can see that the function of two binary coded plates can be combined into one plate having four leads. Note that although the number of plates required has been reduced by a factor of 2, the total number of leads and drivers remains the same. However, if this process of combining the function of two plates into one plate is carried still further (as is the actual case in displays of larger size), the number of leads may increase. This usually occurs when designs having more than eight leads per plate are employed. The advantage of the reduction in the number of plates must then be compared with the additional complexity of additional leads and switching circuitry to obtain an optimum configuration. It should also be noted that as more plates are combined, the total area driven

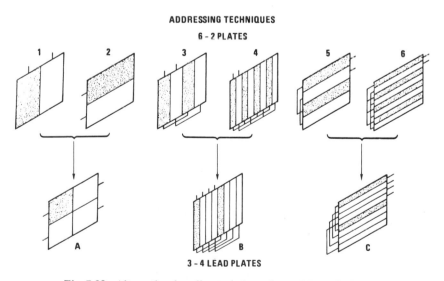

ADDRESSING TECHNIQUES

6 - 2 PLATES

3 - 4 LEAD PLATES

Fig. 7-23. Alternative decoding techniques for an 8 by 8 display.

by the individual switching elements is reduced, thereby reducing the capacitive load and hence reducing switching power. As an example, one effective conductor-coding arrangement developed for a 512 TV/graphic device used only four plates to address the 262,144 beam positions. This design used a total of 96 leads, consideraly less than other matrix displays of that time, which typically required 1,024 leads for a display of similar size.

Devices with a large number of elements could be designed in a multitude of different configurations.[14,69] For example, the basic algorithm for selecting possible decoding schemes is given below:

or

2^{18}: eighteen 2-lead plates
4^9: nine 4-lead plates
8^6: six 8-lead plates
$16^2 \times 32^2$: two 16-lead and two 32-lead plates
64^3: three 64-lead plates
512^2: two 512-lead plates

The decoding scheme of a specific display is then determined by a tradeoff analysis involving per-lead capacitance, desired beam-stepping rate, and practical considerations such as the

$$\text{Number of resolution elements} = 512 \times 512 = n_0 \prod_{i=1}^{N} n_i \quad \text{(Eq. 7.8)}$$

where

n_0 = number of modulated beams
n_i = decoding factor of the i'th switching plate (a half-splitting plate has a decoding factor = 2)
N = number of switching plates

From this relationship, six of the many possible combinations are given as:

$$512 \times 512 = 2^{18} = 4^9 = 8^6$$
$$= 16^2 \times 32^2 = 64^3 = 512^2$$

cost of the plates, making connections to the display, and the number of drivers required.

Conventional monochrome CRT displays normally contain only a single electron gun and therefore have only a single electron beam for writing the information on the phosphor. This device, however, could have as many electron beams as there were channel holes in each plate. In a single-beam device, all but the one desired beam are cut off by the switching plates. One method of converting a single-beam scanner to a four-beam scanner is shown in Fig. 7-24. Note that the functions of plates 3 and 5 shown in

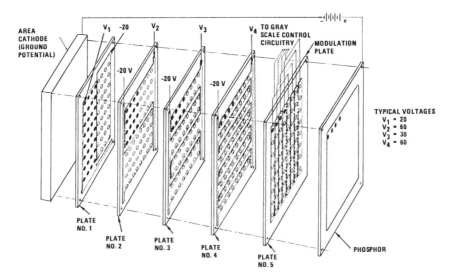

Fig. 7-24. Northrop flat CRT configured for multiple-beam addressing.

Fig. 7-22 (two leads per plate) have been replaced by one four-lead modulation plate (plate 5 in Fig. 7-24). This plate simultaneously modulates the four colinear beams. Several advantages of the multiple-beam technique now become apparent. For a given frame rate, the dwell time of each scanning beam is four times longer in a four-beam scanner that in a single-beam scanner, thus bombarding each phosphor element with four times as many electrons per frame, leading to a corresponding fourfold increase in display luminance (neglecting phosphor saturation). Also, since four beams are available to address the entire raster instead of one, the beam-stepping rate is reduced by a factor of 4. This saves power, as well as allowing the use of slower plate drivers. Figure 7-25 shows the decoding patterns used for a 512 × 512 device written with 32 multiple beams.

The primary shortcoming of the XYZ addressing technique was the cost of fabrication of the large number of plates (typically six plus five spacers) associated with the decoding process. In addition, physical size was limited by the ability to make a large flat vacuum enclosure. Results achieved with this technique will be given in Section 7.10.

X-Axis Control of the Phosphor, Y-Axis Control Grid.   A quite successful variation on the technique using X-axis control of the cathode was developed by ISE Corporation of Japan.[25-33,67] This device (which followed developments by Tung Sol and GE of single-digit versions), often referred to as vacuum fluorescence, has been extremely successful in achieving low-cost alphanumeric displays (up to 300 characters) and medium-resolution graphic displays (up to 256 × 512 was available in 1983). In this technique, a multifilament oxide-coated cathode is used. Figure 7-26 shows a cutaway view of this structure for a segmented eight-character numeric display. The unusual features of this display are:

1. Utilization of an inverse stacking structure, where the cathode is closest to the observer and the phosphor is furthest away.
2. A low-voltage phosphor (ZnO) which operates at about 50 to 100 volts instead of the usual 10 to 20 kV employed in other conventional and flat CRTs.
3. Part of the decoding is done at the phosphor surface.

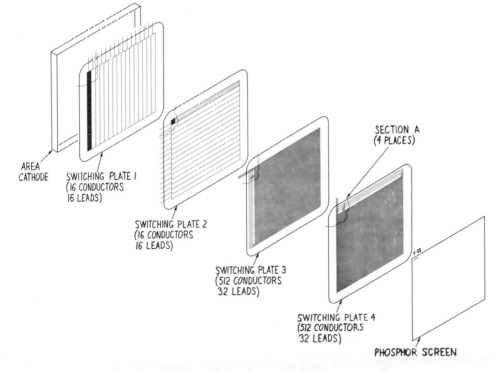

Fig. 7-25. Exploded view of 512 × 512 TV/graphic Digisplay.

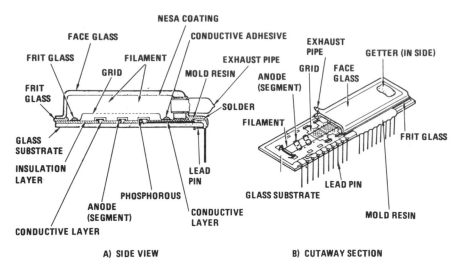

Fig. 7-26. Vacuum fluorescent flat CRT. (Courtesy of ISE)

4. In most dot-matrix character displays, all dots within a character are addressed in parallel.

It is these last two features which are germane to this part of the discussion. To provide a simple and relatively low-cost alphanumeric display, the configuration shown in Fig. 7-27 is used. In this multiline device, the Y-axis control grid, fabricated of a suspended metal mesh, is used to select which column of characters is to be addressed. Then, each dot of the character (35 for a 5 × 7 character) is addressed in parallel by the pattern applied on the glass substrate. The character pattern is defined by the screened phosphor which is applied in the form of the desired character (5 × 7 in this example). If additional rows of characters are desired, they are addressed in parallel. Tubes of this type have been quite successful in commerical applications.

Single Control Grid with Active Matrix. The final addressing technique to be described uses a single grid to control the passage of electrons to the phosphor. This is made possible by the addition of an active matrix to the control grid. Both thin-film transistors (TFTs) and MOS field-effect transistors (MOS FETs) have been used as the active control layer.

In the TFT approach, workers at Arizona State University used the structure shown in Fig. 7-28.[45] The addressing operation is essentially identical to that used by Brody at Westinghouse to address liquid-crystal and electroluminescent displays.[66] These techniques are discussed in detail in Chaps. 2, 8, and 11. Suffice it to say here that one or two TFTs and a storage capacitor reside at each pixel location. The capacitor was used to give a frame-time storage such that the charge supplied to the capacitor during the line-at-a-time addressing pulses was sufficient to keep the control grid in the "on" mode for an entire frame (usually $\frac{1}{30}$ sec). The frame storage allowed very bright displays to be fabricated even with area cathodes (multifilament type) with low output. However, as with nearly all attempts to use TFTs up to 1983, the cost, yield, and complexity of making a useable-size display proved more than could be realized.

In the MOS FET approach, workers at ISE have applied the silicon-transistor approach to vacuum fluorescence[33]. (The basic structure and operation of the MOS FETs are described in Chaps. 2, 8, and 11.) Figure 7-29 shows the structure and the relationship between the cathode and substrate. The phosphor is patterned directly on top of the transistor such that when the element is activated, the phosphor is bombarded by the electrons from the area cathode. In this structure the phosphor and active control layer are located on the same substrate and no control grids are required. The screen shown in Fig. 7-29 is a part of the area cathode struc-

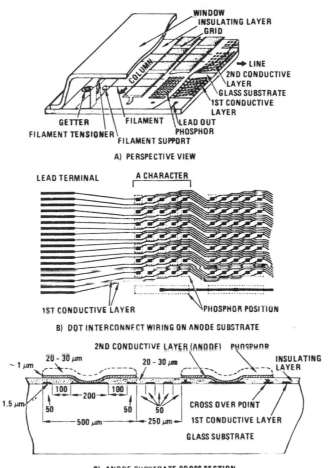

Fig. 7-27. Multicharacter vacuum fluorescent display. (Courtesy ISE)

ture used to improve uniformity and is not used in addressing as in Fig. 7-27. As with other MOS FET structures, most of the difficulty pertains to achieving a defect-free low-cost structure and in extending the use to large sizes.

7.6.3 Modulation Techniques. In the previous discussions of beam-positioning techniques, only brief mention was made of modulation techniques. However, it must be noted that some devices require special techniques to en-

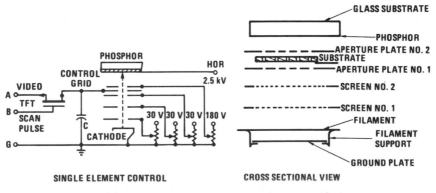

Fig. 7-28. TFT-addressed flat CRT. (After Sirkis, 1976)

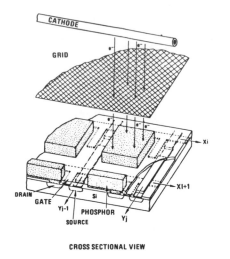 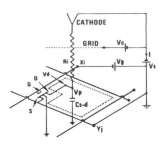

CROSS SECTIONAL VIEW   ADDRESSING CIRCUITS

Fig. 7-29. ISE MOSFET-addressed flat CRT. (After Uemura, 1979)

sure adequate uniformity and performance. Although the following discussion is particularly directed at modulation of flat CRTs, most of the comments apply to all flat-panel displays. Basically, one of three techniques is used to achieve intensity variation in flat CRTs. The special use of on/off operation, as used in alphanumeric and/or graphic-only operation, will not be considered separately as it can be derived as a subset of the video modulation techniques in which a range of gray scale is required. The three techniques are:

1. Amplitude or analog modulation.
2. Pulse-width or digital modulation.
3. Combined amplitude and pulse-width modulation.

In flat-panel CRTs using multiple modulation circuits to control multiple-simultaneous beams hitting the phosphor, special precautions must be taken to ensure that uniform operation results. The two major causes of modulation non-uniformity are:

1. Cell-to-cell variations in the modulation characteristics.
2. Changes in performance from driver to driver.

Selection of a modulation technique requires that both of these factors be considered.

Amplitude Modulation. This technique is the one most commonly used in devices designed to display video signals, as the input signal is already in analog format and the circuits are relatively easy to design to achieve the desired speed and performance. All the deflected-beam approaches described earlier use this approach. Most of the matrix-addressed devices opt to use this technique as a first choice, if it will work with the flat CRT in question. One typical modulation problem which is encountered in many display devices is shown in Fig. 7-30. This figure shows the transfer characteristics of four different pixels. At this point, only pixels with characteristics as shown in curves 1 and 2 will be discussed. At a given drive level V_1, the saturation output currents from the two cells are identical and equal luminance is achieved from the two channels. However, at a lower drive level, V_3, the output is not the same. This may be caused by a number of factors, such as:

1. Different geometries in the holes at different places in the control plate or grid.
2. Differences in the energy or trajectories of the electrons going through different holes, caused by variations in area cathode output.
3. Fabricational misalignments between interrelated parts.

In any event, such variations are not uncommon. Note that other nonuniformities resulting

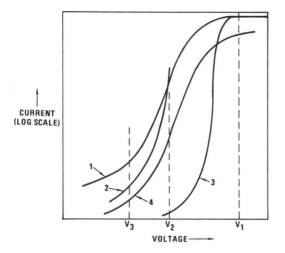

CURRENT
(LOG SCALE)

VOLTAGE ⟶

Fig. 7-30. Modulation characteristics variation – cell to cell.

in differences in output at all drive levels (such as shown in cell 4 vs cell 1) are also likely, but are not considered as part of the modulation discussion. However, it should be noted that with appropriate device feedback monitoring it is possible to implement modulator drivers which compensate for cells with different saturation-level current outputs. As can be seen, if amplitude modulation is attempted in a device having pixel-to-pixel variations as shown in curves 1 and 2 of Fig. 7-30, objectionable nonuniformities will be evident in the darker gray shades. This problem is common to many flat-panel displays, including flat CRTs. If such a characteristic is present, one of the other modulation techniques must be used.

Pulse-Width Modulation. Pulse-width modulation is usually considered as a second choice for most systems. Even in systems using digital refresh memories or digital transmission, it is often easier to D/A convert to get good gray scale than to use pulse-width modulation. One reason for utilizing pulse-width modulation was described above. Another would be if a panel's transfer characteristics appeared as shown by curve number 3 in Fig. 7-30, where the transition from "on" to "off" is so abrupt that it is extremely difficult to achieve the required uniformity using voltage modulation. In these cases, pulse-width modulation is often adopted.

As the name implies, varying shades of gray

are obtained by adjusting the length of time for which a cell is turned "on." Circuits to accomplish such modulation are usually relatively easy to design. The major concern in use of this technique is whether there is adequate time to achieve the desired number of gray-scale steps. In some cases, the maximum number of gray-scale levels achievable with this technique are limited by:

1. The response characteristics of the digital drivers (rise and fall times). Many devices have relatively large capacitances and require substantial voltage swings.
2. The response characteristics of the device cells. Some devices do not respond to very short pulses.
3. In large devices (500 × 500 and up) using something less than line-at-a-time addressing, the maximum full-brightness "on" time is already very short. This places an even heavier burden on the drivers.

For example, suppose the application is to use a flat-panel display to portray conventional 525-line 30-frame-per-second home television pictures. Using line-at-a-time addressing, one has approximately 64 microseconds for each pixel. To avoid any noticeable quantization for good video, it is desirable to have available at least 64 different levels of intensity (6 bits). As the eye essentially operates as a logarithmic detector, it is desirable to have adjacent gray levels differ logarithmically. However, for simplicity, assume the 64 levels to be equally spaced. This allows only 1 microsecond as the minimum pulse "on" time for the darkest gray shade. For good performance, the pulse rise and fall times should be at least five times faster than the desired "on" time—thus requiring drivers with a rise-and-fall response of less than 200 microseconds. As large capacitances and moderate-to-large voltage swings are often required, this presents a formidable design problem. In addition, if the device uses less than line-at-a-time addressing, the problem is even more severe, as the maximum "on" time is reduced correspondingly.

Combined Pulse-Width and Amplitude Modulation. If neither of the above techniques is

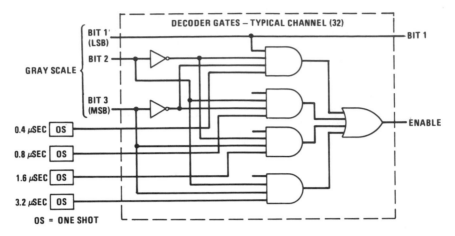

Fig. 7-31. Circuit configuration for combined pulse width and amplitude modulation. (After Goede, 1974)

satisfactory, by itself, it has been shown feasible to use a combination in some devices.[15,68] This requires that at least a portion of the cells transfer characteristics be uniform from pixel to pixel, just as the characteristics of pixels 1 and 2 in Fig. 7-30 were uniform from V_1 to V_2. As an example of how this combined technique might be implemented, assume that because of large-line capacitance characteristics only four levels of pulse-width modulation could be achieved, and that a pixel only has uniform response characteristics ranging from a saturation brightness voltage $\sqrt{2}$ V (corresponding to point V_1 in Fig. 7-30) to brightness voltage V (point V_2). Further assume that eight gray shades are desired, with each shade being $\sqrt{2}$ higher than the previous shade. Note that a linear step difference between shades is undesir-

able as the eye functions much like a log detector of luminance and is not linear (see Chap. 3 for further details).

To achieve additional levels of gray scale over that achievable with either of the individual techniques, the circuit shown in Fig. 7-31 can be used. Here the additional gray levels are obtained by adding a two-state voltage level in the switch. The truth table of this gray level technique is shown in Table 7-8. Although this technique is relatively easy to implement, the cost of the drivers is quite high. As driver costs are already a major problem area, this solution is not without its drawbacks.

7.6.4 Beam Positioning and Modulation Summary. As can be readily seen, a large number of beam-positioning techniques have been tried

Table 7-8.   Gray-Scale Truth Table

Gray Level	Gray-Scale Data (Bits) 3	2	1	Pulse-Width Time in μsec	Amplitude V	Display Brightness Relative
1	0	0	0	0.0	V	0[1]
2	0	0	1	0.4	$\sqrt{2}$ V	1[2]
3	0	1	0	0.8	V	1.4
4	0	1	1	0.8	$\sqrt{2}$ V	2.0
5	1	0	0	1.6	V	2.83
6	1	0	1	1.6	$\sqrt{2}$ V	4.0
7	1	1	0	3.2	V	5.66
8	1	1	1	3.2	$\sqrt{2}$ V	8.0

[1] Actual brightness will be determined by ambient light levels, phosphor scattering, and by leakage currents, if any, from the switching stack.
[2] Arbitrary units—referenced to 1 unit.

in an attempt to achieve a commercially viable flat CRT. At the beginning of the 1980s, only the technique used by the vacuum fluorescence display has been commercially successful. As of this writing in early 1984, the techniques used in the Sony and Sinclair tubes are also beginning to see commercial implementation. However, the extent of their market penetration is unknown. In general, all of the other techniques have failed, due either to high cost caused by either device or driver circuit complexity, or to poor performance particularly in terms of uniformity and brightness, or to inadequate funding. Considerable effort is still under way in some laboratories, and it is possible that some of the beam-positioning techniques described above will be reintroduced as improvements are made in either driver or fabrication techniques.

7.7 BRIGHTNESS-ENHANCEMENT TECHNIQUES

As discussed in the cathode section, several cathode techniques which have been employed in various devices have had inadequate current output to achieve the desired luminance values. Some of the techniques which have been tried to increase luminance, other than improving cathode output, include:

1. Multiple-beam addressing.
2. Electron multipliers.
3. Storage techniques.

7.7.1 Multiple-Beam Addressing. As described in Section 7.6.2, the use of multiple beams is one of the most common techniques used to increase display luminance. This method is easily adapted to all of the techniques previously described except the deflected-beam approaches. The deflected-beam versions usually use no more than three simultaneous beams to achieve color (one beam per primary). In nearly all the matrix-addressing techniques, an area cathode is used which makes possible the use of up to line-at-a-time addressing. As luminance is generally increased nearly linearly with the number of beams used, this is an extremely powerful technique. The only major drawbacks are:

1. In some devices, the number of leads which must be addressed may increase (e.g., XYZ technique).
2. The number of modulator circuits is increased usually with a corresponding decrease in the number of line drivers. Note that modulator drivers are typically more expensive than line drivers as the modulator drivers must control gray-scale levels whereas the line drivers are two-state devices.

On the positive side, in addition to increased luminance, the following advantages are obtained:

1. The speed of the fastest drivers the modulators—is decreased in direct proportion to the number of multiple beams. Thus, less expensive drivers can be used.
2. Switching power is reduced. Reactive power goes as $1/n$ where n is the number of simultaneous beams.

As the advantages of a line-at-a-time addressing (one form of multiple-beam addressing) have been discussed in detail in Chap. 5, it will not be discussed further here.

7.7.2 Electron Multipliers. Electron multipliers have also been extensively used to increase display luminance. In some cases, the electron multiplication process is included as part of the beam-positioning control element section, and in others it is a separate function. One of each of these two classes of electron multiplier approaches will be described. Both types rely on the principle of secondary emission. As several excellent texts describe the basic mechanisms of secondary emission, it is sufficient to note here that when a particle collides with a surface, a certain probability exists that one or more free electrons will be generated.[70,71] The ratio of the number of electrons generated from a single impact is defined as the secondary emission ratio δ. The secondary emission ratio is a function of a large number of factors including: impact energy, angle of impact, and the materials involved. Good

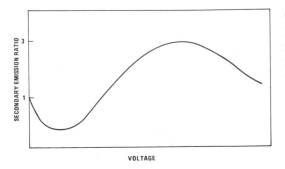

Fig. 7-32. Typical secondary emission curve.

materials have a secondary emission ratio of between 2 and 6 at impact energies of 100 to 300 eV (values of over 20 have been reported under special conditions). A typical plot of secondary emission vs. accelerating voltage is shown in Fig. 7-32.

Electron Multiplication Within Beam-Positioning Elements. In the Northrop XYZ device described previously, early versions used control elements (called dynodes) which performed both beam positioning and electron multiplication.[16] A side view of a single dynode plate is shown in Fig. 7-33. Note that unlike the plates discussed earlier, the inside of the holes are not metallized all the way through. Instead, the inside of the holes are coated with a good secondary emitting material. To prevent the material from accumulating a charge as secondary electrons are drawn off, the material is doped to be slightly conductive. A potential is applied across the hole such that a uniform voltage gradient exists within the hole. This technique has been applied successfully in several nondisplay applications where amplification of low

electron-beam currents are required. The parts performing this amplification have been called microchannel plates and have been used in night-vision goggles and other low-light-level-type products.[72]

Figure 7-34 shows a cross-sectional view of three such plates in which beam multiplication and switching are occurring simultaneously. Although high electron gain could be achieved in such a structure, stable long-life secondary emission materials were never found. In addition, the requirement for an auxiliary current flowing through the resistive secondary emission material within the holes greatly increased power requirements. For these reasons work on this approach was discontinued.

Separate Electron Multipliers. In another approach, independently researched at Northrop, RCA, and Philips, separate metallic electron multipliers were used.[46,52,53,72,73] Figure 7-35 shows a schematic drawing of the Philips device.[72,73] In this structure, the dynodes can be designed for maximum gain without regard for switching performance. In addition, metal dynodes appear to be more stable than the dielectric and semiconducting versions. The Philips type uses material and fabrication processes identical to those used to make the shadow mask in a conventional color CRT. The relatively low cost of this structure compared to previous dynode multiplier attempts holds promise for this technique.

7.7.3 Internal-Storage Techniques. In the beam-positioning section, active matrix-addressing techniques were discussed as one form of matrix addressing. An important benefit of these

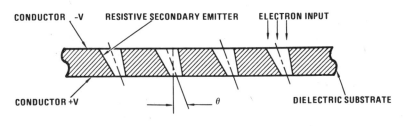

θ = HOLE CANT ANGLE

Fig. 7-33. Multiplier dynode configuration. (After Goede, 1972)

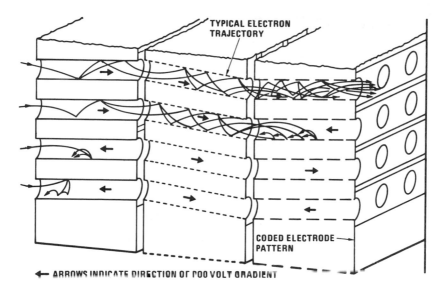

TYPICAL ELECTRON
TRAJECTORY

CODED ELECTRODE
PATTERN

← ARROWS INDICATE DIRECTION OF 700 VOLT GRADIENT

Fig. 7-34. Northrop multiplier dynode plates.

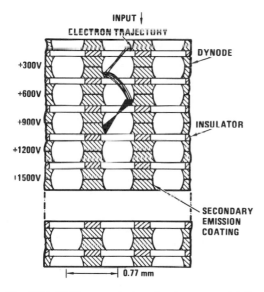

INPUT
ELECTRON TRAJECTORY

DYNODE

+300V

+600V

+900V   INSULATOR

+1200V

+1500V

SECONDARY
EMISSION
COATING

|← 0.77 mm →|

Fig. 7-35. Philips metal dynode electron multiplier. (After Woodhead, 1982)

techniques, using either TFTs or MOS FETs, is the incorporation of a frame storage capacitor. This effectively decreases the duty cycle from $\frac{1}{500}$ for a line-at-a-time-addressed 500×500-element display to a duty cycle of nearly 100%. This increases the luminance by a factor of 500 (assuming no saturation effects). This is an extremely important advantage of this technique, and is one of several reasons that make the active matrix-addressing scheme attractive, de-

spite the tremendous technical difficulty of overcoming the yield and fabrication problems.

Another technique for achieving both brightness enhancement and inherent storage utilizes the addition of a half-tone storage mesh or plate into the structure. This technique borrows heavily on principles developed for conventional CRT storage tube technology.[70] The Northrop Digisplay is one device which attempted to utilize this feature. Figure 7-36 shows a comparison between a conventional CRT storage structure and the flat CRT implementation.[14] The addition of storage to the flat CRT is much simpler than is the case for converting the conventional CRT to storage. This is due to the fact that a uniform area cathode capable of high current output is already available. In addition, the members to support the storage structure already exist.

Test results at Northrop were very promising, showing that normal storage tube operating characteristics could be exceeded. Selective erase (without an erase flash) was also demonstrated and shown to be feasible. As the storage material, typically MgO, could be applied to a switching plate instead of a stretched wire mesh, the structure was also more rugged than a conventional structure. Workers at Northrop were actively pursuing the storage concept up until the time work was discontinued when rights to the

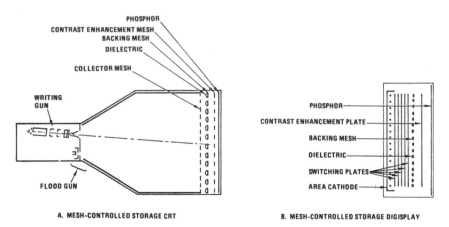

Fig. 7-36. Comparison of CRT conventional storage and Northrop flat storage CRT.

Digisplay technology were purchased by Texas Instruments. At Texas Instruments, the major thrust was to try to adapt the device to compete in the home television market. As the number of gray levels achievable in storage structures of this type are limited and insufficient for television applications, the storage approach was not pursued at Texas Instruments.

A bistable storage target, such as that used in the Tektronix storage tube, could also be used. However, the use of this type of target in a flat CRT has never been reported.

7.8   PHOSPHOR SCREENS

One advantage of flat CRTs is their ability to utilize light-generation materials which are well developed, inexpensive, and very efficient. In Chap. 6, a thorough discussion of phosphor screens was presented, and therefore only brief mention of them will be given here. For monochrome displays, one has a large choice of colors, response times, and other related parameters.[75] This large number of options is an advantage for the flat CRTs as compared to other flat-panel displays which are normally extremely limited in this area.

For color displays, flat CRTs are also able to make use of the technologies developed for conventional CRTs. To date, the only good multicolor displays of any type have had to rely on using phosphor material as the light-generation technique. This is something of a testament to the superiority of this technique

for the generation of multicolor displays. Color flat CRTs have been built which use one following techniques:

1. Three-color primary triad technique.
2. Voltage-sensitive beam-penetration phosphor.
3. Current-sensitive phosphors.

In general, one advantage of the matrix-addressed flat CRT is that perfect registration exists between the beam-positioning structure and the phosphor. Thus, in most cases, no additional shadow mask is required, and the need for convergence is eliminated.

It should be noted that some of the flat CRTs discussed in this chapter are designed to operate at relatively low voltage (4 to 8 kV) as compared to a conventional CRT (10 to 30 kV). At first, this might appear to be a significant advantage. However, phosphors tend to decrease in luminance with use (Coulomb aging). This is strictly a function of how many electrons strike the phosphor. Thus for a given luminance, a device operated at higher voltage will last longer because the current required to achieve that luminance is lower (luminance increases with voltage in approximately the proportion of $V^{1.6}$ to V^2). This coulomb aging effect is described more fully in Chap. 6 under the discussion of Pfahnl's Law. This aging effect applies primarily to insulating type phosphors and not to the very low voltage metallic ZnO

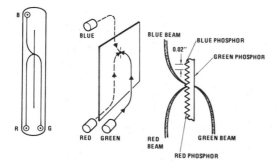

Fig. 7-37. Full-color Aiken Tube using Geer screen. (After Geek, 1964)

phosphor used in vacuum fluorescent displays. These devices typically operate at 50 to 100 V.

One color technique not discussed elsewhere in this book was invented by Geer and was called the Geer Screen.[7] Figure 7-37 illustrates the principle utilized in this device. This was used in an Aiken Tube structure. A transparent internally supported substrate was utilized. One side had a serrated sawtooth pattern as shown in the detail. This side contained the red and blue phosphors on opposite edges of the sawtooth. The other side was flat and contained the green phosphor. Good color purity could be without a shadow mask. Unfortunately, this technique is not applicable to most flat CRT concepts.

7.9 VACUUM ENVELOPE AND PROCESSING TECHNIQUES

A major portion of the effort required to produce a commercially viable flat CRT must be expended in the area of developing a suitable vacuum enclosure, and in developing a material system compatible with whatever processing steps are required to produce long-life reliable vacuum tubes. As this latter point is extremely device specific, the majority of this section will be devoted to the problem common to all the flat CRTs, the vacuum enclosure.

7.9.1 Vacuum Envelope. Virtually all of the structures which have been investigated in the flat CRT category require a flat-face envelope on the side facing the viewer. This single requirement makes it tremendously difficult to develop a general-purpose tube structure capable of being applied to a multitude of display sizes. This is because all such structures are evacuated and must be capable of withstanding atmospheric pressure. In fact, the CRT standard for safety (MIL STD MIL-E-1, ref MIL STD-1311 Method 1141) requires that vacuum envelopes be capable of withstanding a pressure differential of 3 atmospheres or 45 psi (6,500 lb per square foot of surface area)! This huge force has made fabrication of large devices with a flat face a very difficult problem. Although several concepts for overcoming this problem have been proposed and small prototypes fabricated, no commercially viable technique has been verified. To overcome this problem in a conventional CRT, the faceplate is made convex. Glass is strong in compression, but very weak in tension.

A further demand placed on the design of the envelope is that provision for a large number of electrical feedthroughs must be incorporated. In all but the deflected-beam approaches, between 100 and 1,000 leads must be brought out of the envelope for a 500 × 500-element display. The pinouts must not only be reliable and leak-free, but must allow easy and low cost connection. In the following sections, a few approaches to solving these problems will be discussed.

Open-Structure Envelope. The most economical technique for the envelope structure is to utilize commercially available flat glass faceplates. These are available in a number of sizes up to around 25 × 25 cm. Figure 7.38 shows several structures using two such faceplates back to back to form the envelope. In this structure, developed by Northrop, the leads were brought through the frit seal. The lead material was a special thin metal strip referred to in the tube industry as Sylvania alloy #4. The lead patterns were electroformed into a self-supporting lead frame and were wet-hydrogen-fired before frit sealing. This was found to produce reliable leak-free seals. However, connection still had to be individually made to both the external drivers and internal switching plates, a time-consuming and expensive process.

For relatively small sizes, the structure shown

Fig. 7-38. Flat CRT envelopes. (Courtesy of Northrop)

in Fig. 7-26 developed by ISE represents an advantage in that the lead pins shown can be made to plug directly into an integrated-circuit-type socket. Another structure using a glass substrate is shown in Fig. 7-39. Here the substrate, placed between the two conventional CRT faceplates, protrudes on all four sides. In this case the flat substrate represents a convenient surface to make connection to both the internal tube elements and external drivers.[21] By proper design, a connector can be fabricated which plugs directly onto the substrate. In this case, the leads going through the frit glass seal may be either thin or thick film. Special materials are required to ensure leak-free implementation.

All of the structures described in this section

suffer from being unable to be extended to tube sizes larger than approximately 25 × 25 cm, because of the force of the external ambient pressure. To withstand the ambient, it is possible to make thicker glass faceplates. However, there are several negative aspects to this solution. First, as the faceplate gets thicker, the envelope weight increases substantially. In a 25 × 25 cm structure, the envelope weight alone was approximately 5 kg and the faceplate thickness has to be approximately 1.5 cm. In addition to the weight, the thick faceplate creates a contrast and resolution problem due to internal reflections within the glass faceplate. The thicker the faceplate, the bigger the spot-halo and light-spreading problem.

Flat-Face Self-Supporting Envelope. To overcome the size and weight limitations in going to large active area devices, several laboratories have pursued structures which have periodic support elements within the active area. This is commonly referred to as a self-supporting tube structure, as the external ambient pressure is withstood by either pillars or vanes that run from one side of the structure to the other. The faceplate thickness can now be very thin to minimize light spreading and halo. Faceplate

Fig. 7-39. Flat CRT using center mounting plate. (Courtesy of Northrop)

thickness is thus independent of display active area, and tubes of 100 to 150 cm diagonal can now at least be considered. With this structure, however, comes a new set of problems. The first of these is how to fabricate the tube such that the support structure does not affect tube performance. Also, the structure must not be visible to the observer. A similar problem exists in other flat panel displays where maintaining spacing across the active area is important (plasma displays and liquid crystals for example).

Such a structure is particularly difficult to realize if a high-resolution device is required. Fortunately, if one is trying to fabricate a large-screen display (say 75 × 100 cm), the required resolution for a 500-line TV-type display is relatively low (approximately 6 lines/cm).

A structure adopted by RCA of this type is shown in Fig. 7-40 (a and b). In the proposed structure, support vanes would be placed approximately 2.5 cm apart and would run the entire height of the display (approximately 75 cm).[54-56] The vanes would be kept to a minimum thickness and placed such that the channel center-to-center distance was maintained uniformly across the display. To further reduce the visibility of the vanes, an additional black line structure would be introduced with a fine spacing. Thus, instead of trying to eliminate the visibility of the support vane structure, an additional pattern would be introduced into the phosphor to blend with the vane structure. A 5″ × 10″ display consisting of five 1″ × 10″ modules has been demonstrated in both mono and color to verify the module concept (see Section 7.10.8 for further details).

Zenith and Lucitron have also reported working on a somewhat similar structure, also aimed at achieving large display sizes.[46-48] However, a practical structure has yet to be commercially introduced. Conceptually, this idea is very appealing. A cross-sectional view of the Lucitron structure showing the general principles of construction is shown in Fig. 7-41. The malleable metal back, cushion, support vanes, and metal seal are of particular interest.

One problem common to most matrix addressed flat CRTs is the huge amount of internal surface area which results in such structures.

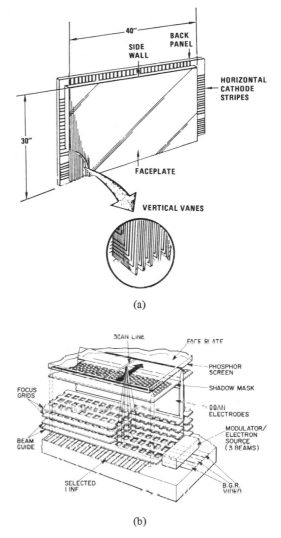

(a)

(b)

Fig. 7-40. RCA self-supporting tube structure. (Courtesy of RCA) (a) Overall structure. (b) Detail.

In a vacuum tube, most materials tend to outgas with time. The emitted gases can affect tube performance, especially the cathode and secondary emission surfaces (if used). The larger the amount of surface area, the more chance of contamination. Appendage pumps or extra getters might be required in such structures. In addition, care must be taken to prevent dielectric surfaces from becoming charged by either an accumulation of free ions or electrons. This charge buildup can deflect the electron beams or pinch off electron channels. Problems such as these have rendered some devices unusable.

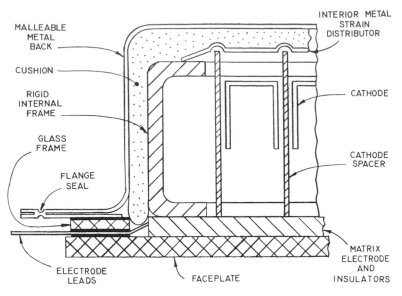

Fig. 7-41. Lucitron self-supporting tube structure. (Courtesy of Lucitron)

7.9.2 Processing. The processing steps required for a flat CRT structure can range from a relatively simple process to one which is very complex. Many of the steps are very device specific and will not be detailed here. However, some general comments might be helpful to understand the design constraints which must be applied. First, to achieve good life, it is probably necessary to use a frit (solderglass) seal. Other seals such as metal-to-metal welds or low-temperature solders can be used, but generally are more costly and not as reliable. These other sealing techniques are used only if the tube components cannot withstand the high temperatures associated with typical frit glass seals (450°C minimum). It is a difficult design task to ensure that all materials can withstand this high temperature, usually performed in an air or other dry gas environment.

Next, another high-temperature bake is usually employed during evacuation in order to drive off water vapor and other undesirable gases. A typical temperature used during evacuation is 350°C. If a large-area oxide-coated cathode is used, this must be activated while the tube is still connected to the vacuum station as large quantities of CO_2 and other gases are given off during the high-temperature activation. The cathode surface is generally heated to approximately 1100°C during activation and operated between 650°C and 850°C. After removal from the vacuum station, a getter is usually flashed inside the vacuum tube. The getter deposits a thin layer of material, usually barium or titanium, over a large surface area. The surface tends to collect gas atoms and serves as a long-lasting zero-power pump to help maintain a good vacuum.

As can be seen, a key element to most tube-processing procedures is the use of high-temperature processing to provide a good seal and high vacuum. In addition, only materials with low outgassing characteristics should be used to ensure good life. Unfortunately, these constraints rule out many otherwise desirable materials from consideration. For example, many of the devices described in this chapter need control grids with a large number of holes. Many materials which are ideally suited for low-cost manufacture of such grids are not suitable vacuum materials. In addition, many potential material systems for the matrix electrodes will not withstand the high-temperature processing for sealing and activation of the cathode.

Expansion coefficients of all materials must also be considered. Many otherwise useful devices have been rendered impractical due to incompatible expansion coefficients of its internal parts. This is particularly a problem for devices of the self-supporting-type tube structure, as parts must maintain alignment and remain intact over large distances, throughout the high-temprature processing steps.

7.10 TECHNICAL ACHIEVEMENTS

Previous sections have described the principles of operation of the various components which comprise a flat CRT. In this section, it will be shown how these various items have been assembled to create a working flat CRT. Emphasis will be on reporting actual accomplishments rather than on techniques, as these have already been discussed. In a few cases, interesting details particular to a specific configuration will be presented. Where possible, devices will be identified by the common name which has been adopted for that technique. However, in many cases, no names other than "flat CRT" have been applied to a particular company device. In these cases, the company name will be adopted. Table 7-9 identifies the displays to be discussed and the techniques which were utilized in the device fabrication. The current status of each technology is also given. As can be seen, seven of the fifteen techniques are believed to be currently in active development. Approximate development dates are also given to allow the reader to trace the history of flat CRT development.

7.10.1 Aiken Tube and Derivatives. The Aiken Tube was first conceived in the spring of 1951 by W. R. Aiken while working at the University of California Radiation Lab, and has had one of the longest and most interesting development histories of any flat display.[4-7g,61-65] The dates given in Table 7-9 trace the development through five of the companies which have worked on this device or variations thereof. Initially the device was designed to compete head on with the CRT for the standard-size color home television receiver. Because of its possible use in home television, many labs have studied the utility of this device.[6] In 1956 a 30 X 30-cm monochrome display was shown at WESCON.[4] Figure 7-42 shows one such tube. Sizes up to 61 cm diagonal and resolutions up to 2,000 lines were demonstrated at Kaiser. A

few versions were also developed for the U.S. Navy which were completely transparent. These were actually tested for use as the first head-up display (HUD) in a T2V jet trainer. Production quantities were desired by the Navy, but quantities were too low to support full-scale development. Work continued at Kaiser until the early 1960s.

About this time, Britain's National Research Development Corp. bought the rights to Kaiser's patents and licensed another British firm, Twentieth Century Electronics Ltd., to extend the Kaiser concept to very large displays.[74] Here the goal was to develop tubes with screen sizes up to 1.2 X 1.8 meters with a depth of less than 30 cm. Tubes with up to 1.2-meter diagonals were built and tested. Apparently, the major problem was in developing a suitable vacuum enclosure. The tubes built used neopreme -0 rings rather than a frit seal. Although life was not reported, this was probably a major problem, and work was discontinued.

Finally, this concept has come full turn and now is being applied to the miniature television receiver market. This occurred around 1978 when Sinclair Electronics took up the approach.[61-65] This device uses a modification of the Aiken concept specifically aimed at small tubes (Fig. 7-11). Unlike the original Aiken tube (Fig. 7-10), which used a series of segmented control electrodes (the vertical electrodes had to switch many kV) to deflect the beam at the appropriate time, this device uses a single set of horizontal and vertical deflection plates very similar to those in a conventional CRT. A separate static field is applied uniformly between the phosphor on the rear interior surface of the envelope and the transparent faceplate. The horizontal deflection plates establish how far the beam will transverse the tube before the static field is able to deflect the beam to hit the phosphor target. A fresnel lens is used to compensate for the nonconventional aspect ratio which results due to the deflection geometry. A single short directly heated cathode similar in nature to those described in the multifilament cathode flood gun section is used to reduce heater power. Figure 7-43 is a photo of early prototypes of this device. Work is still active on this

Table 7-9.   Flat-Panel CRT Summary and Status

Name	Cathode Type[1]	Beam Positioning	Modulation[2]	Brightness Enhancement[3]	Self-Supporting	Multicolor (Demo'd)	Sealed Tubes	Current Status[4]	Years Under Development (Approx.)[5]	Goals of Active Work[6]
Aiken tube										
Kaiser	Electron gun	Electrostatic XY	AN	No	No	No	Yes	D	1953–61	
Video color	Electron gun	Electrostatic XY	AN	No	No	Yes	Yes	D	1960–65	
Twentieth Century	Electron gun	Electrostatic XY	AN	No	No	No	No	D	1965–71	
Sinclair	Electron gun	Electrostatic XY	AN	No	No	No	Yes	A	1978–P	Small TV & M
Sony	Electron gun	X electrostatic Y magnetic	AN	No	No	No	Yes	A	1978–P	Small TV & M
Banana tube	Electron gun	Electrostatic XY	AN	No	No	Yes	Yes	D	1955–63	
Gabor	Electron gun	Electrostatic XY	AN	No	No	No	No	D	1953–68 (?)	
Philips multiplier	Electron gun	Electrostatic XY	AN	Dynodes	No	No	No	A	1978–P	Med TV, M&C
RCA electron guides	Electron gun	X discrete Y electrostatic	AN	MB	Yes	No	No	A	1977–P	VL TV, C
Battelle	Electron gun	Beam addressed matrix	AN	No	No	No	No	D	1974–1978	
Digisplay										
Northrop	Multifilament FG	XYZ control grids	Combined AN & PW	MB & Storage	No	Yes	Yes	D	1963–74	
Texas Instruments	Multifilament FG	XYZ control grids	PW	MB	No	Yes	Yes	D	1973–79	

Name	Cathode Type[1]	Beam Positioning	Modulation[2]	Brightness Enhancement[3]	Self-Supporting	Multicolor (Demo'd)	Sealed Tubes	Current Status[4]	Years Under Development (Approx.)[5]	Goals of Active Work[6]
Vacuum fluorescent	Multifilament FG	X phosphor, Y grid	PW	MB	No	No	Yes	A	1967–P	A/N&G
RCA feedback	Ion & photon feedback	XY control grids	AN	Dynodes	Yes	No	No	D	1971–78	
Zenith	Ion feedback	X cathode, control grid	AN	Dynodes	Yes	No	Unknown	D	1970–76	
Arizona State	Multifilament FG	Active matrix	AN	MB & Storage	No	No	No	D	1970–75	
SRI	Field emission	XY cathode	PW or AN	MB	No	No	No	D	1968–71	
Hybrid Plasma/CRT Lucitron	Plasma	XY self-scan & XY control grids	AN	MB	Yes	No	Yes	A	1977–P	VL AN & G, & VL TV M & C
Siemens/ITT	—	XY self scan & XY control grids	AN	MB	No	Yes	Yes	A	1979–P	AN & G&C

[1] FG = Flood Gun
[2] AN = Analog, PW = Pulse Width
[3] MB = Multip, Beams
[4] A = Active, D = Discontinued
[5] P = Present (1983)
[6] M = Mono, C = Color, VL = Very Large, A/N&G = Alphanumeric & Graphic

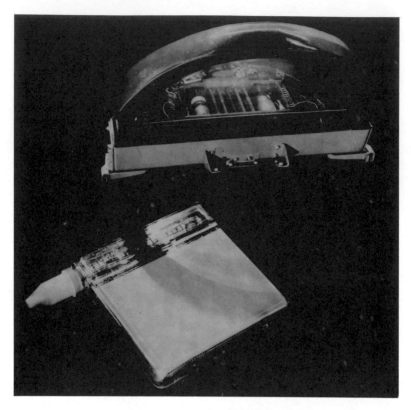

Fig. 7-42. Aiken flat CRT. (After Aiken, 1957)

device. A smaller hand-held version was introduced into the commercial market in late 1983.

The Sony flat CRT is another device which can be linked to the Aiken Tube. As shown earlier in Fig. 7-12, the device is very similar to the Sinclair tube, the major difference being in the substitution of magnetic deflection in the vertical axis for the electrostatic deflection used in the Sinclair tube. The Sony device also eliminated the fresnel lens structure. Figure 7-44 is a photo of the first Sony product introduced as a portable hand-held TV. An improved lower-cost version was introduced in mid-1983.

Fig. 7-43. Sinclair flat CRT.

7.10.2 Banana Tube. This complex system was developed by the British-based Philips division, Mullard Research Labs. It was believed that this concept could achieve lower cost than the competing shadow-mask color cathode ray tube.[8-13] Although a variety of these tubes were made, they were never introduced by Mullard. Problems included noise produced by the mechanical rotating drum which deflected the beams, the odd appearance of the virtual image, and the required large physical display width. Development was discontinued in the early 1960s as the shadow-mask approach was proven superior. Figure 7-18 showed the tube structure. Figure 7-45 is a photograph of one of the actual display systems.

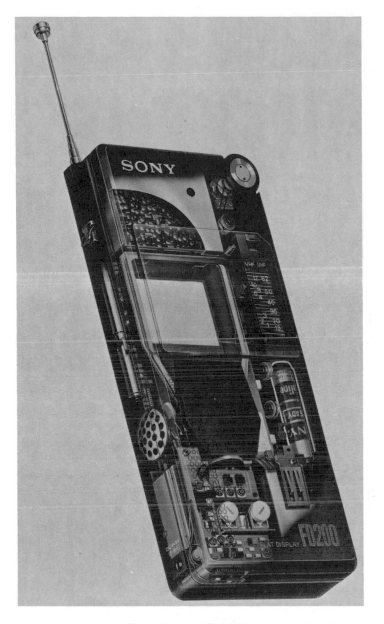

Fig. 7-44. Sony flat CRT.

7.10.3 Gabor Tube.   The Gabor Tube was the first of the flat CRTs and had several extremely novel features.[3-5] As the scanning mechanism was described in detail in an earlier section (see Figs. 7-13 and 7-14), only a few comments about the display system will be given here. Work on the Gabor Tube was carried out almost entirely at the University of London, and no sealed tubes were ever reported. Although this tube had a number of ingenious innovations, the structure was probably too complex to compete with the conventional CRT. The concepts developed for reversing the beam and performing the vertical self-scanning are, however, worthy of note. In addition to its complex structure, perhaps its biggest drawbacks stemmed from only being able to be fabricated in display physical sizes in which it would have to complete head on with the conventional shadow-mask CRT. Because of the relatively

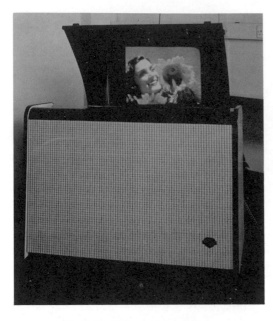

Fig. 7-45. Banana Tube console.

large area required to reverse the beam, it could not be made in small sizes. The requirement of a flat face and an unsupported envelope structure prevented it from achieving large sizes.

7.10.4 Philips Channel Electron Multiplier CRT.
The Philips Channel Multiplier CRT borrows heavily from concepts originally proposed for the Gabor and Aiken Tubes and also used brightness-enhancements techniques similar to those used in other devices.[72a, 73] However, the marriage of these concepts is noteworthy. Figure 7-35, shown earlier, illustrates the multiplier concept. Figure 7-46 is a schematic of how this

Fig. 7-47. Philips CRT image. (Courtesy of Philips)

device works, and Fig. 7-47 is a photo of an image presented on this display. As of early 1984, work on this device was still active.

7.10.5 Battelle Flat CRT.
The Battelle concept used two electron guns to address a matrix of field emission cathodes as shown earlier in Fig. 7-15. Only a few limited experiments were attempted with this technology.[43,44] The primary advantage of this technique was its simplicity of addressing while incorporating the geometrical qualities of a matrix display. As described in the cathode section, the major problem arose from trying to achieve a large-area uniform array of field emission cathodes. Figure 7-48 is a picture of the image on a small 14 × 16-element display. All experiments were conducted in a vacuum chamber. Work was discontinued due to lack of funding.

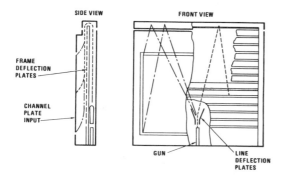

Fig. 7-46. Philips channel multiplier CRT. (Courtesy of Philips)

7.10.6 Digisplay.
The Digisplay also had an interesting development history, being worked

Fig. 7-48. Image on Battelle flat CRT. (Courtesy of Battelle)

on by four different companies (Northrop, Corning, Sylvania, and Texas Instruments).[14-24] In addition, the multilevel XYZ (grid) addressing technique, used to greatly reduce the number of drivers required, has been adapted for use in other flat-panel CRTs and in both plasma and liquid-crystal displays.[69-76] The prime developers of this technology were Northrop (1962-1974) and Texas Instruments (1973-1978).

Table 7-10 lists a few of the display types developed at Northrop and some of the physical and performance parameters for each. Figure 7-49 pictures a 512-character device, and Fig. 7-50 is a photograph from a 160 × 256-element device with gray scale (pulse width). As the photo shows, picture quality was very good. Northrop's motivation was directed towards

alphanumeric and military applications. Two versions of a 512 × 512 TV/graphic display were also fabricated, one at 2.2 lines/mm resolution, the other at 3.1 lines/mm.[18,19] Northrop discontinued work only after being unable to find a commercial partner skilled in mass production of tubelike devices. Texas Instruments bought out rights to this device in 1973.

Texas Instruments continued development on this program from 1973 to 1978. Primary emphasis was to try to develop this display to be competitive with the consumer home color TV. Few details of this intensive effort were made public. The first device described was a 1,920-character alphanumeric-only device.[23] This device had a 5 × 9 font. Spot luminance was 500 fL using 108 drivers and 48 multiple beams. For comparison, it should be noted that an equivalent matrix-addressed display would require 590 drivers. The only other device reported was a full-color video device capable of television display.[74] This device had 75,800 pixels (25,600 color triads) in a 15 × 20-cm active area. Area averaged white luminance was 100 fL using 160 multiple beams and 18 kV phosphor voltage. The total number of drivers was 205 and total tube power was 16 watts. The tube plus driver and electronics power was 20 watts, resulting in a device luminous efficiency of 1.7 lumens/watt. Numerous improve-

Table 7-10. Digisplay Performance Summary

Type	150 × 150	512 Character (5 × 7)	160 × 256	512 × 512	512 × 512
Number of elements	22,500	17,920	40,960	262,144	262,144
Resolution (lines/mm)	2.2	1.5	2.2	3.1	2.2
Active area (cm)	6.9 × 6.9	9.7 × 13.7	7.4 × 11.7	16.3 × 16.3	23.4 × 23.4
Number multiple beams	10	40	40 (20)[1]	32	32
Cathode power (watts)	6.7	12.7	12.0	40	73
Total tube power (watts)	10	14	18	60	100
Luminance (average spot) fL	3500 at 18 kV	240 at 10 kV	300 at 20 kV (150 at 20 kV)	25 at 20 kV	25 at 20 kV
Contrast ratio[2]	50:1	50:1	50:1	50:1	50:1
Number drivers	50	63	84 (60)[1]	96	96
Tube weight (kg)[3]	.8	1.6	1.6	5	15
Envelope size (cm)	10 × 12.8 × 3.2	19 × 14 × 4.4	19 × 14 × 4.4	24 × 24 × 6.4	34 × 44 × 19

[1] Both a 40 and 20 beam version were demonstrated.
[2] No ambient measurements. Contrast limited only by light spreading in the phosphor. Large-area contrast was typically 100:1.
[3] All tubes use flat-faced envelopes except the 512 × 512 at 2.2 l/mm. This tube used a standard 19" V curved-face rectangular faceplate. (Used by a conventional CRT).

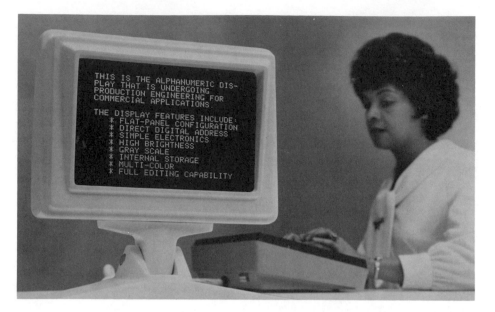

Fig. 7-49. 512-character Digisplay. (Courtesy of Northrop)

ments were incorporated in this device. One of these involved substituting metal for glass in fabrication of the switching plates. This greatly reduced the cost of the switching plates. The other major development involved developing an improved cathode structure to both increase the brightness and decrease the switching voltages. Output current density was increased from about 1 ma/cm² (Northrop) to about 6 ma/cm² (TI). Maximum switching voltage was reduced from approximately 120 V (Northrop) to less than 30 volts (TI). The highest plate capacitance which had to be switched was only 160 pf. The transmission through the switching plates was also increased substantially (from about 25% to over 70%). However, in the attempt to decrease costs to the bare minimum, luminance uniformity and picture quality actually decreased. No photos of the Texas Instruments device could be obtained.

After an intensive effort to decrease costs and improve performance to better than that achievable with a conventional shadow-mask color CRT, the effort was discontinued. This effort probably came closer to meeting these goals than any other flat-panel display to date. However, cost still proved to be the major obstacle. In addition, as with other open-

Fig. 7-50. 160 × 256-element Digisplay. (Courtesy of Northrop)

envelope structures, large sizes were not achievable with this technology.

7.10.7 Vacuum Fluorescence. To date, the most successful of all flat CRT approaches has been the vacuum fluorescent display.[25-32,67] The success has arisen partly from the novel structure using a patterned anode substrate combined with a low-voltage phosphor, and partly by its inventors' realization that such a seemingly complex device could in fact be made less expensively than competing displays.

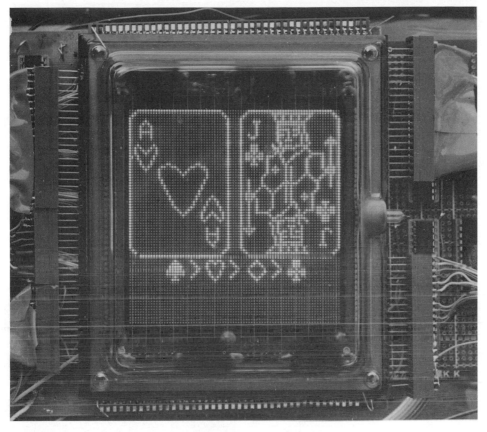

Fig. 7-51. 80 × 80 dot flat CRT. (Courtesy of ISE)

First products of this device were sold in 1967 by ISE Corporation of Japan. Currently a number of companies (ISE, Futaba, CHOA, NEC, etc.) manufacture a wide range of tube sizes. Advantages of this technique include low cost, long life, pleasing appearance, rugged construction, high luminance, and good luminous efficiency. Although nowhere near as low-power as liquid-crystal or other nonemissive displays, it is one of the lowest power and highest luminance light-emitting flat-panel displays currently available.

Figures 7-51 and 7-52 show photos of various tube types. A 240-character version can achieve 220-fL average spot luminance at a duty cycle of 2.2% and anode voltage of 50 volts. Power dissipation for the tube alone is 3.8 watts. Thus, the luminous efficiency of the tube (excluding drivers and electronics) is about 1.3 lumens/watt. Also available in 1983 were 80 × 80, 128 × 128 and 256 × 512 graphics panels. ISE also disclosed that they have demonstrated a MOSFET addressed 241 × 246 element display in a 2.5-cm square active area. This device is early in the experimental stage as of 1983. Currently, a university in China has disclosed success in making VFDs up to 573 × 782 elements.[32] Multi-color panels have also been demonstrated.

The VFD is limited by the large number of drivers required and its inability to be extended to large size because of the open-type envelope structure. In addition, to make devices with greater numbers of elements, one is faced with the typical decrease in luminance caused by the duty cycle problem as discussed in Chap. 5. To circumvent this problem, one is faced with the choice of increasing the number of multiple beams, increasing the phosphor voltage, or adding inherent storage. The pros and cons of some of these techniques are discussed in Section 7.7.

7.10.8 RCA Feedback and Electron Guide Displays. RCA has been pursuing flat-panel CRT

Fig. 7-52. 26 × 256 dot flat CRT. (Courtesy of ISE)

designs since the early 1970s, but the first public disclosures were not made until 1978.[52-57] As discussed earlier, the first types were of the feedback type (Figs. 7-8 and 7-9) using either ion or photon feedback and a cold-type cathode. This work was discontinued in about 1977 in favor of the electron guide versions (Figs. 7-16 and 7-17). Both use the self-supporting-type envelope structure and are intended for use in displays with large diagonals. The electron guide version has been fabricated in single-module sizes of 33 cm long by 2.5 cm wide and in a five-module display. Figure 7-53 is a photo of the five-module display. The advantages of this display are its potential for being made in very large sizes, its readily incorporated color, and its high luminance. However, the structure is relatively complex, and the ability to achieve RCA's goal of a 127-cm diagonal display comprised of a number of modules has yet to be demonstrated.

7.10.9 Zenith Ion Feedback Display. This display and the RCA ion feedback version were nearly identical both in purpose and implementations. The Zenith display was reported first, and was discontinued first as well.[46] Figure 7-54 illustrates the basic structure. No photographs of working devices were ever published. The complexity of the structure, coupled with problems caused by ion-induced sputtering of the cathode and life problems, led to its being discontinued. No new work has been reported.

Fig. 7-53. Modular flat CRT. (Courtesy of RCA)

7.10.10 Arizona State Active Matrix Display. During the period from about 1970 to 1975, Arizona State attempted to incorporate the thin-film transistor technology being developed by Brody and others in a flat CRT.[45,66] The motivation was to develop a very high luminance

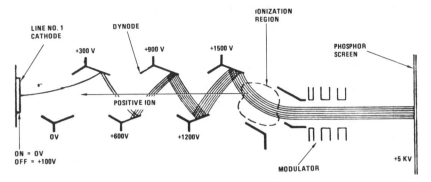

Fig. 7-54. Zenith electron multiplier. (After Schwartz, 1976)

display by achieving nearly a 100% duty cycle. The structure used was shown in Fig. 7-28. Most of the effort was expended on trying to fabricate reproducible, good-quality, thin-film transistors. At the end of the program a 5 × 7 element bell-jar prototype was fabricated with the TFTs internal to the belljar but not on the aperture control plate structure. The work was stopped in 1976 and transferred to Sperry Phoenix, which had funded the early work. Work there was discontinued shortly thereafter. The most pressing problem was in integrating the TFTs and storage capacitors on the control plate. As has been the case with all TFT programs, achieving 100% yield has proven quite difficult. In addition, the workers at the university had difficulty developing a good area cathode. If TFTs become practical, this technique may become useful if coupled with the multifilament cathode.

7.10.11 Stanford Research Institute Field Emission Display.

Workers at SRI developed relatively high performance field emission cathodes in the mid to late 1960s and have been looking for good applications ever since.[37-42] During the early 1970s their use as a matrix cathode for a flat CRT was one such application investigated by SRI. However, severe problems were encountered in achieving cathode-to-cathode uniformity and in fabricating large physical sizes. This led to this aspect of the program being discontinued in the early 1970s. However, work has continued on the field emission-type cathodes and could conceivably be applied to future flat CRTs, if the problems described above are overcome.

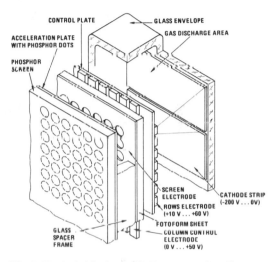

Fig. 7-55. Hybrid plasma/CRT construction Siemens. (Courtesy of Siemens)

7.10.12 Hybrid Plasma/CRT.

As noted earlier, the hybrid plasma/CRT is being developed by both Lucitron and Siemens.[47-51] The construction used in the Siemens version is shown in Fig. 7-55. It is apparently being developed to satisfy the need for a flat, multicolor, modest-size and resolution display. Figure 7-56 is a photo of the display. The Lucitron version is being developed for very large diagonal (>30 in.) displays. Its construction principles are shown in Figs. 7-41, 7-57, and 7-58. It uses a self-supporting structure. A photo of a 88 × 112 element display is shown in Fig. 7-59. Neither of these devices are commercially available as of early 1984. In fact, Siemens appears to have reduced its effort on this program and has teamed with ITT Germany. Lucitron is working on a 256 × 512 element device with a 35-in diagonal, but difficulties encountered in fabricat-

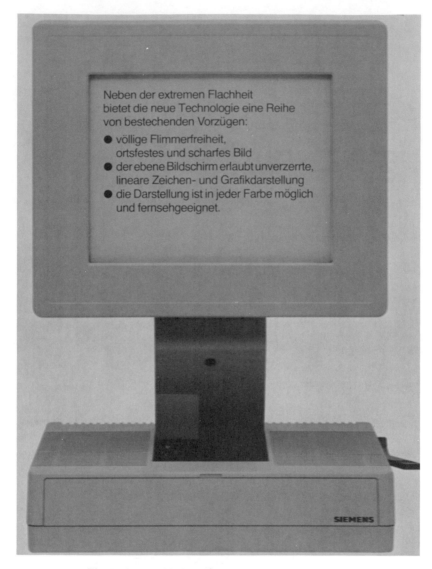

Neben der extremen Flachheit
bietet die neue Technologie eine Reihe
von bestechenden Vorzügen:

● völlige Flimmerfreiheit,
   ortsfestes und scharfes Bild
● der ebene Bildschirm erlaubt unverzerrte,
   lineare Zeichen- und Grafikdarstellung
● die Darstellung ist in jeder Farbe möglich
   und fernsehgeeignet.

SIEMENS

Fig. 7-56. Hybrid plasma/CRT. (Courtesy of Siemens)

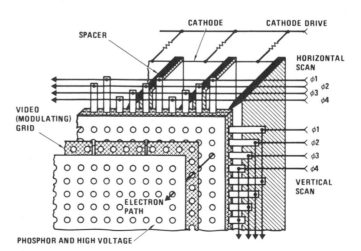

Fig. 7-57. Hybrid plasma/CRT construction–Lucitron. (Courtesy of Lucitron)

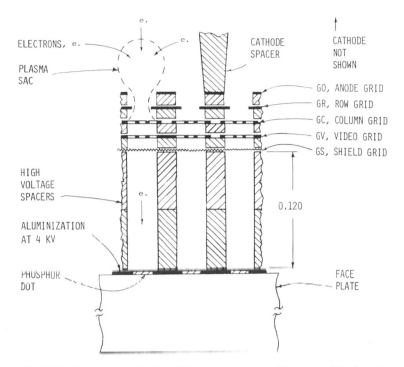

ELECTRONS, e.

PLASMA
SAC

CATHODE
SPACER

CATHODE
NOT
SHOWN

GO, ANODE GRID
GR, ROW GRID
GC, COLUMN GRID
GV, VIDEO GRID
GS, SHIELD GRID

HIGH
VOLTAGE
SPACERS

0.120

ALUMINIZATION
AT 4 KV

PHOSPHOR
DOT

FACE
PLATE

Fig. 7-58. Cross-sectional view of Lucitron structure. (Courtesy of Lucitron)

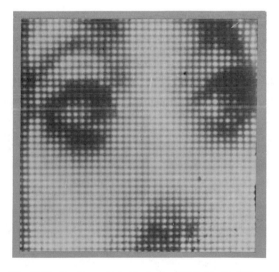

Fig. 7-59. Lucitron hybrid plasma/CRT image. (Courtesy of Lucitron)

ing such a large, complex structure have slowed progress on this device.

7.11 SUMMARY

All of the developments discussed in this chapter arose in response to the investigators' perceived need for an improved display—primarily motivated by the home television market. Some believed that the CRT's bulk was a major disadvantage. However, others saw legitimate shortcomings in terms of performance. In particular, many early researchers viewed the now popular shadow-mask concept for adding color to a CRT as being impractical.

What has proved to be especially disappointing to those involved in these alternative concepts is that, for the most part, they achieved or exceeded the performance goals they established at the beginning of their projects. Many companies found that after several years of relatively good progress, the hope of displacing the conventional CRT had actually diminished. Not because they had failed, but because their competitor—the conventional CRT—had advanced faster than they. As the CRT is still making rather rapid advances in some areas, the would-be inventor should take note that one must not set his goals too low. Rather, he must focus on bettering the performance of the CRT that will exist after his concept has been realized.

In this chapter, the results of over 25 efforts to develop a flat CRT have been summarized. Of these, as of early 1984, at least seven programs are active. However, only one, the VFD, has been successful in establishing a sizable commercial marketplace, this being in the alphanumeric- and digit-type display market. The small hand-held portable television receivers introduced by Sony and Sinclair are also beginning to establish a market. It is still to early to tell how lucrative this new market will be. Many researchers are optimistic that a flat CRT can be successful in the large area video markets as well.

Although flat CRTs are not the envisioned "solid state" panacea, their many practical and highly developed materials and processing techniques give this technology a decided edge over many of the other flat panel displays discussed in this book.

The fact that color is becoming more important leaves many other flat-panel concepts severely lacking. It is hoped that this review of both the successes and failures of flat CRTs will be useful to future researchers.

REFERENCES

1. Gabor, D., et al. "A Fully Electrostatic, Flat, Thin Television Tube." *Proceedings of the IEEE*, Vol. 115, No. 4, April 1968, pp. 467–478.
2. Gabor, D., et al. "A New Cathode-Ray Tube for Monochrome and Colour Television," Institution of Electrical Engineers, May 1958, pp. 581–606.
3. "Gabor Television Tube," *Research*, June 1958, pp. 209–210.
4. Aiken, W. R. "A Thin Cathode-Ray Tube," *Proceedings of the IRE*, Vol. 45, No. 12, December 1957, pp. 1599–1604.
5. Instruction and Maintenance Manual for Kaiser-Aiken Thin Tube Monitor, prepared by Kaiser Aircraft and Electronics, March 1962, 51 pp.
6. Ramberg, E. "Electron-Optical Properties of a Flat Television Picture Tube." *Proceeding of the IRE*, December 1960, pp. 1952–1960.
7. Geer, W. "Color in the Thin Tube." *SID Digest of Technical Papers*, 1964, pp. 278–285.
7a. Kimmy, N. "Electrostatic Focusing Design for Thin C-R Tubes." *Electronic Equipment Engineering*, June 1960.
7b. Aiken, W. R. "Development of the Thin Cathode-Ray Tube." *Journal of the Society of Motion Picture and Television Engineers*, Vol. 67, No. 7, July 1958.
7c. Aiken, W. R. "3-D Data Display." *Journal of ATC*, April 1959.
7d. Kimmy, N. "Transition Lens of the Kaiser-Aiken Thin Cathode-Ray Tube." *Electronic Equipment Engineering*, May 1961.
7e. Aiken, W. R., and Heller, R. E. "Built-In Ion Trap Protects Cathode." *Electronics*, February 1958.
7f. Gruenberg, H. "Discussion of a Deflection System for a Flat Television Picture Tube." *Proceedings of the IRE*, December 1961.
7g. Aiken, W. R. "Flat Tube Monitor Resolves 1000 Lines." *Electronic Products*, March 1962.
8. Schagen, P. "The Banana-Tube Display System. A New Approach to the Display of Colour Television Pictures." *Proceedings of the IEEE*, 1961, Part B 108, pp. 577–586.
9. Eastwell, B. A., and Schagen, P. "Development of the Banana Tube." *Proceedings of the IEEE*, 1961, Part B 108, pp. 587–595.
10. Howden, H. "Mechanical and Manufacturing Aspects of the Banana Tube Colour-Television Display System." *Proceedings of the IEEE*, 1961, Part B 108, pp. 596–603.
11. Freeman, K. G. "Circuits for the Banana-Tube Colour Television Display System." *Proceedings of the IEEE*, May 1961, Part B 108, pp. 604–612.
12. Jackson, R. N. "Colorimetry of the Banana-Tube Colour Television Display System." *Proceedings of the IEEE*, 1961, Part B 108, pp. 613–623.
13. Freeman, K. G., and Overton, B. R. "Appraisal of the Banana-Tube Colour Television Display System." *Proceedings of the IEEE*, 1961, Part B 108, pp. 624–630.
14. Goede, W. "A Digitally-Addressed Flat Panel CRT." *IEEE Electron Devices Transactions*, November 1973, pp. 1052–1061.
15. Goede, W., and Jeffries, L. "Digital Address Thin Display Tube." Final Report, Contract DAAB07-70-C-0110, May 1972.
16. Goede, W., et al. "A Flat Alphanumeric Display Tube." *IEEE Transactions on Electron Devices*, ED-18 No. 9, pp. 692–697.
17. Goede, W. F.; Gunther, J.; and Lang, J. "512 Character Alphanumeric Display Panel." Presented at the 1972 Society for Information Display International Symposium, June 1972, published in conference proceeding and *SID Journal*.
18. Goede, W. "Flat Plate Display." Final Report, Contract F33615-72-C-1581, July 1974, 110 pp.
19. Goede, W. "Digital Address Thin Display Tube." Final Report, Contract #DAAB07-72-C-0776, TRECOM-0076-7, August 1974, NORT 72-280-7, 134 pp.
20. Lang, J.; Goede, W.; and Playter, J. "A 160 × 256-Element Alphanumeric/Graphic Display." Presented at the 1972 IEEE Conference on Display Devices, October 1972, published in conference proceeding, 8 pp.
21. Jeffries, L. "Digitally-Addressed High Brightness Flat Panel Display." *SID Digest*, 1973, pp. 96–97.
22. Elowe, E. FTA-512 Digisplay, Information Display, January/February 1973, pp. 21–27.
23. Holton, W., et al., "Design Fabrication and Performance of a Flat Tube Display." International Electron Device Meeting Proceedings, December 1977, pp. 78–80.
24. Scott, W. C., et al. "Flat Cathode Ray Tube Display." *SID Digest*, 1978, pp. 88–89.

25. Kasano, K., et al. "A 240 Character Vacuum Fluorescent Display and Its Driving Ability." *SID Digest*, 1979.

26. Kasano, K., et al. "A Random Access Memory 26 × 258 Dot Flat Panel Vacuum Fluorescent Display." *SID Digest*, 1980, pp. 74–75.

27. Bylander, E. G. *Electronic Displays*. New York: McGraw Hill Book Co., 1979, Chap. 6.

28. Nakamura, T. *Vacuum Fluorescent Displays and Their Application*. Japanese book, 1977 (in Japanese).

29. Nakamura, T. "Fluorescent Display Tubes May Replace CRT's." *IEE*, January 1980, pp. 67–70.

30. Uchiyama, M., et al. "High Resolution Vacuum Fluorescent Display with 256 × 256 Dot Matrix." *SID Digest*, 1982.

31. Kiernan, R., et al., "High Resolution X-Y Dot Matrix VFD." *SID Digest*, 1982.

32. Ge, S. "Grid-control Matrix Fluorescent Display Panel." *SID Digest*, 1982.

33. Uemura, S., and Kiyozumi, K. "Flat VFD TV Display Incorporating MOS-FET Switching Array." 1980 Biennial Display Research Conference, October 1980.

34. Satu, N. "FIP Technologies Provide for the Future." *IEE*, January 1979, pp. 34–37.

35. Weston, G., and Bittleston, R. *Alphanumeric Displays–Devices, Drive Circuits and Applications*. Granada Publishing Ltd., 1982.

36. "Multicolor Display Tube." *ISE Electronics*, April 8, 1980.

37. Spindt, C. A. "Development Program on a Cold Cathode Electron Gun." Final Report, NASA CR 159570, May 1979, NASA Contract No. NAS 3-20096, SRI Project 5413.

38. Burgess, R., et al. "Corrected Values of Fowler-Nordheim Field Emission Functions v(y) and s(y)." *Physics Review*, Vol. 90, No. 4, May 15, 1953, p. 515.

39. SRI Technical Noted, Submicron Structures Built by SRI's Physical Electronics Group, Technical Note 1, March 1978.

40. Brodie, I. "Microcathode Field Emission Arrays for CRT Application." *SRI Research Brief*, No. 26R, September 1979.

41. Kelly, J. "An Improved Field Emission Cathode for Use in Storage-Tube and Other Electron-Optical Applications. SRI program description—internal document, November 1970, 14 pp.

42. Fowler, R. H. and Nordheim, L. "Electron Emission in Intense Electric Fields." *Proceedings Royal Society* (London), Vol. A119, May 1, 1928, pp. 173–181.

43. Battelle short technical note entitled "Video-Cathodes Display Panel for Portable Oscilloscope or Data Terminal."

44. Battelle short technical note entitled "Video Cathodic Display Panel," #90200.

45. Sirkis, M., et al. "Matrix-Array Cathode Ray Tube." *IEEE Transactions on Electron Devices*, September 1976, pp. 1105–1106.

46. Schulman, R., and Schwartz, J. "A Novel Cath-odoluminescent Flat Panel TV Display." *SID Digest*, May 1976, pp. 134, 135.

47. DeJule, M. et al. "A Gas Discharge Display." *SID Digest*, 1983.

48. Glaser, D. et al. "The Flatscreen Display, Construction and Circuitry." *SID Digest*, 1983.

49. Schauer, A. "A Plasma Electron Excited Phosphor Flat Panel Display." *IEDM*, December 1982.

50. Biancomano, V. "Flat Panel Displays Grow Larger, Focus on Full-Color Performance." *Electronic Design*, November 11, 1982, pp. 79–88.

51. Schauer, A. "Plasma Lights Up 14″ Flat Panel Display." *Electronics*, December 15, 1982, pp. 128–130.

52. Catanese, C. and Endriz, J. "The Multiplier-Assisted Discharge: A New Type of Cold Cathode." *Journal of Applied Physics*, 50 (2), February 1979, pp. 731–745.

53. Endriz, J., et al. "Feedback Multiplier Flat-Panel Television." *IEEE Transactions on Electron Devices*, ED-26, No. 9, September 1979, pp. 1324–1358.

54. Credelle, T., et al. "Cathodoluminescent Flat Panel TV Using Electron Beam Guides. *SID Digest*, 1980, pp. 26–27.

55. Siekanowics, W., et al. "Ladder Mesh and Slalom Electron Guides for Flat Cathodoluminescent Displays." *SID Digest*, 1980, pp. 24–25.

56. "RCA Discloses Technical Details of Concept for a Wall-Mounted, 50-Inch Diagonal Color TV Display." *RCA News*, April 29, 1980.

57. TBD

58. Herrmann, G., and Wagner, S. *The Oxide-Coated Cathode*, Vols. 1 & 2, Chapman & Hall Ltd., 1951.

59. Goede, W. "Investigation of Plasma Area Cathode." Northrop internal memo T 8534-70-24, March 24, 1970.

60. Hatch, A., and Williams, H. *Physics Review*, 100, 1955, p. 1228.

61. Colclough, R. "The Sinclair Side-Window CRT." Internal memo from Sinclair Research, Ltd., 1979.

62. Smith, K. "CRT Slims Down for Pocket and Projection TV's." *Electronics*, July 19, 1979, pp. 67–68.

63. TBD

64. Cohen, C. "Sony Pocket TV Slims Down CRT Technology." *Electronics*, February 10, 1982, pp. 81–82.

65. "Flat Pocket TV." *Popular Science*, May 1982.

66. Brody, T. P. "Integrated Electro-optic Displays." *Non-Emissive Electrooptic Displays*, edited by A. Kmetz. New York: Plenum Press, 1976, pp. 303–341.

67. Morimoto, K. "A High Resolution Graphic Display." *SID Digest*, pp. 218–219.

68. Amano, Y. "A Flat Panel TV Display in Monochrome and Color." *IEEE Transactions on Electron Devices*, ED-22, January 1975, pp. 1–7.

69. Sherr, S. "Research Program on Three Axis

Matrix Addressing Techniques. Contract No. N00014-72-C-0248, AD 751666, August 31, 1972, 75 pp.

70. Kazan, B., and Knoll, M. *Electronic Image Storage*. New York: Academic Press, 1968.

71. Bruning, H. *Physics and Applications of Secondary Electron Emission*. New York: McGraw-Hill, 1954.

72. Wolber, W. "The Channel Photomultiplier—A Photon-Counting Light Detector." *Research/Development*, December 1968, pp. 18–23.

72a. Woodhead, A., et al. "The Channel Electron Multiplier CRT: Concept, Design and Performance." *SID Digest*, 1982, pp. 206–207.

73. Overall, C. et al. "The Channel Multiplier CRT: Table Technology." *SID Digest*, 1982, pp. 208–209.

74. "Flat CRT with Large Screen Around the Bend in Britain." *Electronics*, January 20, 1969, pp. 197–198.

75. *Optical Characteristics of Cathode Ray Tube Screens*. JEDEC Publication No. 16-C.

76. Watts, G., et al. "The Multilayer Gas-Discharging Display Panel." *WESCON Conference Proceedings,* Vol. 26/3, 1974, pp. 1–5.

77. Neale, R. "Towards the Thin CRT." *Electronic Imagery*, February 1983, pp. 36–37.

8

ELECTROLUMINESCENT DISPLAYS

L. E. TANNAS, JR., *Consultant*

8.1 INTRODUCTION

There is today a renewed interest in electroluminescence (EL) as a display medium. This revival has been accelerated by the combination of custom LSI DMOS high voltage drive electronics with matrix-addressed panels of ac thin-film phosphor in a thin-film dielectric sandwich. The primary phosphor of choice is high-purity zinc sulfide activated with manganese. This phosphor is highly nonlinear, with discrimination ratios approaching one million. It has a long life at high luminance and can be made sunlight readable. The panels typically operate over a voltage excursion of 400 volts, peak to peak. A voltage drive suitable for this excursion can be made using double diffusion metal oxide semiconductor field-effect transistor (DMOS FETs) which have very high voltage capability. The DMOS is combined with low voltage CMOS logic on the same substrate to give a highly integrated custom chip for driving a large number of lines. It should be noted that luminance is produced in thin-film ZnS:Mn phosphors by a different mechanism than in other forms of EL such as ac powder EL.

8.1.1 Overview. Over the years, several authors have reviewed the EL technology in some detail. Of note are H. W. Leverenz, 1950;[1] H. F. Ivey, 1963;[2] F. F. Morehead, 1963;[3] E. Schlam, 1973;[4] B. Kazan, 1976;[5] T. Inoguchi and S. Mito, 1977;[6] W. Lehmann, 1980;[7] W. Howard, 1981;[8] and A. Onton and V. Marrello, 1982.[9] During this time period the technology has matured, many misconceptions have been corrected, and many innovations have been made. Because of this long evolution, a detailed historical review is included here to show the continuity of the growth of EL into display products. This should help even those who have worked in the EL field to comprehend the immensity and complexity of its evolution.

Electroluminescence is the emission of light from a polycrystalline phosphor solely due to the application of an electric field. In principle, the construction of such a lamp, or "luminaire," or of an information display, is quite simple. However, the completion of a practical display using the basic phenomenon has stymied the industry for many years. The explanation of the exact mechanisms for light generation is only now beginning to become understood.

The development of EL started in the late 1940s. In the 50s, the initial intended use was for luminaires.[10,11] Enthusiasm was due in part to the prior successes of the fluorescent lamp and cathode-ray tube, which used similar phosphors. Early in the development of the technology, it was quite evident that alphanumeric characters could be made using a segmented font, each segment being simply a flat EL lamp. Examples of an alphanumeric product line typical of the 1960s are shown in Fig. 8-1. The structure for an EL display matrix array was first proposed by W. W. Piper in 1953,[12] and a major industrial pursuit for a TV-on-the-wall was launched.[13,14,15] This enthusiasm finally turned to rejection in the late 1960s because of the inability to solve three technical problems:

- High luminance with
- Long life and
- Matrix addressability.

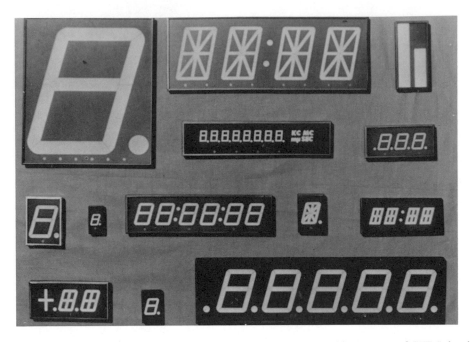

Fig. 8-1. Sylvania hermetically sealed powder EL display produced in 1966 (Courtesy of GTE Sylvania).

The failure to develop successful EL display products prior to 1970 was perhaps the single greatest disappointment in the display-device development community, rivaled only by the fact that the 1980s began without a consumer flat-panel TV in production. EL is now a viable technology made possible by thin-film phosphors and the evolution of custom LSI DMOS to drive the high voltage. It has now been demonstrated that a practical ac thin-film EL flat-panel TV is feasible. Such an experimental high-luminance, long-life, matrix-addressed flat-panel TV (Fig. 8-2) was demonstrated by Sharp of Japan at the 1978 Consumer Electronics Show in Chicago. The 1980s began with companies committed to manufacture EL products.

In the long term, EL could become the leading alphanumeric and vectorgraphic display technology, expanding from there to video, all using ac thin film. This is because of its simplicity of construction, low cost, luminous efficiency, environmental ruggedness and excellent aesthetic performance. This chapter is organized to give the reader a complete grasp of the EL display state of the art. Its current status is placed in historical context by a review of the evolution of EL devices. Principles of operation and performance are then discussed for EL thin films and powders, ac and dc. Included are sections on theory of operation, fabrication techniques, failure mechanisms, color, addressing characteristics, luminous efficiency, and applications.

8.1.2 Configuration Definitions. First, it is necessary to define the various EL configurations. This is best done diagrammatically as shown in Fig. 8-2. (Companies participating in the 1980s are shown in the various categories.) The two major subdivisions are ac and dc, in an analogy to the designation of alternating and direct current. There are further subtleties that require comment to prevent any misunderstanding. Both ac and dc subdivisions are pulse-driven in an actual display application. The main difference is that in the dc configuration the electrons from the external circuit pass through the pixels in the panel. In the ac configuration electrons are blocked from passing through the pixel by encapsulating dielectrics, causing the pixel to be capacitively coupled to the external circuit. The same interpretation of ac and dc applies to gas discharge displays, Chapt. 10.

The technology of EL displays is further subdivided by the two major phosphor classifications, powder and thin-film. The powder phosphors are formed by treating and grinding the phosphor crystals into powder of the proper grain size and then applying them to the sub-

Fig. 8-2. Sharp EL TV shown in 1978 Chicago Consumer Electronics Show (Courtesy of Sharp).

strate by spraying, screening, or doctor blading techniques. The thin-film phosphors are grown from the condensation of evaporants from vacuum vapor depositions, sputtering, or chemical vapor depositions.

A major distinction among EL displays is between those which have intrinsic memory and those which do not. In the memory panels, the image is sustained by ac power applied in parallel across all pixels. Those pixels excited up to the on state remain on and those not excited remain off. To turn an individual pixel on or off, a voltage signal of the appropriate magnitude and phase relative to the ac-sustained power drive is applied to the panel via matrix addressing. The on-pulse is different from the off-pulse.

8.2 HISTORY

The evolution of electroluminescent displays has been long and painful. It is marked by many examples of highly motivated, enthusiastic, and

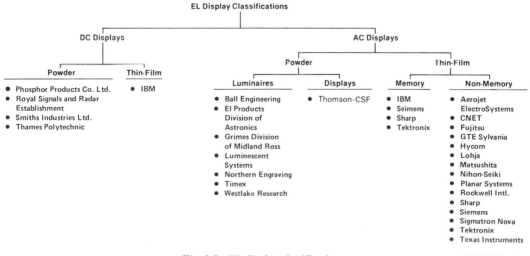

Fig. 8-3. EL display classifications.

dedicated efforts which have fallen short of their aspirations. The delays have resulted in part from a lack of sufficient fundamental engineering, science, and technology; part has been due to underestimation of the magnitude of the technical problems.

This detailed history is included here for two reasons:

- To give a frame of reference for those who have worked with EL in the past.
- To serve as an example of how long it can take for a new display technology to become established.

The reader may skip directly to Section 8.3 and still gain a full appreciation of the technical aspects of EL displays.

8.2.1 Discovery of Electroluminescence in Polycrystalline Films. Electroluminescence in polycrystalline electroded phosphors was first observed in France in June, 1936. A physicist, Professor Georges Destriau of the Faculty of Science, University of Paris, was exploring the general nature of scintillation of phosphors by alpha rays. (Luminescence from crystalline material had already been discovered by H. J. Round [1907] and O. V. Losev [1923].) Destriau was working with a popular photoluminescent phosphor, ZnS activated with Cu, and by accident observed that light was emitted solely because of the application of an electric field. It was first necessary to prove that no UV, IR or corona radiation was partially stimulating the phosphor and that a corona discharge due to the high voltage was not causing the luminance. In 1936, Destriau first published[16] his experimental evidence proving luminescence solely due to an electric field.

Before and after World War II, there were several other papers by Destriau[17] expounding the fact that luminescence in a polycrystalline phosphor could be achieved solely through the application of an electric field. His experimental observation raised considerable skepticism. After operating one of Destriau's samples in his laboratory, H. W. Leverenz,[1] in his book dated 1950, dismissed Destriau's effect as simple corona. Scientific examination and engineering application and dissemination of Destriau's results were delayed becaused of the war. He called the

phenomenon "electrophotoluminescence," and credits Gudden and Pohl[18] with the first observations on the action of electric fields. These authors were the first to note a momentary strengthening of the phosphorescence of certain sulfides which had been previously submitted to ultraviolet rays. Professor Maurice Curie described Destriau's work in the scientific literature under the name of "pure electroluminescence" to emphasize that no photostimulation need be involved. The name "electroluminescence" has prevailed over the name "electrophotoluminescence" used by Destriau and Gudden and Pohl.

The first U.S.-documented discussion of Destriau's work occurred in 1946. A symposium on "Preparation and Characteristics of Solid Luminescent Materials" was held at Cornell University on October 24-26, 1946.[19] During this conference, Destriau's work was discussed only at the concluding-remarks session where it was noted by G. F. J. Garlick of the University of Birmingham, England. Garlick assured the attendees that Destriau had produced luminescence solely by the application of electric fields to the phosphor arranged as a suspension in oil forming the dielectric of a small capacitor cell. The field was at 10,000 volts/cm and oscillating at 50 Hz, where the emission intensity varied with frequency, and increased exponentially with the field strength. The light maxima were out of phase with the field.

Garlick noted that it is difficult to interpret Destriau's results based on the band theory of solids, as the field used (10,000 volts/cm) is not sufficient to raise electrons into conduction levels from luminescent centers or even to produce excitons. He further noted that the optimum efficiency occurs at a much higher copper content in the zinc sulfide phosphor than for ultraviolet excitation, and the electrons taking part in the process do not seem to enter traps. Specimens with a long afterglow to UV excitation show no phosphorescence due to the field excitation. Garlick was obviously not aware of the field concentrating effect of the supersaturated Cu_xS per the Fischer[20] model.

8.2.2 First EL Display Activity. In the U.S. during the 1950s and 1960s there was a great surge of interest in EL as the universal luminaire and display media. In October 1947 Destriau published a widely circulated paper in English in the *Edinburgh and Dublin Philosophical Magazine*.[21] The war was over and scientists and engineers could get back to this new and exciting phenomenon. The possibility of using conventional phosphors in a flat luminaire panel to cover walls and ceilings was mind-boggling. The first practical electroluminescent lamps were made in the Sylvania Laboratories in 1949.[22] GTE Sylvania applied for the first patent (1950),[23] published the first paper (1950),[10] and set up in Salem, Massachusetts, the first plant for their luminaires called "panelescent lighting."[11] They demonstrated their panelescent lighting at a solid state conference in the Spring of 1952 at the Massachusetts Institute of Technology campus.

A solid-state lamp was not new in 1949. The novelty was the use of a low-cost, readily available polycrystalline phosphor as opposed to a single crystal semiconductor (LED). The first application for a patent on an LED lamp had been made by Z. Bay and G. Szigeti in 1939.[24]

A limited degree of success was achieved with phosphor powder, particularly zinc sulfide/zinc selenide mix activated with copper. They made excellent low-light level lamps and displays. Numeric displays of the type shown in Fig. 8-1 were made extensively with this phosphor. They were in 1966 and still are a standard for the industry for an ac powder EL display. However, the displays could not be made with simultaneous high luminance and long life, despite considerable technical effort throughout industry. The problems of EL were best summarized in the introductory paragraph of a paper which S. V. Petertyl and P. R. Fuller[25] of Lear-Siegler's Instrument Division in Grand Rapids, Michigan gave in May 1966 at the National Aerospace and Electronics Conference, Dayton, Ohio:

"Electroluminescent lamps and their promises for use as information displays have historically been plagued with serious shortcomings—low luminance, short useful operating life, poor visibility in normal room light, and no visability under high ambient light. In addition, high voltage control circuitry presented great obstacles, and operation at temperatures much above room temperature caused a rapid deterioration of intensity of emitted light."

The future of EL was definitely in question. Out of this effort led by Petertyl at Lear-Siegler, and sponsored by the Air Force Flight Dynamics Laboratory, came the observation that an ultra-thin, black absorber layer could be used in *front* of the phosphor/dielectric layer to improve the contrast in high ambient illumination.[26]

The Instrument Division of Lear-Siegler developed[27] hermetically sealed alphanumeric displays with the integral neutral density filter for the Apollo vehicle and LM vehicles for NASA's manned lunar program. Each display was sealed in nitrogen in a machined Kovar case with glass and solder seals. A ZnS/ZnSe:CuS green phosphor powder made by RCA Research Laboratory was used in the display. The phosphor was suspended in a resin dielectric (cyanoethylated starch). Eleven display assemblies were used on the Apollo and LM in seven different functions, from event timers to guidance and navigation displays. In 1964 they were sunlight readable and had a life (to one-half luminance) of 2,000 hours. This represented the highest performance point ever achieved by ac powder EL displays. No failures of any of the Lear Siegler EL displays were ever reported during any of the Apollo missions. An example of an Apollo vehicle display is shown in Fig. 8-4.

In terms of readability, the demonstrated re-

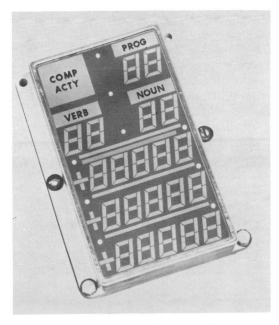

Fig. 8-4. Lear-Siegler hermetically sealed Apollo EL display (Courtesy of S. Petertyl).

sults were excellent as compared to previously fabricated EL displays. Cost of the display was now the major problem. The battle was won but the war was lost.

8.2.3 Thin-Film EL. Several thin-film phosphor screens were made for CRT applications before Destriau discovered EL. J. H. DeBoer and C. J. Dippel[28] were issued a U.S. patent in 1934 for evaporating luminescent phosphor films. In 1947 F. E. Williams[29] used evaporation to produce ZnF_2 films. These films activated with Mn were of good luminescent efficiency based upon a cathodoluminescent application. The ZnF_2 films, however, were not very stable under electron bombardment. Williams also described forming films of a number of other phosphors, ZnS:CdS, ZnO, ZnS:ZnSe, and ZnS. He noted that heating the films improved their luminescence. In 1951 Studer, Cusano, and Young[30] produced luminescent ZnS films by a vapor reaction method. In 1957 Charles Feldman and Margaret O'Hara[31] of the U.S. Naval Research Laboratory reported on their completion of extensive study on thin-film phosphors. They found that the chemistry of the powdered material was applicable to thin films. They also demonstrated the importance of post-heat treatments and found it could affect further crystal growth. Most of the films they made could be either transparent or fogged, depending upon the post-heating. Manganese was the primary activator. Under electron gun excitation, transparent thin film ZnS:Mn was 20% as bright as conventional ZnS:Mn powder and had a 0.89 transmittance compared with 0.28 transmittance with powder.

As early as 1959, W. A. Thornton[32,33] was using thin-film phosphors specifically for EL display purposes. Most of his attention was devoted to dc thin-film EL using copper and silver as activators. The problem with the Cu-and Ag-activated devices was that the conductivity was too high, which made the devices too susceptible to uncontrolled breakdown. Manganese was used only as a co-activator. In 1961 the first real breakthrough came at Servomechanisms, in Santa Barbara, California. A team led by Edwin J. Soxman successfully used thin-film technology to deposit ZnS:Mn phosphor with a dielectric in an ac configuration.[34] In 1962[35] Soxman demonstrated high luminance and matrix addressability using breadboard substrates with a 5 by 6 matrix array.

In 1963, Servomechanisms of Los Angeles wanted to divest themselves of their research laboratory at Santa Barbara, which was then acquired by Martin Reder, Edwin Soxman, and Gordon Steele to form Sigmatron. Soxman and Steele developed a double-sided dielectric sandwich[36] to improve upon the configuration which they had demonstrated to the Office of Naval Research. Their proposed work resulted in a ten-year JANAIR contract[37] for studies on EL. During this time (1965)[37] they completed the first large matrix-addressed EL panels (10 in. by 10 in. with 25.6 lines/in.) and test panels (1.5 in. by 1.5 in. with 33 lines/in.) The test panels achieved luminance values of 500 fL at 400 V rms when excited continuously at 5 kHz. The discrimination ratio was 10^4 to 10^6 at one-third select voltages. A single sided GeO_2 dielectric was used with ZnS:Mn and SnO_2 front electrodes and Al back electrodes. Steele and Soxman went on to develop black layer EL displays.[38] Sigmatron set up a pilot line to fabricate numeric displays[39] which were sold throughout the country.[40] Sigmatron demonstrated[41] their technical accomplishments in New York in September, 1965 at the SID 6th National Symposium on Information Displays. Under the trademark "LEF" (Light-Emitting Film), they showed hardware which included a black layer matrix-addressable EL panel, the test panels, and large panels. An example of the black layer display as demonstrated is shown in Fig. 8-5.

The full impact of the ac thin-film EL was best demonstrated by Sigmatron (mid 1960s) in their numeric twelve-character prototype displays with the black layer *behind* the transparent phosphor, as shown in Fig. 8-6. Sigmatron successfully demonstrated sunlight readability[42] with those prototype displays while emitting only 25 fL. The demonstration was most impressive. It made true believers of all those who ever saw it. The display was the most functional and aesthetically pleasing of any numeric display ever made up to that time and perhaps still today. In 1972, life of up to several thousand hours had been demonstrated.[43] Also, the displays were becoming more matrix-addressable due to a much-improved discrimination ratio.[39]

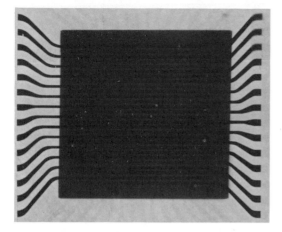

Fig. 8-5. Early ac thin-film EL matrix display with black layer made at Sigmatron and demonstrated at 1965 SID Symposium, N.Y. (Courtesy of E. Soxman).

Sigmatron's inability to manufacture their black layer numeric display for Commodore Computers led to Sigmatron's bankruptcy in September 1973.[40] The assets were acquired in 1975 by investors who set up Sigmatron Nova in Thousand Oaks, Calif., under the technical leadership of W. Essinger.

Others were working on ac thin-film EL. One formulation developed by Kahng and associates[44] in 1968 was called the Lumocen (LUminance from MOlecular CENter). It consisted of SnO_2 (200nm), HfO_2 (300 nm), $ZnS:TbF_3$ (150 nm) and Al electrode, all on a glass sub-

strate. Operating devices were demonstrated, but the life was short. Vlasenko and associates completed devices using $ZnS:Mn$ in 1965.[45,46] Russ and Kennedy reported[47] on $ZnS:Mn$ triple-layer thin-film ac research in 1967.

Sharp Corporation, starting in the early 1950s under the inspired leadership of S. Mito, did considerable research in EL, first with powders and subsequently with thin films. Sharp personnel visited Sigmatron[41,48] in the late 1960s and saw their facilities, devices using ac thin-film EL with $ZnS:Mn$, their black layer display and their matrix-addressed display panel. Sharp improved the technology; they used E-beam evaporation of the $ZnS:Mn$ and applied better dielectrics. They gave two excellent papers at the 1974 SID International Symposium in San Diego.[49,50] In September 1978, they demonstrated a 240 by 320-line thin-film EL monochromatic TV for the first time in the U.S. at the Consumer Electronics Show in Chicago, (Fig. 8-2). The Sharp disclosures gave an "existence proof" of the solution to the second and third problems, i.e., high luminance while preserving long life, and matrix addressability with video for several hundred display lines. Sharp did not use a black layer. They used a diffuse reflective aluminum back electrode and a front circular polarizer or neutral density filter to achieve good contrast in high ambient illumination. The displays, however, were not readable in direct sunlight.

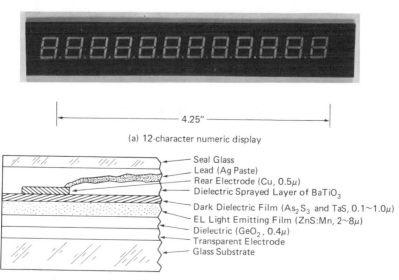

|←——————————— 4.25" ———————————→|

(a) 12-character numeric display

Seal Glass
Lead (Ag Paste)
Rear Electrode (Cu, 0.5μ)
Dielectric Sprayed Layer of $BaTiO_3$
Dark Dielectric Film (As_2S_3 and TaS, $0.1 \sim 1.0\mu$)
EL Light Emitting Film ($ZnS:Mn$, $2 \sim 8\mu$)
Dielectric (GeO_2, 0.4μ)
Transparent Electrode
Glass Substrate

(b) Cross section

Fig. 8-6. Sigmatron sunlight readable numeric display with black layer, ca. 1968 (Courtesy of B. Webb).

Sharp made additional technical advances such as thin-film addressed matrix displays with memory,[51,52,53] TV stop motion,[54] and direct electronic readout capability.[55]

By 1974, almost all of the research and development teams at the various companies throughout the U.S. who had been working on EL during the 1950s and 1960s had been disbanded. Soon after the Sharp disclosures, high-technology companies once more began research and development. Much of this was stimulated by contracts for research and development awarded by the U.S. Army, Ft. Belvoir, and Ft. Monmouth. Of note is the support by B. Gurman, I. Reingold, and E. Schlam, U.S. Army, Ft. Monmouth, New Jersey.

8.2.4 DC Powder EL. In reviewing the history of EL, a note must be made of the fine work conducted at the Thames Polytechnic in London under the leadership of A. Vecht. In using dc powder EL,[56,57] they have demonstrated a wide range of product applications using different colors and demonstrating matrix addressability with high luminance. Short life with high luminance using either ac or dc powders remains a limitation.

8.2.5 TFT Extrinsic EL Drive. During the early 1970s Westinghouse Research Laboratories in Pittsburgh, Pa., and Aerojet ElectroSystems in Azusa, Calif., used thin-film transistors (TFTs) to extrinsically address ac powder EL displays.[58,59] In the hope of achieving longer life, the "hypermaintenance" phosphor powders developed by W. Lehmann[60] at Westinghouse were used in the Westinghouse devices. Two TFTs and a capacitor were used at each pixel to achieve memory and matrix addressability, which was not otherwise possible with ac powder EL.

The application of TFTs to address EL powder extrinsically has several advantages in addition to making the panel matrix-addressable:

- It reduces the line address voltage to CMOS levels (15 volts).
- It adds memory to the display.
- It increases the duty cycle to 100% for higher luminance regardless of the number of rows.

- It separates the power input to the display from the image input, thereby allowing for simple, resonant, power-saving techniques.

However, progress has been slow and difficult.[61] Researchers at Westinghouse and Aerojet Electro-Systems also applied TFTs to ac thin-film EL.[62,63] The ac thin-film is intrinsically matrix-addressable, but the addition of TFTs still gives all the advantages listed above.[64] The life problem of ac powder EL has never proven to be eliminated even with the hypermaintenance approach. This made the substitution of thin-film EL attractive. The resulting structure was very complex for a large-area monolithic poly-crystalline structure, and progress was very difficult.[65]

8.2.6 DMOS Intrinsic EL Drive. Intrinsic matrix addressing of ac thin-film EL using custom LSI DMOS electronics for direct high voltage line drive overtook the extrinsic TFT approach early in the 1980s. This has occurred for several reasons:

- Marked progress on the DMOS technology by Sharp[66] and Texas Instruments[67] using custom LSI MOS with CMOS logic and DMOS high voltage drivers on the same MOS chip.
- High-performance, high-voltage ac thin-film EL fabricated by Sharp, Lohja, Aerojet ElectroSystems, Rockwell International Research Center, and others.
- Slow progress due to the complex structure of the monolithic TFT and powder or thin-film EL structure.

The availability of custom LSI DMOS for the high-voltage line drivers is the second major breakthrough needed to get EL out of the laboratory and into display products. The first was the evolutionary development of high-performance ac thin-film EL using ZnS:Mn phosphor as discussed above. The DMOS technology was developed for the high-voltage power-switching industry independent of display needs. Texas Instruments was the first to develop custom LSI DMOS for displays using their BIDFET[TM] technology. They used bipolar and DMOS transistors on the same MOS chip along with CMOS logic and shift registers. This technology was best exemplified by their now-famous SN75501 TI devices to drive plasma panels. Vacuum fluorescence and EL-display drivers are now available using the same BIDFET^{TM} technology. Sharp was the first to use DMOS custom chips to drive EL displays. They used them in their TV demonstrator shown at the 1978 Chicago Consumer Electronics Show. Sharp did not have the bipolar capability along with the MOS as did Texas Instruments.

It is estimated by this author that over 5,000 man-years of research and development had been spent on EL prior to 1980—and the work goes on.

8.3 THEORY OF OPERATION

8.3.1 Status of Theory. The basic mechanism of operation of EL devices is poorly understood compared with other display technologies. All the EL configurations, ac or dc with powder or thin film, are based as much on experimentation as on theory. The engineering community needs a better theoretical understanding of the principles of operation. This knowledge should help in improving performance and eliminating problem areas. A workshop on the Physics of Electroluminescence was organized by Professor R. Evrard at the University of Liège, Institute of Physics, Belgium in September 1980 and recorded by Professor F. Williams, editor, in a special issue of the *Journal of Luminescence* Vol. 23, in 1981.[68] These two together made significant strides in clarifying high-voltage, ac thin-film ZnS:Mn EL configurations. The earlier theoretical work by Professor A. G. Fischer still holds for the ZnS:Cu$_x$S powder configurations.

EL operates in a materials and voltage realm different from that of any other electronic device. A sound theoretical background is lacking partly because there has not been an industrial need and partly because of the complex nature of the operation of polycrystalline material under the influence of high electric fields. All the EL active layers considered in this chapter are polycrystalline dielectrics that are forced to conduct electrons in a controlled manner. In the displays industry, single-crystal EL is called LED and is discussed in detail in Chapt. 9.

Polycrystalline EL display can be classified

according to their mechanisms of light generation. The three major mechanisms used in displays can be simply described as:

Type 1: Bipolar alternating double-injection luminescence, demonstrable with ac powder ZnS:Cu$_x$S phosphors, or the Fischer Model.[69]

Type 2: Hot electron impact ionization of activator, demonstrable with ac thin-film ZnS:Mn and dc powder ZnS:Cu$_x$S phosphors or the Chen and Krupka Model.[70]

Type 3: Memory effect in ac thin-film ZnS:Mn phosphors first reported by Yamauchi et al. (1974)[52] and described by Onton and Marrello.[71]

Polycrystalline EL is also classified by its external electrical circuit, namely into ac and dc configurations. This classification simply relates to how electrical energy is delivered to the display. In the ac drive method there is a natural ballasting of the display because of the capacitive impedance. This, when properly implemented, prevents current avalanching and ensuing thermal destruction. There is no natural ballasting by-product in the dc drive method. The ballasting used is to pulse drive the display in such a way that before the avalanching current becomes significant the supply is switched off.

8.3.2 Difference in Observed Light Generation Mechanisms. The details of the least resolvable light reveal the variance in the forms of electroluminescence in polycrystalline materials. The difference in the "comets" of ac powder EL (Type 1), the "uniform luminance" (Type 2), and "filaments" (Type 3) of ac thin-film EL is striking. The name "comet" was assigned by Fischer[69,72] to describe the light sources he saw in ac powder EL pixel cross sections (Fig. 8-7). The observation of "uniform luminance" for nonmemory mode and the name "filament" for memory mode was assigned by Marrello, Ruhle and Onton[73,74] to describe the light sources they saw in ac thin-film EL pixels in the viewing plane (Fig. 8-8).

The comets of ZnS activated with Cu$_x$S and coactivators have been observed in the cross section and described by Fischer[69,72] as a pair

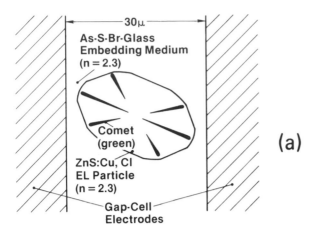

Fig. 8-7. "Comet" light source in ac powder EL display (Courtesy of A. G. Fischer).

Fig. 8-8. Individual light filaments in ac thin-film memory display (Courtesy of A. Onton and V. Marrello).

of mirror image comet-shaped luminescent volumes occurring in the interior of a crystal of cubic ZnS along a fracture filled with supersaturated Cu_xS. In the presence of an alternating electric field, the fracture fill materials glow at a relatively low field strength. As the field is increased, the glowing volume increases in both size and luminance. Their orientation is fixed. Their direction is determined by the fracture orientation. The highest luminance occurs from those parallel to the field lines of force; the lowest, from those perpendicular. The relationship holds true as the field is rotated. The volume of a comet is typically from 4 μm long by 1 μm in diameter to 10 μm long by 2 μm in diameter in a 20-μm-diameter ZnS grain.

Figure 8-7 is drawn from mircroscopic observations of well-crystallized cubic ZnS:Cu, Cl particles observed at 1200 X in the plane of the ac-field lines of force.[69,72] Disturbing reflections due to irregular particle surfaces are eliminated by embedding the particle in an insulating, low-melting glass (As-S-Br) of the same refractive index as ZnS (n = 2.36). Comets always exist in pairs, and the one next to the anode always lights first. During aging, the comets become fainter and fainter and only dim light comes from comet heads. During aging, the field-emitting tips of the comets become blunted due to copper diffusion. The double comets follow main crystal axes. They are associated with copper-decorated edge dislocations at the boundaries of stacking fault planes.

The light filaments in the ac thin-film EL of ZnS activated with Mn have been observed in end view and described by Marrello, Ruhle and Onton[74,75] as single light filaments of less than one μm in diameter. The display is made up of an ensemble of the uniform filaments roughly uniformly distributed as seen in Fig. 8-8. The filaments appear to move about somewhat randomly and repel each other when close together. The filaments are sometimes initiated at a discontinuity point called a "source" and appear to move outward from the source as the field is increased and retreat to the same source and disappear when the field is reduced. The filaments move about as if controlled by nucleating sites and mutual repulsive forces and migrate as if controlled by film morphology or thermal effect. The end of the filament is less than one μm in diameter, below the resolving power of precision microscopes. The higher luminescence is achieved by the higher density of filaments. Each filament acts in a bistable manner. The luminance performance curve of a display is due to the hysteresis of an ensemble of filaments.

Filaments move; comets do not. Furthermore, filaments are bistable; i.e., they are either on or off, and comets luminesce proportionally to the one-half power of applied voltage.

The filaments are associated only with the memory mode of ac thin-film EL with ZnS:Mn, and only when the manganese concentration is greater than one mole percent. When the concentration is below one mole percent, memory does not occur and filaments do not occur. Non-memory EL (Type 2) does not exhibit any optically resolvable sources of light. The luminance is due to a uniform haze of evenly distributed, unresolvable light sources.[76] It is characterized by a very high discrimination ratio of 1×10^6 for one-third select voltages.[77] These differences among Types 1, 2, and 3 EL are outlined in Table 8-1.

Many other experimental differences will be presented in the balance of this report. It is important to note these fundamental differences. Observations and conclusions about one EL configuration do not generally apply to another configuration.

8.4 AC THIN-FILM EL

8.4.1 Physical Structure. The basic pixel structure is shown in Figs. 8-9 and 8-10. This configuration has been extensively researched[78] and has been experimentally proven to have long life,[79] high luminance,[77] and video display capability.[79] There are several approaches in the selection of dielectrics and seals. The phosphor of zinc sulfide activated with less than one mole percent manganese has thus far been the best performer in terms of discrimination ratio, luminous efficiency, life, and luminance. The color is a pleasant yellowish orange centered at 585 nm, with a discrimination ratio of one million, a luminous efficiency of as high as 6 lumens/watt of panel power,[80] life of over 10,000 hours and luminance appropriate for sunlight readability.

The front electrode is arbitrarily selected as the column electrode, which must be as transparent as possible so as not to block the emitted light. The fringe field or light-emission fringe is

Table 8-1   Differences in Observed Light Mechanisms in Electroluminescence

Characteristic	Type 1	Type 2	Type 3
Activator	Cu_xS and others	Mn < 1 mole%	Mn > 1 mole%
Light characteristics	Comet pairs	Haze	Ensemble of filaments
Motion	Stationary	Undetectable	Move randomly; mutually repelling; speed of motion increases with excitation frequency
Light amplitude with voltage	Analog $\sim V^{1/2}$	Threshold with 10^6 discrimination ratio	Hysteresis; bistable
Size	1 by 4 μm to 2 by 10 μm when viewed in plane of field	Continuous	Point < 1 μm when viewed in plane perpendicular to field equipotential surface
Aging	Continuous reduction in luminance to < 10 fL	Voltage shift in performance curve	Voltage shift in hysteresis curve and narrowing of hysteresis loop

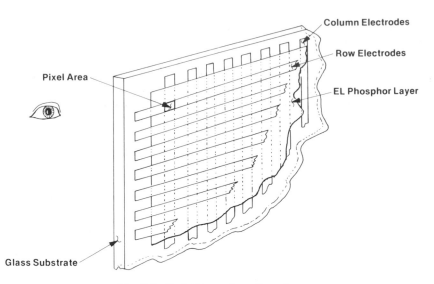

Fig. 8-9. Basic EL panel configuration.

insignificant. Light is emitted only from the area where the two electrodes overlap. The best front electrode has been found to be indium–tin oxide (ITO), which is a transition metal oxide semiconductor with resistance of approximately 5 ohms per square and 90% transmittance at a thickness of 600 nm. A ratio of 10 mole percent indium is optimum.[81]

The best phosphor has been experimentally determined to be ZnS as a host with Mn as the activator. Coactivators of Cu, Cl, or Ga of much lower concentrations slightly improve performance by increasing luminance,[77] accelerating aging,[82] and reducing operating voltage,[9] respectively. The significance of the Cu coactivator

with Mn in increasing the persistence can be seen in Fig. 8-11. This Cu is at one-tenth the concentration of Mn and should not be confused with the phosphors where Cu is the dominant activator and is at or above the Cu saturation level in ZnS. Gallium initially reduces the operating voltage, but after stabilization its advantage is lost. Zinc selenide mixed with ZnS has also been used[83] as a host with the intent of lowering the operating voltage due to the lower band gap. This has been done, but not without lowering the operating luminance. Concentrations of Mn above approximately one mole percent are necessary for a hysteresis effect in memory flat-panel and memory CRT displays[84,85] and for a

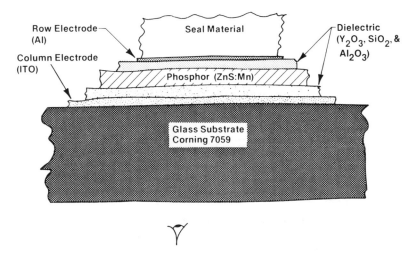

Fig. 8-10. Basic EL pixel structure showing thin-film sandwich layers.

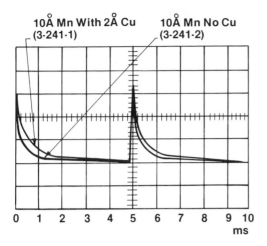

10Å Mn With 2Å Cu (3-241-1) **10Å Mn No Cu** (3-241-2)

Fig. 8-11. Oscilloscopic trace of light pulse persistence, showing ac thin-film ZnS·Mn with and without Cu coactivation.[77]

photo-optic effect appropriate for light pen write and read applications.[55] This higher concentration of Mn results in lower luminance and efficiency. However, if the device has memory, then 100% duty factor can be used to regain the higher average luminance associated with the optimum Mn concentration used for nonmemory displays.

The biggest variance in the structure of Fig. 8-10 has been in the selection of the dielectric materials. From first principles, the dielectric should be of high dielectric strength to prevent electrical breakdown of the thin-film stack and should have a high dielectric constant to cause most of the electric field to be concentrated across the phosphor layer. The dielectric-phosphor interface is the probable location of trapped charges fundamental to the operation of ac thin-film EL. The choice of dielectric does affect the efficiency, luminance, and voltage performance of the panel. For example, no memory has been reported with a dielectric Al_2O_3-phosphor interface.

The most-used dielectrics include Y_2O_3, Al_2O_3, Si_3N_4, and amorphous $BaTiO_3$. Dielectric layers can be double-layer composites of dissimilar materials, a procedure used in thin-film technology to minimize the occurrence of pinholes in the films. A selection of the most-often used dielectrics and phosphors are tabulated in Table 8-2, with their major physical properties. The dielectrics are the weakest layers in the series of thin films in the EL sandwich; it has been

impossible to completely eliminate pinhole electrical breakdown in them. Better solutions are continually being sought. Sharp is using layers of silicon-oxy-nitride next to the ZnS:Mn in their production panels.[86] Texas Instruments has experimented with SiAlON with promising results.[87] Amorphous films generally have dielectric properties with fewer defects and are less sensitive to contamination and stoichiometry than crystalline films. However, dielectrics will usually not remain amorphous when subjected to a 550°C phosphor anneal. There is not the same unanimity in the selection of the best dielectrics as there has been with the phosphor. It is inevitable that both better dielectrics and phosphors will be discovered.

There are several ways to seal a display. There is no disagreement that the phosphor layer must be protected from humidity. Any water molecules in the sandwich will hydrolyze under the influence of the electric field and delaminate the film. The water can get into the sandwich during fabrication or through pinholes after fabrication. Sharp has used a silicon oil dielectric fluid contained within a second glass substrate and an organic fillet seal. The author has used a Y_2O_3 thin-film passivation layer, a solid polymer with metal cladding for seal, protection and cooling, and an organic fillet as shown in Fig. 8-12. Lohja has used a thin-film Al_2O_3 passivation layer with a polymer-adhered glass plate.

The commonly used substrate is drawn or ground low-alkali borosilicate glass. The low alkaline content minimizes the potential of contamination[88] by alkali ions diffusing into the phosphor. The borosilicate glass is desirable because of its high-temperature stability, needed during the annealing of the phosphor, typically at 550°C. LE-30 glass from Hoya Electronics Co. in Japan and 7059 glass from Corning Glass Works in Corning, NY are two glasses used for EL displays. Soda-lime glass has also been used.

The preferred row or back electrode is almost universally aluminum. The aluminum electrode survives better than other electrodes because it fuses to an open circuit in the immediate vicinity of any pinhole breakdown that may occur. Indium tin oxide is the next most common back electrode; however, its fusing properties are not as good as those of aluminum, particularly if the Al is made very thin—20 to 50 nm.

Table 8-2   Dielectrics Used in AC Thin-Film Panels

Dielectric	Refractive index	Dielectric constant, ϵ/ϵ_o	Dielectric strength MV/cm	Maximum electric displacement, $\mu C/cm^2$	Reference/Remarks (see footnotes)
Al_2O_3	1.98	7.9	4	3	(A)
TiO_2	2.586–2.741	3.0–4.0	0.04	0.01	
ZnO_2		20	4	7	(A)
SiO		5.1	2.95	1.4	(A)
SiO_2	1.470–1.484	3.8	10	3.5	(B)
Y_2O_3		11–13	4.5	4.5	(A)
Ta_2O_5		20–23			(C)
Sm_2O_3		15.5			(C)
ZnS	2.356–2.378	10	100	100	(D)
Y_2O_3		12	1–3	1–3	(E) Moisture sensitive
Ta_2O_5		25	2–3	4–6	(E) Reaction with ITO
Si_3N_4		7	5	3	(E) Poor adhesion, stress
Al_2O_3		10–11	1–3	1–3	(E) Reaction with ITO
SiAlON		8–9	8	6	(E) Moisture Barrier
Y_2O_3		12	3–5	3–5	(F)
Ta_2O_5		23	1.5	3	(F)
$BaTa_2O_6$		22	3.5	7	(F)
$SrTiO_3$		140	1.5–2	19–25	(F)
$PbTiO_3$		150	0.5	7	(F)
$BaTiO_3$		14	3.2	4	(G)

(A) Thin Solid Films, Physics Abstracts 1970.
(B) *Handbook of Chemistry and Physics*
(C) Kozawaguchi, et al. *SID Proceedings,* Vol. 23/3, 1982.
(D) Hanays and Campbell. *Dielectric Properties of Thin Films.*
(E) Tiku, et al (Texas Instruments). *SID 83 Digest.*
(F) Fiyita, et al (Matsushita). *Japan Display 83 Digest.*
(G) Alt, IBM.

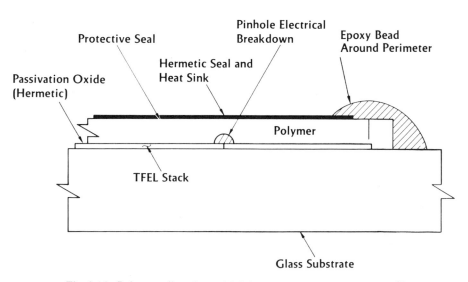

Fig. 8-12. Polymer-adhered metal foil for hermetic seal of EL display.[92]

8.4.2 Performance. Each EL pixel gives two pulses of light during each refresh frame, giving the designer more flexibility in controlling flicker and luminance. During refresh the panel is addressed one line at a time using matrix addressing signals of one voltage polarity. At the end of the refresh frame, during a period equivalent to CRT fly-back time, the entire panel is reset by a pulse having the opposite voltage polarity, completing the ac cycle. Pixels selected during the first pulse cycle emit light during both pulses. A one- or two-frame transient exists in turning a pixel on or off, depending on the voltage amplitude.

The characterization of a knee-type inflection is used in evaluating EL luminance vs. voltage performance. The definition in Fig. 8-13 would permit the quantitative comparison of results. By this definition, the knee of the curve is close to the optimum luminous efficiency and the nominal operating point. The movement of this point in life studies and aging studies is important. The point is easily found graphically by simply sliding a straightedge (with a slope of one decade in luminance per 100 volts peak voltage) aligned to the graph paper until it be-

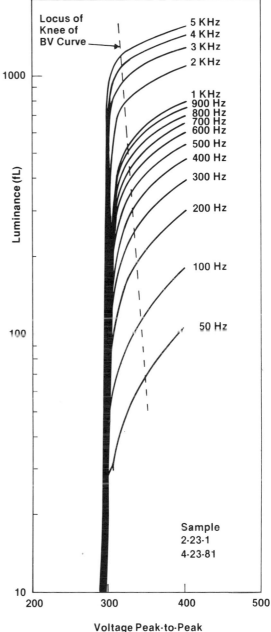

Fig. 8-14. EL performance as a function of excitation frequency.

comes tangent to the luminance vs. drive voltage curve. This definition is used to define the knee in Fig. 8-14 and the results are plotted out in Fig. 8-15.

The luminance of the display is directly controllable by varying the refresh frequency, as shown in Figs. 8-14 and 15. In certain panels it

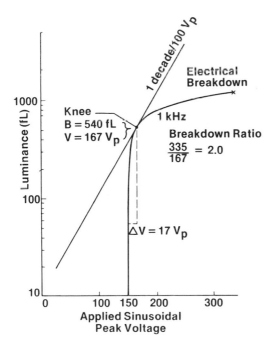

Fig. 8-13. Typical luminance vs voltage performance curve depicting the definition of the knee of the performance curve.

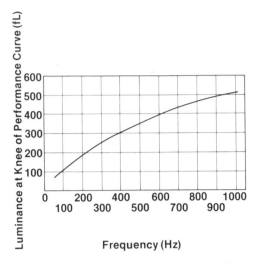

Fig. 8-15. EL luminance performance at knee of curve as a function of excitation frequency.

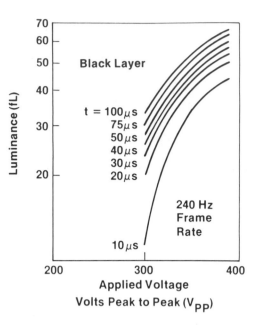

Fig. 8-16. Influence of excitation pulse width at a constant frame rate.

can also be controlled by pulse-width modulation, as in Fig. 8-16.

The luminance at the knee of the curve is very important. Along with duty factor, it limits the display matrix size or number of rows when using line-at-a-time addressing. The better-performing ac thin-film EL formulations such as those shown in Fig. 8-17 have discrimination ratios at one-third voltage of better than a million. The excess discrimination ratio can be used to simplify the electronic drive. The difference between the on-pixel voltage and off-pixel voltage can be reduced to 50 volts or lower, and the nonaddressed row lines can be allowed to float electrically.

The performance of an EL device can vary widely depending on the materials or methods used in its fabrication.[89] However, the peak luminance performance of ZnS:Mn from three different laboratories on three different continents has thus far proven to be quite similar. A comparison is given in Fig. 8-18. These three laboratories, located at Sharp, Lohja and Aerojet ElectroSystems, used three different deposition techniques and three different dielectric formulations. As indicated above, the luminance is also a function of the refresh frequency and pulsewidth, and to a lesser extent, of pulse shape.

The importance of rigorously documenting and defining the parameters associated with the reported luminance vs. voltage performance curve should be discussed. A luminance vs.

voltage performance curve is not interpretable without the luminance units, refresh frequency, voltage wave shape, aging, environments, and duty factor, or pulsewidth.

Very high luminance EL was achieved using the sandwich structure sequence outlined in

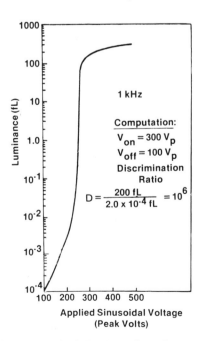

Fig. 8-17. Discrimination ratio performance.

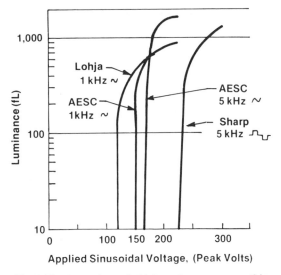

Fig. 8-18. Examples of high performance ac thin-film EL.

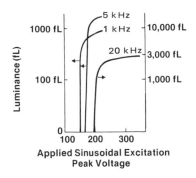

Fig. 8-19. High-performance EL panel.

Table 8-3. In Fig. 8-19, the high frequency performance is shown. This performance is attributed to high purity materials, low moisture content in chamber, precise doping with activator and coactivator, substrate heating during deposition, and proper post annealing. The ZnS was thermally evaporated from a resistance boat and coactivated with 0.4 At% (0.8 mole%) Mn and 0.04 At% (0.08 mole%) Cu.[77] The copper coactivator influence is shown in Fig. 8-11. Since the Cu is well below its solubility level in ZnS, it is not a factor in the phosphor conductivity. The technique for evaporating the phosphor is

of secondary importance. What is important is that a dense film be deposited with minimal pinholes. Several other formulations are available in the literature.[49, 78]

Several other researchers have obtained super-high luminance comparable to that obtained at Aerojet ElectroSystems,[77] and shown in Fig. 8-19. In 1972 Sigmatron[37] reported on an experimental panel with a luminance of 3,800 fL at 20 kHz sinusoidal excitation at a peak voltage[90] of approximately 500 V. The discrimination ratio of this Sigmatron El film was much lower than that shown in Fig. 8-19. IBM has made but not reported on films with a luminance of 2,000 fL using a 5 kHz square wave excitation.[91]

The luminance data as reported here and by most authors is for aluminum back electrodes which are ideal reflectors. The method used to measure EL performance is shown in Fig. 8-20.

Table 8-3 Process Steps for Fabrication of Low-cost E L Display Panel

Layer	Material	Approximate thickness in nanometers	Kovar mask	Deposition
Edge connection leads	ITO:Au	100:160	Yes	Sputter
Transparent electrode	ITO	200	Yes	Sputter
First dielectric	Y_2O_3	80	Yes	E-Beam
	Al_2O_3	80	Yes	E-Beam
	Y_2O_3	80	Yes	E-Beam
Phosphor	ZnS:Mn, Cu	500	Yes	Thermo
Second dielectric	Same as first	Same as first	Yes	E-Beam
Black layer	—	—	—	—
Step edge	Ti:Al	20:100	Yes	Sputter
Back electrode	Al	20 to 50	Yes	Sputter
Passivation	Oxide	200 to 500	Yes	E-Beam
Protective seal	Polymer	50,000	No	None
Hermetic seal	Metal	25,000	No	None

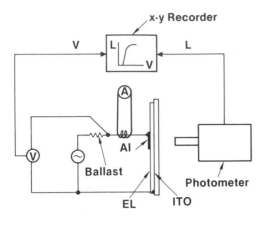

Room Ambient Darkened, 20 °C

Fig. 8-20. Test setup for characterization of EL display panels.

The back reflector doubles the luminance by reflecting half the flux from the isotropic EL radiation forward. A back reflector is normally unacceptable in a display application because it also reflects ambient illumination. The ambient illumination can be counteracted with a black back layer between the phosphor and aluminum electrode, or with a front circular polarizer or transparent back electrode and black back absorber. All of these approaches nominally reduce the effective display luminance by half. However, for comparison purposes, the setup shown in Fig. 8-20 is normally used when measuring the performance of the basic sandwich. Methods to enhance the display's readability vary—e.g., filters, polarizers, absorbers, or antireflecting coatings. They cloud the picture when measuring the performance of the phosphor.

8.4.3 Discrimination Ratio. The matrix addressability is of key importance. This means that only row and column drivers are needed. A thin-film transistor array or other nonlinear element does not need to be added at each pixel.

The degree of matrix addressability is a function of discrimination ratio, which again is defined as the ratio of the luminance of a nonarrayed pixel in response to the applied "on" voltage, to the luminance of that same pixel in reponse to the voltage that would occur due to the one-third-select phenomenon of matrix-addressed displays when the pixel is intended

to be "off." For arrays with more than ten rows when using line-at-a-time addressing, the contrast ratio equation (neglecting ambient illumination—see Chapt. 5) reduces to:

$$\text{Pixel contrast ratio} = (L_{on}/NL_{off}) + 1$$
$$= (D/N) + 1$$

where L_{on} is the luminance of the on pixel and L_{off} is the luminance of the off pixel based on the luminance performance curve of a single pixel, N is the number of rows, and D is the discrimination ratio. Therefore, the discrimination ratio must be equal to the number of rows for a contrast ratio of 2, which is not acceptable when considering that we have not taken into account ambient illumination. A discrimination ratio of ten times the number of rows is more appropriate. The discrimination ratio of ac thin-film EL is one million, as can be computed from the example shown in Fig. 8-17.

The highly nonlinear nature of ac thin-film as depicted in Fig. 8-17 gives the designer more flexibility in selecting the drive voltage conditions. For optimal discrimination in a matrix-addressing scheme, the off voltage is one-third the on voltage. The off voltage can never be less than one-third. However, one-third for off is difficult to implement electronically. An off voltage of one-half the on voltage is easier, and more than one-half is easier than that. As the difference in on to off voltage decreases, the discrimination ratio decreases to the point where contrast is lost, as can be computed from the above equation. For a drive voltage difference of 50 V and an on voltage of 275 V, and using the curve of Fig. 8-17, the usable discrimination ratio is 10,000, as depicted in Fig. 8-21. Using this relationship for a display with 1,000 rows, the contrast ratio is 11.

8.4.4 Panel Size. The number of addressable rows for ac thin-film EL such as those shown in Figs. 8-17 and 8-19 is limited not by the discrimination ratio but by the frame rate and duty factor. For most display technologies, including EL powders, the available discrimination ratio limits the number of row lines. A finite amount of time is needed to electrically excite each row. This leads to a minimum duty factor required

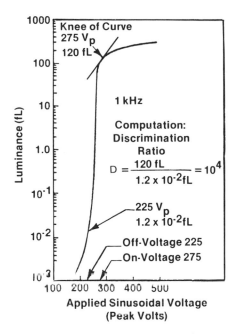

Fig. 8-21. EL performance curve showing definition of discrimination ratio.

for acceptable luminance. For the ac thin-film EL example of Fig. 8-16 where a luminance of 50 fL is desired, this is approximately 20 μsec at 240 Hz frame rate. The effect of frame rate can be scaled from the slope of Fig. 8-15. For example, it would take 4 msec (200 × 20 μsec) to write one frame on a 199-row display allowing one row time for the reset pulse. The columns are addressed in parallel, and therefore do not affect the addressing time. In this case, for a 4-msec frame time there is enough time to refresh the display at a maximum frame rate of 250 Hz. Since only 240 Hz is required to give the desired 50 fL, we have a satisfactory solution. An EL display of this size refreshed at 200 Hz frame rate has been tested and found to be sunlight readable.[92]

The physical size of an EL panel is limited by the transparent ITO conductor resistance in whatever axis it is used. As EL panels become larger, the line resistance of the front transparent conductor increases. The electrode line acts like a transmission line where each pixel adds line capacitance and pixel-to-pixel ITO adds line resistance. The line width is of no consequence because the resistance decreases in proportion to the capacitance increase, so the RC (resistance times capacitance) time constant

stays the same. In 1983 the best ITO layer that could be made was 5 Ω/square at 90% transmittance in a 200-nm thin film. Thicker ITO can lead to lower line resistance but causes fabrication problems as the upper layers must replicate the ITO step edge. It has been experimentally determined that this is adequate for up to 5-inch lines driven from only one end, or a 10-inch line driven from both ends. Conductor improvements are needed for larger panels.

8.4.5 Temperature Dependence. The luminance and electrical characteristics of ZnS:Mn ac thin films are quite insensitive to temperature. From −269°C to about 77°C, only minor changes are seen in luminance and electrical properties such as threshold voltage and discrimination ratio.[93] Above 77°C there is a general reduction in the threshold voltage and discrimination ratio. For memory panels with manganese activation above one mole percent, there is a gradual falloff in luminance with increasing temperature. Burn-in[6] is accelerated by orders of magnitude in time at temperatures around 200°C and operation above the knee of the luminance curve. The pinhole breakdown failure rate of weak thin-film sandwiches is greatly accelerated at higher temperatures. Weak thin-film dielectrics will conduct and heat the panel, and at the high operating temperatures with internal heating, the panels deteriorate rapidly.

The temperature insensitivity of an ac thin-film ZnS:Mn panel is a very significant advantage and greatly enhances its potential range of practical applications. It also gives some clues as to the light-generating mechanism: it eliminates any thermal process.

8.4.6 Life, Aging, and Burn-In. The life of the phosphor in the ac thin-film sandwich has been reported by Sharp to be over 20,000 hours[6] with no aging effect observed at all. The samples on which they report the 20,000 hr life were in a controlled environment. Sharp has also reported[94] that "large panels" typically have 30% reduction in luminance after 10,000 hr, which is more consistent with that observed in cathodoluminescent phosphor.

A life test on a panel at Aerojet ElectroSystems in a quasi-controlled environment is shown in Fig. 8-22. The panel was operated at 100%

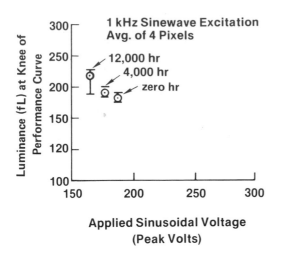

Fig. 8-22. Aging of EL panel as exhibited by motion of knee of performance curve.

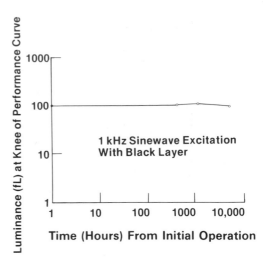

Fig. 8-23. Life test results for 320 × 192 (4 in. × 6 in.) panel.

duty cycle, with sine wave excitation of one kHz above the knee of the performance curve. The seal was a polyimide polymer cured on the EL sandwich. There were no line or pixel failures out of 900 pixels during 12,000 hours of operation. The substrate was made as shown in Table 8-3. Cathodoluminescent phosphor devices typically go through an initial burn-in which may take several hours to tens of hours. This is not necessarily always true of ac thin-film EL. During burn-in, the EL performance curve (Fig. 8-13) may shift to the right or rotate clockwise or counterclockwise about the knee of the curve. This motion is thought to be due to the formation of deep traps.[95] It may be due to the motion of atoms in the crystal, diffusion, or chemical reactions under the influence of the electrical field. Burn-in motion is very dependent on the fabrication process, and the speed of stabilization is very dependent on the temperature and electrical stress.

Aging and burn-in properties are important to the display designer. He must design controls to account for burn-in and aging panels if changes are significant. If initial stabilization effects are significant, the panel must be burned in before installation or the panel may show image ghosting. After burn-in, the longer-term effects are called "aging." From numerous observations, EL burn-in is over before the first 100 hours of operation. An adjustment in voltage may be needed to keep this display operating at an optimum point relative to the knee of the performance curve over the rest of its life. Optimum

efficiency occurs just below the knee, and optimum uniformity just above the knee. Each panel manufacturer will have his own performance data and criteria.

It appears from the author's experience and from published data, that luminance and voltage changes are less with EL than with CRT phosphors. Examples of life test data showing burn-in and aging for a prototype EL vectographic display application without gray shades are given in Fig. 8-23. The complete display is shown in Fig. 8-24.

8.4.7 Light-Generating Process Model. The ac thin-film EL light mechanism is fairly well understood and agreed upon at the measurable level. But this is not comprehensive enough to allow one to synthesize improved systems. For ac thin film, the best performing materials are ZnS:Mn. This phosphor and configuration had been in experimental evolution at several companies for over twenty years (starting with Soxman's work at Servomechanisms) before a production product was achieved. The combination of ZnS host with Mn as the activator is unique, and works well only in the ac thin-film configuration.

The equivalent electrical circuit (Fig. 8-25) is modeled based on the measured results. The electro-optical response is measured without knowledge of why or how it works. These models are used to help hypothesize the solid-state phenomena explaining the mechanism of light generation.

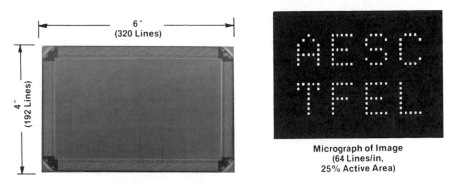

**Micrograph of Image
(64 Lines/in,
25% Active Area)**

Fig. 8-24. AC thin-film EL display with black layer for U.S. Army, developed by Aerojet ElectroSystems.

The Liège Workshop was devoted to the task of explaining luminescence resulting from a high electric field. The paper "Modeling AC Thin-Film Electroluminescent Devices" by Smith[95] of Tektronix summarizes results of several workers in this field including his own. This model was first proposed by Chen and Krupka[70] of Bell Telephone Labs. The principal features of this model are:

- Trapping of carriers at localized sites at or near the ZnS/dielectric interface.
- Tunneling of carriers from the interface states to the conduction band with the application of a high electric field.

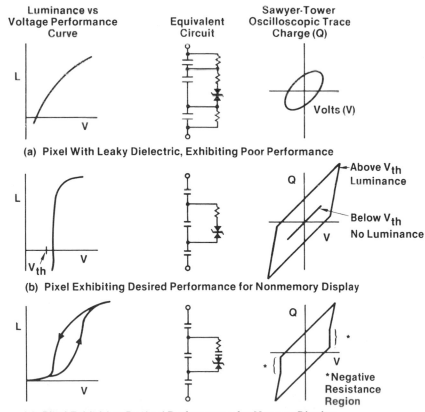

(a) **Pixel With Leaky Dielectric, Exhibiting Poor Performance**

(b) **Pixel Exhibiting Desired Performance for Nonmemory Display**

(c) **Pixel Exhibiting Desired Performance for Memory Display**

Fig. 8-25. AC thin-film EL measurable properties for basic dielectric phosphor sandwich.

- Impact excitation of localized centers by hot electrons.

The band diagrams for this model are shown in Fig. 8-26. The first diagram is for the zero trapped charge case, the second is for the trapped charge case. It has been observed for a long time that a start-up condition occurs during the first half cycle or more in the light-generating process in an ac EL device. Therefore at least two band diagrams are necessary, one for the initial conditions and one for the steady-state operating conditions. The trapped charge boundary conditions are necessary in a display application to get a sufficiently high internal field. When the external voltage is reversed with the trapped charge, the internal field is doubled.

The trapping and tunneling of carriers explains the very steep performance curve characteristics shown in Fig. 8-17. Fig. 8-26b shows the high density of deep levels from which carriers can tunnel into the conduction band. The filled traps on one side and empty traps on the other side create an internal field. When an external potential is applied with the same polarity as the internal field, the internal and external fields combine, giving an even larger internal field. The electrons are now pulled into the conduction band and are accelerated by the field to an energy level sufficient to excite the electrons of the Mn atom. On reaching the opposite side, the electrons are retrapped at the phosphor-dielectric interface until the next opposite polarity voltage is applied. The emitted light spectrum corresponds to that associated with the energy level of the $3d^5$ shell of Mn. The cross section of Mn in ZnS is rather low as compared to Cu. This is made up for by the fact that Mn can be put into ZnS at rather high concentrations without forming precipitates which cause concentration quenching, which in turn would degrade electrical properties.

Tunneling from a Dirac well would account for the emission rate associated with the very steep function of field. The steepness of the curves would be greatly influenced by impact ionization and space charge. The device's characteristic insensitivity to temperature seems to preclude the possibility of a thermally induced process such as Poole-Frenkel conduction. No conduction component of the current is nor-

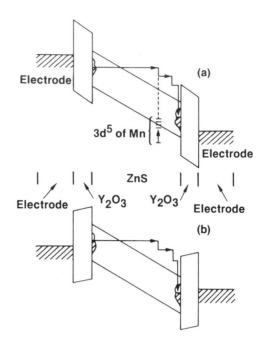

Fig. 8-26. Band diagram of ac thin-film EL (after D. H. Smith).

mally seen in the trailing edge of the pulse, which indicates that the charge which flows with the voltage pulse is trapped. The polarization associated with the trapped charge is real and can be measured. It will add algebraically to the next voltage pulse.

One of the things not explained by the model is the broadband low-level blue flash which Marrello and Onton report observing at the leading edge of the luminance.[73] The blue emission can be seen in ZnS without the Mn. It may be due to the activation of some impurity, of which there are many at the few-parts-per-10-million level, or from the ZnS itself. One part per million of impurity is normally the minimum threshold for activation in photoluminescent phosphors.[19]

The hysteresis effect is difficult to explain in a detailed sense with this model. It may be due to band bending from space charge in the phosphor.[95] The change in threshold voltage and initial stabilization effects may also be due to changes in interface states due to hot electron bombardment.[95]

8.4.8 Fabrication Techniques for AC-Thin-Film EL. The performance of EL is very process-dependent. There is some black art and there are some tricks, but in general, good thin-film technique and laboratory technology will suffice. The high electric fields require high breakdown strength in the ac configuration and good current conduction mechanisms in the dc configuration, both of which require the highest of quality in film fabrication. Also, when compared to other electronic devices, a display is an ultra large-area substrate; uniformity is often a problem.

There are four techniques in use in fabricating thin-film ac or dc EL. They are vapor deposition, sputtering, atomic layer epitaxy, and chemical vapor deposition.

The differences between ac and dc thin-film lie simply in the activators and in dc devices' lack of dielectrics. The following description will be for ac thin-film.

Vapor Deposition and Sputtering. The typical basic sandwich is shown in Fig. 8-10. Each layer is deposited by any one of the deposition techniques. In general, a deposition can be divided into three parts:

- Vaporization and disassociation of the source material
- Transportation of material to the substrate
- Condensation of material on the substrate

The last is by far the most critical in obtaining a high-quality thin film. The major parameters are substrate cleanliness, substrate temperature, substrate angle and position with respect to the source, deposition rate, chamber pressure, and background gases. It is not too critical how the phosphor is evaporated so long as it does not spatter and the net ratio of the activator to phosphor is preserved. The phosphor is best evaporated from crystals or hot-pressed pellets. ZnS absorbs water and gases, which will disperse the crystals unless removed slowly. The substrate temperature is a critical chamber parameter. For ZnS depositions the substrate should be held at a uniform temperature between approximately 125°C and 225°C. Stoichiometry is guaranteed within this range, as zinc or sulfur will not stick but will reevaporate unless the reactant is formed. The next most critical parameter is background gas. It is essential that water vapor be reduced below 10^{-7} torr. In general, the phosphor and dielectrics should be deposited at as high a temperature as possible to cause Type II or Type III columnar crystal growth unless amorphous films are desired. The substrate should be perpendicular to the vapor stream to achieve as smooth a morphology as possible. The roughness of each layer is often amplified up through the sandwich. See Section 8.4.9.

Evaporation method is not as critical as substrate temperature, background gas, and control of contaminants. Each laboratory has its particular mix of equipment and operator skills. A process used in one laboratory may not be the best for another. One typical evaporative process is outlined in Table 8-3.

The etching or in-contact masking used to define the ITO lines is critical. The edges of the lines must be rounded to prevent damage from electrical breakdown at sharp edges and voids along the ITO edges. The electric field is concentrated at the edge, and voids in the thin films along this edge will break down first.

Annealing of the phosphor is usually required, one hour at 550°C being sufficient. This annealing usually destroys metal electrodes and therefore should be done after the phosphor is in place but before any metal deposition. The glass substrate will normally shrink during the anneal temperature cycling, and if this is a problem the substrate should be preshrunk. It is best to complete the deposition of the dielectric-phosphor-dielectric in one pumpdown to minimize water contamination of the phosphor, and then to anneal afterwards.

Atomic Layer Epitaxy (ALE). Atomic layer epitaxy is a process developed by Suntola[96] of Lohja. The process is carried out by separate surface reactions between the surface to be grown and each of the components of the compound serving as reactants. The temperature of the surface is kept high enough to prevent the condensation of each reactant. Accordingly, only one atomic layer is able to make a compound-bond with the original surface, resulting in a self-controlled growth of one atomic layer at a time. The reactant can be an element of the compound in vapor phase or a volatile compound of the element.

As a further difference from CVD, the reactants in ALE act alternately in the reaction chamber, which fully eliminates the harmful gas-phase reactions typical in CVD. Moreover, local differences of the reactant vapor-pressure concentrations don't have harmful effects on growth uniformity. For example, ZnTe, ZnSe, and ZnS films have been grown by this technique but ZnO cannot. Dielectrics such as Al_2O_3, Ta_2O_5, $InSn_xO_y$, TiN, TiO and $TiAlO_3$ can be grown by ALE.

In the deposition of ZnS phosphor thin films, the substrate first passes through Zn_x vapor, where a monolayer of Zn adheres to the substrate surface by a surface reaction or chemisorption until the surface is fully covered. A second layer of Zn will not form because all the reaction sites are occupied and the substrate is too hot to thermodynamically accept a Zn-to-Zn bond. Therefore, only a monolayer of Zn is left on the surface.[96] Next, the surface is subjected to S_x vapor, which reacts with the Zn to form a ZnS bond. Again, once all the Zn bonds are complete, no further S will stick, as an S-S bond is not stable at the temperature used. The process is repeated, alternating between Zn_x vapor and S_x vapor until the desired film thickness, one monolayer at a time, is achieved. The sequence cycle is repeated every two or three seconds, which gives a net growth rate of 0.1 μm/hr, similar to the growth rate of molecular beam epitaxy.

The process is used to make EL displays of the ITO, Al_2O_3, ZnS:Mn, Al_2O_3, ITO or Al, and Al_2O_3 sandwich configuration. The final Al_2O_3 is a passivation and seal layer. Very high quality EL films have been grown by Lohja. The EL efficiency is on the order of six lumens per watt, and the EL performance curve is excellent. The film is very close to being a single crystal in the vertical direction, and the phosphor is hexagonal as determined from X-ray diffraction analysis. The sandwich may be grown on soda-lime glass, as annealing is not necessary.

Chemical Vapor Deposition (CVD). The first CVD films grown for EL purposes were reported in 1982 by Cattel, et al, of the Royal Signal and Radar Establishment, England. They used an organometallic chemical vapor deposition (OMCVD) technique[97] and incorporated it into both ac and dc coupled EL devices. The resulting phosphor gave good luminance at a low duty factor.

To accomplish OMCVD[98] a substrate is placed upon a graphite holder heated by RF induction to the desired temperature of between 325°C and 400°C. A continuous flow of either hydrogen or helium carrier gas passes over the substrate. Added to the carrier are small quantities of dimethyl zinc and hydrogen sulfide, which decomposes when heated by the substrate to form a ZnS thin film. The activator Mn is vaporized in TCM (tricarbonylmethylcyclopentadienyl manganese). The carrier gas is bubbled through a stainless steel vessel containing the TCM. The concentration is controlled by varying the TCM container temperature. A cadmium stannate transparent conductor and Si_3N_4 transparent dielectric were sputtered, and an aluminum back electrode was vapor deposited.

8.4.9 Transmission Electron Microscopy (TEM) of EL Thin Films.[99]

For the assessment of device performance, Theis, Oppolzer, Ebbinghaus, and Schild of Siemens investigated the microstructure of thin cross sections of electroluminescent ZnS:Mn films using TEM. Examples of ZnS:Mn films prepared by thermal or e-beam evaporation (EBE), by atomic layer epitaxy (ALE), and by sputtering were examined and are discussed and exhibited here.

Summarizing, EBE films are cubic, always show twinning structure, and exhibit a very fine-grained layer of initial nucleation growth upon which cubic, columnar, conical grains grow in the direction of the materal flow, depending on the substrate temperature. ALE films, on the other hand, are hexagonal, show no fine-grained region and no twinning structure, and exhibit comparatively large columnar grains, most of which extend from the bottom to the top of the film. In spite of large microstructural differences, the overall electro-optic performance (threshold behavior and luminance characteristic) of EBE and ALE samples is surprisingly comparable. The efficiency of ALE samples is higher and not explainable based on microstructure analysis alone.

Since it is expected that the microstructure of polycrystalline films strongly influences the device performance, Theis and associates used transmission electron microscopy (TEM) of cross-sectional specimens to explore the differences. This method allows one to visualize the microstructure of the ZnS thin films directly as a function of deposition thickness. Comparatively little has been reported on the microstructural properties in electroluminescent thin-film display devices. Samples from three different laboratories were investigated:

- Siemens Research Labs (EBE, sputtering) by Theis.
- Tektronix, Inc. (EBE) by King and Smith.
- Lohja Corp. (ALE) by Skarp and Törnqvist.

Cross-sectional TEM specimens were stuck front-to-front in pairs with epoxy resin, ground normal to the surface down to a thickness of 50 μm, and finally thinned by ion-beam etching. The investigations were performed at 100 kV beam voltage.

Electron Beam Evaporation (EBE). Generally, ZnS EBE films deposited on glass or an insulating layer to form an ac thin-film EL structure always exhibit the same basic microstructure features. In the first stage of film growth, a very fine-grained region develops adjacent to the substrate. With increasing thickness, large columnar grains begin to grow. Qualitatively, the mean grain diameter in the fine-grained region, as well as the tendency of the columnar grains to become larger and cone-shaped are related to the substrate temperature during deposition. In all cases a high density of twinning structures in the grains is observed. Figure 8-27 shows typical cross-sectional TEM images of ZnS films grown in the Siemens Research Laboratories on amorphous BaTiO$_3$ layers at three different temperatures. The grain size immediately at the interface is not significantly altered by the substrate temperature. In all three films, two different growth regions can be distinguished. In the first region up to deposition thickness of 200 nm, the increase in grain diameter is essentially due to the growth of larger grains with no preferred orientation, whereas in the second, upper region the columnar or conical shape of the grains results in a retarded increase of surface-parallel grain diameter with film thickness. The size of grain growth increases in both regions with increased substrate temperature.

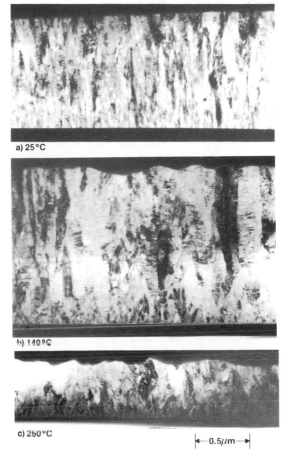

a) 25°C

b) 140°C

c) 250°C

|← 0.5μm →|

Fig. 8-27. TEM images of cross-sections through ZnS films (from Siemens Labs.) deposited on amorphous BaTiO$_3$ by electron-beam evaporation at different substrate temperatures: (a) 25°C, (b) 140°C, (c) 250°C (Courtesy of D. Theis).

Figure 8-28 shows cross-sectional TEM images of EBE samples grown in the Tektronix laboratories. Comparing them to Siemens samples grown at higher temperatures (Fig. 8-27c), it is clear that the basic microstructural features of EBE films grown at different laboratories and on different substrates remain unchanged. It is noteworthy that there is no significant difference observable between films grown on amorphous SiO$_x$N$_y$ layers (Fig. 8-28b) and those grown on fine-grained polycrystalline Y$_2$O$_3$ layers (Fig. 8-28a). The mean grain diameter for the Tektronix samples is quite similar to the Siemens samples.

According to X-ray diffraction analysis performed by Theis and associates, the preferred orientation of all films grown above 130°C,

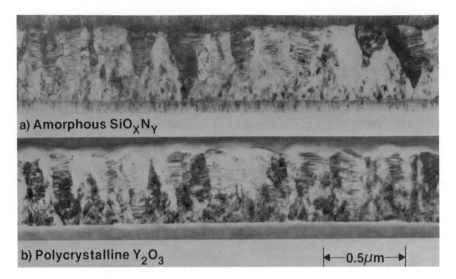

Fig. 8-28. TEM cross-section of ZnS:Mn films (from Tektronix Labs.) deposited by electron-beam evaporation at $T_s = 250^{\circ}C$: (a) polycrystalline Y_2O_3, and (b) amorphous SiO_xN_y (Courtesy of D. Theis).

which all showed good luminescent properties, is $\langle 111 \rangle$. Lattice constant of the cubic unit cell was 0.5421 ± 0.0002 nm.

Atomic Layer Epitaxy (ALE).   The microstructure of ALE samples from Lohja laboratories observed in cross-sectional TEM images is considerably different from EBE samples. There is no very fine-grained region in the initial growth stage of the films. Figure 8-29 shows typical cross sections of ALE films deposited on amorphous Al_2O_3. Very pronounced columnar grain

growth occurs, and many grains extend from the bottom to the top of the layer with a minimum of conical grain growth. Selected area diffraction patterns showed that the hexagonal phase of ZnS had formed due to the high substrate temperature of about $550^{\circ}C$. No twinning structure in the grains is observed as in the EBE films. The high substrate temperature will not explain all the differences between ALE and EBE.

Sputtered EL Films.   Sputtered films were initially comparable to EBE films. A very fine-

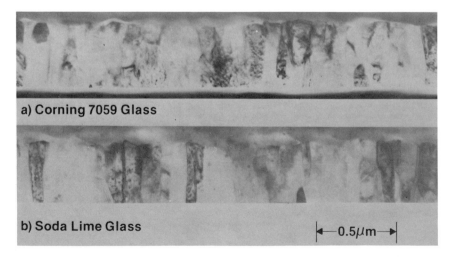

Fig. 8-29. TEM images of ZnS:Mn films (from Lohja Labs.) grown by atomic layer epitaxy and deposited on amorphous Al_2O_3: (a) Corning 7059 glass substrate, and (b) soda lime glass substrate (Courtesy of D. Theis).

Fig. 8-30. TEM images of ZnS:Mn film (from Siemens Labs) deposited by sputtering (Courtesy of D. Theis).

grained region extending up to deposition thickness of 60–70 nm is exhibited in both techniques. The mean grain diameter parallel to the film plane at a deposition thickness of 200 nm is about 30 nm and in sputtered films increases only up to 40 nm at the top interface. Although the increase in grain diameter with deposition thickness is very similar to that of the EBE film grown at room temperature (Fig. 8-27a), the shape of the columnar grains is much more regular in the sputtered film. The sputtered ZnS film of Fig. 8-30 also exhibits a high density of twinning structure in the grains.

Correlation of Microstructure to Electro-Optical Properties. The microstructures do not as yet correlate with all the relevant electro-optical properties. There were no significant differences in the microstructure of EBE samples before and after annealing at 550°C for three hours; however, the electroluminescent characteristics improved considerably. In spite of significant differences in microstructure of EBE and ALE films, the electro-optical properties of these two devices do not differ from one another as much as one would expect from TEM and X-ray analysis alone. A sharply defined voltage threshold for light emission, a steep luminance-voltage relation, and a saturating current-voltage relation are common to both types of samples (see Fig. 8-30). As far as external efficiency is concerned, ALE films are superior. For comparison, measurements of efficiencies were made for several films in portions adjacent to those which were then prepared for TEM analysis. The efficiency values obtained were reported as follows: For the Siemens EBE samples (Fig. 8-27), the efficiency was 1.5 lm/W; for the Tektronix EBE samples (Fig. 8-28), 1.1 lm/W; and for the Lohja ALE samples (Fig. 8-29), efficiencies of 3 lm/W corrected for Al back reflective electrode.

8.4.10 Electrode Definition. The line definition of the ITO transparent electrode is a particular problem. To get low resistance, the ITO must be typically 200 nm thick. This leaves a sizable step edge for the 1000-nm thin film of dielectric and phosphor to replicate. Further, when etching by conventional resist techniques, there are no simple means to avoid the line step edge.

The step edge can be avoided by depositing through an in-contact mask and etching off or removing the mask. One successful technique used is to make the mask of double Ni-clad Kovar, magnetically chuck the mask, deposit through it, and then lift off the magnet and mask. Such a mask is shown in Fig. 8-31. This eliminates the need for any wet chemistry or photolithography in defining EL display electrodes. It also greatly reduces process fabrication cost, time, and equipment.

8.4.11 Memory Mode. When properly fabricated, the ac thin-film EL using ZnS:Mn phosphor has a luminance-vs.-voltage hysteresis.[52,53] When properly addressed electronically,[51] the hysteresis can be used to give an EL panel memory analogous to the ac plasma panel (see Chap. 10). The hysteresis can also be used to freeze a frame of video when properly addressed.[54] Images stored or frozen on the panel can be read out to a memory device.[55] Also, the hysteresis panel is near-UV photosensitive. The UV energy lowers the hysteresis on-threshold voltage and can be used to optically write and erase an electrically sustained panel.[100]

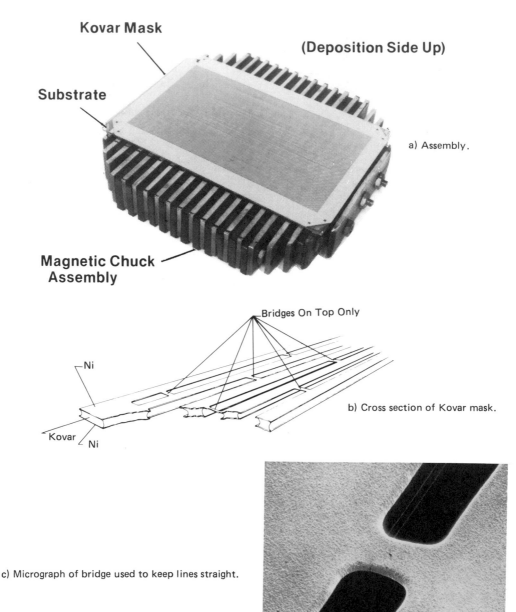

Kovar Mask

(Deposition Side Up)

Substrate

a) Assembly.

Magnetic Chuck Assembly

Bridges On Top Only

Ni

b) Cross section of Kovar mask.

Kovar   Ni

c) Micrograph of bridge used to keep lines straight.

Fig. 8-31. Kovar mask used to define row and column electrode lines.[92]

EL panels can be written on with an electron beam[101,102] as in a cathode-ray tube. One very attractive application of the EL memory mode is for a faceplate to a storage CRT. The EL memory configuration would be used over the entire faceplate and continually operated at the sustain- ing voltage. No matrix array is needed, and the back electrode is thin enough to be penetrated by the electron beam. Addressing the panel with an electron beam in a conventional manner would lower the on-threshold voltage to the ac sustain- ing voltage, turning on the addressed area. The image would remain on as long as the sustaining voltage is applied. The image is bulk-erased by turning off the sustaining voltage. The application of the memory mode to CRTs has been studied extensively at IBM[103] and Tektronix[102] and reported elsewhere.

The fabrication of a good memory ac thin-film device is governed by five factors:

- The ZnS deposition process including substrate temperatures (typically 175°C for ZnS).
- The Mn activator concentration and its method of inclusion in the ZnS (typically one mole percent).
- The absolute and relative thickness of each thin film in the sandwich (typically 0.5 μm for ZnS).
- The dielectric and its interface with ZnS.
- The substrate anneal sequence and temperature (typically 500°C).

The goal is to achieve wide stable memory margin with good luminance.

One of the motivating factors for memory is the ability to stimulate the entire panel in parallel, using 100% duty cycle excitation regardless of the number of rows or columns. Addressing with display information is done at a lower rate by conventional matrix addressing techniques. With nonmemory panels, the duty factor cannot be greater than the reciprocal of the number of rows, assuming line-at-a-time addressing. Therefore, the maximum panel luminance diminishes inversely with the number of rows.

The memory effect has great potential in advanced display devices. Further improvements need to be realized in memory margin and long term stability.

8.4.12 Acoustical Noise from AC Thin-Film EL Panels.

Panels sometimes make so much noise they are dubbed "the talking displays." The glass substrate makes an excellent sounding board. The sound is from an elastic deformation in the thin films because of electrically induced strain. The field is high enough to cause significant electrostriction and piezoelectric effects. "Electrostriction" is a property of all dielectrics where the induced strain polarity is independent of the field direction. The strain is proportional to the dielectric constant, the elastic compliance, and the square of the electric field. The "piezoelectric effect," on the other hand, occurs only in piezoelectric materials such as ZnS and is proportional to the field, changing sign with field reversal.

The sound from a 2.5 in. by 2.5 in. ac thin-film EL test panel was measured by the author with an accelerometer mounted at the center of the substrate, its sensitive axis normal to the substrate. The simultaneous trace of the accelerometer signal and display excitation signal on a dual-trace oscilloscope proved conclusively that piezoelectric force was the dominant effect causing the sound. The accelerometer signal was not only in phase and alternating with the excitation signal, it was also linear with the EL excitation signal in amplitude and independent of any light generation. The test panel clamped along the edge was in perfect resonance with the excitation signal at 7.7 kHz. A null in both sound and accelerometer signal occurred at one-half the resonant frequency.

Two obvious observations regarding this particular experiment may be useful in designing mounting fixtures for EL displays: (1) Peak accelerometer signals occurred at resonance and a multiple of resonance, and (2) minimum accelerometer signals occurred at one-half resonance, and near minimums occurred at numerous other frequencies.

In conclusion, the sound from EL displays is due to a piezoelectric-driven panel resonance and has no connection with the light-generating process. The sound can be made minimal by designing the panel and mounting to be resonant at twice the excitation frequency.

8.4.13 Failure Modes.

Several failure mechanisms have slowed down the development of ac thin-film EL displays. These failures fall into three general classifications:

- Electrical breakdown.
- Thin-film delamination.
- Field-induced chemical reactions.

A scenario for each of these failure modes is given with pictures of examples and solutions.

The basic validity of ac thin-film EL as a viable display technology has been demonstrated by several leading laboratories confirming the existence of long life at high luminance with matrix addressability. The more difficult problem is to demonstrate yield and reproducibility appropriate for manufacturing at low cost.

Basic Configuration. The typical cross section of an ac EL pixel is shown in Fig. 8-32. Within this configuration, several different dielectric systems are used by different laboratories to minimize the failure mechanisms (see Table 8-2).

Several observations should be made before discussing the failure modes. First of all, the entire sandwich is made of a series of thin films. Each film has its characteristic growth sequence of nucleation, coalescence, channel stage, continuous film, and columnar growth. As with all polycrystalline and amorphous films, defects and voids form as the film grows. These defects and voids are inherent regardless of the surface quality and contamination level. The structure is schematically shown in Fig. 8-32. Amorphous films would not have the grain structure shown.

Second, it is generally agreed that the light emission from ZnS:Mn is due to hot electron excitation of the Mn atom, and that the electrons are held in traps at the phosphor-dielectric interface after each excitation voltage pulse.[95]

Third, the electric field in the thin film is 2 MV/cm, as can be calculated directly from Fig. 8-32. It exceeds this level in either the dielectric or phosphor when the trapped charge at the dielectric-phosphor interface is taken into account.

Fourth, the device is operated in a manner different from any other electronic device. That is, the phosphor, a wide-band gap dielectric, is forced to conduct while the encapsulating dielectrics are required to stop the conduction (hot electrons) from reaching the electrodes. When the conduction does complete the circuit from electrode to electrode, electrical breakdown usually occurs. The trapped electrons at the interface serve to self-limit the conduction in the phosphor due to their counteracting field. At the same time, they increase the field in the dielectric until the externally applied field is reversed, whereupon the field in the phosphor is increased.

Modes of Failure. The failure modes of electrical breakdown, thin-film delamination, and field-induced chemical reactions would be anticipated directly from this model. Any voids or contamination in the film will result in elec-

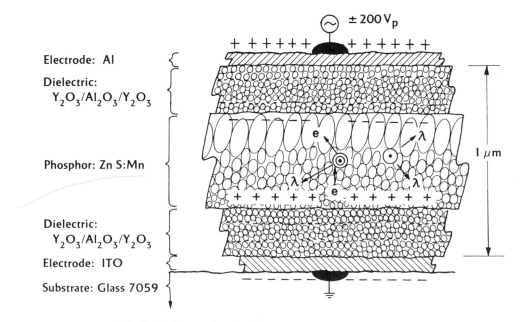

Fig. 8-32. First-order physical model of ac thin-film EL process.

trical arcs between the electrodes and thermal destruction. Any weakened area caused by poor adhesion in the thin film will be ripped apart by the electrostatic forces between the trapped charge and the applied voltage charge. Any H_2O will be ionized in the film by the high field.

In spite of these problems, it is unquestionably possible to make the EL structure of Fig. 8-32 good enough for display purposes.[104] There is no particular secret to making good ac thin-film electroluminescent display products. The preparation must include an ultraclean surface and very densely packed polycrystalline films made at elevated temperatures in a moisturefree deposition process. The phosphor and dielectric materials must be of high purity, and the thickness of the layers and activation levels must be precisely controlled.

Difficulties arise in eliminating failures when detailed reproducibility is not maintained. The failure analysis is further complicated by compound mechanisms simultaneously at work. Further, there is no prior art to which to refer, as no other solid state electronic device is designed to operate beyond its dielectric breakdown voltage, as is the ZnS phosphor. The failures discussed here generally occur within the first few minutes of operation; all occur within the first twenty hours of operation.

Electrical Breakdown. In large area polycrystalline thin films, electrical breakdowns in the form of pinhole electrical arcs are unavoid-

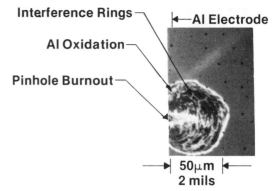

Fig. 8-33. Stabilized pinhole burnout after 12,000 hours of operation. (See Fig. 8-22.)

able.[43,77,80,94,105] As voltage is increased, these breakdowns occur at some typical voltage, which in capacitor device technology is called the "clearing voltage." In this case, the circuit opens around the pinhole, stopping the arcing. An example of such a pinhole breakdown is shown in Fig. 8-33. These burnouts are to be expected and tolerated. They typically are less than 50 μm in diameter and stable. They leave an injury through which moisture can penetrate and lead to hydrolysis and subsequent delamination. Also, if they become too numerous they can limit the light output or open up an entire line. A density of fewer than two pinhole breakdowns per square inch is achievable, and if they do not destroy a pixel or line this is permissible.

When the failure propagates as shown in Fig. 8-34, the breakdown can consume the entire device. The propagation is usually the result of

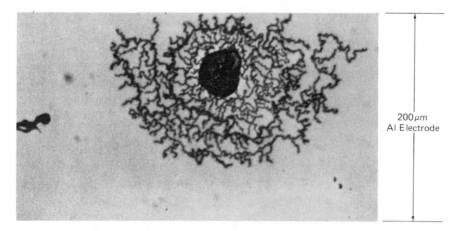

Fig. 8-34. Aluminum electrode breakdown with propagating tributaries typical of failure mode in ac thin-film EL displays.

a thermal runaway fueled by the power input. When the dielectric becomes hot or melts, it becomes conductive. Evidence of the heat are the color of the arc, the beads of formerly molten film material, and the occurrence of thermal shock fractures, called "Hackle lines", in the glass surface. Dielectric conduction is evidenced in the propagation path of the tributaries. They run away from the top electrode, through the dielectric, along a path to reach the bottom electrode, thus completing the electric circuit as shown in Fig. 8-35.

Each pinhole burnout is followed by a thermal transient which may either be self-limiting or develop into a runaway meltdown. At the instant of the pinhole burnout, the dielectric in the immediate vicinity melts, as can be seen by an SEM examination. The temperatures exceed the melting point of Y_2O_3, Al_2O_3, and all known dielectrics. The color temperature of the arc is in the red to blue range. As the temperature rises and the heat propagates outward from the pinhole, one of two things can happen:

1. The Al electrode can melt ($660°C$) or oxidize and fuse open in a self-limiting manner, as seen in Fig. 8-33.
2. The failure propagates by melting the film in an uncontrolled manner (Fig. 8-34 and 8-35).

It is hypothesized that the propagation is due to excessive ohmic heating initiated from a pinhole breakdown and driven by a low impedance power source and exothermic oxidation of the Al electrode. The conductivity of the dielectric increases exponentially with temperature.[106] The thermal gradient across the dielectric is very large due to the heat sinking of the dielectric to the glass and oxidation of the Al. Consequently, the flow of current (and possibly of ions) is in the plane of the display sustaining the molten dielectric state ahead of the Al electrode. The molten dielectric acts as an extension of the Al electrode. The propagation is supported by the reduction in dielectric strength,[106] which is approximately 0.5 MV/cm per $100°C$ increase in temperature. The propagation tributaries shown in Fig. 8-35 support this theory. Figure 8-35 shows the channels filled with molten dielectric before cooling to ambient and beading up along the tributary edges.

Four things must be done to eliminate the propagation:

1. Use high density contamination-free dielectric films.
2. Ballast the power supply.
3. Use resistive electrodes.
4. Use thin Al electrodes (20 to 50 nm).

The pinhole breakdowns like those seen in Fig. 8-33 were controlled in this manner.

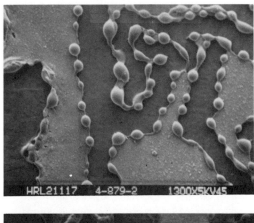

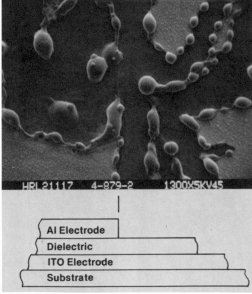

Fig. 8-35. Molten dielectric conduction in thin-film dielectric sandwich.

Thin-Film Delamination. The coulomb forces from the trapped charge in the high field are in the direction which would tend to separate the thin films at the time of external field polarity reversal. If there are any weaknesses in adhesion, the films will separate.

The separation in the film shown in Fig. 8-36 was induced by including air as a background gas during dielectric depositions. The water entrapped in the film was ionized, and the increased volume initiated the separation. The separation can be confirmed by the existence of interference rings in the film. Delamination can be controlled by using state-of-the-art techniques for improved adhesion.

Field-Induced Chemical Reactions. During operation in air, the high field anodizes the back electrode. An example of this is shown in Fig. 8-37. The reaction is induced by the electric field, as evidenced by the fact that it occurs only where the electrodes overlap. The oxidation starts at the edge of the electrode and moves inward until the electrode is completely oxidized over the entire pixel area. This does not occur when the display is operated in a vacuum. The oxidation is confirmed by the fact that the thickness of the Al is doubled in the oxidized region.

Oxidation can be stopped by blocking the source of oxygen with a thin-film passivation layer of, or by encapsulating with, oxygen-free material.

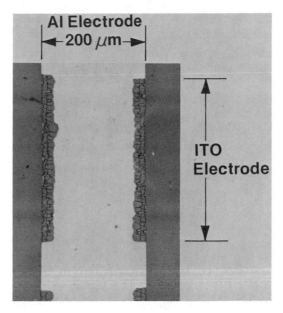

Fig. 8-37. Aluminum electrode anodization over pixel after one hour of operation exposed to room air.

All the problems cited can be controlled, as exhibited by the substrate of Fig. 8-33, assuming that a panel product is appropriately sealed. This substrate has been operated in a dry box for 12,000 hours at 1 kHz sine wave excitation. The substrate has had no loss of pixels nor loss in luminance. It has had a loss in active area of 4.5% due to pinhole burnouts in 10% of the pixels. The life test results are shown in Fig. 8-22.

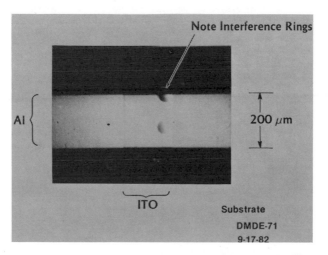

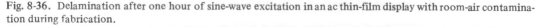

Fig. 8-36. Delamination after one hour of sine-wave excitation in an ac thin-film display with room-air contamination during fabrication.

8.4.14 Color. A wide variety of colors has been demonstrated with thin-film[107, 108] and powder[109] EL. The colors from powder EL are discussed later. One is further encouraged by the large set of rare earths and metals available as color activators whose performance has already been demonstrated in cathodoluminescence. There is an almost infinite set of combinations of phosphors, activators, and coactivators that have not even been tried. Some of the rare earth activators such as Tb change color with concentration.

A summary of rare earth and high performance phosphor activators is given in Table 8-4. Mn is used extensively because of its high efficiency and high luminance. However, the luminance of ZnS:Tb is very good and approaches 50% of the luminance of ZnS:Mn.[110] A comparison is given in Fig. 8-38 of the spectral response where both devices were made by the process detailed in Table 8-3. The 1931 CIE

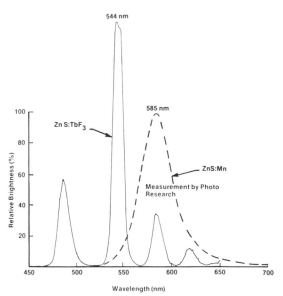

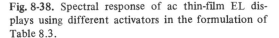

Fig. 8-38. Spectral response of ac thin-film EL displays using different activators in the formulation of Table 8.3.

Table 8-4   Color Performance in AC Thin-Film EL

Phosphor Coactivator	Color	Excitation Frequency, kHz	Luminance fL	cd/m^2	Reference (see footnotes)
ZnS:Mn/TbF$_3$	Red	0.06	4	14	(A)
ZnS:TbF$_3$	Green	5	700	2,400	(B)
ZnS:TmF$_3$	Blue	5	0.5	1.7	(B)
ZnSe:Mn	Yellowish-orange	1	30	100	(C)
ZnS$_{0.4}$Se$_{0.6}$:Mn	Yellowish-orange	1	150	800	(C)
ZnS:Mn	Yellowish-orange	1	300	1,000	(C)
ZnS:Mn	Yellowish-orange	1	600	2,000	(D)
ZnS:Mn	Yellowish-orange	2.5	1,000	3,500	(E)
ZnS:TbF$_3$	Yellowish-green	2.5	300	1,000	(E)
ZnS:PrF$_3$	White	2.5	60	200	(E)
ZnS:DyF$_3$	Yellow-white	2.5	20	70	(E)
ZnS:Mn/Cu	Yellowish-orange	5	1,700	6,000	(F)
ZnS:TmF$_3$	Blue	5	2	7	(G)
ZnS:ErF$_3$	Green	5	60	200	(G)
ZnS:HoF$_3$	Green	5	70	240	(G)
ZnS:TbF$_3$	Green	5	500	1,700	(G)
ZnS:DyF$_3$	Yellow	5	140	480	(G)
ZnS:SmF$_3$	Red	5	200	700	(G)
ZnS:NdF$_3$	Red	5	6	20	(G)
ZnS:Tb/P	Green	5	1,000	3,500	(H)
SrS:CeF$_3$	Blue	1	45	150	(I)

(A) Yamauchi, et al. *IEDM 74 Digest.*
(B) Hale, et al. (Rockwell). *IEDM 80 Digest.*
(C) Miura and Sato (Fujitsu). *IEDM 80 Digest.*
(D) Suntola, et al. (Oy Lohja). *SID 80 Digest.*
(E) Yoshida, et al. (Sharp). *SID 80 Digest.*
(F) Tannas (Aerojet). *SID 81 Digest.*
(G) Okamoto. Osaka. U. thesis, 1981.
(H) Tohda, et al. (Matsushita), Private communications, 1983.
(I) King, et al. (Planar) Private communications, 1984

coordinates are as follows:

$$x = 0.534 \quad \text{and} \quad y = 0.466 \text{ for Mn/Cu}$$
yellowish orange

$$x = 0.310 \quad \text{and} \quad y = 0.600 \text{ for TbF}_3$$
yellowish green colors

The transparency of thin-film EL gives it a unique advantage in making multicolor displays. Normally in any color display, except penetration phosphor CRTs, resolution or active area has to be given up to make room for the other color areas in the plane of the display. Because of the high transparency of EL, colors can be added in series. Monolithic color planes can be made eliminating parallax.

A dual-color monolithic thin-film EL breadboard was made by Tektronix demonstrating this principle using ZnS:TbF$_3$ for green and ZnS:Mn for yellowish orange.[111, 112]

Three colors can be displayed with only two monolithic EL stacks with common row electrodes, as shown in Fig. 8-39. The width of the column electrodes can be adjusted to give equal pixel luminance for each color. For example, the widest column electrode may be used for the color blue. The red and green colors share the same host phosphor. The phosphor is alternately activated in stripes for red and green.

Each row electrode is subdivided to give gray shade of each color.

It is interesting to note that the ZnS:Mn spectrum is invariant to activator concentration, and that at concentrations above the optimum (approximately one mole percent) the luminance and luminous efficiency decrease.[113] At high concentration a red band appears.[114] The phosphor ZnS:TbF$_3$, on the other hand, has a spectrum that is very concentration-dependent.[111] The color can vary from saturated green[115] at 542 nm to yellowish-green, 80% saturated at 555 nm. Red can be achieved by combining Mn and TbF$_3$ as coactivators.[115] Depending on the concentrations of Mn and TbF$_3$, it is possible to generate any color from green to red. As might be expected, the fabrication of blue is the most difficult, due in part to the low sensitivity of the eye and the wide bandgap required of the materials. Unactivated ZnS gives a faint blue emission.

8.5 AC POWDER EL

AC powder EL is made from ZnS and ZnSe phosphor, and activated with supersaturated copper sulfide and other coactivators such as Mn, Cl, and Ag for color. It is prepared in such a way that a small excess of copper sulfide re-

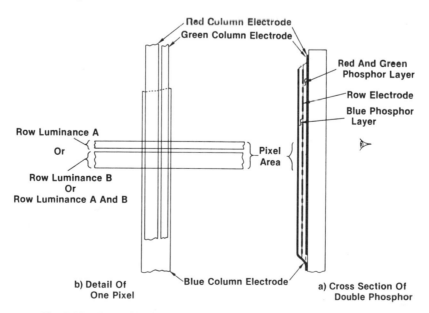

Fig. 8-39. Three-color EL monolithic display concept (Tannas, 1981).

mains as a separate phase not dissolved in the ZnS lattice. This is sometimes called Destriau-type EL as it is like the phosphor in which he made his original observations. The ac powder EL is essentially the same today as it was in the 1950s. The primary changes have been in particle size, quality control, dielectrics, and fabrication techniques.

The key ingredient in the ZnS host is the copper sulfide. Ac powder EL is made with many different host combinations and coactivators to achieve different colors, but $ZnS:Cu_xS$ must always be present. Copper sulfide has a fairly high electrical conductivity. The Cu_xS is embedded in the ZnS particles in voids, cracks, and dislocations inside the particles. The Cu_xS is a good conductor, while the ZnS is a good dielectric. An applied electric field is highly deformed by conducting Cu_xS needles where the local field strength can easily be 1,000 times higher than the average field.

In the model suggested by Fischer, the high field in the ZnS causes injection of electrons and holes from the opposite ends of Cu_xS needles. The electrons then excite the Zn, Cu, or coactivator to emit light. Also, holes are trapped on the Cu recombination centers, and upon field reversal the emitted electrons may recombine with the trapped holes to produce light.

To prevent arc-over and thermal destruction of the device, the powder is embedded in a dielectric between the electrodes as shown Fig. 8-40.

8.5.1 Fabrication Technique. For those skilled in the trade, the preparation of EL powder phosphors is simple and inexpensive.[116] Briefly, the ZnS and ZnSe host powders are typically ground with the Cu_2S and coactivator compounds and fired at a temperature of 1000°C and higher in an H_2S atmosphere. The hexagonal phase forms, then is cooled to the cubic structure phase. The dissolved copper forms a sulfide. After cooling, it exceeds the solubility limits and precipitates on defects resulting from the hexagonal to cubic transformation. Embedded Cu_xS conducting needles are formed by this process. The phosphor is then ground to the proper size ranging from 1 to 20 μm depending on the application. Surface Cu_xS causes ohmic dc conduction, which is not desirable. The surface conduction prevents the high field from forming. The excess copper is removed from the grain surfaces by washing in cyanide such as hot KOH-KCN. The cyanide does not reach the internal embedded Cu_xS as it is protected by the ZnS.

The substrate may be a flexible plastic or glass. The front conductor may be tin oxide, indium-tin oxide, or a transparent thin-film metal such as 10 nm of gold. The phosphor powder prepared as above is then applied to the substrate held in a dielectric such as cyanoethyl-cellulose or low-melting glass to a thickness of approximately 20-40 μm. The slurry of phosphor and dielectric is applied by spraying, screening, or doctor blading. A second layer of dielectric is applied to give further protection against arc-over and destruction. The substrate is then fired if on glass, or cured if organics are used. The back electrode is normally a thin film of vapor deposited aluminum. To complete the device, a final seal of polymer or other suitable material is applied.

8.5.2 Applications. AC powder EL devices are primarily used for transillumination of panels, keyboards, and other displays such as ac plasma

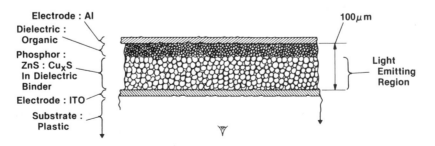

Fig. 8-40. Basic structure of an ac powder EL lamp.

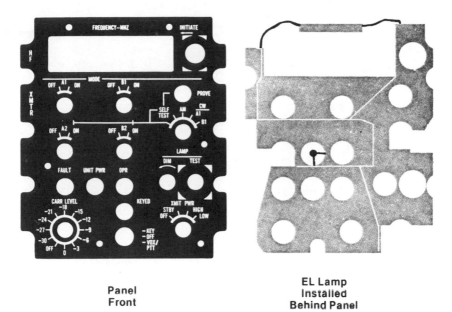

**Panel
Front**

**EL Lamp
Installed
Behind Panel**

Fig. 8-41. AC powder EL lamp application (Courtesy of Grimes Division of Midland Ross).

panels and twisted nematic liquid crystals. They may be used anywhere a continuous low-luminance film is needed. An example of a modern ac powder EL lamp for an aircraft control panel is shown in Fig. 8-41. The application of an EL lamp for transilluminating a liquid-crystal display is shown in Fig. 8-42.

The use of ac powder EL lamps has been on the increase since 1980 for several reasons:

- The availability of low cost solid state miniature electrical power inverters tailored to drive EL lamps.
- The need for transillumination (backlighting) for night reading of new products such as liquid-crystal displays and membrane switches.
- The shift to solid state electronic input and output devices for automobiles, airplanes, boats, etc. where both sunlight and nighttime readibility are important.
- The thrust for miniaturization.

The solid-state inverter was developed by Endicott to convert from voltages such as 12 V dc to 170 V ac at 750 Hz. The inverter may be located right at the lamp, eliminating the need for having high voltage lines throughout the equipment. The transillumination of an LC display turns it from a nonemitter to an emissive display without altering its sunlight readability or electronic addressing characteristics.

The construction of a modern ac powder EL lamp is shown in Fig. 8-43. The low cost construction is rugged, drip proof, flexible, efficient, and long-lived at low luminance (less than 7 fL). The life characteristics are shown in Fig. 8-44. The standard and long-life phosphors differ in the quality control and burn-in time. Luminescent Systems makes a tungsten white-color lamp for aircraft cockpit applications with a performance of 7 fL plus or minus 10% for 10,000 hours.[117]

Applications of ac powder EL devices as an information display are very limited for several reasons:

- Low discrimination ratio.
- Short life when at moderate to high luminance.
- Low contrast ratio.

The low discrimination ratio is due to the fundamental physics of the light generating mechanism. The shortened life with high luminance is an experimental fact. It results from the migration of the copper atoms when they are exposed to the higher field required to get the higher

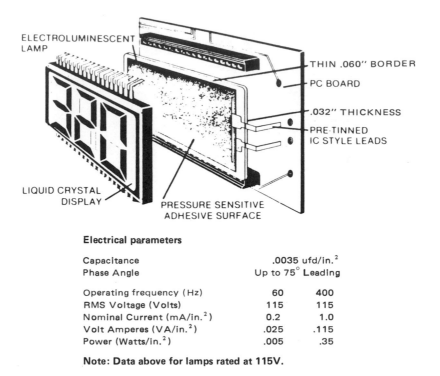

ELECTROLUMINESCENT LAMP

THIN .060" BORDER

PC BOARD

.032" THICKNESS

PRE-TINNED IC STYLE LEADS

LIQUID CRYSTAL DISPLAY

PRESSURE SENSITIVE ADHESIVE SURFACE

Electrical parameters

Capacitance	.0035 ufd/in.^{2}	
Phase Angle	Up to 75° Leading	
Operating frequency (Hz)	60	400
RMS Voltage (Volts)	115	115
Nominal Current (mA/in.^{2})	0.2	1.0
Volt Amperes (VA/in.^{2})	.025	.115
Power (Watts/in.^{2})	.005	.35

Note: Data above for lamps rated at 115V.

Fig. 8-42. Back-lit LC display using ac powder EL lamp (Courtesy of Luminescent Systems).

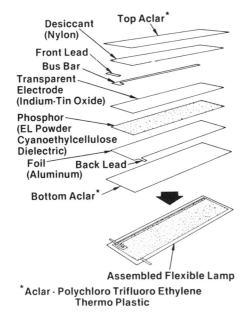

Desiccant (Nylon)

Top Aclar*

Front Lead

Bus Bar

Transparent Electrode (Indium-Tin Oxide)

Phosphor (EL Powder Cyanoethylcellulose Dielectric)

Foil (Aluminum)

Back Lead

Bottom Aclar*

Assembled Flexible Lamp

*Aclar - Polychloro Trifluoro Ethylene Thermo Plastic

Fig. 8-43. Modern ac powder EL lamp (Courtesy of Luminescent Systems).

luminance. The low contrast ratio in moderate to high ambient illumination is due to the high reflectivity of the whitish phosphor powder itself. Filtering can be of some help, though the additional luminance required to compensate for the filter reduces the useful life. When ac powder displays are operated in the 50 fL range as would be expected in a nominal industrial application, the time to half-luminance (maintenance) is typically 1,000 hours. The typical performance characteristics of high quality production ac powder EL alphanumeric displays by GTE Sylvania in 1966 are given in Table 8-5. These displays have been out of production since 1970.

The reduction in luminance with operating time can be compensated by increasing the voltage. This is fine for a lighting application. It does not work for a matrix display, as image ghosting or lower luminance of the pixels used more often would occur in a short time without any means of compensation.

8.5.3 Mechanism of Light Generation. A detailed model of the ac powder EL process has been developed and verified by Fischer.[69,72,118] The model has been scrutinized by Lehmann[119] and others and is generally considered accurate. The Fischer model is not applicable to the ac thin-film ZnS:Mn configuration and is most

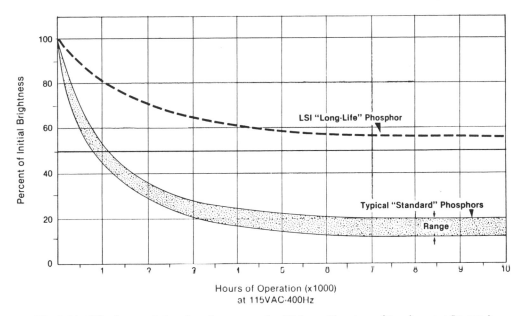

Fig. 8-44. Life characteristics of modern ac powder EL lamp (Courtesy of Luminescent Systems).

Table 8-5 Characteristics of Sylvania Hermetically Sealed AC Powder EL Alphanumeric Display Products of 1966

Operating Conditions:
Temperature range: −55° to +71°C
Peak voltage: 300
Peak transient voltage: 350
Maintenance at 115 V RMS, 400 Hz: 55% of original intensity at 1,000 hr
Original intensity: 12–18 fL

Optical:
Peak output wavelength: 510 ± 20 nm
Color: Blue-green (400–800 Hz)
Brightness uniformity: +20% max between digits
Contrast ratio: 0.20 at 100 fc ambient
0.06 at 300 fc ambient
Reflectance: 65% at 100 fc ambient

likely not applicable to any configuration without supersaturated Cu_xS and possibly Ag_xS.

According to Fischer's model, the applied field is distorted by the imbedded Cu_xS conducting needles. The needles permeate the insulating ZnS particles. The ZnS particles are mainly of cubic zincblende structure with residual traces of the hexagonal wurtzite structure causing linear defects. The field concentrates at the sharp end tips of the invisible needles of Cu_xS. The local high-field regions around the Cu_xS tips cause electrons and holes to be alternately injected into the surrounding phosphor. The injected holes are thought to be trapped by the copper in deep luminescent centers, forming an activated luminescent volume at the dislocation edge. The electrons remain mobile. Upon field reversal, the electrons recombine with the hole-charged volume of luminescent centers and radiate light.

The microscopic luminescent volume has been named a "comet" because of its bright cometlike shape with the head toward an electrode and the tail back toward a mirror imaged comet as seen under microscopic inspection.[69] Most observations on Cu_xS supersaturated EL powder can be explained by this model. With aging the comets diminish in volume or break up into two or more comets each. The comet is the lowest resolvable light center seen in EL powders. It is very bright, making further resolution by the microscope-aided eye impossible. A special preparation of index matching transparent materials is needed to couple the microscope into the high-index (2.3) ZnS phosphors.

8.5.4 Life and Aging. The exponential decay in luminance of ac powder EL appears to be an insurmountable problem. However, although performance typically decays to one-half luminance after 1,000 hours of operation, this is only true when the powder film is driven at the high luminance appropriate for alphanumeric display. At a low luminance of 1 fL, its life is near infinite. AC powder EL nightlights have been known to operate 100% of the time for over ten years at 100 V rms, 60 Hz.

Aging has several typical characteristics:[118]

- The higher the initial luminance, the shorter the time to half-luminance (called "maintenance" or "aging").
- The higher the temperature, the faster the aging; luminaires operating at liquid nitrogen temperatures do not age.
- The higher the frequency of operation, the faster the aging.
- Nonoperating luminaires do not age.
- Humidity accelerates deterioration of unsealed luminaires.

The aging is due to localized heating in the comets,[118] which causes material changes, which in turn cause the deterioration of their light generating properties. The main material change is the copper sulfide segregation and diffusion inside the ZnS powder particles.[118] A comet head operating at 10 kHz has been measured[120] to emit 10^5 fL. From the light emitting volume, it is estimated[118] that the charge carriers that flow along the comet-conducting Cu_xS defect approach 10^4 A/cm^2. At these current levels, there can be no question that atoms can be swept along and that the local ohmic heating can be immense. This model is consistent with the experimental observations above.

Several solutions have been proposed[118,119] to minimize aging. These include high-pressure sulfurization to reduce copper diffusion, surface passivation by forming inert skins around each particle to prevent surface electrolysis, use of aluminum coactivation in place of the potentially corrosive halogens, and hypermaintenance[121,122] (a phosphor preparation technique developed for long life by Lehmann).

8.5.5 Color in AC Powder EL Lamps. The EL lamp can provide all the specialized colors required for aircraft panels and commercial and military applications. Of special note are (1) equal energy white:

$$x = 0.330 \pm 0.030; \quad y = 0.330 \pm 0.030,$$

(2) USAF white MIL-STD 7788E:

$$x = 0.440 \pm 0.020; \quad y = 0.405 \pm 0.020,$$

and (3) aviation red, green, and blue in accordance with Mil-C-25050.

A set of standard colors available from Luminescent Systems is shown in Table 8-6.

Some colors are best made with converters to improve efficiency, color stability, and chrominance.[117] Aviation red is often made with a daylight fluorescent overlaid on an EL lamp with a matched emission peak. A final application of white paint produces a red shift due to multiple internal reflections. The resultant color will meet the stringent requirements of MIL-C-25050 for aviation red. White can be made with a blend of blue, green, and red phosphors or by a dye conversion. The conversion uses a fluorescent dye in the lamp dielectric which gives excellent internal conversion of luminous energy.

8.6   DC POWDER EL

An EL display can be made in a dc powder configuration. It has been developed mainly at the Thames Polytechnic in London, and at Phosphor Products Co., Ltd., Dorsett, England, under the leadership of A. Vecht.[123,124,128] The basic structure of the panel is shown in Fig. 8-45.

8.6.1 Fabrication. The basic composition of the best performing dc EL powder phosphor is small grain ZnS host material activated with Mn and coated with Cu_xS for conduction. A wide range of colors[126] has been achieved with alkaline earth activators in place of the Mn. The phosphor grains of micron size are in a dielectric binder such as nitrocellulose. The mixture can be doctor-bladed, sprayed, or screened onto the substrate. Screening is best for production control and to achieve the patterning required for electrical isolation for electronic addressing. The dc phosphor is conductive, and to prevent shorting from row to row the phosphor must be

Table 8-6   AC Powder EL Colors—Nominal Characteristics*

Color	Frequency, Hz	Luminance, fL	Luminance, cd/m²	Radiance, W/ft²	Spectral Peak Emission, nm	Chromaticity x	Chromaticity y
Blue Green	60	4	13.70	.010	515	.200	.500
	400	20	68.52	.062	505	.185	.420
Aviation Green	60	4	13.70	.009	530	.250	.550
	400	20	68.52	.051	520	.230	.500
Yellow Green	400	18	61.67	.036	540	.375	.575
Yellow Orange	400	15	51.39	.036	580	.520	.450
Blue	60	2	6.85	.012	470	.175	.215
White	400	12	41.11	.040	N/A	.330	.350
Blue White	400	12	41.11	.037	N/A	.270	.350
Pink White	400	12	41.11	.036	N/A	.390	.380
Tungsten White	400	4.38	15	.011	N/A	.540	.430
Red	400	4	13.70	.019	620	.666	.333

*Courtesy of Luminescent Systems.

dimensioned like the row conductors for electrical isolation.

8.6.2 Theory of Operation. The light is emitted from a thin layer formed in the phosphor next to the anode. The anode is made the transparent conductor located on the viewing side. The cathode is located to the rear and is made from evaporated Al, Au, or Ag.

A dc powder EL display does not emit light until after the forming sequence. During forming, the Cu_xS conductor migrates from the anode, leaving a thin region from which most of the light is emitted. Initially, each particle of phosphor is covered with Cu_xS, which can schematically be represented as a PNP semiconductor sandwich:

$$Anode\,|\,P\text{-}Cu_xS/N\text{-}ZnS\text{:}Mn/P\text{-}Cu_xS\,|\,Cathode$$

where the Cu_xS is P-type and the phosphor is an N-type semiconductor. The dielectric bonds the particles together between the anode and cathode. When voltage is applied to a virgin device, a high current will flow due to its low resistance. The Cu_xS is the primary conductor, and the Mn activator is not excited. As forming continues, the resistance increases due to the ionic motion of Cu, mostly near the anode. After several hours of forming and continual increases in voltage, the display will begin to luminesce.

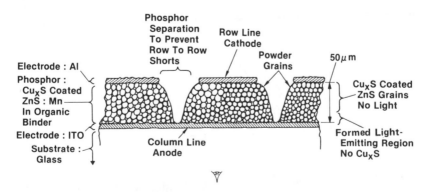

Fig. 8-45. Basic structure of a dc powder El matrix display.

$$\text{Anode} \left| (\text{N-ZnS:Mn}) \right\| (\text{P-Cu}_x\text{S/N-ZnS:Mn/P-Cu}_x\text{S}) \left| \text{Cathode} \right.$$

- Thin lumi-
  nescing
  region
- N-type
- High
  resistance

- Extended cathode
- P-type
- Low resistance

The schematic representation of the display cross section[127] is now as above.

Reversing the device polarity does not reverse the process. Instead, a new thin luminescing region is created at the new anode, formerly the cathode. The overall luminance is now lower since the applied voltage has to be divided between two localized high-resistance regions.

The electrical conduction is best described by different laws at different stages. The conduction process at low voltages is due to the Poole-Frenkel effect. At the higher voltages after forming, it is due to Zener tunneling across the heterojunction interface. After several hundred hours of operation, conduction is governed by space charge accumulation.[127]

8.6.3 Aging and Failure Mechanisms.

Because of the continuous ionic movement in the phosphor film, voltage must be increased continuously to maintain luminance until electrical breakdown mechanisms begin to destroy the film.

The thin luminescent layer of phosphor at the anode is approximately 1.5 μm thick and depleted of copper after forming. It has been demonstrated that a dc powder EL display can be made by placing a thin layer of ZnS:Mn phosphor powder without Cu_xS next to the anode and avoiding forming.[119] The electromigration moves all the copper out of the high field region near the anode. As the width of the region grows, the field is reduced, rendering the process stable and insuring uniformity. As the width increases, the field must be increased to cause electrons to traverse the dielectric gap. The Cu_xS conducts the electrons from the cathode up to the edge of the remaining Cu_xS, and then it is believed they tunnel out of the Cu_xS. They are then accelerated by the high field and impact-excite the Mn^{++} luminescent centers. This process is quite different from ac powder EL Type 1 of Section 8.3.1, as there

are no embedded needles of Cu_xS, only surface Cu_xS. The impact excitation of Mn or other activators is more akin to ac thin-film EL, Type 2 of Section 8.3.1.

8.6.4 Applications.

Of the powder EL sandwiches, the dc is more readily matrix addressable than is the ac. The discrimination ratio is higher, and the luminance at a low pulsed duty factor is higher. For example, when a suitably prepared panel is excited with 120-V, 3-μsec pulses at a duty factor of 0.5%, the luminance is 50 fL.[128] The luminance is proportional to approximately the third power of voltage for ac powder EL and the sixth power for dc pulsed powder EL.[129] This advantage greatly improves the contrast ratio for dc over ac powder in a matrix-addressed panel.

A breadboard dc powder EL flat-panel TV display was made and reported upon in 1973. The 13-inch diagonal display has been described by Kawarada and Ohshima.[129] It is driven at 200-volt pulses using line-at-a-time matrix addressing techniques at 60-Hz frame rate showing live TV pictures with no interlace. The panel has 224 rows by 224 columns. The display achieved 10:1 contrast ratio at 10 fL in a low ambient light setting. Because of the scattering and reflecting properties of all powder phosphors, the contrast ratio at this luminance would be unacceptable in a normal office lighting environment. The overall efficiency of the display was low, and the life to half luminance was 300 hours.

The dc powder EL using ZnS:Mn has been used commercially in 80- and 256-character readout displays,[109] automotive panels, and similar configurations. These same devices have been made in various colors using rare-earth phosphors on a breadboard basis.

8.6.5 Colors with DC Powder EL Phosphors.

A wide range of colors have been achieved using

Table 8-7   Color Performance in DC Powder EL

Phosphor	Color	Peaks in Emission Wavelength, nm	Continuous Drive		Pulsed* Drive	
			fL	Volts	fL	Volts
Alkaline earth sulfide:						
CaS:Ce	Green	580 & 520	300	70	175	110
CaS:Er	Yellowish-green	580 & 530	100	80	25	120
CaS:Tb	Yellowish-green	550	5	80	15	120
CaS:Eu	Red	650	30	50	5	120
SrS:Ce	Bluish-green	520 & 480	120	70	60	110
SrS:Mn	Yellowish-green	550	80	120	–	–
SrS:Cu, Na	Yellowish-green	540	80	80	15	120
Ternary sulfides:						
Ba$_2$ZnS$_3$:Mn	Red	630	10	100	5	120
NaYS$_2$:Tb	Yellowish-green	550	5	100	5	120
SrGa$_2$S$_4$:Eu	Green	–	5	100	5	120
Other phosphors:						
ZnS:Te, Mn	Redish-orange	–	150	70	50	70
ZnS:Mn, Cu	Yellowish-orange	585	–	–	–	–

*Pulsed panels operated at 10 to 20 μs and 1 to $1\frac{1}{4}$% duty factor.

alkaline earth sulfides activated with rare earths.[109,126,130] Calcium sulfide host has a better discrimination ratio than ZnS:Mn and with some activators, such as cerium, a higher luminance. Zinc sulphide with manganese is still the most durable phosphor.

Numerous colors have been demonstrated as outlined in Table 8-7.[130] The persistence of the rare earth activated alkaline sulfides is very short compared with ZnS, microseconds compared to milliseconds.[109]

8.7 DC THIN-FILM EL

The dc thin-film EL is potentially the simplest structure of all EL configurations. It is the oldest configuration next to the ac powder EL,

and was studied by Thornton[32] and others as early as 1959, and yet it has borne the least fruit.

The thin films of phosphor at high fields tend to suddenly spark and catastrophically break down. The EL operating point is above the capacitor clearing point. This can be minimized by loading up the thin film with copper. Such a film will conduct and form analogous to the Vecht-type dc powder EL. The dc conduction necessitates the isolation of the phosphor between pixels for matrix addressing as with the powder EL. The basic structure is shown in Fig. 8-46.

The most recent best performing dc thin-film EL panel was completed at the Centre National D'Etudes Des Telecommunications,

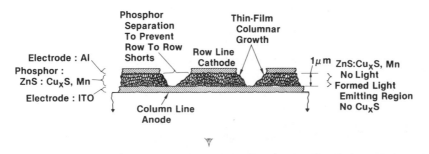

Fig. 8-46. Basic structure of a dc thin-film EL matrix display.

Bagneux, France by Abdalla, et al.[131] A small matrix addressable panel (20 rows by 20 columns) was made from ZnS:Mn,Cu where the islands of phosphor for each pixel were deposited through a Mo mask. A test device achieved a life of 7,300 hours operating at one kHz repetition frequency with a duty cycle of 0.5% at 45 volts. The initial luminance was 10 fL and the final luminance was 7.1 fL.

The main attractions of dc thin-film EL have been its relatively low operating voltage and its simplicity of construction. However, the construction is perhaps more difficult than ac thin-film EL because of the requirement for isolating the phosphor at each row line or each pixel. Also, the necessity for low voltage has disappeared with the availability of high-voltage DMOS drivers in custom LSI chips with integral low-voltage CMOS logic.

8.8   LUMINOUS EFFICIENCY

The luminous efficiency of all emitting display media is very important. Efficiencies above one lumen per watt are desirable but not sufficient for good contrast and luminance ratios. High luminous efficiency can be defeated by high reflectivity of the display media. For example, if the display medium is highly reflective, as powder EL phosphors are, the display must be operated at a much higher luminance level to achieve the same contrast ratio. The transparent nature of the thin-film phosphors is a definite advantage in this regard.

The display-material efficiency for ac thin-film EL is as high as 6 lumens/watt,[80] near the optimum point as measured in the conventional manner for displays using the test setup of Fig. 8-20. However, much of the light generated in the phosphor is trapped due to the high index of refraction for ZnS (2.3) and light piped to the edge of the display. A display made by the formulation of Table 3 was operated in an integrating sphere, and by using the method of Fig. 8-20, the ratio of total light output (including light trapped) to usable display light was determined. The ratio was found to be 4:1. That is to say, only 25% of the light generated is used for display purposes. The basic material efficiency in this example is 24

lumens per watt. This is a very respectable number when one considers that the efficiency of a 60-watt incandescent lightbulb for home use is only 14 lumens per watt when operated at a voltage appropriate for 1,000 hour life.

The luminous efficiency of ac powder EL is also very high. However, due to scattering in the powder, light is not trapped and light-piped to the edges of the device as occurs in thin-film EL. The light generation and scattering is affected by crystal grain size as reported by Lehmann.[132] For 6-micron grain size powder EL, the optimum efficiency was 14 lm/W. It was estimated by Lehmann that 18 to 19 lm/W were actually generated in the material.

The luminous efficiency for the ac thin-film EL display sample of Fig. 8-14 is shown in Fig. 8-47. Note that the maximum efficiency is at a luminance below the knee of the luminance-vs.-voltage curve. To achieve high luminance

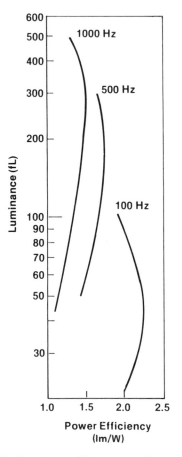

Fig. 8-47. Luminous efficiency vs. frequency in ac thin-film EL panel.

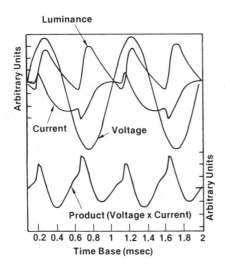

Fig. 8-48. Voltage, current, and product waveforms, with luminance signal from photometer, at 1 kHz.

and uniformity from pixel to pixel, a display would be operated in saturation at or above the knee of the luminance curve.

Measurement of the EL efficiency is complicated by the very high capacitance. The real current-causing luminance is small compared to the displacement current. The oscillographic wave traces are shown in Fig. 8-48. The small perturbation on the sinusoidal current trace corresponds to the charge conduction in the phosphor which generates electroluminescence. The circuit and instruments used to generate these curves are shown in Fig. 8-49.

The product waveform of Fig. 8-48 was produced using a Tektronix 7854 waveform processing oscilloscope with a 7A26 preamplifier and 7B53A dual time base. The current waveform was acquired, averaged over 1,000 cycles, and then stored in memory using 1,024 intervals per period. The voltage waveform was similarly processed. Both waveforms were then recalled and interval-by-interval multiplication was performed, resulting in the product waveform. The luminous efficiency curves of Fig. 8-47 were calculated from power values obtained by the above-outlined method and luminance measurements from a Photo Research 1980B Pritchard Photometer, using the test setup of Fig. 8-20. The luminous efficiency for a Lambertian emitter in English units is:

$$\text{Efficiency} = \frac{AL}{1/t \int (V \times I)\, dt}$$

where A is the Lambertian emitting area in ft^2, L is luminance as measured on the photometer in fL, and t is time duration in seconds of the power integration of the voltage V and current I product as integrated by the Tektronix 7854 processing oscilloscope.

The problem is in measuring the electrical power where the real current occurs over a very brief time and is embedded in a larger reactive power component. Another method of obtaining the power is to use the Sawyer-Tower circuit[95] and graphically integrate the area in the charge-vs.-voltage loop, which gives power per cycle or average power.

8.9 CONCLUSION

The evolution of EL into practical cost-effective display products has been very slow. This slowness was partly because of the inability to make ac powder have long life with high luminance and matrix addressability. In spite of a massive technical effort, the inability of engineers and scientists to solve this problem led

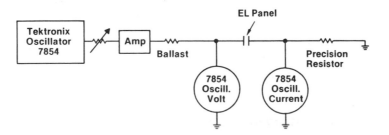

Fig. 8-49. Test setup for power consumption measurements of an EL panel.

to its disfavor in the late 1960s. The technologies of gas discharge, light-emitting diodes, vacuum fluorescence, and liquid crystals overtook EL at that time.

In 1961 a new form of EL, ac thin-film EL, was being invented, using a different phosphor: $ZnS:Mn$ without Cu_xS. By 1962, high luminance matrix addressable breadboard devices were made but not reported in the open literature. It was not until the Sharp papers were given at the 1974 SID International Symposium that the technical community took the discovery seriously, and it was not until 1983 when Sharp began to operate a production plant and Grid Systems put the display into a commercial product that the industrial community took the display seriously. Twenty-one years from breadboard device to production is a long time.

Electroluminescence is now a viable display technology manufactured by Sharp and others. There are two developments which have made this so:

1. Switching to ac thin-film using $ZnS:Mn$ phosphor.
2. The independent development of custom LSI driver chips with high voltage DMOS to drive the display.

The ac powder EL has found a product application of its own in low-light-level luminaires (less than 10 fL) for nightlights, backlighting, and transillumination. All the conventional colors are in production, the cost is low, the environmental properties are excellent, and life span is effectively infinite at low luminance. The luminaires are being used to backlight plasma panels, liquid-crystal displays, and keyboards.

There is no reason to believe that ac powder EL will ever lead to matrix-addressable displays of the complexity achievable with ac thin-film EL. It is now obvious that these two approaches as used in display devices are fundamentally different. The ac powder EL is an electron injection process, and ac thin-film EL is an electron impact process. The electron impact mechanism has a very high discrimination ratio and is temperature insensitive. The transparency

of the thin-film phosphor is a tremendous advantage in achieving sunlight readability.

The dc form of EL has inherent limitations because of ionic migration and chronic electrical breakdown, which appear to be inevitable consequences of dc operation. To refute this, there would need to be independent technical data from several laboratories that devices could be built with long life at high luminance. In matrix-addressed displays, the dc configuration will always be more costly because of the requirement for segmenting the phosphor in matrix arrays to prevent cross-coupling. One of the motivations for pursuit of dc EL is that they can be operated at a low voltage. This motivation has been neutralized due to the availability of custom DMOS high voltage-drivers.

The first application of ac thin-film EL will be in portable systems such as computers, communicators, instruments, and vehicle and aircraft panels. This is because of their inherent low power, compact size, excellent environmental properties, and sunlight readability. An example of a compact design for an EL display with drivers, high-voltage circuitry, and high-voltage power supply is shown in Fig. 8-50. A Sharp EL display in a Grid personal computer is

Fig. 8-50. Sharp EL display of 320 columns and 240 rows at 68 lines per inch, packaged with DMOS drivers, high-voltage circuitry, and high-voltage power supply (Courtesy of Sharp Corp.).

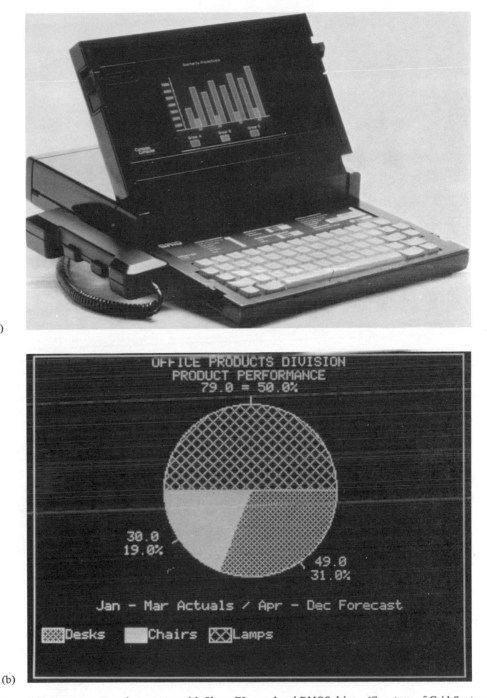

(a)

(b)

Fig. 8-51. Grid Systems personal computer with Sharp EL panel and DMOS drivers (Courtesy of Grid Systems).

shown in Fig. 8-51. AC thin-film EL is also finding applications in large alphanumeric terminals, as shown in Fig. 8-52.

The technical maturity of ac thin-film EL is behind that of plasma and vacuum fluorescent technologies in accumulated production volume and production experience, user experience, and associated test and engineering data. Also, the fundamental knowledge of the EL physical phenomena is very immature when compared

a. Terminal Installation b. Character Module

Fig. 8-52. (a) Helsinki International Airport Terminal arrival and departure board using an array of ac thin-film El characters made by the ALE method. (b) Character module (Courtesy of the FINLUX Division of Lohja Corp).

with that of plasma (gas discharge) and vacuum fluorescence (cathode-ray tubes). However, all of this will be overcome because of the decided advantages which are evident in Table 8-8. Further, because of its simplicity in construc-tion, EL is bound to have lower production cost than competing technologies.

Areas of interest for future ac thin-film EL research and development are larger size, multi-color, memory effect, and improved dielectrics.

Table 8-8 Weight, Power, and Volume Comparison of Leading Flat-Panel Emissive Displays in 1983
(Values include display panel and drive electronics except as noted.)

	Vacuum Fluorescent Display*	Plasma Display†	ACTFEL Display‡
Weight, lb	NA	3 to 4	1.32
Power, W	15	15	8
Volume, in.^{3}	NA	132.19	55.11
Width, in.	NA	12.50	7.03
Height, in.	NA	4.70	5.85
Depth, in.	NA	2.25	1.34
Volume (panel only), in.^{3}	33.25	20.08	5.02
Width, in.	7.05	11.88	5.70
Height, in.	6.46	3.38	4.40
Depth, in.	0.73	0.50	0.20
Resolution:			
Characters	1196	1148	1196
Columns	320	576	320
Rows	240	128	240
Lines/in.	64	60	68

*Based on data from Noritake Electronics, Inc., Los Angeles; product announced 1983.
†Based on data from Electro Plasma, Inc., Millbury, Ohio, Product available.
‡Based on data from Sharp on their model LJ-320U01. Product available from SEC, Paramus, NJ. Includes high-voltage power supply and case.

REFERENCES

1. Leverenz, H. W. *An Introduction to Luminescence of Solids.* New York: Wiley, 1950.
2. Ivey, H. F. *Advances in Electronics and Electron Physics*, Supplement 1. New York: Academic Press, 1963.
3. Morehead, F. F. "Electroluminescence." In *Physics and Chemistry of II-VI Compounds*, edited by M. Aron and J. S. Prener. New York: Wiley, 1963.
4. Schlam, E. "Electroluminescent Phosphors." *Proceedings of IEEE* 61, 1973, pp. 894.
5. Kazan, B. "Electroluminescent Displays." *Proceedings of SID* 17, No. 1, 1976, pp. 23–29.
6. Inoguchi, T., and Mito, S. "Phosphor Films." In *Electroluminescence*, edited by J. I. Pankove. Berlin: Springer-Verlag, 1977.
7. Lehmann, W. "Electroluminescent Large-Area Image Displays." *IPC Business Press Ltd.*, April 1980, pp. 29–38.
8. Howard, W. "Electroluminescent Display Technologies and Their Characteristics." *Proceedings of SID* 22, 1981, pp. 47–56.
9. Onton, A., and Marrello, V. "Physical Phenomena in AC Thin-Film EL Phosphor." *Advances in Image Pickup and Display* 5, edited by Ben Kazan. New York: Academic Press, 1982.
10. Payne, E. C.; Mager, E. L.; and Jerome, C. W. "Electroluminescence–A New Method of Producing Light." National Technical Conference of the Illuminating Engineering Society, 21–24 August 1950, Pasadena, CA. *Illuminating Engineering*, November 1950, pp. 688–693.
11. *Panelescent Lighting.* Sylvania Technical Brochure TR-124, Salem, MA. ca 1950.
12. Piper, W. W. "Phosphor Screen." U.S. Patent 2,698,915, filed 28 April 1953, issued 4 January 1955.
13. Kilburn, M. A.; Hoffman, F. R.; and Hayes, R. E. "An Accurate Electroluminescent Graphical-Output Unit For a Digital Computer." *IEEE Paper* 2441M, October 1957, pp. 136–144.
14. Yoshiyama, M.; Kawarada, H.; and Sato, T. "Electroluminescent Flat-Panel Display." *Proceedings of IEEE International Computer Group Conference*, Washington, D.C., June 16–18, 1970, pp. 261–69.
15. Kawarada, H., and Oshima, N. "DC EL Materials and Techniques for Flat-Panel TV Display." *Proceedings of IEEE* 61, 1972, pp. 907–15.
16. Destriau, G. "Research into the Scintillations of Zinc Sulfides to Alpha Rays." *Journal de Chimie Physique et de Physico-Chimie Biologiques* 33, 1936, pp. 587–625.
17. Destriau, G. *Transactions of Faraday Society* 35, 1939, pp. 227.
Destriau, G., and Loudette. *Academie des Sciences, Paris. Annuaire. Comptes Rendus Hebdomadaires des Seances* 208, 1939, pp. 891.
Destriau, G., and Saddy, J. *Journal de Physique et de Radium* 6, 1945, pp. 12.
Destriau, G., and Mattler, J. *Journal de Physique et de Radium* 6, 1945, pp. 227 and 7, 1946, pp. 259.
Destriau, G., and Ivey, H. F. "Electroluminescence and Related Topics." *Proceedings of the IRE* 43 No. 12, December 1955, pp. 1911–38.
Destriau, G. "Phosphor." U.S. Patent, filed 13 May 1947, issued 6 September 1960.
18. Gudden and Pohl. In *Zeitschrift für Physik* 2, 1920, pp. 192.
19. *Preparation and Characteristics of Solid Luminescent Materials*, Cornell Symposium of the American Physical Society, October 24–26, 1946. New York: Wiley, 1948, pp. 441.
20. Fischer, A. G. "Electroluminescent Lines in ZnS Powder Particles, 11. Models and Comparison with Experiments," *Journal of the Electrochemical Society* 110, July 1963, pp. 733.
21. Destriau, G. *Edinburgh and Dublin Philosophical Magazine*, Series 7, Vol. 38, No. 285, October 1947, pp. 700–37.

22. Sentementes, T. J. Senior Engineering Specialist, GTE Lighting Products Business, Danvers, MA. Private communication, November 1980.

23. Mager, E. L. "Electroluminescent Lamp." U.S. Patent 2,566,349, filed 38 January 1950, issued 4 September 1951.

24. Bay, Z., and Szigeti, G. "Electric Source of Light." U.S. Patent 2,254,957, 2 September 1941. Hungarian Patent 140,547.

25. Petertyl, S. V., and Fuller, P. R. "Improved Contrast and Visibility for Electroluminescent Displays." 18th Annual National Aerospace Electronics Conference, Dayton, Ohio, 16–18 May 1966.

26. Petertyl, S. V., et al. "Development of High Contrast Electroluminescent Displays." Air Force Flight Dynamics Laboratory Report AFFDL-TR-66-183, March 1967.

27. Petertyl, S. V. Private communication, February 1983.

28. DeBoer, J. H., and Dippel, C. J. U.S. Patent 1,954,691, 10 April 1934.

29. Williams, F. E. *Journal of the Optical Society of America* 37, 1947, pp. 302.

30. Studer, Cusano, and Young, J. *Journal of the Optical Society of America* 41, 1951, pp. 559.

31. Feldman, C., and O'Hara, M. "Formation of Luminescent Films by Evaporation." *Journal of the Optical Society of America* 47, April 1957, pp. 4.

32. Thornton, W. A. "Electroluminescent Thin Films." *Journal of Applied Physics* 30, January 1959, pp. 123–24.

33. Thornton, W. A. "DC Electroluminescence in Zinc Sulfide Films," *Journal of Applied Physics* 33, No. 10, October 1962, pp. 3045.

34. Soxman, E. J. *Electroluminescent Thin Films, Army/Navy Instrumentation Program, Summary Report, 1961.* Servomechanisms, Inc. Report SMIR 62-3. Contract: NONR-1076 (prime: Douglas Aircraft Company, Inc), 5 January 1962.

35. Soxman, E. J. *Army/Navy Instrumentation Program Phosphor Research, Final Report.* Servomechanisms, Inc., Report SMIR 62-7. Contract: NONR 1076(00) (prime: Douglas Aircraft Company, Inc.), December 1962.

36. Soxman, E. J., and Steele, G. N. JANAIR Report AD 437-866 on Contract NONR 4165(00), 4 November 1963.

37. Soxman, E. J., and Ketchpel, R. D. *Electroluminescent Thin Film Research, Final Report.* JANAIR Report AD 754-781, July 1972. Also following interim JANAIR reports: EL-1, AD 475-700L, August 1965; EL-2, AD 800-992L, August 1966; EL-3, AD 815-950L, January 1967; EL-4, AD 682-547, July 1967; EL-5, AD 704-536, April 1969; EL-6, AD 704-537, May 1969.

38. Steele, G., and Soxman, E. J. "Dark Field High Contrast Light-Emitting Display." U.S. Patent 3,560,784, filed 26 July 1968, issued 2 February 1971.

39. Ketchpel, R. D. "Light-Emitting Film Display System with Two-Axis Time-Multiplex Circuitry." *SID 72 Digest,* June 1972.

40. Webb, Bob. Private communication and product bulletins, February 1983.

41. Steele, Gordon. Luxel Corp., 515 Tucker Avenue, Friday Harbor, Washington 98250. Private communication, February 1983.

42. Gurman, B. S. U.S. Army, Avionics R & D Activity. Private communication, February 1983.

43. Lippman, M. E. "Physics and Failure Analysis of High-Reliability LEF Displays." *SID 72 Digest,* June 1972.

44. Kahng, P. *Applied Physics Letters* 13, No. 210, 1968.

45. Vlasenko, N. A. *Optical Spectroscopy* 18, No. 260, 1965.

46. Vlasenko, N. A., and Yaremko, A. M. *Optical Spectroscopy* 18, No. 263, 1965.

47. Russ, M. J., and Kennedy, P. I. *Journal of the Electrochemical Society* 114, No. 1066, 1967.

48. Mito, S. Japan. Private communication, October 1983.

49. Inoguchi, T., et al. "Stable High Brightness Thin-Film Electroluminescent Panels." *SID 74 Digest,* 1974, pp. 84–95.

50. Mito, S., et al. "TV Imaging System Using Electroluminescence," *SID 74 Digest International Symposium,* 1974, pp. 86–87.

51. Takeda, M., et al. "ZnS:Mn Thin Film EL Panel with Inherent Memory Function." *SID 75 Digest* Paper 7.8; not bound in *Digest,* 1975.

52. Yamauchi, Y.; et al. "Inherent Memory Effects in ZnS:Mn Thin Film EL Devices." *IEDM Technical Digest,* 1974, pp. 348.

53. Yoshida, M., et al. "The Mechanism of Inherent Memory in Thin Film EL Device." *Japan Journal of Applied Physics* 17, Sup. 17-1, 1978, pp. 127–33.

54. Kabo, N., et al. "EL TV Display with Stop Motion." *SID 78 Digest,* 1978.

55. Suzuki, C., et al. "Direct Electrical Readout from Thin-Film EL Panel." *SID 1978 Digest,* 1978.

56. Vecht, A. "Electroluminescent Displays." *Journal of Vacuum Science Technology* 10, 1973, pp. 789.

57. Vecht, A. "EL Powder Technology for the Eighties." *SID 81 Digest,* April 1981.

58. Brody, T. P., et al. "A 6 × 6 in. 20 lpi Electroluminescent Display Panel." *IEEE Transactions on Electron Devices,* September 1975, pp. 739.

59. Kramer, G. "Thin-Film Transistor Switching Matrix for Flat-Panel Displays." *IEEE Transactions on Electron Devices,* September 1972, pp. 733.

60. Lehmann, W. "Hyper-Maintenance of Electro-

luminescence." *Journal of the Electrochemical Society* 113, No. 1, January 1966, pp. 40.

61. Cresswell; M. W.; et al. *Manufacturing Methods and Engineering for TFT Addressed Displays*, Final Report. U.S. Army ERADCOM Contract DAAB07-76-C-0027, February 1980.

62. Fugate, K. O.; "High Display Viewability Provided by Thin-Film EL, Black Layer, and TFT Drive." *Proceedings of SID* 18, No. 2, 1977, pp. 125–133.

63. Kun, Z. K.; Luo, F. C.; and Murphy, J. "Thin-Film Transistor Switching of Thin-Film Electroluminescent Display Elements." *SID 79 Digest*, May 1979.

64. Tannas, Jr., L. E. "Fabrication and Application of Thin-Film Transistors to Displays." U.S. Army '80 ERADCOM Hybrid Microcircuit Symposium, Ft. Monmouth, New Jersey, June 1980.

65. Tannas, Jr., L. E.; Helm, W. J.; and Bassi, D. L. *Integrated Thin-Film Transistor Display*, Final Report. U.S. Army ERADCOM Contract DAAB07-77-C-0583, October 1982.

66. Awane, K., et al. "High-Voltage DSA-MOS Transistor for Electroluminescent Display." *IEEE Electronics Conference Record*, San Francisco, 1977.

67. Spencer, I. "The High Voltage IC and Its Future." *SID 82 Digest*, 1982.

68. Organized by Professor R. Evrard and recorded in special issue of *Journal of Luminescence* 23 (1981), edited by Professor F. Williams, North-Holland Publishing Company.

69. Fischer, A. G. "Electroluminescent Lines in ZnS Powder Particles, I. Embedding Media and Basic Observations." *Journal of the Electrochemical Society* 109, November 1962, pp. 1043.

70. Chen, Y. S., and Krupka, D. C. *Journal of Applied Physics* 43, 1972, pp. 4089.

71. Onton, A., and Marrello, V. "Physical Phenomena in AC Thin-Film Electroluminescent Phosphors." In *Advances in Image Pickup and Display* 5, edited by Ben Kazan. New York: Academic Press, 1982.

72. Fischer, A. G. "Electroluminescent Lines in ZnS Powder Particles, II. Models and Comparison with Experiments." *Journal of the Electrochemical Society* 110, July 1963, pp. 733.

73. Marrello, V., and Onton, A. *Applied Physics Letters* 34, 1979, pp. 525.

74. Marrello, V.; Ruhle, W.; and Onton, A. "The Memory Effect of ZnS:Mn AC Thin-Film Electroluminescence." *Applied Physics Letters* 31, pp. 452–54.

75. Ruhle, W.; Marrello, V.; and Onton, A. "AC Thin-Film Electroluminescence, Filamentary Emission and its Memory Effect." *Journal of Luminescence* 18/19, 1979, pp. 729–38.

76. Marrello, A. IBM, San Jose, CA. Private communications, March 1983.

77. Tannas, Jr., L. E. "Thin-Film Electroluminescent Emitter." *SID 81 Digest*, May 1981.

78. Hurd, J. M., and King, C. N. "Physical and Electrical Characterization of Co-Deposited ZnS:Mn Electroluminescent Thin-Film Structures." *Journal of Electronic Materials* 8, No. 6, 1979.

79. Suzuki, C.; Inoguchi, T.; and Mito, S. "Thin Film EL Displays," *SID Information Display Journal*. Spring 1977.

80. Suntola, T. "Performance of Atomic Layer Epitaxy Devices." *SID 81 Digest*, May 1981.

81. Tueta; R., and Braguier, M. "Fabrication and Characterization of Indium Tin Oxide Thin Films for Electroluminescent Application." *Thin Solid Films* 80 (Netherlands), 1981, pp. 143–148.

82. Kun, Z. K., et al. "The Influence of Chlorine on the Crystal Structure and Electroluminescent Behavior of ZnS:Mn Films in Thin Film Electroluminescent Devices." *Journal of Electronic Materials* 10, No. 1, 1981, pp. 287.

83. Miura, S., and Sato, S. "Thin Film AC $ZnS_x Se_{1-x}$:Mn EL," *IEDM Technical Digest*, December 1980, pp. 715 ff.

84. Alt, P. M.; Howard, W. E.; and Sahni, O. *IEEE Transactions on Electron Devices* 26, 1979, pp. 1850.

85. Smith, D. H. *IEEE Transactions on Electron Devices* 26, 1979, pp. 1850.

86. Takeda, M., et al. "Commercial Production Version of Thin Film Electroluminescent Display." *SPIE Proceedings* 386, paper 386-07, January 1983.

87. Tiku, S. K.; Smith, G. G.; and Johnson, M. R. "SiAlON—A New Composite Dielectric for ACTFEL Displays." *SID 83 Digest*, 1983.

88. Faulkner, E. K.; Whitney, R. K.; and Zeman, J. E. "Alkali Extraction as a Determinant in the Selection of a Glass for Displays," *International Display Research Conference Record*, October 1982, pp. 171.

89. Miller, M., et al. "Characterization of Electroluminescent Display Devices." *SID 82 Digest*, May 1982.

90. Ketchpel, R. D. Rockwell International Research, Thousand Oaks, California. Private communication, 1981.

91. Alt, P. M. IBM Research, Yorktown Heights, New York. Private communication, 1981.

92. Tannas, Jr., L. E. *High Contrast EL Display Interim Report*, U.S. Army ERADCOM Contract DAAK20-80-C-0313, October 1982.

93. Yang, K. W., and Smith, D. H. *IEEE Transactions on Electron Devices,* 28, 1981.

94. Takeda, M. et al. "Practical Application Technologies of Thin-Film Electroluminescent Panels." *SID 80 Digest*, May 1980.

95. Smith, D. H. "Modeling AC Thin-Film Electroluminescent Devices." *Journal of Luminescence* 23, 1981, pp. 209–35.

96. Suntola, T., et al. "Atomic Layer Epitaxy for Producing EL-Thin Films." *SID 80 Digest*, 1980, pp. 109.

97. Cattell, A. F., et al. "Electroluminescence from Films of ZnS:Mn Prepared by Organometallic Chemical Vapour Deposition." *International Displays Research Conference* 25, October 1982.

98. Wright, P. J., et al. *Journal of Crystal Growth* 56, 1982.

99. Theis, D., et al. "Cross-Sectional Transmission Electron Microscopy of Electroluminescent Thin Films Fabricated by Various Deposition Methods." *Journal of Crystal Growth*, 63, No. 1, September 1983, pp. 47–57.

100. Suzuki, C., et al. "Optical Writing and Erasing on EL Graphic Displays." *SID 77 Digest,* April 1977.

101. Howard, W. E., and Alt, P. M. *Applied Physics Letters* 31, 1977, pp. 399.

102. Dunham, M. E., et al. "A Storage CRT Using a Thin-Film Electroluminescent Screen." *IEDM* Paper 27.4, December 1980.

103. Sahni, O., et al. "Device Characterization of an Electron-Beam-Switched Thin-Film ZnS:Mn Electroluminescent Faceplate." *IEEE Transactions on Electron Devices* 28, 1981, pp. 708–19.

104. Takeda, M., et al. "Commercial Production Version of Thin Film Electroluminescent Display." *SPIE Proceedings* 386, Paper 386–07, January 1983.

105. Tannas, Jr., L. E. "Failure Modes in ACTFEL Panels." *SID 83 Digest*, 1983.

106. Maissel and Glang, *Handbook of Thin Film Technology*. New York: McGraw-Hill, 1970, p. 19-5 and p. 16-10.

107. Yoshida, Masaru, et al. "AC Thin-Film EL Device that Emits White Light." *SID 80 Digest*, May 1980, pp. 106.

108. Miura, S., and Sato, S. "Thin-Film AC $ZnS_x Se_{1-x}$:Mn EL." *IEDM*, December 1980, pp. 715.

109. Vecht, A., et al. "DC EL Dot Matrix Displays in a Range of Colors." *SID 80 Digest*, 1980.

110. Hale, L. G., et al. "Multi-Color Thin-Film Emitters." *IEDM*, December 1980, pp. 719.

111. Coovert, R. E., and King, C. N. "A Dual-Color AC Thin-Film Electroluminescent (TFEL) Display." SPIE, Los Angeles (Jan. 1983), Paper 386-12.

112. Coovert, R. E.; King, C. N.; and Tuenge, R. T. "Feasibility of a Dual-Color ACTFEL Display." *SID 82 Digest*, May 1982, pp. 128.

113. Menn, R., et al. "Thin Film Electroluminescent Devices: Influence of Mn Doping Method and Degradation Phenomena." *1982 International Display Research Conference Record*, October 1982, pp. 38–41.

114. Alt, P. M. IBM Research, Yorktown Heights, NY. Private communication, 1983.

115. Yamauchi, Y., et al. "Red Electroluminescence from ZnS:Mn-F Thin Film." *IEDM Technical Digest*, 1974, pp. 352.

116. Hegyi, I. J., et al. *Journal of the Electrochemical Society* 104, 1957, pp. 717.

117. Fleming, G. R. "Design Guide, Electroluminescent Lighting." Company Bulletin, Luminescent Systems, Inc., Lebanon, NH, 1982.

118. Fischer, A. G.; Hoger, K.; Herbst, D.; and Knufer, J. "Advances in AC-Electroluminescent Powder Layers." *Proceedings of the 3rd European Electro-Optic Conference*, SPIE 99, October 1976.

119. Lehmann, W. "Electroluminescent Large Area Image Display." *Displays*, IPC Business Press Limited, April 1980, pp. 29.

120. Fischer, A. G. Chapter 10 in *Luminescence of Inorganic Solids*, edited by P. Goldberg. New York: Academic Press, 1966, pp. 541–602.

121. Lehmann, W. *Journal of the Electrochemical Society* 102, 1960, pp. 20.

122. Lehmann, W. *Journal of the Electrochemical Society* 113, 1966, pp. 40.

123. Vecht, A.; Waite, M. S.; Highton, M. G.; and Ellis, R. *Journal of Luminescence* 24/25, 1981, pp. 917.

124. Vecht, A. "EL Powder Technology for the Eighties," *SID 81 Digest*, 1981, pp. 32.

125. Mears, A. L.; et al. "An Operating 36-Character DC Electroluminescent Alphanumeric Display." *SID 73 Digest*, 1973, pp. 30.

126. Higton, M.; Vecht, A.; and Mago, J. "Blue, Green and Red DC EL Powder Display Developments." *SID 78 Digest*, 1978.

127. Abdalla, M. I.; Brenar, A.; and Noblanc, J. P. "Electrical Conduction and Degradation Mechanisms in dc EL ZnS:Mn, Cu Powder Displays." *Biennial Display Research Conference*, 1980, pp. 174.

128. Vecht, A., et al. "Direct Current Electroluminescence in Zinc Sulfide: State-of-Art." *Proceedings of the IEEE* 61, 1973, pp. 902–07.

129. Kawarada, H., and Ohshima, N. "DC EL Materials and Techniques for Flat-Panel TV Displays." *Proceedings of the IEEE* 61, 1973, pp. 907–15.

130. Vecht, A. "Developments in Electroluminescent Panels." Department of Material Science Report, Thames Polytechnic, Wellington Street, London, SE 18 6PF, 1982.

131. Abdalla, M. I., et al. "Performance of DC EL Co-Evaporated ZnS:Mn, CU Low Voltage Devices." *Biennial Display Research Conference*, 1980, pp. 165.

132. Lehmann, W. "Particle Size and Efficiency of Electroluminescent Zinc Sulfide Phosphors." *Journal of the Electrochemical Society* 105, No. 10, October 1958, pp. 585.

9

LIGHT-EMITTING DIODE DISPLAYS

M. GEORGE CRAFORD, *Hewlett Packard*

9.1 INTRODUCTION

The intent of this chapter is to give an overview of issues in LED technology that are important to an engineer considering the use of LEDs for display applications. An attempt has been made to make this chapter self-contained. Due to space limitations, however, many topics are only briefly discussed. Additional information on various portions of LED technology is available in a number of books and review articles[1-9] and the reader is referred to these publications if he wishes to pursue the topic further.

We begin with a brief history and an overview of basic LED technology. Following sections include a review of the status of LED performance, a discussion of LED display devices indicating the types of devices that are available commercially, a discussion of performance parameters and drive requirements for LED displays, and a section discussing the materials and processes used to fabricate LED display devices. The chapter concludes with a summary in which future applications of LED devices for the 1980s are discussed in comparison with alternative display technologies.

9.2 History of LED Display Devices. Light-emitting devices were observed as early as 1907, when yellow light was produced by passing current through a silicon carbide detector,[10a] and were studied in some detail prior to 1928.[10b,c] However, the history of today's commercial LED displays can be traced to the observation of light emission from p-n junctions in compound semiconductors only twenty years ago and to the fabrication of visible injection lasers in gallium arsenide phosphide (GaAsP),[11] Following this discovery, several laboratories began the development of GaAsP for LED display applications. This work resulted in the commercial introduction of LED displays in 1968 by Monsanto and Hewlett Packard.

In parallel with the GaAsP research of the 1960s, development of GaP LEDs was also being pursued at a variety of laboratories including Bell Laboratories and IBM. This work utilized, for crystal growth, a liquid-phase epitaxial technique (LPE) instead of the vapor-phase epitaxial technique (VPE) used in the production of GaAsP devices. (The LPE and VPE techniques are described in Section 9.7 of this chapter.) Both red and green GaP liquid-phase epitaxial devices were developed. For red-emitting devices, GaAsP became the dominant commercial technology, although substantial quantities of red GaP devices have also been sold. In the early 1980s, red-emitting gallium aluminum arsenide (GaAlAs) devices have also become important commercially. In the case of green-emitting devices, GaP grown by both the LPE and VPE techniques has been extensively employed, but in the late 1970s LPE became the dominant technology.

A key breakthrough occurred in 1966 when workers at Bell Laboratories found that the addition of nitrogen to GaP substantially improved the performance of green-emitting devices.[12,13] In 1971 nitrogen doping was utilized in the GaAsP system and the fabrication of high-performance red, yellow, and green devices using VPE technology was made possible.[14-17]

Through the 1970s and early 1980s the performance of LED devices has continued to improve and the cost of LED devices has progressively decreased. The first LED display devices were designed to replace the vacuum-tube displays which through the late 1960s and early 1970s dominated the instrument market. These devices were followed by smaller numeric devices utilized in hand-held calculators. The calculator market grew rapidly and was important in establishing LED technology as a significant commercial enterprise. In the 1973–1975 time period, the market became highly competitive, with LED displays sharing the market with vacuum fluorescence and liquid-crystal displays (LCD) with the later ultimately becoming the dominant technology. When the market for LED calculator displays decreased, it was replaced by a rapidly growing market for LED watch displays. This market generally switched to LCD technology in the 1976–1978 time frame, but has been replaced by a broadly based and steadily growing market for LED devices which encompasses consumer, industrial, and military applications. New applications for LEDs are continually being developed, many of which are closely coupled to the microprocessor evolution which is continuing with no apparent end in sight.

9.3 BASIC LED TECHNOLOGY

9.3.1 Radiative and Nonradiative Recombination.

An LED, consisting of a single p-n junction, is in principle a simple semiconductor device. A schematic diagram of an LED structure is shown in Fig. 9-1a. Light is emitted when the junction is forward biased so that minority-carrier injection and electron-hole recombination occur. The recombination can be either radiative or nonradiative, as indicated in Fig. 9-1b. The radiative recombination often occurs through bound states associated with impurity atoms, resulting in a photon energy, $E_p = E_g - E_I$, where E_g is the semiconductor energy gap and E_I is the combined binding energy of the impurity levels involved. The energy of the emitted photon is inversely proportional to the wavelength, λ, according to the relationship $E_p = hc/\lambda$ (if the recombination occurs through shallow levels $E_I \ll E_g$ so that $\lambda \approx hc/E_g$). Thus, the color of the emitted light can be con-

trolled by choosing appropriate combinations of semiconductor material and impurities. The semiconductor energy gaps for materials important to LED technology today are shown in Table 9-1. The peak emission wavelength is also shown assuming $\lambda \approx hc/E_g$. It can be seen that the materials presently utilized in LED technology are all III-V compounds—that is, they are composed of atoms from columns III and V of the periodic table. These compounds are used because energy gaps corresponding to the visible spectral region can be produced, and because it has been possible to establish a materials technology such that high-quality p-n junctions can be fabricated. Table 9-1 shows both binary (two-element) and ternary (three-element) compounds. Ternary coumpounds have the property that the energy gap can be adjusted by changing the composition "x" of the alloy. Quaternary (four-element) compounds such as GaInAsP, which are being studied for infrared-emitting devices, may also become important for visible LED technology in the future.

The primary objective in LED technology is to maximize the light output by increasing the probability for radiative recombination and decreasing the probability for nonradiative recombination. Nonradiative recombination can occur through a variety of shunt paths such as a deep trap state, as shown in Fig. 9-1b, Auger recombination, or photon emission. The origin and physics of the nonradiative centers that limit emission efficiency are not fully understood, although powerful new analytical techniques are now being utilized which could result in major breakthroughs in the future.[18,19]

Today's commercial LED devices typically require more than 100 injected electrons for each photon that is emitted, so there is substantial room for improvement.

9.3.2 Direct-Indirect Transition.

Before proceeding further, it is important to review the concepts of direct and indirect energy gaps in semiconductors. A schematic representation of direct and indirect semiconductors is shown in Fig. 9-2. In single-crystal materials, the periodicity of the electrons results in a relationshp between the energy and momentum of electrons, which can be quite complex. In semiconductor materials, the conduction band and valence

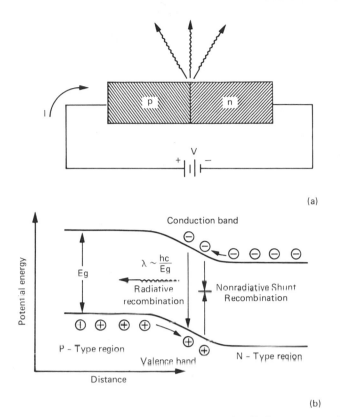

(a)

(b)

Fig. 9-1. Schematic of an LED and forward-biased LED junction. Both radiative processes, which result in photon emission, and nonradiative processes, which result in the generation of heat, are illustrated.

band are separated by the energy gap in which, in a perfect crystal, no states exist. The energy difference between the valence and conduction band is a function of the momentum of the electrons.

In a direct semiconductor, such as GaAs or InP, the lowest energy conduction band state occurs at the same momentum or "k" value as the highest energy valence band state. In an indirect semiconductor, such as GaP, AlAs (and

Table 9-1. Materials Important for Visible LEDs

Material System	Energy Gap, eV, at 300°K	Peak Emission Wavelength, Å
GaAs	1.43 direct	8670 infrared
GaP	2.26 indirect	5485 green
AlAs	2.16 indirect	5740 greenish yellow
InP	1.35 direct	9180 infrared
$GaAs_{1-x}P_x$ $x_c = 0.49$	1.43–2.03 direct 2.03–2.26 indirect	8670–6105 infrared → red 6105–5485 red → green
$Al_xGa_{1-x}As$ $x_c = 0.43$	1.43–1.98 direct 1.98–2.14 indirect	8670–6525 infrared → red 6525–5790 red → yellow
$Ga_xIn_{1-x}P$ $x_c = 0.62$	1.35–2.18 direct 2.18–2.26 indirect	9180–5685 infrared → yellow 5685–5485 yellow → green

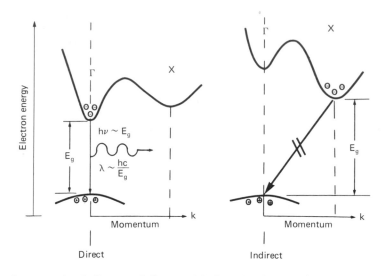

Fig. 9-2. Schematic energy band diagram of direct and indirect bandgap semiconductors. Electron energy is plotted versus momentum, k. In the direct bandgap material, radiative transitions are allowed. In the indirect bandgap material, they are forbidden.

silicon), these band extrema occur at different "k" values. Due to momentum conservation requirements, the change in momentum of an electron and hole involved in radiative recombination must equal the photon momentum. A photon has negligible momentum so that the momentum of the electron hole involved in electron-hole recombination must remain essentially unchanged. Consequently, on an energy-versus-momentum diagram, such as Fig. 9-2, only "vertical" transitions are allowed. As indicated in Fig. 9-2, there is a high probability of "vertical" transitions in a "direct" semiconductor, which explains the early success of GaAsP.

In the case of an indirect semiconductor, the electron must make a "diagonal" transition requiring a substantial change in momentum. This can only be accomplished by scattering of the electron from, for example, a lattice vibration or an impurity atom potential that permits the conservation of energy and momentum. There is generally a low probability that this will occur, since an extra "participant" must be involved in the recombination process. Consequently, nonradiative recombination usually dominates in indirect semiconductors.

The importance of the difference between direct and indirect bandgap semiconductors becomes readily apparent when LEDs are fabricated using different compositions of a ternary alloy system in which the energy gap changes from direct to indirect at a certain composition. The $GaAs_{1-x}P_x$, $Ga_{1-x}Al_xAs$, and $In_{1-x}Ga_xP$ ternary alloys are all examples of systems of this type. The band structure of the $GaAs_{1-x}P_x$ system, which is utilized for a substantial fraction of commercial LEDs, is shown schematically in Fig. 9-3. $GaAs_{1-x}P_x$ is a solid solution of GaAs and GaP. GaAs (x = 0) is a direction bandgap semiconductor since the direct minimum in the conduction band is lower in energy than the indirect minima. Therefore, the electrons in the conduction band collect in the direct energy minimum. GaAs is an efficient material for the fabrication of LEDs. Since the energy bandgap is approximately 1.4 electron volts at room temperature, the radiation is emitted near 9,000 angstrom (Å) units and is in the infrared spectral region.

As x is increased by the addition of phosphorus, the energy bandgap increases and the band structure changes. The energy gap as a function of alloy composition is shown in Fig. 9-4. The direct energy bandgap, E_Γ, increases more rapidly than the indirect energy bandgap, E_x, so that at x = 0.49 the energy minima have reached the same energy. Beyond x = 0.49 the indirect minima become the lowest in energy, and thus the "physics" of radiative recombination change since the electrons now collect in

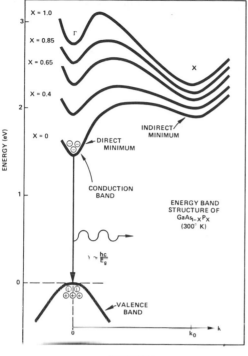

Fig. 9-3. Schematic energy band diagram of the $GaAs_{1-x}P_x$ material system. For $0 < x < 0.49$, the energy bandgap is direct. For $x > 0.49$, the energy bandgap is indirect. The alloy compositions shown correspond to red ($x = 0.4$), orange ($x = 0.65$), yellow $x = 0.85$), and green ($x = 1.0$). (After Reference 3).

the indirect minima and, in the case of a pure crystal, have a low probability for radiative transitions. It has been found, however, that high-efficiency LEDs can be fabricated from indirect materials such as GaP and $GaAs_{1-x}P_x$ with $x > 0.49$, by adding nitrogen to the crystal.[12-17]

9.3.3 Nitrogen Doping in $GaAs_{1-x}P_x$. Figure 9-5 shows a plot of the quantum efficiency of $GaAs_{1-x}P_x$ LEDs plotted as a function of the mole fraction x. This curve shows that high-efficiency devices without nitrogen can be fabricated for $x \leqslant 0.4$. For $x > 0.4$ the efficiency falls off by two to three orders of magnitude, due to the proximity of the direct-indirect bandgap transition which occurs at $x = 0.49$. With nitrogen doping, however, the quantum efficiency for $x > 0.49$ can be increased by more than two orders of magnitude. Thus,

nitrogen doping has the effect of substantially "softening" the direct-indirect transition and making high-performance devices possible for all alloy compositions. Figure 9-6 shows a plot of brightness as a function of peak emission wavelength and energy for $GaAs_{1-x}P_x$ devices with and without nitrogen doping. We can see that high-performance devices can be made with nitrogen doping, from the red through the green spectral regions. The relatively constant LED performance from red through green is explained by the fact that the CIE relative luminosity function (also shown in Fig. 9-6) is increasing with peak emission energy (and alloy composition) in the same region as the device efficiency, shown in Fig. 9-5, is decreasing. The product of the CIE curve and the efficiency curve results in the luminous performance shown.

The emission energy which is obtained in $GaAs_{1-x}P_x$ devices doped with nitrogen is a function of nitrogen concentration as well as alloy composition. The relationship between peak emission energy and composition shown in Fig. 9-4 corresponds to an N density between 10^{17} and $10^{19}/cm^3$, which is the concentration range yielding maximum efficiency for most alloy compositions. However, in the case of GaP:N, nitrogen doping in the range of $10^{20}/cm^3$ has been found to give high-performance devices emitting in the yellow spectral region instead of the green.[20] This change in wavelength is due to the fact that the heavy N doping results in the occurrence of a substantial number of traps in which N atoms form pairs on adjacent lattice sites. The "NN pair" levels are lower in energy than single N levels and give yellow emission.

Nitrogen is basically a different type of impurity than is commonly used in semiconductor devices. It is dubbed an isoelectronic impurity because it comes from column V of the periodic table and therefore has five valence electrons, as do arsenic and phosphorus which it replaces in the crystal. As a consequence, nitrogen, unlike a donor or acceptor, introduces no charge carriers in the crystal lattice. Unlike conventional shallow impurities, however, nitrogen does provide a strong radiative recombination center, particularly for indirect bandgap materials. The

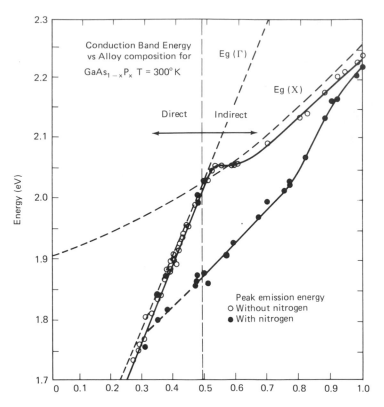

Fig. 9-4. Peak emission energy and energy gap versus alloy composition for $GaAs_{1-x}P_x$ (after Reference 15). Dashed lines indicate the direct and indirect energy gaps. The direct energy gap, $Eg(\Gamma)$, increases more rapidly than the indirect energy gap, $Eg(x)$, resulting in a direct-indirect transition at $x = 0.49$. The peak emission energy for LEDs fabricated with and without nitrogen doping is also shown. In the case of devices without nitrogen doping, the peak emission energy closely follows the energy gap. It can be seen that with nitrogen doping the peak emission is substantially lower in energy. Thus, the color of an LED made with a given alloy composition will be somewhat different, depending upon whether the device is nitrogen-doped.

reason for this is the unique way in which carriers are bound at the nitrogen site. The physics of binding at a nitrogen site has been studied theoretically,[17,21] but the basic concepts can be rationalized using arguments based on the Heisenberg uncertainty principle.

In the case of normal donors and acceptors, charge carriers are bound to the shallow impurities by weak coulombic forces, which are proportional to $1/r$ and which result in a relatively large spread in position in any direction. We know from the uncertainty principle that the uncertainty in position, Δr, is inversely proportional to the uncertainty in wave number, Δk. Consequently, since Δr is large for weak coulombic binding, this implies that Δk is small and that the momentum of a carrier bound to a shallow impurity is closely localized in momentum space in the vicinity of the impurity.

In the case of nitrogen doping, since nitrogen is isoelectronic and has no net charge, coulombic binding is not operative. Instead, charge carriers are bound at nitrogen centers by a much shorter range attraction which results from a combination of the difference in the electronegativity between the nitrogen atom and the column V atom it replaces, and the hydrostatic deformation of the lattice around the nitrogen site.[17,21] As a consequence, Δr is smaller for the isoelectronic trap binding and Δk is larger. The result is schematically shown in Fig. 9-7, where the shaded portion illustrates that the electron bound to the nitrogen impurity is widely distributed throughout momentum space. The electron density is highest near the X minima, but there is also a significant probability that the electron can be located anywhere between Γ and X with a comparatively high probability of being also

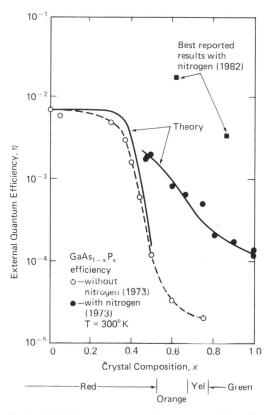

Fig. 9-5. Efficiency versus alloy composition for GaAs$_{1-x}$P$_x$ with and without nitrogen. The efficiency without nitrogen drops sharply in the composition range $0.4 < x < 0.5$ because of the proximity of the direct-indirect transition at $x = 0.49$. The efficiency with nitrogen is higher in the region for $x > 0.49$, but still decreases steadily with increasing x because of the increasing separation between the direct and indirect bandgap. The dashed line is experimental, and the solid lines represent theoretical calculations described in Reference 17. The theoretical curves shown the relative efficiency as a function of alloy composition and are normalized to fit the 1973 data from References 16 and 17. The best reported results in 1982 (Reference 45) are substantially higher, but the relative performance of the 1982 data points are consistent with the theory.

at $k = 0$. It can be shown that the probability of finding the electron at $k = 0$ is related to the separation between the direct and indirect minima, $E_g(\Gamma) - E_g(X)$, and that this probability increases as the separation between the direct and indirect minima decreases.[17,21] The fact that an electron bound to a nitrogen impurity has a high probability of being at $k = 0$ permits the nitrogen-doped indirect-energy-gap crystal to behave as if it were a direct-energy-gap crystal

and has "vertical" transitions. In other words, the momentum conservation rule can be obeyed even in an indirect crystal, with the result that nitrogen doping results in a high radiative recombination efficiency for these materials. This has been important in filling the color spectrum between red and green, as we shall see below.

Since the probability of an electron being at $k = 0$ decreases as the separation between the energy minima increases, we would expect the radiative recombination probability and hence the quantum efficiency to decrease as we move from GaAs towards GaP. This is what occurs as shown in Fig. 9-5. With the addition of nitrogen, the quantum efficiency is high (compared to devices without nitrogen) and decreases as the crystal composition is shifted toward GaP, due to the increasing separation between the direct and indirect energy bandgap minima.

9.3.4 GaAlAs. The properties of the GaAs$_x$-As$_{1-x}$ alloy system are in many ways similar to those of the GaAs$_{1-x}$P$_x$ system. This is not too surprising, since the binary compounds involved are quite similar and thus the alloy that is based on the two binaries. For $x = 0$ we have GaAs in both cases. For $x = 1$ we have GaP and AlAs for the two systems. These two binary compounds both have indirect energy gaps with similar energies as shown in Table 9-1. The respective ternaries both have a direct-indirect transition between $x = 0.4$ and $x = 0.5$, with GaAlAs occurring at a slightly lower composition, consistent with the slightly lower energy gap of AlAs.

The GaAsP system has been more important commercially for visible LEDs, primarily because it can be grown using a well-developed vapor-phase epitaxial technology as discussed in Section 9.7. On the other hand, GaAlAs requires liquid-phase or OM-VPE growth technology, the latter only now being developed. GaAlAs has the advantage, however, that Ga and Al atoms are much more similar in size than Ga and P. As a result, in GaAsP the atoms do not fit together as well as do the GaAlAs atoms, and there are less strain and dislocation-related defects in GaAlAs than GaAsP. This makes it possible to grow very high-quality devices using the GaAlAs system.[46] Historically it has been difficult to

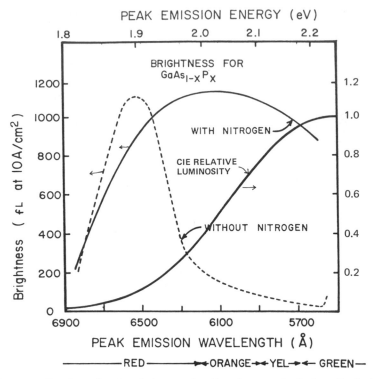

Fig. 9-6. Brightness as a function of peak emission wavelength and energy for GaAs$_{1-x}$P$_x$ devices with and without nitrogen doping. The brightness is measured in foot-lamberts with a drive current of 10 mA or 10 A/cm^2, corresponding to a device area of 10^{-3} cm^2. The data shown correspond to the 1973 efficiency data shown in Fig. 9-5. Brightness data for the best 1982 devices whose efficiency is shown in Fig. 9-5 are not available but would be expected to be in the 10,000-fL range at 10 mA. The CIE relative luminosity curve is also shown.

grow commercial quantities of good GaAlAs material because of the lack of a viable production technology. This has changed in the early 1980s, and high-quality visible GaAlAs devices are now available commercially. Only red-emitting devices with compositions in the direct-energy-gap region (x ~ 0.35) have been developed.

9.3.5 GaP:Zn,O. Red-emitting GaP devices utilize a deep trap state which is similar in many ways to the N state. In this case, the trap is formed by having Zn and O on nearby lattice sites in the crystal.[1] As in the case of N doping, this Zn,O state has a short-range potential in real space, behaving as an isoelectronic "molecule" or "pair" with a large uncertainty in k space and a high recombination efficiency. Since the Zn,O level has a binding energy of ~0.5 eV, the transition energy is approximately 1.8 eV, which corresponds to red emission.

9.3.6 Optical Coupling Efficiency. The external quantum efficiency of an LED chip can be expressed as the product of the internal quantum efficiency and the optical coupling efficiency of the device. The internal quantum efficiency, or the amount of radiative recombination divided by the sum of the radiative and nonradiative recombination, is largely determined by the material system, the material quality, and the type and concentration of impurities that are incorporated in the crystal. These issues have been discussed in sections 9.3.1 through 9.3.4. The optical coupling efficiency is the percentage of radiative recombination that escapes from the crystal. A substantial amount of radiation is reabsorbed internally before it can escape because of internal reflection at the chip surface. Since the index of refraction of LED material is typically 3.2, the critical angle for total internal reflection, θ_c ($\theta_c = \sin^{-1} 1/n$), is 18 deg. Thus, only light which strikes the

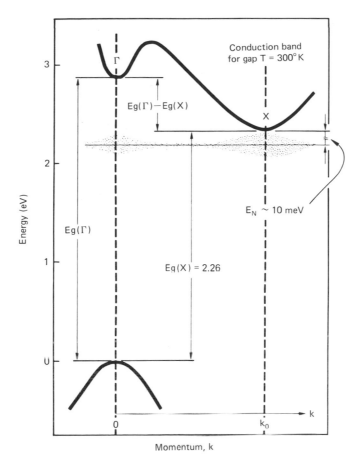

Fig. 9-7. Energy band structure for GaP showing the effect of nitrogen doping. The probability distribution for an electron band at a nitrogen site is indicated by the shaded region. The maximum probability is at k_0 beneath the indirect (x) minimum, but there is a relatively high probability that the electron will be at k = 0 beneath the direct (Γ) minimum. (After Reference 22).

surface within an 18-deg cone of perpendicular will escape from the crystal. The remainder of the light, which strikes at a lower angle, will be internally reflected. As a consequence, it can be shown that less than 2% of the light will escape from the crystal without reflection. As a result of this, the type of substrate used is a critical issue.

If the substrate is GaAs, as in Fig. 9-8a, the light that is generated at the junction and proceeds downward into the crystal is absorbed when it reaches the substrate because GaAs has a smaller band gap than the GaAsP epitaxial structure. As a consequence, the maximum external quantum efficiency that can be achieved for a visible device on a GaAs substrate is only a few percent even if the internal quantum effi-

ciency is 100%. The external quantum efficiency can be increased by a factor of 2–3X by encapsulating the chip in epoxy which has an index of refraction of 1.6 and therefore increases the critical angle for internal reflection.

If the substrate is GaP, as in Fig. 9-8b, then the light that passes downward into the crystal will be transmitted through the substrate to the back side of the chip and reflected upward, assuming that the back side of the chip is covered with a reflecting contact. GaP is transparent to the emitted radiation since the energy gap of GaP is larger than the energy of the emitted photon. Because of the possibility of multiple internal reflections of the type illustrated in Fig. 9-8b, it can be shown that the external coupling efficiency for a properly designed chip

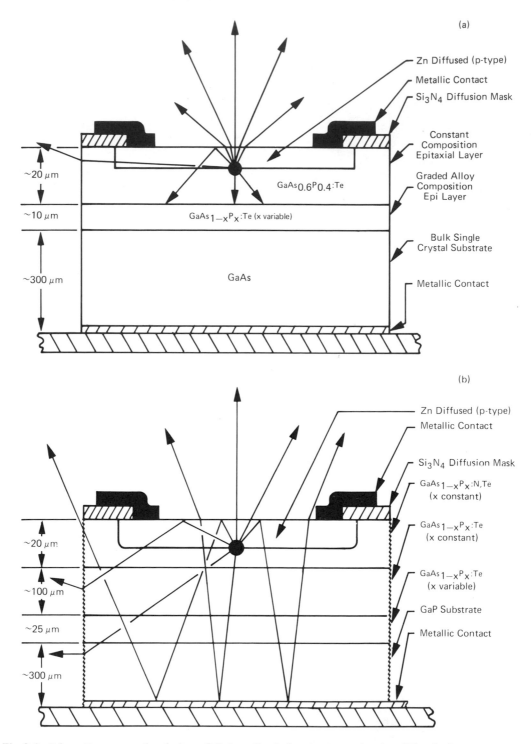

Fig. 9-8. Schematic cross-sectional view of $GaAs_{1-x}P_x$ device structures. A red-emitting device structure on a GaAs substrate is shown in (a), and an orange, yellow, or green nitrogen-doped structure on a GaP substrate is shown in (b). Light that is emitted downward in the GaAs-based structure is absorbed, whereas that emitted downward in the GaP-based structure can be reflected from the metallic contact and emitted from the edges or top of the chip. (Reference 3).

with a GaP substrate is approximately 12% and, if the chip is encapsulated, can be as high as 25%.[23] As a consequence, GaP substrates are generally used for higher-performance LEDs even though the cost of GaP is three times the cost of GaAs. The exceptions to this are red-emitting $GaAs_{0.6}P_{0.4}$ and $Ga_{0.65}Al_{0.35}As$, which generally utilize GaAs substrates. The GaAsP devices of this type are used in applications where minimum cost is a critical parameter or where large areas of LED material are required. Examples are watch displays and calculator displays, as discussed in Section 9.5.

9.4 LED PERFORMANCE—STATE OF THE ART

9.4.1 Performance Characteristics of Different Materials Technologies.

The performance status for LEDs that are important commercially is summarized in Table 9-2. The epitaxial growth technique used to fabricate the devices is indicated in the second column of the table where LPE (liquid-phase epitaxy) indicates that the epitaxial film is grown from a liquid gallium solution and VPE (vapor-phase epitaxy) means that the epitaxial layer is grown by deposition from a gaseous source. These growth techniques are further discussed in Section 9.7. Table 9-2 also indicates the LED color, peak emission wavelength and energy, luminous efficacy, quantum efficiency, and luminous performance. Luminous efficacy is the number of lumens emitted per watt of *emitted* radiation. This is a measure of the degree to which the LED spectral distribution is matched to the human eye, and is not by itself an indicator of LED performance.[7] Quantum efficiency is the ratio of the number of photons emitted to the number of electrons injected into the device. The LED quantum efficiency for the best reported devices and typical commercial devices of each material type are given in columns 6 and 7, and the luminous performance is indicated in columns 8 and 9. The luminous performance of the device is the lumens emitted, divided by the current input to the device (lumens/amp). The luminous performance can be calculated as follows:

Peak emission energy (column 4) $\times$
luminous efficacy (column 5) $\times$

quantum efficiency (column 6 or 7) = luminous performance (column 8 or 9)

The *luminous efficiency*, which is the *lumens emitted* per *watt input* (input current $\times$ input voltage) is not shown in Table 9-2 because the input voltages have not been reported in many cases. The input voltage is approximately 2 volts at the relatively low currents used (usually ~10 mA) in Table 9-2. Thus, the luminous efficiency can be estimated from the Table by dividing the luminous performance by 2 volts. Thus, the best reported GaAlAs LPE device, which has a luminous performance of 21 lumens/amp, would have a luminous efficiency of approximately 21/2 = 10.5 lumens emitted per watt input. (The actual value for this AlGaAs device is 11.3 lumens per watt since the forward voltage is 1.85 volts.)

Brightness values are often used to compare the performance of different display technologies. The surface brightness of commercial LED chips ranges from 150 foot-lambert/A/cm² for red-emitting chips without N doping to over 1000 foot-lambert/A/cm² for GaAlAs. The surface brightness of a chip can easily be tens of thousands of foot-lamberts when operated at maximum direct-current-drive conditions, and even higher when operated in a pulsed mode. However, since in most LED display applications the chip is magnified and the light is scattered by light-diffusing epoxy, the actual display surface brightness is different than the brightness of the chip itself. As a consequence, chip surface brightness is of limited utility in comparing LED displays to other display technologies. A complete methodology for comparing technologies is given in Chap. 2.

Generally, LEDs are specified in terms of on axis intensity, using units of candelas. Candelas are equivalent to lumens per steradian. For example, if an LED is a lambertian emitter, which is often approximately the case, then a device with a 1.0 lumen/amp luminous performance (from column 8 or 9 of Table 9-2) would have an on-axis intensity of $1/\pi$ candelas/ampere. Since LEDs are typically operated at current inputs of 10 mA, the user would obtain $(10/\pi) \times 10^{-3}$, or 3.2 mcd from such a device.

Table 9-2.   Performance Characteristics of Different Types of Commercial LEDs

LEDs	Epitaxial Growth Technique	Peak Emission		Luminous Efficacy	External Quantum Efficiency (Percent)		Luminous Performance (Lumens/Ampere)	
		Wavelength, Å	Energy, eV	Lumens/Radiated Watt	Best	Typical Commercial	Best	Typical Commercial
$GaAs_{0.6}P_{0.4}$	VPE	6490 (red)	1.91	75	0.5	0.1–0.3	0.72	0.14–0.43
$GaAs_{0.35}P_{0.65}$	VPE	6350 (red)	1.95	135	2.0	0.7–1.2	5.3	1.8–3.1
$GaP:Zn, O$	LPE	6990 (red)	1.77	20	15.0	2.0–5.0	5.3	0.7–1.8
GaAlAs/GaAs	LPE	6650 (red)	1.86	36	6.6	2.2–6.0	4.4	1.5–4.0
GaAlAs (GaAs substrate removed)	—	6550 (red)	1.89	58	19.0	—	21.0	—
GaP:N	LPE	5650 (yellowish green)	2.19	640	0.7	0.1–0.4	9.9	1.4–5.6
GaP:N	VPE	5650 (yellowish green)	2.19	640	0.1	0.05–0.1	1.4	0.7–1.4
GaP	LPE	5570 (green)	2.23	650	0.19	0.05–0.15	2.8	0.7–2.2
$GaAs_{0.15}P_{0.85}:N$	VPE	5850 (yellow)	2.12	540	0.37	0.1–0.3	4.2	1.1–3.4
GaP:NN	VPE	5900 (yellow)	2.10	~450	0.1	~0.05	~0.9	~0.5
GaN	VPE	4900 (blue)	2.53	~170	0.12	~0.03	~0.5	~0.1

Intensity is usually used to specify commercial LEDs instead of lumens/amp or lumens/watt, partly because intensity is easier to measure. Unfortunately, it also has the disadvantage that it is very package-dependent and does not give a true indication of the quality of the device unless the package geometry is carefully specified. A relatively minor change in the shape of the lens of an LED can focus more light on axis, at the expense of viewing angle, and thereby increase the intensity by more than an order of magnitude. For example, narrow-beam GaAlAs red-emitting devices are commercially available that have intensities of 1000 mcd. Typical discrete LED lamps are more focused than a lambertian emitter but less focused than the 1000-mcd narrow-beam device. Intensity values typically range from 1 mcd to 100 mcd at 10 mA depending on the package size and shape, and whether or not the encapsulating epoxy contains a colored dye and/or a diffusant. The particular package parameters are chosen based on the desired appli-

cation. For this reason, when comparing the performance of different types of LED materials, as in Table 9-2, we find it more meaningful to talk in terms of luminous performance.

Figure 9-9 shows emission spectra for various types of devices that are important commercially.

Red-Emitting Devices. Table 9-2 indicates that good luminous efficiencies in the red spectral region can be obtained using GaAsP:N, GaP:Zn, O, and GaAlAs devices. The GaP:Zn,O devices can be quite efficient at low current densities, but as the current is increased the GaP:Zn,O devices saturate and the efficiency decreases. The opposite is true for the other types of red-emitting LED devices, where the efficiency increases with increasing current density until the average power dissipated in the device gives rise to thermal degradation. This effect is illustrated in Fig. 9-10, which shows luminous output as a function of current for different types

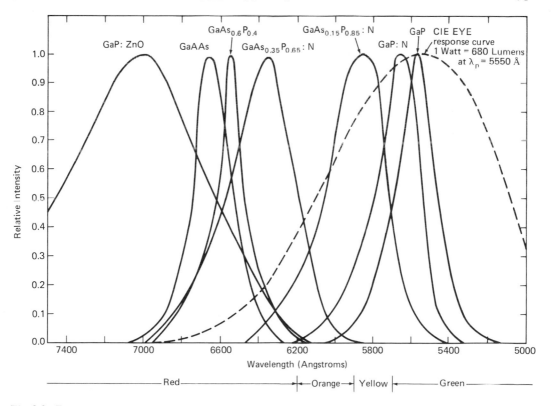

Fig. 9-9. Emission spectra for commercially important LED types plotted as a function of emission wavelength. The CIE eye response curve is also shown. It can be seen that the red emitters overlap weakly with the CIE curve, compared to the yellow and green devices, which explains the much higher luminous efficacy for yellow and green emitters.

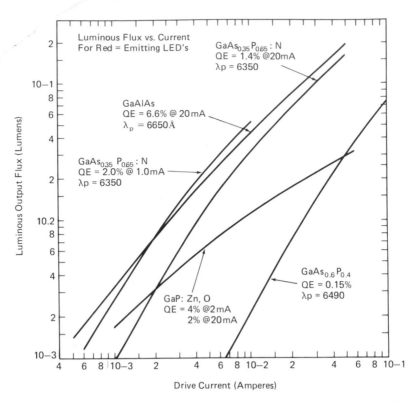

Fig. 9-10. Luminous output flux as a function of drive current for four types of red-emitting LEDs. Two $GaAs_{0.35}P_{0.65}$:N devices are shown; one device (2% QE @ 10 mA) is optimized for low current operation, and the other for higher current and pulsed applications. The GaAsP:N and GaAlAs devices shown represent state-of-the-art performance. The GaP:Zn,O device is the best available commercial device. The GaAlAs device performs well at all current densities. The GaAsP:N devices increase in efficiency at high current densities as a result of saturation of nonradiative defect centers. The GaP:Zn,O device decreases in efficiency at high currents as a result of saturation of radiative recombination centers. The $GaAs_{0.6}P_{0.4}$ standard red emitter shown represents typical commerical performance.

of devices. The performance of the GaP:Zn,O device becomes increasingly inferior to GaAsP and GaAlAs devices as the current increases.

Two GaAsP:N devices are shown, one optimized for low current operation and another designed for high current multiplexed operation. The low current device has a smaller junction area, which means that for a given drive current the current density is higher, resulting in increased efficiency. The increase is efficiency with current is believed to be due primarily to the saturation of nonradiative traps at high current density. The large junction GaAsP:N device is relatively inefficient for current below 10 mA, but performs well for higher currents, either pulsed or dc. The GaAlAs device performs well at all current densities, apparently because the density of defects in these devices is relatively low.

The best reported GaAlAs device, shown in Table 9-2 which has a performance of 21 lumens/amp, is not shown in Fig. 9-10 because it is not yet commercially available. The performance of this device would be approximately for four times better than the AlGaAs device shown.

For some LED applications, particularly discrete LED indicator lamps, low current can be utilized and GaP:Zn, O devices are suitable. These devices dominate the red LED market in Japan. For the majority of LED numeric or alphanumeric display applications, multiplexing is used, and the LEDs are driven with current pulses in the range of tens to hundreds of milli-amps on a low duty cycle basis. Because of this, $GaAs_{0.6}P_{0.4}$ and $GaAs_{0.35}P_{0.65}$:N dominate the domestic commercial market, with the choice between these technologies based primarily on cost/performance considerations.

$GaAs_{0.6}P_{0.4}$ is less expensive, but $GaAs_{0.35}P_{0.65}:N$ is more efficient.

The other red-emitting device that is becoming important commercially is GaAlAs. The performance of these devices is excellent. They behave similarly to GaAsP in that the efficiency increases with increasing current density. The devices have had a significant commercial impact only since the beginning of the 1980s, primarily for discrete LED applications. The extent to which these devices will penetrate the red-emitting LED market will depend on evolving cost, performance, and reliability comparisons with the other technologies. A key issue will be whether devices with the GaAs substrate removed (similar to the 21 ℓ/A device) can be produced commerically.

Green-Emitting Devices. GaP:N is at the present time the clearly dominant technology for green emission. The only question has been whether to use the VPE or LPE growth technique. In the 1970s, the majority of green-emitting commercial LED displays used VPE material. However, higher-performance LPE devices are now available from most suppliers and these devices dominate the commercial market.

GaP devices are also commercially available that do not use nitrogen doping. As indicated in Table 9-2, these devices have a peak emission energy of 5570 Å instead of 5650 Å for the nitrogen-dopant devices. Although this would seem to be an insignificant shift in color, the eye is extremely sensitive in the yellow and green spectral regions. As a result, these devices have a "pure" green color instead of the "yellowish green" characteristic of the GaP:N devices. This color is appealing to some users. Unfortunately, the GaP devices are two to three times less efficient than the GaP:N devices, so more power is required to achieve comparable luminous output.

Yellow-Emitting Devices. In the yellow spectral region, the dominant technology is based upon $GaAs_{0.15}P_{0.85}:N$. The only competitor for this technology indicated in Table 9-2 is GaP:NN, GaP with high nitrogen doping ($N \sim 10^{20}$), which results in yellow emission

because of nitrogen atoms on nearest neighbor sites in the crystal. High N doping has generally been found to be difficult to control reproducibly, and devices of this type are not a significant factor commercially.

The technology used for yellow devices is similar to that used for the other GaAsP devices and for the VPE GaP:N green-emitting devices. The capability of a single technology to provide viable devices of all three colors accounts in large part for the commercial success of this technology. The performance obtained with these devices can be expected to continue to improve.

Blue-Emitting Devices. The technology required to produce blue-emitting devices is much more difficult than that for red through green. It is necessary to go to a totally new material system, either SiC[30] or GaN.[31] It is difficult to grow high-quality crystals, and even more difficult to form a good p-n junction. As a result, viable blue-emitting LEDs have been a tantalizing but elusive goal for a variety of laboratories around the world since the 1960s. In 1981 both SiC and GaN devices were announced commercially, and the work on the GaN devices has been described in the technical literature.[47]

The performance of the blue-emitting devices remains an order of magnitude lower than that of the other colors, and the cost is generally substantially higher. The reliability is unproven, and the drive voltage is roughly two times higher. As a consequence, blue devices have not had a significant commercial impact and will probably tend to be used only for applications where the blue color is essential.

It would be desirable to have viable blue devices so that a truly full-color LED flat-panel display could be achieved. Unfortunately, unless there are additional breakthroughs in blue LED technology, devices of this type will not be practical in the foreseeable future.

Variable-Hue Devices. It is possible to combine the red and green GaP technologies to make LED chips which change color with drive current.[24] If a p-n junction is grown with Zn,O doping in the p-type layer and N doping in the n-type layer, the device will be red at low currents where the GaP:Zn,O process is highly

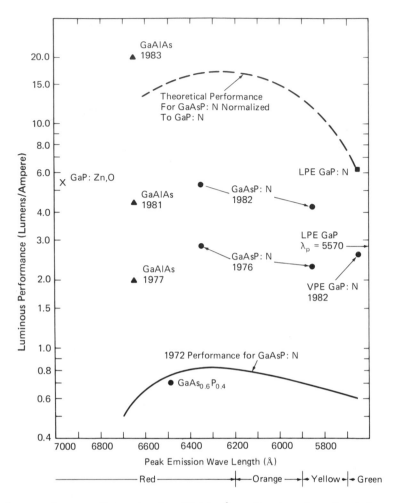

Fig. 9-11. Luminous performance (lumens/amp) at 10 A/cm^2 as a function of wavelength for a variety of types of LEDs. The data shown represent best achieved results for the years shown. It can be seen that the performance for most types of LEDs has been increasing steadily, and from a theoretical point of view a further improvement of severalfold should be possible.

efficient and will turn green at higher currents where the superlinearity of the green emission process begins to dominate and the GaP:Zn,O. emission saturates. Chips of this type have been used to fabricate prototype multicolor x-y arrays but, due in part to difficult process control problems, the devices are not important commercially.

When variable hue is desired, a generally more effective approach is to utilize red and green LED chips mounted side by side. The hue can be changed from red through yellow to green by appropriately changing the relative drive current to the chips.

9.4.2 Future Performance Improvement. Figure 9-11 is a plot of luminous performance (lumens/amp) as a function of wavelength for various types of LEDs. The fact that LEDs have been continuing to improve is indicated by the maximum performance that has been achieved in 1972, 1976, and 1982 for GaAsP devices with N doping and in 1977 and 1982 for GaAlAs devices. The performance achieved in GaAsP:N does not appear to be near a fundamental limit, because LPE GaP:N green devices have much higher performance than VPE green devices, as indicated in Fig. 9-11. If the reason for the difference in performance between VPE and LPE devices could be understood and eliminated, the VPE devices could improve by a factor of approximately two times. One would expect the red and yellow devices with N doping to move up in proportion to the increase in performance of the VPE green-emitting devices, because the relative performance of GaAsP:N red, yellow, and green devices can be predicted theoretically and is in general agreement with the observed device performance.[17] This can be seen in Fig. 9-11 by comparing the relative performance in 1972, 1976, and 1982 with the predicted theoretical relative performance shown by the dashed curve, which is normalized to the best observed LPE GaP:N device.

Even the best LPE green-emitting device performance probably does not represent a fundamental performance limit. It is simply the best demonstrated performance for green-emitting GaP:N devices; i.e., it is a feasibility demonstration. It is quite possible that performance levels significantly higher than those achieved to date using LPE may be achieved with either VPE or LPE if the performance-limiting parameters can be identified and eliminated. A substantial amount of effort has been aimed at understanding the mechanisms or defects responsible for limiting LED performance. The possible mechanisms include native defects such as vacancies or vacancy clusters, dislocations, Auger recombination, and deep levels related to trace amounts of unintentional and undetected impurities. Any or all of these mechanisms, or a combination of mechanisms such as impurities associated with vacancy complexes, may contribute significantly to nonradiative recombination.

Within recent years it has become possible to do more than speculate upon the nature of these nonradiative recombination centers. The emergence of analysis techniques such as deep level transient spectroscopy (DLTS) has enabled researchers to begin to correlate the relationships between the energy of deep level traps, specific impurity types, and device performance. As an example, DLTS has been used to determine that sulfur is responsible for a complex deep level in GaAsP which strongly degrades LED performance.[25] It has not yet been established whether sulfur is present in commercially available LEDs or contributes significantly to the variability or performance limitations of these devices. As DLTS and other similar techniques continue to be developed and utilized, it is reasonable to expect that the key nonradiative recombination mechanisms will be identified. Then, depending on whether or not the dominant mechanisms are fundamental properties

of the crystal system, it may be possible to substantially reduce or even eliminate the key mechanisms, which would result in major improvements in LED performance.

It has already been shown, for example, that high dislocation densities degrade performance of GaP green-emitting devices.[26, 27] This has resulted in an effort by technologists to produce substrates with reduced or even zero dislocation densities.[28] Whether this can be achieved for large diameter GaP ingots remains an open question. In silicon technology it has been possible to develop techniques to grow zero dislocation ingots 5 in. and larger in diameter. There is no fundamental reason that this cannot also be achieved with GaP, although the technical problems involved are substantially more difficult.

The GaAlAs red-emitting devices have also shown substantial performance improvement over the past several years. The performance of these devices was approaching a fundamental limit of around 6 lumens/amp as long as absorbing GaAs substrates were utilized, because of internal absorption effects. Recently, however, a transparent substrate technology has been developed for GaAlAs, and the performance of these devices has improved to 21 lumens/ampere. The technology is very complex, and it remains to be seen whether devices of this type can be produced in commercial quantities.

The GaP:Zn,O devices have not improved for more than five years. The performance of commercial devices is already lagging behind the GaAlAs and GaAsP:N devices, as shown in Table 9-2. It appears that this technology will decline in importance in the 1980s.

Other material systems and new growth techniques are also being investigated in an attempt to improve LED performance. These systems include the gallium indium phosphide (GaInP) ternary system, which has a direct-indirect crossover in the yellow emission range, and the gallium indium aluminum phosphide (GaInAlP) quaternary system, which also has a direct-indirect crossover in the yellow emission range. This alloy system has the advantage that it should grow well on GaAs substrates since, like the GaAlAs system, the epitaxial film can be "lattice matched" to the substrate. By using the extra degree of freedom afforded by a four-component instead of three-component alloy system, the device designer can, for certain alloy-substrate combinations, construct the epitaxial film so that it has the desired energy gap and the appropriate spacing between the atoms to match the spacing between the substrate atoms, or "lattice match" the substrate, thus resulting in high-quality material with low defect densities. Utilizing lattice-matching ternary and quaternary systems also gives technologists another advantage, since they can, by adjusting the alloy composition, create epitaxial structures with changes in the energy band gap that can be utilized to confine carriers within the active region close to the junction.[29] This could result in increased quantum efficiencies without the need to pulse the devices to high current densities. For infrared-emitting devices, this epitaxial "heterostructure" technique has resulted in high-performance injection lasers that have made long-range communication utilizing fiber optics viable. If these heterostructure techniques can be applied to LEDs, it may be possible to achieve similar substantial improvements in LED performance.

In the early 1980s, injection lasers have been further improved by utilizing increasingly complex epitaxial structures in which the individual layer thicknesses are as thin as 10 Å.[48] These "quantum-well" lasers have exhibited performance superior to that of conventional heterostructure lasers at wavelengths as short as 6500 Å.[49] These techniques may or may not lead to substantially improved LEDs, but it is likely that during the 1980s visible injection lasers with power outputs of several hundred milliwatts will become available.[50,51] Devices of this type could replace LEDs in a variety of applications requiring high intensity.

Another complex thin-layer epitaxial structure has been recently proposed that could affect the future of visible LEDs. This structure, called a superlattice, consists of alternating layers of semiconductor material with different alloy compositions. If the layers are thin enough so that quantum mechanical effects dominate, the resulting "composite" material can have properties substantially different than either of the alloy compositions used in the alternating layer structure. It has been predicted that structures of this type can be grown using alternating layers of *indirect*-energy gap semiconductor material such as GaP and $GaAs_{0.4}P_{0.6}$, which will result in a *direct*-energy gap material.[52] If this can be demonstrated and a high-performance device developed, it could in principle be possible to obtain yellow and green LEDs with quantum efficiencies comparable to the direct-energy-gap GaAlAs. Because of the higher luminous efficacies for these devices, luminous performance levels in excess of 20 ℓ/A are conceivable. Key issues that remain to be resolved are whether sufficiently high-quality epitaxial structures can be grown, and whether these thin-layer structures will be stable enough for reliable commercial implementation.

As mentioned earlier, work on blue-emitting devices is continuing at a modest level, primarily in Japan. In addition to silicon carbide and gallium nitride devices which have been announced commercially, other material systems such as zinc selenide are being investigated. In order for blue LEDs to be a viable cost-effective product, a major technical breakthrough will be required. All of the material systems that have an energy gap large enough to produce blue light are poorly developed and more challenging to work with. The problem is further complicated by the fact that the luminous efficacy for blue is three to four times lower than for green devices. Thus, in order to be comparable to green LEDs, the blue devices must be three to four times higher in efficiency, using materials which are more difficult to produce.

9.5 LED DISPLAY DEVICES

9.5.1 Overview. There are a variety of types of LED displays, the simplest of which is the discrete LED indicator lamp. An extension of this is the bar-of-light LED display family which can be used to illuminate legends and for a variety of annunciation applications. Next on the scale of complexity are the seven-segment LED numeric displays and alphanumeric displays. Numeric and alphanumeric displays can also be purchased in clusters with from two to forty characters. Many of these displays can be obtained with on-board decoding drive circuitry. More complex matrix-addressable LED displays have been fabricated on a research or prototype basis, or to satisfy specialized instrumentation requirements. Large-area flat-panel LED alphanumeric displays are generally not available on the commercial market. In this section the different types of LED displays will be discussed, ranging from the discrete emitter to the complex large-area flat-panel LED displays.

9.5.2 Discrete Emitter.

The simplest LED display device is a discrete emitter shown in Fig. 9-12. Devices of this type are generally used as indicator lamps. For certain applications, discrete emitters have been fabricated into arrays to form large-area flat-panel alphanumeric displays. Discrete LEDs can be obtained in red, yellow, and green colors with a variety of intensity distribution patterns. Figure 9-12 pictures the key elements of the discrete LED emitter, namely the lead frame, the semiconductor chip, and the epoxy encapsulation lens. The epoxy lens is a critical part of the device. A small change in the position of the lead frame within the epoxy lens can have a major impact on the distribution of the light emitted by the device. The epoxy is generally colored with dye in order to enhance the on/off contrast ratio of the device. Another common variation is to mix glass particles in the epoxy that scatter the light such that when the device is turned on the entire

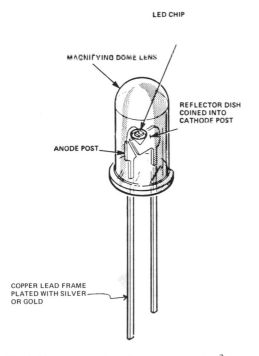

LED CHIP

MAGNIFYING DOME LENS

REFLECTOR DISH COINED INTO CATHODE POST

ANODE POST

COPPER LEAD FRAME PLATED WITH SILVER OR GOLD

Fig. 9-12. Construction features of a T-1 $\frac{3}{4}$ discrete LED emitter. (After Reference 2).

lens appears to light up uniformly. The majority of LED lamps are sold in either the T1 ($\frac{1}{8}$-in dia.) or T1 $\frac{3}{4}$ ($\frac{7}{32}$-in. dia.) sizes.

9.5.3 Bar-of-Light Displays.

In some applications where more than a single dot of light is required, an LED light-bar module can be effectively utilized. The cross section of one type of LED light-bar module is shown in Fig. 9-13. This device is basically a row of discrete LEDs connected together within a common reflector cavity. The cavity is filled with epoxy containing particles of glass. The result is that when the LEDs are turned on, the bar lights up uniformly. Different numbers of LEDs can be combined in a light-bar module to form a variety of bar lengths and widths. As an example, four LEDs can be combined to form a bar 0.15 in. × 0.17 in. Bar-of-light displays can simply be used as large rectangular indicators, or they can be used in conjunction with a legend as indicated in Fig. 9-14. Another variation of the bar-of-light display is to maintain a separation between the reflecting cavities of the different LED chips forming a linear array of small emitters. Arrays of this type can be used individually or stacked end-to-end to form bar-graph displays.

High-resolution bar-graph displays are also available in which the light-emitting elements are directly viewed instead of being in reflecting cavities. Devices of this type can be obtained with as many as 101 elements in a single device. They are particularly well suited to industrial process control systems as a status or position indicator for controller panels. They are also well suited for instrument and panel meter applications. A device of this type is shown in Fig. 9-15.

Bar-graph displays can also be fabricated using different colors of LED chips in a single display so that the color can change, from red to yellow to green for example, as more of the bar is progressively illuminated. For special applications, bar-graph displays could also be fabricated with two different colors of chips at each position so that the color of the entire display could change, from red to green for example, with drive level.

9.5.4 Numeric Displays.

The simplest numeric display is a single-character seven-segment dis-

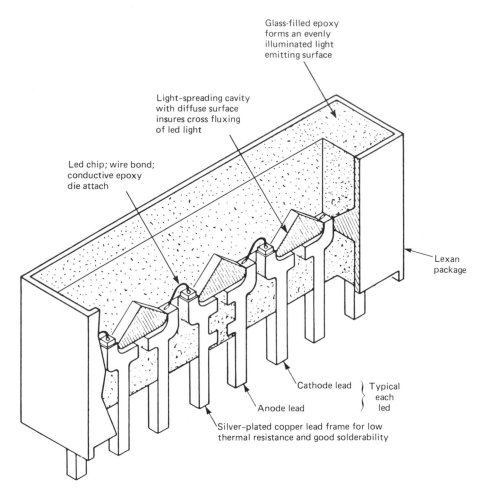

Glass-filled epoxy forms an evenly illuminated light emitting surface

Light-spreading cavity with diffuse surface insures cross fluxing of led light

Led chip; wire bond; conductive epoxy die attach

Lexan package

Cathode lead } Typical each led

Anode lead

Silver-plated copper lead frame for low thermal resistance and good solderability

Fig. 9-13. Construction features of a plastic-encapsulated LED light-bar module. A legend may be placed directly onto the light-emitting surface using a transparent adhesive, as shown in Fig. 9-14. Alternatively, the reflector cavity can be segmented so that a row of separate light-emitting regions are obtained, resulting in a display that can be used for bar-graph applications. (After Reference 2).

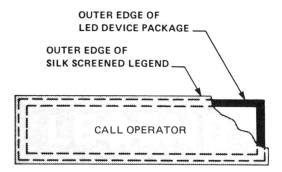

OUTER EDGE OF LED DEVICE PACKAGE

OUTER EDGE OF SILK SCREENED LEGEND

CALL OPERATOR

Fig. 9-14. LED light-bar module, shown in Fig. 9-13, with legend attached. (After Reference 2).

play. For applications where a small numeric display on the order of 50–150 mils high is desired, as in watch and calculator applications, the display is fabricated using a single piece of GaAsP as illustrated in Fig. 9-16. A single piece

of n-type GaAsP is masked photolithographically such that seven bar-shaped p-type regions can be diffused into it, forming the seven segments of the display. Metal is deposited on the p-type regions and on the back side of the display to form the anode and cathode contacts respectively. Displays of this type can be assembled in a variety of packages with different cost-performance tradeoffs depending on the application.

Probably the most common package for monolithic devices is the type used in calculator displays where typically 8 to 12 displays are die-attached and wire-bonded to a circuit board and then covered with a plastic magnifying lens. The size of the monolithic digit and the power of the magnifying lens can be adjusted to give the desired brightness, character height, and

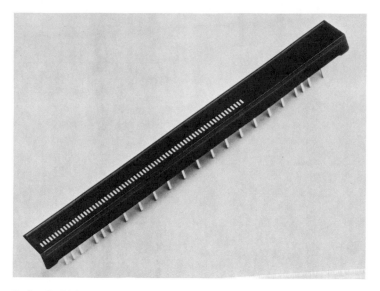

Fig. 9-15. Hewlett Packard's–high-resolution 101-element linear LED array. The array, which is 10.58 cm (4.16 in.) long, is a rugged, easy-to-read, high-resolution alternative to mechanical meters.

viewing angle. The light intensity varies as m^2 and the viewing angle as $1/m$, where m is the lens magnification. Only red-emitting monolithic displays are available commercially because of the high cost of the yellow and green LED material that is grown on the more expensive GaP substrates. An additional reason is that optical crosstalk occurs between segments if the substrate is transparent, as it is when GaP is used.

When LED numerics larger than 150 mils high are desired, the cost of a monolithic device becomes excessive due to the large area of GaAsP involved. For large numeric displays, a hybrid assembly technique is used in which small LED chips, generally identical to those used in discrete emitters, are mounted in a reflecting cavity

such that the chips appear as uniformly illuminated bars of light. A cross section of a device of this type is shown in Fig. 9-17. With seven small LED chips, one can form seven-segment displays with character heights ranging from 0.3 to 0.8 in. high.

The maximum size that can be fabricated is determined by the efficiency of the LED chip. In order to diffuse the light uniformly over increasingly longer segments, it is necessary to add more glass particles to the epoxy with which the cavity is filled. This results in increased absorption in the package. In addition, larger segments have more surface area that the LED chip has to illuminate. Since the LED chips degrade when the current density exceeds a

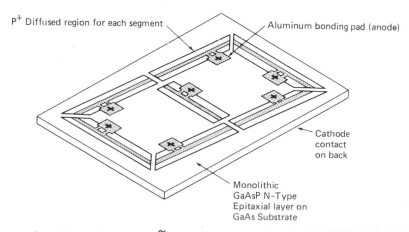

Fig. 9-16. Construction of a small (typically, $\lesssim 0.1$ in. high) seven-segment monolithic chip LED display of the type commonly used in watch and calculator displays.

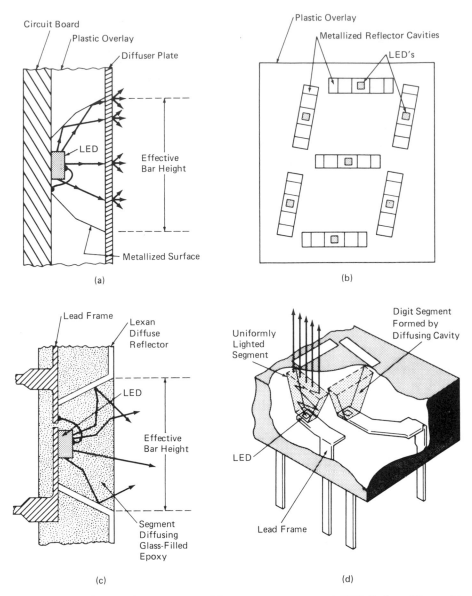

Fig. 9-17. Construction features for two types of large-area, seven-segment LED displays. Figures (a) and (b) illustrate a construction technique in which the LED chip is mounted on a printed circuitboard and the light is reflected from a metallized surface and scattered by a diffusing plate. Figures (c) and (d) show an epoxy-filled display mounted on a lead frame. (After References 2 and 3.)

critical value, it has been found that, with the performance levels achievable at the present time, the maximum display size which can be fabricated is ~0.8 inches for a device with a single LED chip per segment. If more than one chip is used per segment, then larger numerics can obviously be fabricated. However, utilizing more than one chip per segment increases the display cost and the power required to drive the display. As higher-performance LED devices

are developed it will be possible to fabricate larger numeric displays utilizing a single chip per segment.

In order to reduce the costs for LED numeric displays, in some applications, it is possible to group several LED numerics on a single substrate as illustrated in Fig. 9-18. In the case of this clock display, 30 LED chips are attached to a single circuitboard. After the die-attach and wire-bond steps are completed, a single plastic

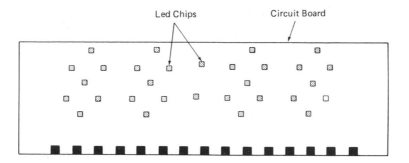

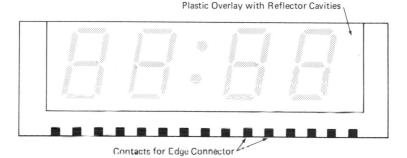

Fig. 9-18. Schematic diagram of multiple-digit LED display array. In this example, a clock display is fabricated by attaching 30 LED chips to a single circuitboard, adding a plastic overlay containing reflecting cavities, and encapsulating. This technique is an extension of the technique shown in Fig. 9-17.

piece containing 30 reflector cavities can be snapped over the LED chips. Different techniques can be used to complete the display assembly, such as encapsulating the display in epoxy or attaching a diffusing plate to the surface of the display. The appropriate assembly technique will generally be determined by cost considerations and by the environment in which the display will operate.

9.5.5 Alphanumeric Displays. There are two general types of LED alphanumeric displays, segment and dot matrix displays. A variety of fonts used for LED numeric and alphanumeric displays are shown in Fig. 9-19. The 14- or 16-segment alphanumeric fonts C and E respectively are used for LED displays less than 150 mils high. A display chip of this type, shown in more detail in Fig. 9-20, is fabricated in the same manner as the monolithic seven-segment numeric display. However, this device requires more than twice as many wire bonds as the seven-segment display and therefore is somewhat more costly. The 16-segment devices can display the entire alphabet, as indicated in Fig. 9-20, however, the shape of some of the

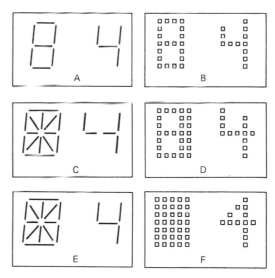

Fig. 9-19. Display fonts used for segmented and dot-matrix LED displays.

characters is found to be undesirable by many viewers.

The dot matrix display is generally found to be a more desirable display font. It gives more flexibility in shaping characters, is capable of generating a variety of symbols, and has a lower probability of being misread in case of a display

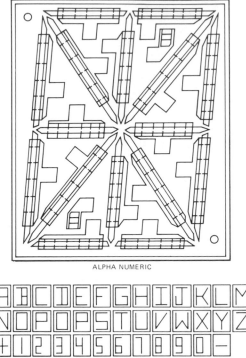

ALPHA NUMERIC

TYPICAL DISPLAY PATTERN

Fig. 9-20. Sixteen-segment alphanumeric LED display and typical display characters. This small (typically ≲0.1 in. high) display is fabricated using the same technology as the seven-segment display shown in Fig. 9-16. (Reference 3).

failure. However, the 5 X 7 display requires a larger number of wire bonds, resulting in increased display complexity and cost. Dot matrix LED displays have been fabricated using a variety of techniques. For large alphanumeric displays, hybrid assemblies of discrete LED emitters can be utilized. For displays 0.3 to 1 inch in height, hybrid assemblies of 35 discrete chips can be utilized. This requires 35 die-attach and wire-bond steps per display.

For alphanumeric displays between 0.1 and 0.3 inches in height, the sliver display technology illustrated in Fig. 9-21 is generally employed. Seven bars, each containing five junctions, are die attached and the bars are stitch bonded vertically. Handling seven bars is a substantial improvement over handling 35 discrete chips, but sliver displays are still complex devices to assemble. It would be desirable for small alphanumeric displays to use monolithic techniques.

Monolithic displays have been fabricated by several research groups.[32,33,34] One approach that has been used to fabricate matrix-addressable displays, illustrated in Fig. 9-22, has been to grow a p-type layer on an n-type substrate and follow this with the growth of an n-type layer. An isolation diffusion is then driven

Circuit Board With Traces Leading to Adjoining Characters

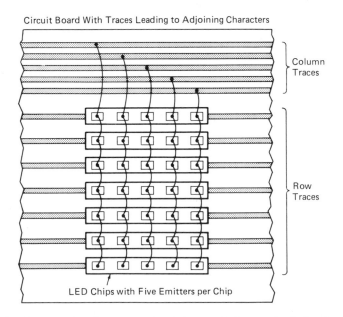

Column Traces

Row Traces

LED Chips with Five Emitters per Chip

Fig. 9-21. Construction features of a sliver-type LED alphanumeric dot-matrix display. Each sliver contains five junctions forming rows that are stitch bonded together vertically to form the column contacts. Larger displays (not shown) use discrete chips instead of slivers.

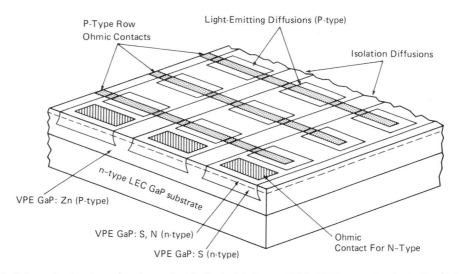

Fig. 9-22. Schematic drawing of a planar electrically isolated monolithic LED array structure. The isolation is accomplished by a deep ZN diffusion through the upper (~50 mm thick) n-type layer of an n-p-n structure. The light is generated at shallow (5-10 μm) junctions diffused into the upper n-type layer. (After Reference 32.)

through the n-type epitaxial layer, and finally the p-type anode region is diffused. In order to complete the array, a complex multilevel metallization scheme is required.[32] In order to make displays of this type viable economically, major improvement will have to be made in LED technology. It will be necessary to have LED wafers with quality similar to those achieved in today's silicon technology, several inches in diameter, and nearly defect-free. It will also be necessary to reduce substantially the cost per square inch of the material. While improvements of this type are in principle possible, it does not seem probable that monolithic matrix-addressable LED displays will be economically viable by the mid 1980s.

A further complication in the case of monolithic LED alphanumerics is illustrated in Fig. 9-23. There is a substantial optical crosstalk problem from segment to segment if monolithic displays are fabricated from materials grown on transparent substrates. This is illustrated in the bottom portion of Fig. 9-23 where light generated at the left-hand junction travels laterally through the substrate, bounces off the metal contact at the back, and appears to be emitted by the right-hand junction. Thus, the junction at the right side of the diagram appears to be "on" when the junction at the left is turned on and vice versa. This is not true of a display fabri-

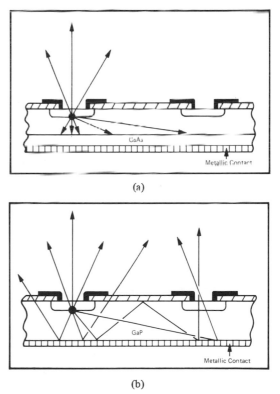

Fig. 9-23. Schematic cross-sectional view of a multiple-junction device illustrating optical crosstalk. In the case of a GaP-based device (b), the light emitted downward is reflected by the contact or back surface of the chip and after multiple reflections can be "emitted" in the vicinity of another junction, giving the appearance that the junction is "on." (Reference 3.)

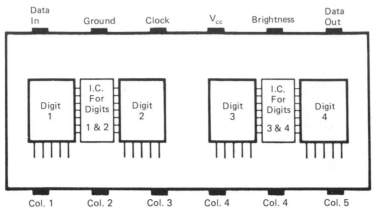

Fig. 9-24. Package layout for four-character LED alphanumeric display with on-board shift register and drive circuitry. The device contains two integrated circuits, one for each pair of characters. (After Reference 4.)

cated of material with an opaque substrate such as GaAs, as illustrated in the top portion of Fig. 9-23. Since high-performance yellow and green devices use transparent GaP substrates which not only have optical crosstalk but also cost three times as much as GaAs substrates, the monolithic alphanumeric technology is best suited to red-emitting LEDs only.

9.5.6 Numeric and Alphanumeric Displays with On-Board Integrated Circuits. A variety of LED displays can be obtained with on-board integrated circuit (OBIC) decoding capability. Figure 9-24 schematically illustrates the layout of an OBIC LED display. The incorporation of the integrated circuit into the display greatly simplifies the interfacing and reduces the number of interconnections required to activate the display. Devices of the type illustrated in Fig. 9-24 are available commercially, as shown in Fig. 9-25 which illustrates a Hewlett Packard yellow alphanumeric diode display. More complicated LED displays with on-board microprocessor-based control circuitry can also be obtained, as illustrated in Fig. 9-26 which shows a 40-character segment-type alphanumeric display system. A 32-character dot matrix-type alphanumeric display system using devices of the type illustrated in Figs. 9-24 and 9-25 can be obtained. Many numeric and alphanumeric LED displays can also be obtained in hermetic packages.

9.5.7 Large-Area x-y Addressable LED Arrays. Large-area flat-panel hybrid LED displays have been fabricated by a number of research labo-

ratories. These displays are not available as standard commercial products. A recent example of a green-emitting display is that fabricated by workers at Sanyo who made a TV size (12 cm × 16 cm) prototype using 38,400 discrete LED chips wire-bonded into an array with a resolution of 20 lines per cm.[35] The display surface brightness with 40 fL for the 38,400-element array, but the brightness could be increased to 70 fL if the resolution was reduced to 12.5 lines/mm and reflector caps were placed around the LEDs. The Sanyo displays have been used to display a TV picture which has acceptable viewing characteristics.

A flat-panel green-emitting display designed for aircraft applications has been announced and is shown in operation in Fig. 9-27. The display was built by Litton Systems (Canada) and Optotek Limited and was developed under a joint program between the U.S. Air Force and the Canadian Department of Industry, Trade, and Commerce.[36] The display has a resolution of 64 lines per inch, is 4 in. by 3 in. in size, and contains over 49,000 light-emitting junctions. The display is composed of four 1 in. × 5 in. modules (with 1 in. × 3 in. display sections) which are edge stackable. The modules have an average brightness of 30 fL and require a total of 49.7 watts of power including drive circuitry assuming a 20% "on" condition. The modules each consist of a mosaic of monolithic $\frac{1}{4}$ in. × $\frac{1}{4}$ in. GaP LED arrays, with each array containing 16 × 16 light-emitting junctions. The arrays are x-y addressable, with the "row" isolation obtained by sawing through the epitaxial layer, as shown in Fig. 9-28. The "column" contacts

Fig. 9-25. Hewlett Packard yellow alphanumeric light-emitting diode display with on-board deciding. The devices are end stackable.

are made by stitch bonding from row to row. This process can be contrasted with the planar process shown in Fig. 9-22 which eliminated the sawing and wire bonding, but involved more complicated materials growth and wafer processing technology.

Several large-area LED displays have been fabricated by Litton Data Systems.[37] The Litton group has assembled arrays using mosaics of monolithic arrays as well as assemblies of discrete LED chips. One of the more impressive devices is a multicolor display utilizing both red and green discrete LED chips. The display is designed to be read through a tactical military map, and the development was sponsored by the U.S. Army. Each data point of the display

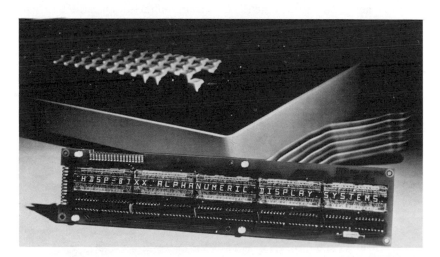

Fig. 9-26. Hewlett Packard 40-character alphanumeric segment display system with on-board microprocessor-based control circuitry.

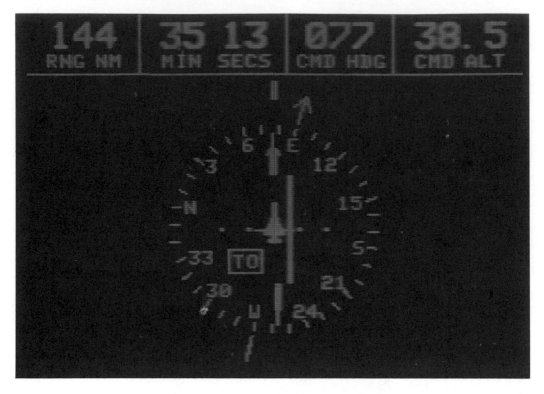

Fig. 9-27. Green-emitting x-y addressable aircraft display.

has both a red and a green chip, and, by adjusting the relative current through the chips, any color from red through green can be realized. The display is composed of edge-stackable modules each of which is 1.5 in. X 3.0 in. and contains 2,048 display points (4,096 LED chips) with a resolution of 22 lines/inch. The module requires 4 watts of power at 10% display loading. Display systems containing as many as eight modules have been fabricated, and ultimately the plan is to assemble 1 meter X 1 meter systems. The Litton group has also

assembled single-color systems, with 33 lines/inch resolution, which are mounted in modules similar to the multicolor displays. Figure 9-29 is a photograph of a Litton display module, and Fig. 9-30 illustrates the module components. Figure 9-31 shows the Litton display system in operation.

LED displays are suitable for aircraft and other military applications in part because of their high reliability. The mean time between failure of an LED flat-panel display is estimated at 10,000 hours, and the LED display generally

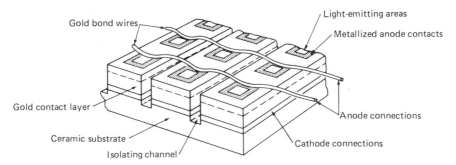

Fig. 9-28. Construction detail of display shown in Fig. 9-27. LED isolation is achieved by saw cuts through the epitaxial layer. (Reference 36.)

Fig. 9-29. Litton Data Systems' multicolor flat-panel LED array module, which has a resolution of 22 lines/inch.

fails gradually, as opposed to the catastrophic failure which can occur with tube displays. LED arrays utilizing modular construction also have the advantage that, by edge-stacking modules, arrays of arbitrary size and shape can be constructed. The major technical disadvantage of flat-panel LED displays is their higher power consumption.

The LED displays described in this section are outstanding technical achievements and represent the state of the art in hybrid LED display technology. They clearly demonstrate the technical feasibility of large-area LED displays. Nevertheless, it is likely that the cost of display systems of this type will remain high enough that their use will be limited to demand-

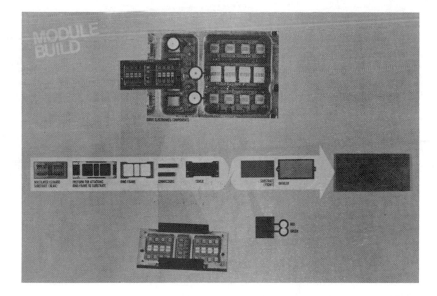

Fig. 9-30. Construction detail of the Litton display shown in Fig. 9-29. The display is constructed of 1.5 in. × 3 in. modules, each containing an LED panel that consists of a hybrid assembly of 4,096 discrete LED chips, half red and half green, resulting in 2,048 data points (32 × 64 lines).

Fig. 9-31. Litton's large-surface display utilizing LED modules of the type shown in Figs. 9-29 and 9-30. Modules are viewed through the military map and can display alphanumeric and graphical information used in planning and coordinating military maneuvers.

ing applications and environments requiring the unique advantages that LEDs have to offer.

9.5.8 Large-Area Displays Using a Combination of LED Products. One advantage of LED dis-

plays is that there are a variety of types, colors, shapes, sizes, and fonts available, and these devices can be combined in any desired arrangement to form a large-area information display for a variety of applications. One example is the

Fig. 9-32. LED automotive display module utilizing a variety of types of LED display products, resulting in a large-area multicolor display which with appropriate filtering is visible in the high-ambient automotive environment.

automotive display shown in Fig. 9-32. This display combines numeric displays, bar-of-light displays, and bar-graph displays with a variety of colors and legends to form a large-area multi-color display with extremely good viewability and reliability.

9.6   LED PERFORMANCE PARAMETERS

The current-voltage characteristics for typical standard red-emitting LED lamps and for high-efficiency red, yellow, and green LED lamps with N doping grown on GaP substrate are shown in Fig. 9-33. It can be seen that both types of devices have typical diode character-istics and that the standard red devices begin to conduct at slightly below 1.5 volts, while the N-doped devices conduct at approximately 1.6 volts. This difference is due to the higher energy gap present in the N-doped devices, which is necessary for shorter wavelengths. An-other difference is that the N-doped devices on GaP substrates have significantly higher dynamic resistances than the red emitters on GaAs sub-strates. Typically, the dynamic series resistance for a red lamp on GaAs is on the order of 3 ohms, whereas the dynamic series resistance for the GaP-based lamps is in the range of 20 to

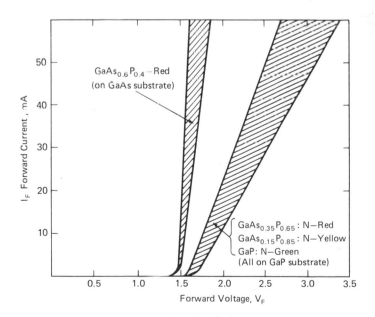

Fig. 9-33. Typical current-voltage characteristics for red, yellow, and green-emitting LEDs. The red-emitting $GaAs_{0.6}P_{0.4}$ on GaAs substrate has a typical dynamic resistance of 2-6 ohms. The nitrogen-doped devices on GaP substrates have typical dynamic resistances of 20-30 ohms.

30 ohms, because GaP and indirect-bandgap GaAsP alloys have lower electrical conductivity than GaAs. In a high-current or pulsed operating mode, this can result in increased drive requirements for the GaP-based lamps. Generally, however, all types of LEDs are TTL-compatible for both current and voltage and can be driven with a wide range of commercially available interface circuits.

Reverse currents are low for all types of LEDs. Typically, the reverse current is $<10\ \mu A$ for reverse voltages of <10 volts. Some types of devices will withstand reverse voltages of >50 volts with leakage currents less than $0.01\ \mu A$. The response speed of LEDs is also high, typically in the 10–200 nanosecond range, and there is practically no duty cycle limitation. All of the factors cited above contribute to making the LED highly suited for multiplexing and x-y matrix addressing.

It is desirable to use the maximum peak forward drive current possible consistent with system requirements and with the limitations of the LED chips because, as discussed in Section 9.4, and indicated in Figs. 9-10 and 9-34, the relative efficiency increases with increasing peak segment current. Figure 9-34 is drawn for a GaAsP:N red display. Similar behavior is also characteristic of yellow, green, and $GaAs_{0.6}P_{0.4}$ and GaAlAs red displays. It can be seen that the efficiency is increased by more than 60% by operating the display at a current of 50 milliamps per segment instead of 5 milliamps per segment. The efficiency increases even more rapidly with current if the effects of heating are eliminated by using short pulses and low duty cycles, as shown by the dashed curve in Fig. 9-34. Peak brightnesses in excess of 70,000 fL can be obtained for currents on the order of 1 amp/segment. However, this case is imprac-

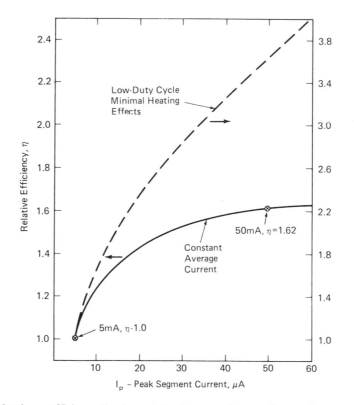

Fig. 9-34. Relative luminous efficiency (luminous intensity per unit current) vs. peak current per segment for a $GaAs_{0.35}P_{0.65}$:N high-efficiency red-emitting display. The lower curve is obtained if the duty cycle is adjusted to maintain a constant average current in the display; it is applicable for most display applications. The upper (dashed) curve is obtained if narrow pulses and low duty cycles are utilized to eliminate heating effects. An increase in efficiency with increasing current is also characteristic of yellow, green, and standard red ($GaAs_{0.6}P_{0.4}$) displays.

tical for display applications because the duty cycle must be very short and the average luminous intensity is too low. The efficiency increase can be defined by the equation

$$\eta = \eta_0 \left(\frac{I}{I_0} \right)^x$$

where η and I are the final efficiency and current, η_0 and I_0 are the initial efficiency and current, and x is empirically found to be in the range 0.2–0.4 for most situations.

The improvement that can be obtained in the time-average luminous intensity per segment is illustrated in Fig. 9-35, where time-average luminous intensity per segment is plotted as a function of the average current per segment. It can be seen that by maintaining an average of a 6 mA per segment, but strobing with a $\frac{1}{6}$ duty factor, a 560-μcd intensity is obtained as compared to 380 μcd for d-c operation. Or, considering the situation from the standpoint of saving power rather than increasing intensity, we see that if an intensity of 560 μcd is required, the display can be operated at an average of only 6 mA per segment instead of a 8.2 mA per segment if dc operation is utilized. This results in a power dissipation per segment of 14.1

mW per segment for the dc case as compared to only 10.3 mW for the strobed case, a substantial savings in power. Typical average drive currents for LED displays are 1–2 mA per segment for small red GaAsP monolithic displays, 5–10 mA per segment for large stretched segment displays using GaAsAs or GaAsP:N, and 10–20 mA for stretched segment displays with red GaAsP material.

The improved materials performance described in Section 9.4 has resulted in new families of devices designed to operate at increasingly lower currents. Red, yellow, and green devices can now be obtained that have intensities of 2–5 mcd for drive currents in the 2–3 mA range. Devices of this type can substantially reduce power requirements and drive circuitry costs, as indicated in Fig. 9-36. Most CMOS and low-power TTL circuits use an external driver to power a 10-mA LED because they can drive only 1–3 mA. Low-current devices can be driven directly from the gate. Since there is no need for an external driver, $0.05 can be saved in high-volume component costs, plus extra space on the circuitboard.

The increased performance of LED displays has also made it possible to effectively utilize LEDs for a variety of high-brightness ambient applications such as avionics and automotive. LED displays with optimized combinations of package reflectance and appropriate contrast enhancement filters are now being used in ambients up to 107,000 lm/m² (10,000 footcandles). In these bright ambients, key factors that affect the readability of the display are luminance contrast, chrominance contrast, and front surface reflections. Chrominance contrast, which is the color difference between the illuminated LED and the background, is a critical factor that must not be neglected when comparing LEDs to other technologies. In many applications, the chrominance contrast, which is strongly affected by the filter utilized, can have a several-times-larger effect in determining display readability than luminance contrast. The designer should evaluate the total display discrimination index, which is a combination of luminance and chrominance contrast. This issue is discussed more fully in Reference 2.

If it is desired to drive LEDs to high current levels for applications in high-brightness ambi-

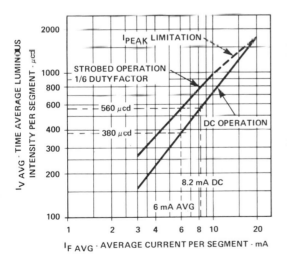

Fig. 9-35. Comparison of time-average luminous intensity as obtained from strobed and dc operation with 6-mA, time-averaged current in both cases for a GaAs$_{0.35}$P$_{0.65}$:N high-efficiency red display. Strobed operation results in higher light output for the same average current or equivalent light output at lower average current. (Reference 2.)

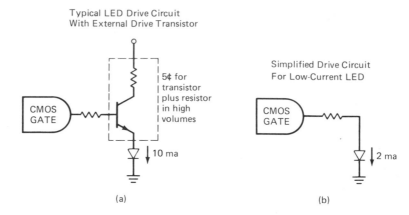

Fig. 9-36. The typical LED drive circuit shown in (a) which utilizes an external drive transistor and extra resistor. This type of circuit has usually been necessary because the LED current requirements exceeded the capability of the CMOS or low-power TTL integrated circuit gates. With the improved low-current LED products now available, simplified and less expensive drive circuits like that shown in (b) can often be substituted.

ents, or where multiplexing is appropriate, one needs to understand the device limitations. Increased current densities cause increased heating in the device, which results in degradation of the light output. The relationship between efficiency and temperature for several types of LEDs is shown in Fig. 9-37. Typically, the light output decreases by approximately 1% for each degree centigrade the temperature is increased. Even if the duty factor is adjusted to keep the average power constant, excessive heating can occur with increased peak current if the pulses are too wide. The colors of LED displays also change with temperature due to the fact that the energy gap decreases. Typically, the emission shifts to longer wavelengths at the rate of ~0.1 nm $°C^{-1}$. Application problems of this type are discussed in detail in Reference 2.

In general, LED displays are more tolerant of temperature variations and extremes than

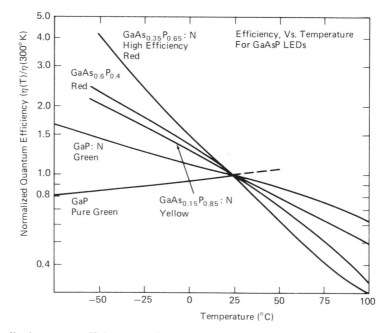

Fig. 9-37. Normalized quantum efficiency as a function of temperature for a variety of types of LEDs. The efficiency is normalized to the room-temperature value. The devices, without nitrogen doping except for the green GaP device, all decrease in performance with increasing temperature. The rate of decrease varies from about 0.5%/°C to 2%/°C depending on the color of the device.

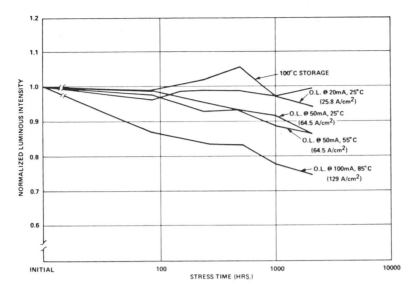

Fig. 9-38. Normalized luminous intensity vs. operating life (O.L.) and high-temperature storage for GaAs$_{0.6}$P$_{0.4}$ red-emitting devices. (Reference 2.)

most other technologies. The maximum operating temperature is typically 70°C to 85°C and the minimum operating temperature −40°C to −20°C. LEDs are also outstanding with regard to thermal shock and temperature cycling.

LEDs are also extremely rugged, and insensitive to most hostile environments. They are excellent with regard to vibration, mechanical shock, and impact resistance. They therefore provide clearcut advantages for such applications as automobiles, trucks, farm equipment, military and commercial avionics, and military ground equipment. Many LED displays are available in hermetic packages for extreme operating requirements.

The typical failure mode for LEDs is gradual degradation as opposed to catastrophic failure. The gradual-degradation characteristic of LEDs is a function of both temperature and drive-current density. Typical LED degradation curves for standard red devices and for GaAsP:N and GaP:N red, yellow, and green devices are shown in Figs. 9-38 and 9-39 for a variety of operating conditions. It can be seen that devices degrade

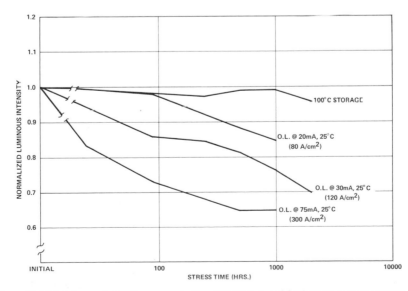

Fig. 9-39. Normalized luminous intensity vs. operating life (O.L.) and high-temperature storage for typical N-doped red, yellow, and green LEDs. (Reference 2.)

more rapidly for increased current densities and at higher temperatures. It can also be seen that the devices with N doping degrade more rapidly than the standard red emitters. However, at reasonable operating current densities, all the devices have extrapolated lifetimes to half-brightness in the range of 10^5 hours or greater. The mechanisms responsible for LED degradation are not well understood, nor is the reason for the difference between the degradation rates of standard red and N-doped devices.

For most applications, LED lifetime is more than adequate, and when this is coupled with the ruggedness and environment-resistant packaging that LEDs have to offer, the overall reliability of LED display products is excellent compared to other technologies.

9.7 MATERIALS AND PROCESSES

9.7.1 Substrate Preparation. In this section, techniques used to grow and process materials for LED fabrication will be summarized. The preparation of materials for LEDs begins with the growth of the single-crystal substrate. The substrates utilized for the GaAsP and GaP material systems are GaAs for direct bandgap GaAs $_{0.6}P_{0.4}$ and GaAlAs for red emitters, and GaP for GaAs$_{0.35}P_{0.65}$:N (red), GaAs$_{0.15}P_{0.85}$ (yellow), and GaP (green) emitters. GaP is employed for the latter three types of devices both because it offers a better lattice match than GaAs for these materials, and because it is transparent to the emitted radiation so higher efficiency can be achieved.

GaP is grown using the liquid-encapsulated Czochralski (LEC) technique illustrated schematically in Fig. 9-40.[38] This technique is similar to the Czochralski technique utilized to pull single-crystal silicon for integrated circuits except that in order to prevent the escape of the volatile phosphorus, a boric oxide layer must be used to cover the molten GaP and the crystal pulling must be carried out at elevated chamber pressures. Utilizing this technique, it is possible to routinely grow ingots two inches in diameter or larger.

GaAs is commonly grown using the LEC technique, similar to GaP, or the gradient freeze technique. The latter technique involves imposing a temperature gradient across the melt, then lower-

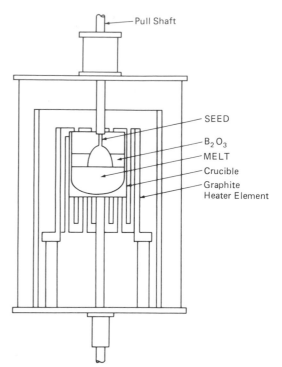

Fig. 9-40. Schematic diagram of liquid-encapsulated Czochralski crystal-pulling system. The crystal is pulled up through the molten boric oxide (B_2O_3) layer that covers the GaP melt. The B_2O_3 prevents the volatile phosphorus from escaping from the melt.

ing the temperature. The single crystal starts to freeze at a seed placed at one end of the melt and the freezing interface passes across the melt as the temperature is lowered. The cost of GaAs is approximately three times lower than GaP since the growth can be carried out at lower pressure and thus requires equipment which is less complicated and expensive.

Germanium is sometimes used instead of GaAs as a substrate for GaAs$_{0.6}P_{0.4}$. Germanium is grown using the Czochralski technique, and can be grown inexpensively, in large diameters, and with low defect densities. However, if trace quantities of germanium are incorporated into the growing epitaxial film, poor LED performance is obtained. This necessitates a sealing coat on the back of the wafers which complicates the epitaxial deposition step and is the primary reason that GaAs remains the dominant substrate material.

9.7.2 Epitaxial Deposition. After the ingot is grown, it is sliced and polished and placed in an

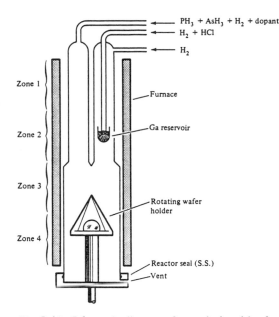

PH$_3$ + AsH$_3$ + H$_2$ + dopant

H$_2$ + HCl

H$_2$

Zone 1

Furnace

Ga reservoir

Zone 2

Zone 3

Rotating wafer holder

Zone 4

Reactor seal (S.S.)

Vent

Fig. 9-41. Schematic diagram of a vertical multiwafer reactor for the vapor-phase epitaxial (VPE) deposition GaAs$_{1-x}$P$_x$.

epitaxial reactor for the growth of the epitaxial film which constitutes the active portion of the device. The epitaxial growth can be carried out using either the vapor-phase epitaxial (VPE)[7,39] or liquid-phase epitaxial (LPE) techniques.[40,41] The VPE technique is illustrated in Fig. 9-41 and the LPE technique in Fig. 9-42.

In the case of the VPE technique, the substrate wafers are placed on a holder in the hot zone of the reactor, and, for the case of GaAsP, the reactant gases containing gallium, arsenic, and phosphorus, as well as the doping gases, are passed over the wafers. The gallium is transported by passing HCL gas over molten gallium, forming a gallium chloride. When the gas mixture reaches the substrate surface, a single-crystal epitaxial layer will be grown if the appropriate gas flows and temperatures are provided. Growth rates are

typically in the range of one micron per minute. The technology used for vapor-phase growth shares much in common with the technology used to grow silicon epitaxial layers. This has helped GaAsP VPE become established as a viable commercial process. Generally speaking, adequate control of layer thicknesses, surface quality, and carrier concentration can be routinely obtained.

In the late 1970s an alternate VPE technology, organometallic VPE (OM-VPE), which is also called metal organic chemical vapor deposition (MO-CVD), became important in the fabrication of injection lasers using the aluminum gallium arsenide (AlGaAs) ternary material system.[42,43] The organometallic technique uses trimethyl gallium instead of gallium chloride to transport gallium. The OM-VPE system has the advantage that the quartz reactor walls are cold and the substrate is hot, generally heated by RF induction heating. As a consequence, aluminum compounds can be grown. In the case of the gallium chloride transport system, aluminum compounds cannot be easily grown because the reactor walls are hot and if aluminum is present it reacts with the reactor walls, resulting in destruction of the reactor and the incorporation of impurities in the growing film. Using OM-VPE it may be possible to extend the excellent infrared results which have been obtained into the visible spectral regions, making red-emitting AlGaAs an even more serious contender for the commercial market. In addition, the OM-VPE technique may yield improved results for GaP:N or for a variety of other binary, ternary, and quaternary systems which have heretofore been difficult to grow and/or have yielded disappointing results.

In the case of LPE growth, as illustrated in Fig. 9-42, the substrate wafer is placed in con-

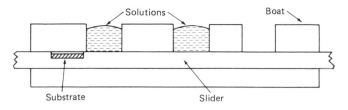

Solutions

Boat

Substrate

Slider

Fig. 9-42. Schematic diagram of a liquid-phase epitaxial (LPE) growth system of the type used for the growth of GaP red- and green-emitting LEDs. The substrate wafer (or wafers) is mounted so as to be brought into contact with gallium solutions that contain the appropriate dopants for achieving growth of the p-n junction.

tact with a gallium melt that is saturated with GaP and the appropriate dopants. When the melt in contact with the wafer is cooled, growth is nucleated on the substrate surface and an epitaxial film is formed. This growth technique has been widely utilized in a variety of laboratories around the world and has resulted in LEDs with some of the highest performance values to date.

On a small scale, the LPE growth technique is easier to utilize than the VPE technique since the handling of toxic gases and the precision control of gas flow is not required. In addition, the gallium melt acts as a "getter" to collect trace amounts of unintentional impurities which may be present in the growth chamber and which, if incorporated in the epitaxial film, could give rise to nonradiative recombination centers. It is straightforward to establish a basic LPE growth system, and LEDs with reasonable performance can generally be produced with a nominal amount of process optimization.

VPE systems, on the other hand, are more complicated, requiring the handling and precise control of a variety of highly toxic gases, and, since no gallium melt is in contact with the substrate to act as a "getter," the cleanliness of techniques utilized in assembling and operating VPE systems are critical. However, once established and properly maintained, VPE growth is a proven high-volume commercial technique. LPE systems have been, until recent years, primarily utilized for research. However, diligent effort in a variety of laboratories has resulted in LPE growth systems capable of producing substantial quantities of LED material which is now commercially available in the form of red- and green-emitting devices, chips, and wafers.

Two significant differences between LPE and VPE material are surface quality and junction formation technique. VPE surfaces tend to be smoother than the LPE grown surfaces, permitting high-resolution photographic fabrication techniques. This is generally not critical for LED devices, although it could be a factor if larger-area monolithic arrays become important. The junction formation for high-efficiency LPE devices is accomplished by growing rather than diffusing the p-n junction. Typically an n-type LPE layer is grown first, followed by the growth of a p-type LPE layer. The p-n junction covers the entire surface of the wafer. In the case of VPE

devices, only an n-type epitaxial film is grown. The p region is formed by diffusion into the n-type region, and the shape of the diffused region can be controlled by photolithographic techniques. The ability to control the shape of the junction using photolithographic techniques is not critical for discrete emitters but is important for monolithic devices. If monolithic devices are fabricated using LPE-grown junctions, it is necessary to etch mesas or saw grooves to define the desired p-region geometry. Etching and sawing are difficult processes to control and result in nonplanar surfaces which have historically given rise to metallization continuity problems.

Cross sections of chips formed from VPE wafers are shown in Fig. 9-8. The VPE epitaxial structure illustrated in Fig. 9-8 consists of the substrate, followed by the growth of a GaAsP taper region in which the alloy composition is changed continuously from that of the substrate to the alloy composition required to achieve the color of emission desired. Following the growth of the taper region, a 10–40-micron thick region of constant composition is grown. The growth of the taper region is required because the arsenic and phosphorus atoms have different sizes so that the GaAsP lattice spacing is different than the GaAs or GaP substrate. The stress in the lattice associated with this difference must be accommodated for by gradually changing the crystal lattice composition or poor LED performance is obtained. In the case of nitrogen-doped devices, nitrogen is introduced during the growth of the final 20 microns of the epitaxial structure, as shown in Fig. 9-8b.

The wafer fabrication process includes diffusion, backlapping, metallization, and dicing. The diffusion step is not required for LPE devices.

9.7.3 Wafer Fabrication. Following the growth of the epitaxial structure, a silicon nitride film is deposited over the surface of the wafer and the areas to be diffused are defined photolithographically. After removing the nitride from the regions to be diffused, the wafer is placed in a diffusion furnace. The source is usually metallic Zn or zinc arsenide. Typically the diffusion is carried out at a temperature of $800°$ for a period of an hour or more. The diffusion step is critical in the formation of high-performance and high-reliability devices. The light output can be lower by more than an order of magnitude if inappropriate diffusion parameters are employed. After diffusion, the wafers are removed from the furnace, lapped to the appropriate thickness, typically 6 to 8 mils, and metallized on both sides.

In the case of devices on a transparent GaP substrate, the metallization on the back side of the wafer is important because it is desirable to have a reflecting contact, as discussed in Section 9.3.5. The reflecting contact can be formed by depositing an SiO_2 film on the back side of the wafer, photolithographically cutting holes in the SiO_2, and evaporating Au-Ge alloy on the SiO_2 and into the holes. This metal film is then alloyed into the back side of the wafer. Typically the holes in the SiO_2 constitute approximately 20% of the back side. These alloyed regions are non-reflecting, but the 80% of the back side that is coated with SiO_2 and Au-Ge alloy is reflecting. The p-type contact is formed by evaporating an Au-Ge or Au-Zn film. This must also be alloyed to obtain good ohmic contact.

In the case of GaAsP devices on GaAs substrates, the contacting process is less complicated. In this case, the back contact need not be reflecting, so an Au-Ge film may be evaporated over the entire back surface of the wafer and be alloyed. The p-type contact for GaAsP is generally aluminum. The reason that Au-Ge or Au-Zn is used on GaP devices is that it is more difficult to form a good ohmic contact to wide-bandgap semiconductor material and aluminum does not form a good ohmic contact on GaP or phosphorus-rich GaAsP alloy compositions unless special techniques are employed to increase the p-type surfaces doping concentration.

After the wafer fabrication is complete, the wafers are tested and then cut into individual chips. The dicing operation is accomplished using a scriber, dicing saw, or laser depending upon the device type involved. Generally, the diced wafers are attached to an expandable membrane which facilitates high-speed device assembly.

In the past LED chips were assembled into devices using labor-intensive techniques. However, this is changing rapidly as a higher degree of automation becomes increasingly more desirable and feasible. Reasons for this include labor rates, higher volumes, increased standardization of LED products, and the availability of improved automated assembly equipment and techniques for semiconductor devices. Automated assembly can be expected to result in lower costs and increased product uniformity and reliability.

9.8 SUMMARY AND CONCLUSIONS

Many types of LED displays are currently available commercially ranging from the simple discrete indicator lamps to 40-character alphanumeric displays with on-board electronics. In the future, it can be expected that LED performance and reliability will increase, resulting in larger and more complex displays with improved specifications. Continued development of new types of LED displays can be anticipated particularly in the alphanumeric area. Interfacing with microprocessors is a key issue. It is not clear yet what amount of on-board electronics is most suitable, but a variety of devices can be expected to be developed in an effort to capture a segment of this rapidly growing market.

It is clear that different applications will require different levels of on-board decoding electronics in the display. For consumer-oriented products produced in high volume, it is likely that a minimum of electronics will be desired on the display. Instead, specialized interface electronics will be designed by the user to do a specific task and to minimize the component count and cost. However, for many applications with intermediate volume, a display which can be directly interfaced with a microprocessor with a minimum of engineering will be desired, since the product volume will not justify the engineering cost of designing a specialized display interface.

In competition with other technologies, LEDs have a variety of advantages and disadvantages which are summarized in Tables 9-3 and 9-4. In the long term, it is not clear what percent of the display market LED devices will occupy as LED and other display technologies continue to improve. It is clear that the display market is increasing as microprocessors continue to permeate society and as new applications for displays are developed. As a result of this growth of the display market and the inherent advantages of LED displays for many applications, it is likely that the market for LED display devices will con-

Table 9-3. Advantages of LED Displays

Looks/Format

Active, light-emitting
Color range (red, orange, yellow, green)
Size range (2–25 mm)
Excellent viewing angle compared to liquid crystals
Flexible format
Multiple colors within a single device
Sunlight viewability with appropriate filter

Performance

Good efficiency of 1–6 L/A at $V_F \cong 2V$
Large dynamic range (1–300 mA) without saturation
High response speed (10–20 nsec)

Interface Electronics

Diode characteristics simplifies matrix addressing
TTL-compatible for current and voltage
Wide range of available interface circuits
Practically no duty cycle limitation
Direct drive from CMOS for some applications

Reliability/Environmental Range

Rugged, not fragile
Generally gradual noncatastrophic failure
Wide temperature range
Long operating life
Hermetic (optional)

tinue to grow. Future trends in LED technology are summarized in Table 9-5.

LED displays, which are an "interconnect technology" requiring a die attach and wire bond for each picture element, compete most effectively for applications in which the information density is relatively low, below approximately 100 picture elements/cm². As the information density increases, the "area technologies" such as CRTs and LCDs generally become more cost-effective. It seems unlikely that LED displays will compete effectively on a cost/performance basis for a major segment of the large-area flat-panel display market. However, high-performance LED displays of this type have been successfully fabricated. In applications where cost is not a critical parameter and the special advantages of LED displays such as high brightness, reliability, and multicolor capability are desired, LEDs can be the most effective technology. One past and present limitation of high-density flat-panel LED displays is that the heat dissipation of the panel is relatively high. As the LED material performance continues to improve, it should be possible to obtain excellent brightness with substantially less heat buildup. The panel cost should also decrease as LED materials technology continues to evolve and automated assembly techniques are developed.

An example of the continuing decrease in LED costs is shown in Fig. 9-43, where the average high-volume market price of discrete LED lamps is plotted as a function of time. Lamp volume is also shown, which is steadily increasing. Lamps are plotted instead of more complex LED displays primarily because a good data base exists and the product configuration is relatively invariant. In addition, a discrete lamp represents one pixel of information, so lamp costs can be used as a rough, and generally worst-case, estimate of the cost/pixel for more complex LED display products such as an x-y addressable matrix display. Each lamp requires one chip and the die-attach and wire-bond steps which are required for each pixel of a larger display. In addition, each lamp requires a complete scaled package.

The range of prices in 1982, as indicated by the error bar, goes from 5 cents to 25 cents. The 5-cent price represents a standard lamp with minimal specifications in very high volume. The 25-cent price typically represents a high-performance device with a variety of possible package

Table 9-4. Disadvantages of LEDs

High power consumption (compared with passive displays)

High peak currents for long digit strings

Restricted color range (no commercially viable blue)

Cost of complex assembly (one die attach and/or wire bond per display element)

Table 9-5.   Trends in LED Display Technology

Higher brightness and lower power requirements for
  all colors

Improved reliability

Reduced cost due to automated assembly

More products with on-board electronics for microprocessor
  interfacing

Improved product uniformity due to better-controlled
  materials growth and device assembly techniques

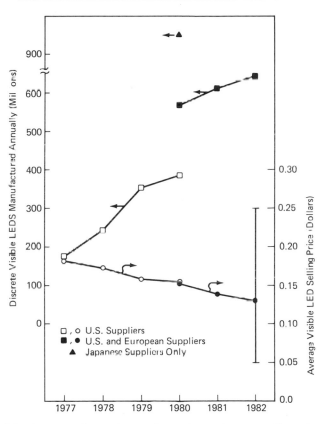

Fig. 9-43. Discrete visible LEDs manufactured annually vs. time and average selling price vs. time. The U.S. and European data are obtained from the Semiconductor Institute of America (SIA) and represent worldwide distribution. The 1980 Japanese data is from the Electronic Industries Association of Japan (EIAJ). In 1982 the selling price averaged 13.2 cents but ranged from 5 cents to 25 cents, depending on device type and quantity. The general trends are increasing volume and decreasing price. More units are produced by Japanese suppliers than by U.S. and European suppliers combined.

requirements. Hermetic lamps would be still higher in price.

Based on Fig. 9-43 and other considerations, it is reasonable to expect that LED x-y matrix displays can be manufactured and sold for 5 cents/picture element. Since some economic advantage can be realized by assembling a large number of LED chips in a single package, the price could go lower. There is also a trend towards using smaller, and hence less expensive, chips as the automated equipment becomes more sophisticated. As a result, LED x-y matrix displays could become increasingly cost-effective for special applications where the other advantages of LEDs are required.

REFERENCES

1. Bergh, A. A., and Dean, P. J. *Light Emitting Diodes*. Oxford: Clarendon Press, 1976.

2. Gage, S. R.; Evans, D. L.; Hodapp, M. W.; Sorenson, H. O.; Jamison, R. E.; and Krause, R. L. *Optoelectronic Applications Manual and Supplement*. New York: McGraw-Hill Book Co., 1977.

3. Craford, M. G. "Recent Developments in Light-Emitting-Diode Technology." *IEEE Trans. Electron Devices*, 24(7), 935–943, July 1977.

4. Haitz, R. "Trends in LED Display Technology." *Proc. Electronic Components Conf.*, Washington, D.C., 2–10, May 1974.

5. Duke, C. B., and Holonyak, H., Jr. "Advances in Light Emitting Diodes." *Physics Today*, 26, December 1973.

6. Bhargava, R. N. "Recent Advances in Visible LED's." *IEEE Trans. Electron Devices*, 22 (9), 691–701, September 1975.

7. Craford, M. G., and Groves, W. O. "Vapor Phase Epitaxial Materials for LED Applications." *Proc. IEEE*, 61, 862–880, July 1973.

8. Craford, M. G. "Properties and Electroluminescence of the $GaAs_{1-x}P_x$ Ternary System." *Progress in Solid State Chemistry*, 8, 127–165, 1973.

9. Craford, M. G., and Holonyak, N., Jr., "The Optical Properties of the Nitrogen Isoelectronic Trap in $GaAs_{1-x}P_x$," *Optical Properties of Solids–New Developments*, edited by B. O. Seraphin, pp. 187–253. North Holland Publishing Company, 1976.

10. (a) Round, T., *Electrical World*, 309, 1907.
    (b) Lossev, O. V., *Telegraphia i Telefonia*, 18, 61, 1923.
    (c) Lossev, O. V., *Phil Mag.*, 6, 1024, 1928.

11. Holonyak, N., Jr., and Bevacqua, "Coherent (Visible) Light Emission from $GaAs_{1-x}P_x$ Junction." *Appl. Phys. Lett.*, 1, 82, 1962.

12. Thomas, D. G., and Hopfield, J. J. "Isoelectronic Traps Due to Nitrogen in Gallium-Phosphide." *Phys. Rev.*, 150, 680–689, 1966.

13. Logan, R. A., White, H. G., and Wiegmann, W. "Efficient Green Electroluminescence in Nitrogen-Doped GaP p-n Junctions." *Appl. Phys. Lett.*, 13, 139, 1968.

14. Groves, W. O.; Herzog, A. J.; and Craford, M. G. "The Effect of Nitrogen Doping on $GaAs_{1-x}P_x$ Electroluminescent Diodes." *Appl. Phys. Lett.*, 19, 184–186, September 1971.

15. Craford, M. G.; Shaw, R. W.; Groves, W. O.; and Herzog, A. H. "Radiative Recombination Mechanisms in GaAsP Diodes With and Without Nitrogen Doping." *J. Appl. Physics*, 43, 4075–4083, October 1972.

16. Craford, M. G.; Keune, D. L.; Groves, W. O.; and Herzog, A. H. "The Luminescent Properties of Nitrogen Doped GaAsP Light Emitting Diodes." *J. Electron. Mater.*, 2, 137–158, January 1973.

17. Campbell, J. C.; Holonyak, N., Jr.; Craford, M. G.; and Keune, D. L. "Band Structure Enhancement and Optimization of Radiative Recombination in $GaAs_{1-x}P_x$:N (and $In_{1-x}Ga_xP$:N). *J. Appl. Phys.*, 45, 4543–4553, October 1974.

18. Lang, D. V. "Deep-Level Transient Spectroscopy: A new method to characterize traps in semiconductors." *J. Appl. Phys.*, 45 (7), 3023–3032, July, 1974.

19. Henry, C. H. "Some Recent Fundamental Advances in Radiative and Nonradiative Transitions in Semiconductors," *J. of Luminescence*, 12/13, 47–56, 1976.

20. Nicklin, R.; Mobsby, C. D.; Lidgard, G.; and Hart, P. B. "Efficient Yellow Luminescence from Vapor Grown GaP with High Nitrogen Content," *J. Phys. C.*, 4, L344–L347, 1971.

21. Faulkner, R. A. "Toward a Theory of Isoelectronic Impurities in Semiconductors." *Phys. Rev.* 175, 991–1009, November 1968.

22. Holonyak, N., Jr., Campbell, J. C., Lee, M. H., Verdeyen, J. T., Johnson, W. L., Craford, M. G., and Finn, D. "Pumping of $GaAs_{1-x}P_x$:N (77°K, X < 0.53) by an Electron Beam from a Gas Plasmas." *J. Appl. Phys.*, 44, 5517–5521, December, 1973.

23. Bachrach, R. Z.; Joyce, W. B.; and Dixon, R. W. "Optical-Coupling Efficiency of GaP:N Green-Light-Emitting Diodes." *J. Appl. Phys.*, 44, 5458–5462, December 1973.

24. Rosenzweig, W.; Logan, R. A.; and Wiegmann, W. "Variable Hue GaP Diodes." *Sol. St. Electron.*, 14, 655, 1971.

25. Craven, R. A., and Finn, D. "The Sulfur-Related Trap in $GaAs_{1-x}P_x$." *J. Appl. Phys.*, 50 (10), 6334–6343, October 1979.

26. Brantley, W. A.; Lorimar, O. G.; Dapkus, P. D.; Haszko, S. E.; and Saul, R. H. "Effect of Dislocations on Green Electroluminescence Efficiency in GaP Grown by Liquid Phase Epitaxy." *J. Appl. Phys.*, 46 (6), 2629–2637, June 1975.

27. Stringfellow, G. B., and Kerps, D. "Green-Emitting Diodes in VPE GaP." *Sol. St. Electron.*, 18, 1019, 1975.

28. Rokshoer, P. J.; Huijbregts, J. M. P. L.; Van De Wiggert, W. M.; and DeKock, A. J. R. "Growth of Dislocation-Free Gallium-Phosphide Crystals from a Stoichiometric Melt." *J. of Crystal Growth*, 40, 6–12, 1977.

29. Holonyak, N., Jr.; Stillman, G. E.; and Wolfe, C. M. "Compound Semiconductors." *J. Electrochem. Soc.*, 125 (12), 487c–499c, December 1978.

30. Matsunami, H.; Ikeda, M.; Suzuki, A.; and Tanaka, T. "SiC Blue LED's by Liquid Phase-Epitaxy." *IEEE Trans. Electron Devices*, 24 (7), 958–961, July 1977.

31. Pankove, J. I. "Low-Voltage Blue Electroluminescence in GaN." *Proc. SID*, 16, 140–143, 1975.

32. Keune, D. L.; Craford, M. G.; Graves, W. O.; and Johnson, A. D. "Monolithic GaP Green-Emitting LED Matrix Addressable Arrays." *IEEE Trans. Electron Devices*, 20, 1074–1077, 1973.

33. Frescura, B. L.; Luechinger, H.; and Bittman, C. A. "Large High-Density Monolithic x-y Addressable Arrays for Flat Panel LED Displays." *IEEE Trans. Electron Devices*, 24, 891–897, 1977.

34. Barnett, A. M.; Galginaitis, S. V., and Huemann, F. "GaP Planar Monolithic Matrix-Addressable Displays." *IEEE Trans. Electron Devices*, 18, 638–641, 1971.

35. Nioka, T.; Koroda, S.; Yonei, H.; and Takesada, H. "A High-Brightness GaP Green LED Flat-Panel Device for Character and TV Display." *IEEE Trans. Electron Devices*, 26, 1182–1186, August 1979.

36. Burnette, K. T., and Melnick, W. "Multi-Mode Matrix (MMM) Flat-Panel LED Varactor-Graphic Concept Demonstrator Display." *Proceedings of the Society for Information Display*, Vol. 21:113–126, 1980.

37. Obright, N. A., Litton Data Systems. Private Communications.

38. Grabmaier, B. C., and Grabmaier, J. G., "Dislocation-Free GaAs by the Liquid Encapsulation Technique." *J. Crystal Growth, 13/14*, 635–639, 1972.

39. Burd, J. W. "A Multi-Wafer Growth System for the Epitaxial Deposition of GaAs and $GaAs_{1-x}P_x$." *Trans. Met. Soc.* AIME, 245, 571–576, March 1969.

40. Niina, T., and Yamaguchi, T. "An Improved Liquid Phase Epitaxial Growth Method for Mass Production of GaP Green LED's." *IEEE Trans. Electron Devices*, 24, 946–950, 1977.

41. Gillessen, K., and Marshall, A. J. "Correlation Between Location of Light Emission and Spatial Nitrogen Distribution in GaP LED's." *IEEE Trans. Electron Devices*, 26, 1186–1189, August 1979.

42. Dupuis, R. D., and Dapkus, P. D. "Room Temperature Operation of $Ga_{(1-x)}Al_xAs/GaAs$ Double-Heterostructure Lasers Grown by Metalorganic Chemical Vapor Deposition." *Appl. Phys. Lett.*, 31 (7), 466–468, 1977.

43. Dupuis, R. D., and Dapkus, P. D. "$Ga_{(1-x)}Al_x As/Ga_{(1-y)}Al_y As$ Double-Heterostructure Room-Temperature Lasers Grown by Metalorganic Chemical Vapor Deposition." *Appl. Phys. Lett.*, 31 (12), 839–841, 1977.

44. Moon, R. L. "Liquid Phase Epitaxy," *Crystal Growth*, 2nd ed., edited by B. R. Pamplin. Elmsford, NY: Pergamon Press, 1980.

45. Kellert, F. G. "EL Lifetimes in VPE $GaAs_{1-x}P_x$ and LPE GaP:N LED's." Presented at the 1982 IEEE Specialist Conference on Light Emitting Diodes and Photodetectors, Ottawa-Hull, Canada.

46. Nishizawa, J., and Suto, K. "Minority-Carrier Lifetime Measurements of Efficient GaAlAs p-n Heterojunctions." *J. Appl. Phys.*, 48(8), 3484–3445, 1977.

47. Ohki, Y.; Toyoda, Y.; Kobuyasi, H.; and Akasaki, I. "Fabrication and Properties of a Practical Blue-Emitting GaN m-i-s Diode." Conference of GaAs and Related Compounds, Oiso, Japan, 1981, p. 479.

48. Holonyak, N., and Hess, K. "Quantum-Well Heterostructure Lasers," in *Synthetic Modulated Materials*, edited by L. L. Chang and B. C. Giessen. New York: Academic Press, 1983.

49. Camras, M. D.; Holonyak, N., Jr.; Hess, K.; Coleman, J. J.; Burnham, R. D.; and Scifres, D. R. "High Energy $Al_xGa_{1-x}As$ (ocxco. 1) Quantum-Well Heterostructure Laser Operation." *Appl. Phys. Lett.*, 41(4), 317–319, 1982.

50. Burnham, R. D.; Scifres, D. R.; and Streifer, W. *Electron. Lett.*, 18, 507–509, 1982.

51. Scifres, D. R.; Burnham, R. D.; and Streifer, W. *Appl. Phys. Lett.*, 41, 118–120, 1982.

52. Osbourn, G. C.; Biefeld, R. M.; and Gourley, P. L. "A GaAsP/GaP Strained-Layer Superlattice." *Appl. Phys. Lett.*, 41(2), 172–174, 1982.

10

PLASMA DISPLAYS

LARRY F. WEBER, *University of Illinois*

10.1 INTRODUCTION

Plasma displays utilize the physical phenomena of the gas discharge (they are also called *gas-discharge displays*). They have found widespread commercial success in sizes ranging from small seven-segment numeric indicators to large graphics displays with two million pixels and diagonals of 1 meter (39 inches). In early 1983, plasma displays were the dominant large flat-panel display technology on the market. This success is due to many desirable properties of gas discharges that lend themselves to flat-panel displays and that made plasma displays one of the earliest technologies to achieve practical matrix addressability. Each of these attributes will be briefly discussed in this introduction.

Figure 10-1 shows the wide range of successful products available on the market in early 1983. The maunfacturers listed include only those who fabricate plasma display devices. Excluded are those who buy the device from one of the listed manufacturers and fabricate the electronics to sell complete display systems. Also excluded are the many instututions doing only research and development on plasma displays.

Plasma displays are generally divided into two major families—ac and dc, as seen in Fig. 10-1. The reason for this division is the method of current limiting, which will be covered in Section 10.5. Section 10.6 will cover dc displays and Section 10.7 will cover ac displays. Some recently introduced displays use both ac and dc current limiting and are presented in Section 10.8. The number of different ways of configuring gas discharges to make plasma displays is too large to cover in this chapter. Sections 10.6, 10.7, and 10.8 will filter out the less significant technologies by emphasizing those that have made it to the marketplace with a successful product.

While all of the commercial success of plasma displays has been in computer and information displays, a large amount of research and development has been performed on image plasma displays. Section 10.9 will cover techniques for making plasma displays with gray scale and color. It will also cover the impressive efforts to make flat-panel color television using gas discharges.

To understand the many aspects of plasma display technology it is desirable to have a firm understanding of the fundamental gas-discharge properties. The basic electro-optical characteristics of discharges are covered in Section 10.3. Since no suitable elementary text on gas discharges for plasma displays currently exists, Section 10.4 covers the important factors of gas-discharge physics.

The purpose of this chapter is to review the factors of the plasma display technology that contribute to its success. Since the chapter is by no means a complete treatment of the subject, the reader is encouraged to read some of the excellent review articles available (see References 1 through 14).

Plasma Display Attributes. The remainder of this section will briefly review the fundamental reasons why plasma displays have been so successful as flat-panel displays.

A: Very Strong Nonlinearity. To achieve intrinsic matrix addressability, the display element must have a nonlinear characteristic. The greater this nonlinearity, the greater the number of lines

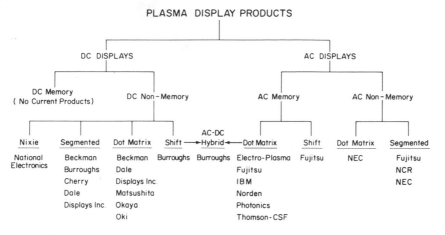

Fig. 10-1. Family tree of plasma display products available in early 1983.

that can be displayed. The gas discharge has very strong nonlinearity. For voltages below the firing voltage it emits practically no light. This nonlinearity is so strong that the detailed nature of it is almost never considered by the plasma display engineer. The ratio of light from a full selected cell to that of a half selected cell does not significantly influence the contrast ratio of a plasma display even with as many as 1,000 pixels per line.

B: Memory. Plasma display cells can be operated in the memory or refreshed mode. The advantage of the memory mode is the increased brightness achievable for large displays over that obtained for the low duty cycle of the scanned refreshed displays. The advantage of a memory display of not requiring an external bit map memory to hold the image may or may not be significant depending on the relative cost of the display and the external memory.

C: Discharge Switching. Many matrix displays have the disadvantage of requiring a large number of drive circuits to address each of the matrix lines. The gas discharge exhibits physical properties that allow it to act as a switching element. This switching action allows logical functions to be performed internal to the display which can drastically reduce the number of drive circuits required. Two examples of this are the shift register type ac plasma display and the dc Self-Scan™ display that eliminate the need for individual column address circuits.

D: Long Lifetime. It is not uncommon to achieve 50,000 hours lifetime in plasma displays. The failure mode is not catastrophic but rather a slow degradation in display quality, with the pixels that have been discharged the longest becoming the weakest. For most display applications, the duty cycle of the average pixel is 10% or less, which means that in the average application the panel lifetime will be even greater. It is not unusual for a plasma display to outlast the entire life of the product.

E: Good Brightness and Luminous Efficiency. Commercially available plasma displays have maximum brightnesses in the range of 10 to 500 foot lamberts. The luminous efficiency of the neon-based gas is between 0.1 and 0.5 lumens per watt depending on the gas mixture used.[15,16] Although these numbers are not as good as in some other display technologies, plasma displays compete favorably since they can achieve high contrast ratios because the display media is transparent, which gives low ambient-light reflectivity. The brightest commercial plasma displays have good contrast ratio when viewed outdoors in direct sunlight without light filters. Laboratory plasma cells have achieved peak luminance values of 100,000 foot-lamberts at more than 5 lumens per watt.

F: Low-Cost Materials–Simple Structure. Plasma displays are usually made of glass and metal in a simple structure that is amenable to mass fabrication. The panel materials are very low-cost. The basic structure of many plasma displays could hardly be simpler, being only slightly more complex than that necessary for the minimal matrix display—an array of x and y electrodes on glass. Custom shapes and sizes of displays can easily be designed and fabricated with quick turnaround time.

G: Rugged Self-Supporting Structure. Plasma displays are usually as rugged as the sheet of glass they are made of. With suitable mounting they will operate in the high shock and vibration military environments. This is because the structures are self-supporting, having no massive parts other than the substrate glass, with all other parts usually directly bonded to the glass substrate. This structure makes large-area plasma displays practical.

H: High Resolution and Large Size. Plasma displays are available on the market with resolutions as high as 39.4 lines per centimeter (100 lines per inch). Laboratory panels with resolutions of 49.2 lines per centimeter (125 lines per inch) have been reported.[17] The one-meter (39-inch) diagonal panels first appeared on the market in 1981[18] (see Fig. 10-34).

I: Transparent Display Media with n = 1. The gas discharge is transparent, and because it is a gas it has an index of refraction of 1. In many other display devices the index of refraction is considerably greater than 1, which causes significant problems. Because of total internal reflection, light generated in a high-index material is trapped, which can severely limit the efficiency of the device. Also, the reflectivity of the high-index displays to ambient light is high, which reduces the contrast ratio.

J: Does Not Scatter Ambient Light. Plasma displays can be made so that they do not scatter ambient light. This allows for an increase in the contrast ratio. For instance, CRT phosphors are made in powder form so that total internal reflection does not limit efficiency. However, the ambient-light reflectivity becomes very high. In many plasma displays, the ambient light is reflected mostly form the metal electrodes on the substrate. This reflection can be conveniently

filtered with a circular polarizer arrangement, which gives excellent contrast ratios in high ambient illumination. This polarizer does not work on the CRT because the phosphor scatters and depolarizes the ambient light.

K: Tolerant to Harsh Environments and Temperature Extremes. To operate with reasonable lifetimes, plasma displays must be hermetically sealed with a high-temperature ($> 400°C$) glass frit seal. This means that plasma panels will operate under high-humidity conditions or in the presence of reactive gases. For many plasma displays the limiting factor on the operating temperature is determined by the drive circuits and not the display device. For design purposes, the characteristics of the ac plasma displays shown on the right branch of Fig. 10-1 can safely be considered invariant with temperature. The dc displays on the left branch of Fig. 10-1 have a more restrictive temperature range because they use mercury in the gas, for reasons discussed in Section 10.6.1.

L: Reasonable Impedance Characteristics. The dielectric constant of the gas discharge is very close to 1, and therefore the capacitance of plasma displays is low compared to some other matrix display technologies such as thin-film electroluminescence. This means that the drive circuits require less current capability.

Although the high voltage of plasma displays is frequently looked upon as a disadvantage, it is in one way quite favorable. If the voltage were to be lower, then for the same pixel brightness and efficiency, the current would have to be higher. Thus, the plasma displays require reasonable currents, which is a distinct advantage when using thin-film conductors in avoiding voltage drops.

M: Diffuse Glow. The spacial distribution of light from a gas discharge is a diffuse glow which extends beyond the electrode that excites the discharge. This adds great flexibility to the display design because thick opaque electrodes having high electrical conductivity can be used without significant penalty in light output. This is especially important for large matrix displays, since transparent electrodes would have too much resistance.

N: Transparent for Back Projection. Some plasma displays can be made transparent so that other images can be projected on top of the plasma image. This is quite useful for map-type displays where the fixed high-resolution map is projected and the plasma display is used to indicate acitivity at the various regions of the map. By placing a solid green, thick-film ac electroluminescent panel behind the transparent orange plasma display, one can obtain a display with a pleasing color contrast.[19]

O: Natural Color Capability. By depositing phosphors in a plasma display the full range of colors can be achieved. Flat color television displays with diagonals as large as 16 inches and excellent picture quality, have been made in laboratories[20, 21] (see Fig. 10-76). The brightness and efficiency have not yet approached that of the CRT, but, steady progress is being made.

P: Flat-Panel Display. Plasma displays have all of the advantages of flat-panel displays, including low volume, slim profile, and freedom from display distortion.

10.2 HISTORY

10.2.1 History of Gas Discharges. The early history of gas discharges follows closely the early history of electricity. A very significant synergism existed between these two disciplines. This was because the gas discharge exhibited many startling and brilliant luminous phenomena that enticed the early experimenters to study electricity. Unlike those of solids or liquids, the discharges in gas afforded a wide range of intriguing effects that were used to great advantage in the investigation of the fundamental properties of electricity. Likewise, many of the significant gas-discharge discoveries followed shortly after the major discoveries in electricity. In the current solid-state revolution it is easy to forget the very significant role that gas-discharge studies played in our modern understanding of electronics.

The first gas discharge in an evacuated vessel was discovered in 1675 by the astronomer Jean Picard, who is most famous for his celestial observations that confirmed Newton's laws of mechanics.[22] The discharge was observed in a mercury barometer made by filling a glass tube with mercury and then inverting it into a cup of mercury so that the falling mercury would evacutate the top of the tube. When the mercury was agitated, static electricity built up and caused a gas discharge in the mercury vapor, which gave the intriguing weak flash of light. Picard had no idea what caused this unusual light, but he did communicate and publish his results.

In 1705 Francis Hauksbee performed a series of impressive experiments directed toward understanding Picard's results.[22] This work culminated in the remarkable machine shown in Fig. 10-2. This had an evacuated glass ball that could be rotated at high speed by a hand-cranked lathe arrangement. To the delight of all observers, the gas inside the glass ball would glow when the operator's hand touched the rotating ball. Hauks-

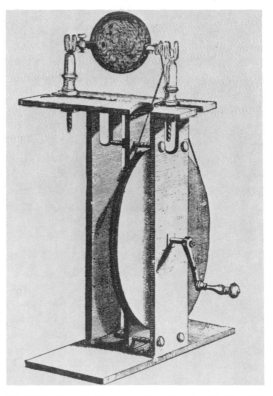

Fig. 10-2. Gas-discharge machine developed by Francis Hauksbee in 1705. (Reference 22 copyright John Wiley & Sons)

bee realized that all of the phenomena known at the time concerning static electricity were closely associated with the mysterious glow. Machines similar to this were used throughout the eighteenth century to generate static electricity for a wide variety of experiments. For instance, the Leyden jar (a primitive capacitor discovered in 1745) was usually charged up by such a rotating glass static-electricity machine.[22]

Continuous gas discharges had to await the announcement of the battery by Volta in 1800. In 1801 Humphry Davy began experiments with continuous electric arcs between carbon electrodes. In 1808 he demonstrated an impressive carbon arc lamp that was powered by the largest electric battery of its day, comprised of 2,000 cells.[23] Although the brilliance of this new lamp exceeded anything at the time, its application had to wait a half century for a more practical power source: the dynamo. With this new power source, the first practical use of a gas discharge occurred when Frederick Holmes installed an arc-lamp system in the lighthouse at South Foreland, England, in 1858.[24] During the last quarter of the nineteenth century, arc lamps were in common use for many types of illumination.

In 1838 Michael Faraday did the first experiments on a gas discharge between two metal electrodes placed in an evacuated glass bulb.[25] With this apparatus he observed the luminous regions of the gas discharge (see Fig. 10-7). He observed that the cathode was covered with a glow and that the region associated with the positive electrode (the positive column) could be any length. He also discovered the dark region between these two luminous regions which is now called the Faraday dark space.

In 1854 Heinrich Geissler, a glassblower at the University of Bonn, began making evacuated-gas-discharge tubes in large numbers.[24] These were shipped to many schools, colleges, and other laboratories where they were used for demonstrations and general entertainment. The tubes were made in a wide variety of shapes. Most were filled with 5 torr of air, but other gases were used. Some tubes used a special glass having fluorescent salts that gave added color. These tubes, forerunner, of the fluorescent lighting tubes of today, made gas-discharge phenomena widespread common knowledge.

In 1875 William Crookes experimented with a gas-discharge tube that could achieve unprecedentedly good vacuum.[26] He was able to observe all of the standard Geissler tube phenomena over a wide range of pressures. At very low gas pressures, the various luminous regions began to disappear because the electrons were making virtually no collisions with the gas atoms. Crookes noted the fluorescence of various substances when the electrons in this gas discharge hit them. He called these electrons cathode rays. This is the early ancestor of the cathode-ray tube (see CRT History, Section 6.2). By using a similar gas-discharge device, Wilhelm Roentgen discovered the X-ray in 1895.[26] By applying magnetic and electric fields to the Crookes tube, J. J. Thomson discovered the electron in 1897.[26]

In 1887 Heinrich Hertz discovered radio waves.[26] In his experiments he used a gas arc switch to initiate the oscillations in his transmitter. His receiver consisted of a wire loop with a small spark gap. Hertz detected the radio signal when a small gas discharge occurred across this gap.[27] Others improved on this detector by using a Geissler tube instead of the spark gap to make a much brighter discharge. In 1892 L. Zehnder devised a further improvement with a four-electrode Geissler tube.[27] Two of the electrodes were connected to a battery with a voltage just below the sparking potential. The other two electrodes were connected to the receiving antenna so that the presence of the signal would trigger a bright discharge between the battery electrodes. This is perhaps the first development of the principles of triggering and amplification.

Radio transmitters continued to use gas-discharge switching until the late 1920s. Continuous-wave transmitters were installed with powers as high as 500 kW that used the negative-resistance characteristic of the arc discharge to generate the powerful CW oscillations in the transmitter LC circuit.[28]

In 1901 Peter Cooper Hewitt commercialized the low-pressure mercury-discharge lamp.[23] This was a 1.25-meter (50-inch) long, 2.5-centimeter (1-inch) diameter tube having a small pool of mercury. In 1902 he developed the mercury-arc rectifier. In 1904 D. McFarlan Moore began installing in commercial stores gas-discharge lighting tubes which used the light from the positive column.[23] These tubes used N_2 or CO_2 and were made to lengths of 60 meters (200 feet). Georges Claude developed a technique for obtaining low-cost neon gas and used it to make the first commercial neon sign in 1910.[29]

In subsequent years, numerous other gas-discharge devices have been commercialized. These include the high-pressure sodium and mercury-arc lamps used for street lighting, the Geiger counter, the cold-cathode voltage reference tubes, the modern fluorescent lamp, the neon and xenon flash tubes, the thyratron, the radar T-R tube, and the cathode glow bulb.

10.2.2 History of Plasma Displays. In a very early flat-panel display, the Bell System used a receiver based on a gas-discharge display for their 1927 demonstration of live television.[30] The signals were sent between Washington, D.C., and New York City via cable and between Whippany, N. J., and New York via wireless. The display, shown in Fig. 10-3, was an array of 50 by 50 gas-discharge cells with a display area approximately 60 by 75 centimeters (2 by 2.5

feet). This was intended for viewing by an auditorium audience. The display panel was constructed of a neon-filled glass tube bent back and forth fifty times in a zigzag manner, and having 2,500 electrodes cemented to the outside of the tube, each being excited by an ac voltage. This display demonstrated a full-grayscale television image at 16 frames per second.

The drive electronics sat behind the display as shown in Fig. 10-4. It consisted of a motor-driven distributor having 2,500 output wires that connect to each of the gas-discharge cells. These wires were arranged so that as the distributor brush rotated, the discharge cells would light up in a raster scan sequence. The distributor made a complete revolution for each display frame. The video signal was applied to the brush so that each selected cell would glow at the proper intensity. As with many of today's flat panels, the state of the art of the drive electronics was the major limiting factor for the viability of this impressive early flat-panel technology. This display did succeed in generating a great

Fig. 10-3. Early gas-discharge television display developed by the Bell System in 1927. (Reference 30 reprinted with permission from *The Bell System Technical Journal.* Copyright (1927), AT&T)

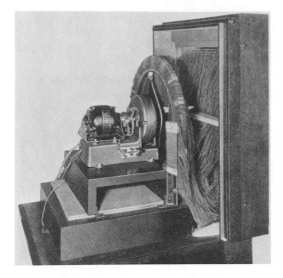

Fig. 10-4. Drive electronics positioned behind the gas-discharge television display consisting of a motor-driven distributor. (Reference 30 reprinted with permission from *The Bell System Technical Journal.* Copyright (1927), AT&T)

deal of interest in the concept of television, which was so aptly accomplished by the cathode-ray tube in the early 1930s.

In the 1940s and 1950s a number of gas-discharge devices were developed for use in computer switching and logic. One of these that developed into a rather successful product was a counting device called the Dekatron.[31] This had multiple cathodes arranged in a ring along with a common central anode. The appropriate voltage pulses would make the gas discharge jump from one cathode to the next in a shift-register fashion. This counting device doubled as a display since the state of the counter could be determined by observing which cathode was discharging.

In the early 1950s the NIXIE tube was developed at Burroughs Corporation.[32,33] This was the first commercially successful electronic numeric indicator. It had the ability to rapidly display changing numbers with unprecedented clarity. It opened up many new product possibilities, one of which was to provide a better display for the Dekatron. It operated on the principle of making each cathode electrode the shape of a different numeral (see Fig. 10-15). This technology gained immediate success and is still manufactured in 1983 by National Electronics.

The Burroughs engineers noted that the life of a NIXIE tube could be greatly increased by adding mercury to the tube.[34] This inhibited the destructive effects of sputtering which would erode away the cathode and deposit a mirrorlike film on the inside of the tube that obscured the display. To this day (1983) all commercial dc plasma displays are still manufactured with mercury.

The potential for a variable-format display having an array of x and y electrodes and a display element at each intersection was realized in the late 1950s and early 1060s.[35-38] These experiments led to the development of arrays of gas discharges that could be made to emit light efficiently. However, certain problems existed. No more than one cell could be on in a line because once a single cell discharged it would hog current and lower the voltage on that electrode and prevent other discharges from igniting. It was then realized that if the memory of the gas discharge was to be exploited in a dot matrix panel, a resistor was needed in series with every display element. Unfortunately, the value of this resistor had to be in the 1-meg-ohm range and it is difficult to integrate resistors of this value in a display panel.

A key breakthrough came in 1964 when Bitzer and Slottow, working at the University of Illinois, realized that a capacitor could replace the resistor.[39] Capacitors of the proper impedance were easily fabricated by placing a dielectric layer between the electrodes and the gas discharge. As a bonus, the capacitor could store charge, and when combined with the negative-resistance characteristics of the gas discharge, memory could be achieved when excited by the proper ac voltage. The inventors named their new device the plasma display panel.

The structure of the ac plasma display panel developed at the University of Illinois was very fragile, and the panel would only operate for a few hours. A new breakthrough occurred in 1968 when Owens-Illinois developed the rigid-substrate open-cell structure that was capable of the high-temperature bakeout necessary for long operating lifetime.[40] With this development, the ac plasma display panel emerged as a rugged practical device that was mass-producible.

The excitement of the ac plasma panel invention stimulated Holz and Ogle at Burroughs Corporation to explore new techniques for making gas-discharge displays. This work resulted in the invention of the Self-Scan™ display that was introduced as a product in 1970.[41] This display successfully used the internal-logic ability of the gas discharge to greatly reduce the number of circuit drivers by making the display act as a shift register. Many of the principles of the Dekatron tube were incorporated in this design.

In 1972 Umeda and Hirose of Fujitsu presented the development of an ac plasma display that acted as a shift register.[42] Like the Self-Scan™ this display has the advantage of a low number of circuit drivers. It also has the advantage of internal memory.

In late 1972 researchers at Zenith and Bell Labs demonstrated high-quality real-time television pictures on a Self-Scan™ display.[43,44] The excellent performance of these systems stimulated a great deal of research into flat-panel television using gas discharges during the next decade.

During the 1970s and early 80s, plasma displays moved from the pilot production stages to full-scale production in numerous plants throughout the world. This is indicated by the large number of manufacturers with commercial products in 1983, as shown in Fig. 10-1. This activity has made plasma displays the dominant flat-panel technology for large-area displays.

10.3 BASIC ELECTRO-OPTICAL CHARACTERISTICS OF THE GAS DISCHARGE

The fundamental characteristics of the gas discharge will first be discussed here. This will lay a foundation for a more complete discussion of the gas-discharge physics in Section 10.4.

10.3.1 I-V Characteristic.
The I-V characteristic of a gas discharge typical of that used in plasma displays is seen in Fig. 10-5. Note that the current is plotted on a log scale which spans 9 orders of magnitude. For this large current range, numerous different physical phenomena become dominant over different current ranges. This accounts for the various labeled regions

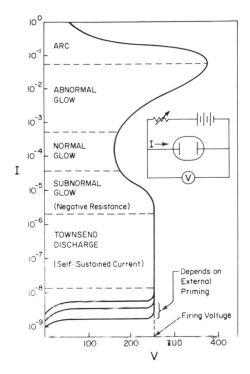

Fig. 10-5. The I–V characteristic of a gas discharge typically found in plasma displays.

of the curve seen in Fig. 10-5. A complete discussion of these regions and their physical mechanisms appears in Section 10.4.4. In this section we will accept this characteristic as a given and discuss how the plasma display engineer might use this characteristic.

Probably the most striking feature of this curve is the extreme nonlinearity exhibited by the gas discharge. Note that the current is plotted on a log scale and that for voltages near the firing voltage of 250 volts the current can change by more than 3 orders of magnitude for a very small incremental voltage change. Since the light output of the display is roughly proportional to the current, this extreme nonlinearity is very useful for making matrix displays with a large number of display electrodes, as discussed in Section 5.3. This strong nonlinearity is one of the major reasons why plasma displays are successful as flat-panel matrix displays.

Note that the very-low-current regions of the discharge characteristic depend on external priming. Gas discharges need some energetic particles to get them started, and the generation of these particles is called priming. These particles can be electrons, ions, photons, or other excited species. Without external priming, a gas discharge will not occur even with very high applied voltages. This phenomenon is used to great advantage in plasma displays to perform internal gas-discharge switching operations.

The I-V characteristic exhibits a negative resistance which can be used to make memory devices if the proper external resistance and power supply voltage are used as discussed below.

Figure 10-6 shows the I-V characteristic of a plasma display gas discharge plotted on a linear current scale. This scale covers the subnormal-glow (negative-resistance) region, the normal-glow region, and part of the abnormal-glow regions of Fig 10-5. Since the current of the gas discharge can reach very large levels, as seen in Fig. 10-5, some external means of current limiting is usually used to prevent the device from the destructive effects of the high currents. The most common way of limiting the current is to place a resistor in series with the gas discharge.

10.3.2   Resistor Load-Line Technique.   To determine where on the I-V characteristic the resistor–gas-discharge circuit will operate, one can use the load-line technique shown in Fig.

10-6. Here a resistor load line is drawn over the gas-discharge characteristic. Differing resistor values have different slopes; also, the load lines intersect the voltage axis at the power supply voltage. Figure 10-6 shows four different load lines for four different supply voltages, 175, 225, 275, and 300 volts. The operating points of the discharge are determined by the intersections of the load line with the gas-discharge characteristic. These operating points can be either stable or unstable depending on the slopes of the load line and the gas-discharge characteristic. For instance, the 500 kΩ load line for a power supply of 225 volts has three operating points: two stable and one unstable. This load line is configured to operate in the memory mode with two stable states corresponding to the two stable operating points. One is the off state and the other is the light-emitting on state.

To determine which of the two stable operating points will be active, one must know some of the previous history of the circuit. One usually forces the circuit in one state or the other by adjusting the power supply voltage. If the voltage is lowered to 175 volts, then the load line can only intersect the gas-discharge characteristic at one stable point, which is an off state as seen in Fig. 10-6. Similarly, if the voltage is raised to 275 volts, the load line will only allow a stable on state to exist. If the voltage is first raised to 275 volts and then lowered to 225 volts, the stable on state will exist at 225 volts. If the voltage is then lowered to 175 volts, the discharge will go off, and it will stay off when the voltage is raised back to 225 volts.

For the characteristic of Fig. 10-6 and a resistance of 500 kΩ, there is no way that a steady state discharge can operate in the negative-resistance region of the curve, since all of the possible operating points in this region will be unstable. One might ask how this region can be measured if all operating points are unstable. The answer is that one must use a much greater resistance such as 5 MΩ, which corresponds to the load line that intersects the voltage axis at 300 volts in Fig. 10-6. This load line intersects the negative-resistance region. Because of the relative slopes, this operating point is stable. Thus, with this large resistance, one can

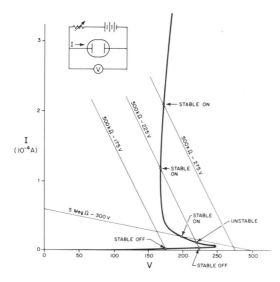

Fig. 10-6. Linear plot of a typical gas-discharge I-V characteristic showing various resistor load lines. For clarity, the current values near the horizontal axis have been exaggerated.

measure the entire I-V characteristic by varying the supply voltage and measuring the current and the voltage across the gas discharge. The data of Figs. 10-5 and 10-6 could be taken using this technique.

10.3.3 External Current-Limiting Requirement.
The discharge can operate at very high current levels, as demonstrated by the arc region of Fig. 10-5 which has a negative-resistance region where the applied voltage can be lowered considerably and still maintain the arc. This is a very strong negative resistance which, once entered, will certainly mean destruction of the device. This destruction comes from the extremely large power dissipation which literally melts or vaporizes the gas-discharge electrodes. The plasma display engineer must use a suitable method of limiting the current to prevent destructive operation in the arc mode or in the high-current regions of the abnormal glow. These current-limiting techniques will be discussed in Section 10.5.

10.3.4 Luminous Regions of a Gas Discharge.
Plasma displays generally operate in the normal-glow or the abnormal-glow regions of the I-V characteristic of Fig. 10-5 when they are emitting light. In these regions the light output of the gas discharge has the spacial distribution seen in Fig. 10-7. This distribution is due to the complexities of the various distributions shown below the luminous intensity. The physical mechanisms behind these distributions will be discussed further in Section 10.4.5.

For now we will note that there are two important luminous regions: the negative glow and the positive column. The light from these two regions has considerably different characteristics. For instance, it may have different colors and shapes. The negative glow usually covers the cathode, whereas the positive column frequently takes the shape of the glass tube that contains it. Examples of a positive-column discharge are the common fluorescent lamp used for lighting and the neon sign used for commercial advertising. Both of these devices illustrate the manner in which the positive column takes the shape of the glass tube. All of the plasma displays currently on the

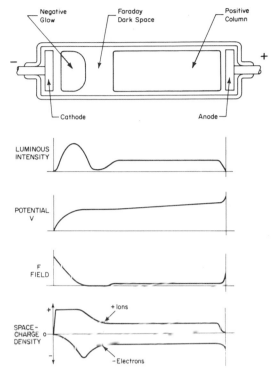

Fig. 10-7. The important regions of a gas discharge typically found in plasma displays.

market use the light from the negative glow. However, a number of experimental plasma displays have been designed around the positive-column light, as discussed in Section 10.9.3. To determine which of the two light-emitting regions will dominate, the engineer controls the parameters of the gas discharge such as the geometry, the gas pressure, and the applied voltage.

10.3.5 Wavelength Distribution.
Figure 10-8 shows the wavelength distribution for a typical plasma display. This is the light from the negative glow. It is the classic line spectrum of the gas discharge. This is the spectrum of neon gas, the only luminous gas used in the plasma displays on the market today (1983). The major reason for the selection of neon is that it exhibits the highest luminous efficiency among the noble gases. This spectrum gives the familiar neon orange color emitted by most plasma displays. Other colors can be achieved by introducing phosphors into the display or by using other gases as discussed in Section 10.9.2.

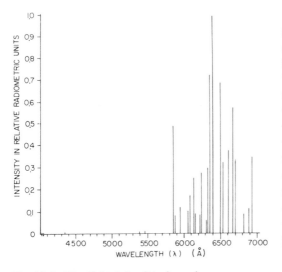

Fig. 10-8. The light intensity dependence on wavelength for a plasma display having neon gas. (Reference 4 courtesy of Academic Press)

10.4   GAS-DISCHARGE PHYSICS

At this point this text will leave the simplified characterization of the gas discharge and turn to the details which explain why the discharge exhibits the characteristics presented in Section 10.3.

10.4.1 Gas-Discharge Reactions. Figure 10-9 shows a simplified view of the most important reactions that occur in the gas discharge. Numerous other reactions occur that are beyond the scope of this discussion; for a more complete treatment of these reactions, the reader should consult one of the texts on gas discharges.[45-48] Even this simplified view is rather complex so each of the reactions will be covered in this section. This example assumes that the gas is a mixture of predominantly neon with a very small percentage of argon (typically 0.1% Ar). This is a common gas used in plasma displays, called a Penning mixture.[49]

The reactions can be divided into gas volume reactions and cathode surface reactions. The gas volume reactions include ionization (I), excitation (E), metastable generation (M), and Penning ionization (P). The cathode surface reactions are concerned with ejection of electrons from the cathode by ions, metastable atoms, or photons. The details of each of these reactions is discussed below.

The gas discharge has numerous species that are transported by differing mechanisms. Each of the species found in this discharge example is listed in the lower left corner of Fig. 10-9. The charged species, which include the electrons and the positive neon and argon ions, are transported mainly by the field-induced drift, with diffusion transport of secondary importance. The neutral species include the neutral neon and argon gas atoms, the excited neon atom, and the metastable neon atom. These species are not influenced by the electric field, and thus their transport is governed solely by diffusion. The photons of course obey the propagation laws for electromagnetic radiation.

Neon Atom Energy Levels. Figure 10-10 shows the allowed energy levels for the outer electron of the neon atom. These energy levels are important because the gas-discharge reactions discussed below are strongly dependent on these levels.[50] One major consequence of these levels is the spectrum seen in Fig. 10-8 which is due to the photons emitted when an excited neon atom makes a transition between the 2p energy levels and the 1s levels. Figure 10-10 shows the visible photon transitions and wavelengths for some of the brightest spectral lines.

Ionization. Probably the most important reaction of a gas discharge is ionization. When an electric field is placed across the gas, the electrons that may be in the gas volume are accelerated by that field. These electrons are continually colliding with the neutral gas atoms. Since Fig. 10-10 shows that the neon atom does not have any allowable energy levels between 0 and 16.6 electron volts, most of these collisions will be elastic, which means that during the collision the electron does not gain or lose energy. After many such collisions, the electric field will have accelerated some of these electrons to energies greater than the 21.6-electron-volt ionization limit of the neon atom seen in Fig. 10-10. The collisions between these energetic electrons and the neon atoms can cause an electron to be ejected from the atom. This will create a positive neon ion and a new free electron. As seen in Fig. 10-9 the ionization

GAS DISCHARGE REACTIONS

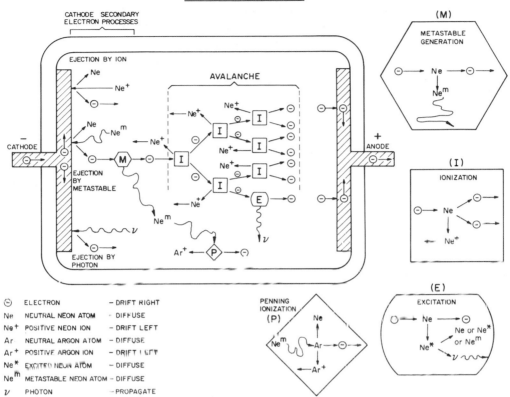

Fig. 10-9. Model of important gas-discharge reactions.

NEON ATOM

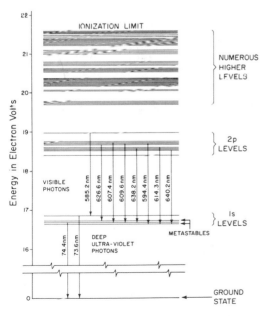

Fig. 10-10. Energy-level diagram of the neon atom.

reaction can be written as:

$$Ne + e^- \longrightarrow Ne^+ + 2e^-$$

The electrons drift to the right and the ion drifts to the left. Because the positive ion is much larger and more massive, it drifts roughly 100 times slower than the electrons.

Excitation. Some of the electrons will gain sufficient energy to excite a neutral neon atom during a collision. This reaction occurs at electron energies in the range of 16.6 to 21.6 electron volts. Energies less than this will cause only elastic collisions; energies greater begin to favor ionization. The excited atom can only remain excited for a relatively short time (roughly 10^{-8} seconds) before it radiates a photon and returns to the ground state. Thus the excitation reaction seen in Fig. 10-9 can be written as:

$$Ne + e^- \longrightarrow Ne^* + e^-$$

$$Ne^* \longrightarrow Ne \text{ (or } Ne^*, \text{ or } Ne^m) + \nu$$

Many of the photons that are due to the transitions between the 2p and the 1s levels of Fig. 10-10 are visible and are responsible for the spectrum seen in Fig. 10-8. These transitions leave the neon atom in an excited 1s state that will eventually be de-excited by some later reaction. Figure 10-10 shows that the transitions between the 1s levels and the ground state are deep ultraviolet photons. These cannot pass through the glass envelope of the gas discharge, but they can be important for cathode surface reactions discussed below in this section.

Metastable Generation. Not all of the excited states of the neon atom are allowed to emit a photon. Two of the four 1s levels of the neon atom shown in Fig. 10-10 cannot radiate photons and are called metastable levels. These metastable atoms can be generated by an electron with greater than 16.6 electron volts colliding with a neutral neon atom, as shown in Fig. 10-9. The reaction is:

$$Ne + e^- \longrightarrow Ne^m + e^-$$

Metastables can also be generated as a by-product of the excitation reaction. For instance, many of the radiative transitions shown in Fig. 10-10 end on one of the two 1s metastable levels. Thus, metastable generation can occur by the two-step process whereby an 18.5-volt or greater electron excites a neutral neon atom to a 2p state and that atom radiates a photon and places the neon atom in a 1s metastable state. The two-step reaction is written as:

$$Ne + e^- \longrightarrow Ne^* + e^-$$

$$Ne^* \longrightarrow Ne^m + \nu$$

Metastable atoms are simply excited neon atoms that do not radiate a photon and thus are not charged. This means that their transport is only by the slow process of diffusion. Metastables move much slower than do the positive ions which are of virtually the same mass. The metastable atoms survive in the discharge for relatively long times. In plasma displays, metastable lifetime decay time constants are usually in the 1-to-10-microsecond range. Metastables usually do not decay naturally but are de-

excited by a reaction with some other body. Among other reactions, metastables can be de-excited by colliding with the discharge chamber walls and by the Penning ionization process discussed below.

Penning Ionization. The metastable atoms have about 16.6 electron volts of energy, as seen in Fig. 10-10. Since these atoms have a relatively long lifetime, they can make a large number of collisions with other atoms in the discharge. If a metastable collides with an argon atom, there is a high probability that it will ionize the argon atom. This is because the 16.6-electron-volt metastable energy is greater than the 15.8-electron-volt ionization energy of argon. Thus the reaction is:

$$Ne^m + Ar \longrightarrow Ne + Ar^+ + e^-$$

This reaction is of great importance to plasma displays because it generates additional ionization beyond that produced by the ionization reaction. As discussed in Section 10.4.3, this additional ionization allows plasma displays to operate at a lower voltage. The amount of this additional ionization is strongly dependent on the argon atom concentration. The mixture that gives the largest amount of ionization per applied volt was found by F. M. Penning in 1937 to be Ne plus 0.1% Ar.[49] This is a common mixture used in plasma displays and is called a Penning mixture. Other Penning mixtures include neon plus small amounts of xenon, krypton, or even certain noninert gases.

Cathode Surface Reactions. A number of important reactions occur at the surface of the gas container. The most important reactions are at the cathode electrode surface. Figure 10-9 shows the three reactions that can eject electrons form the cathode surface. This ejection can be stimulated by cathode collisions from positive ions, metastables, and photons. The ejection of electrons is of critical importance to a gas discharge because these cathode electrons initiate the volume reactions and thus participate in the gas-discharge feedback equation which determines the firing voltage of the discharge, as discussed in Section 10.4.2.

The most important cathode surface reaction for plasma displays is the electron ejection due to positive ions. The neon ion has 21.6 electron volts of energy, and the argon ion has 15.8 electron volts. During collision with the cathode, these ions capture an electron from the surface and they become neutralized. In this process the ions give up their energy to the surface. This energy is more than enough to allow an electron to escape the work-function energy of the cathode surface, which is usually in the 3-to-10-electron volt range. This large amount of excess energy means that there is a high probability that an ion hitting the cathode will cause an electron ejection. The factor, coupled with the fact that all positive ions produced in the gas volume are directed by the field to drift toward the cathode, makes this ejection reaction dominant in plasma displays.

Photoemission also is a significant mechanism for electron ejection. Since the work function of the cathode is generally greater than 3 electron volts, only ultraviolet photons will have significant photoemission. Figure 10-10 shows that the transitions between the 1s levels and the ground state will generate these photons. There will be a very large number of these photons generated, but they will have random directions and only a small fraction will be directed toward the cathode.

Metastable neon atoms have 16.6 electron volts of energy that can be given to the cathode surface to eject an electron and de-excite the metastable. Although the metastable has about the same probability of ejecting an electron as an ion, the metastable is not nearly as important as the ion since the metastables diffuse in random directions at a very slow rate compared to the drift of the ions toward the cathode.

Avalanches. The ionization reaction is initiated by one electron and results in two free electrons and a positive ion. These two electrons can then go on to create two more ionizations and so on, resulting in an avalanche as shown in Fig. 10-9. The electron that initiates the avalanche can be one that was ejected from the cathode. As the avalanche progresses toward the anode, the number of ionizations increases exponentially in space and time. In plasma displays, the number of ionizations occuring in an avalanche is in the range of 10 to 300.

10.4.2 Gas-Discharge Feedback Model. Figure 10-11 shows a feedback model of the gas discharge. The model is quite useful for understanding a number of gas-discharge characteristics such as the firing voltage, the current growth rate, and the influence of priming currents on the discharge growth.

The avalanche shown in Fig. 10-9 is represented in the model of Fig. 10-11 as an amplifier with gain $e^{\pi V} - 1$, where π is the number of ion pairs generated per volt and V is the voltage between the anode and the cathode. The value of π depends on the gas mixture, gas pressure p, and the electric field E. Figure 10-12 gives the data for π for two gases commonly used in plasma displays. The π for neon + 0.1% argon is much larger than that for pure neon because of the Penning reaction discussed in Section 10.4.1. It should be clear that the avalanche gain increases as the voltage is increased.

The avalanches create ions, metastables, and photons which can find their way back to the cathode and eject secondary electrons as shown in Fig. 10-9. Since these secondary electrons can now start new avalanches, this process can be represented by the feedback path in Fig. 10-11. The delay shown represents the transit time for the ions, metastables, or photons between their point of generation and the cathode. Once these particles reach the cathode, there is a probability γ that the particle will cause an ejection of an electron. This γ coefficient multiplies the feedback term as shown.

The secondary electrons that are ejected from the cathode must be added to the electrons due to the priming current I_o. The priming current arises from some external means such as photoemission from photons generated by an independent nearby gas discharge. The mechanisms for priming are very important for plasma displays and are fully discussed in Section 10.4.6. The sum of the priming and the secondary currents represent the flow of electrons that can start new avalanches and so they are fed back into the avalanche gain amplifier. Note that this results in a positive feedback.

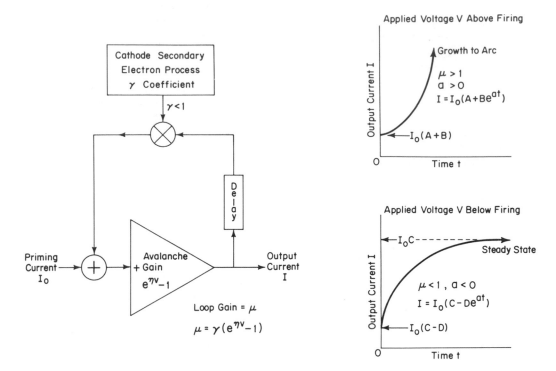

Fig. 10-11. The feedback model for the gas discharge.

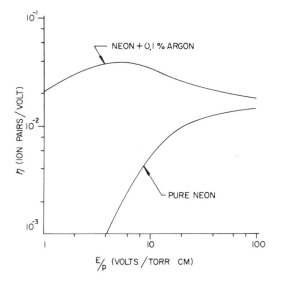

Fig. 10-12. Graph of the ionization coefficient η as a function of the electric field divided by the gas pressure (E/p). (Reference 4 courtesy of Academic Press)

Loop Gain. The model in Fig. 10-11 is relatively easy to analyze. Because of the positive feedback, it will become unstable if the loop gain is greater than 1. The loop gain μ is equal to the product of the avalanche gain and the secondary electron γ coefficient:

$$\mu = \gamma(e^{\eta V} - 1)$$

The condition $\mu = 1$ represents the point of instability. This condition defines the firing voltage of the gas discharge. Note that the value of μ is strongly dependent on V. For V above the firing voltage V_f, $\mu > 1$. Also for V below V_f, $\mu < 1$.

Current Growth. The upper right side of Fig. 10-11 shows how the output current behaves for V above firing. The current grows exponentially with a positive time constant a which depends on the value of μ and the feedback delay time. The coefficients A and B are constants that are independent of time. Note that the output current I is proportional to the priming current I_o. Once started, the output current will continue to grow until some other physical mechanism limits it. If some current-limiting mechanism does not take over, the current will grow until an arc occurs that will usually destroy the device.

The lower right of Fig. 10-11 shows the current growth for V below firing. Here the exponential time constant a is less than 1, so that the output current eventually reaches a low steady state value. The coefficients C and D are independent of time. Again the current I is proportional to the priming current I_o.

Priming Requirement. Because of the proprotionality of the output current to the priming current I_o, it is clear that a gas discharge will not occur without some external priming current. This priming can be as little as a single external electron that is necessary to get the first avalanche started. However, if this single electron does not exist, then even voltages much greater than the firing voltage will not cause a gas discharge. The importance of priming to the initiation of a gas discharge impacts the design of plasma displays to a great degree. The mechanisms of priming will be covered more extensively in Section 10.4.6.

10.4.3 Paschen Curve. The display engineer is frequently concerned with the dependence of the firing voltage on the gas-discharge cell design. This is conveniently characterized with the Paschen curve shown in Fig. 10-13. This curve gives the firing or breakdown voltage as a function of the product, pressure (p) × gas gap distance (d). It can be derived by noting the voltage where $\mu = 1$ in the feedback model of Fig. 10-11. A different curve is arrived at for different gas mixture or different cathodes, since each will have differing values of π and γ.

The important principle seen from this curve

is that the firing voltage will remain unchanged for differing cathode-anode distances as long as the pressure is changed so that the product pd is constant. This fact is used to great advantage by display engineers. In order to reduce drive circuit costs it is frequently desirable to design the firing voltage to be as low as possible. The vlaue of d is usually determined by other constraints such as the required resolution of the display. Thus the engineer has the freedom to choose p for a given d in order to minimize the firing voltage.

The Paschen curve in invariably U-shaped. This means that for a given gas mixture and cathode material there is a minimum firing voltage. It is not possible, for instance, to achieve a lower firing voltage by shortening the anode-cathode gap distance d. Although this would increase the electric field across the gas, it would also cause the firing voltage to increase from the minimum, since the pd product will be smaller, and push operation up the left side of the Paschen curve.

The physical reason for this higher firing voltage with smaller d and thus higher E can be understood by considering that if d gets very small there will not be enough collisions between the electrons and the gas atoms so that the number of ionizing collisions, $\pi \times V$, will be low. Thus a higher voltage will be needed to achieve the $\mu = 1$ firing condition. If the pressure is increased at this smaller d, the number of electron-atom collisions will increase, as will pd, and thus the firing voltage will then approach the minimum.

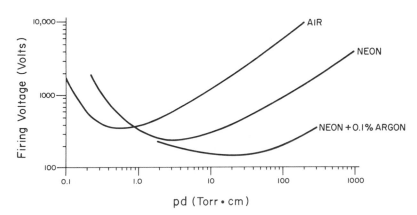

Fig. 10-13. The Paschen curve showing the dependence of the breakdown voltage on the product of gas pressure, p, and the cathode-anode separation, d.

On the right side of the Paschen curve minimum, d is too large, so that the E field is too low and the electrons do not gain sufficient energy between collisions to efficiently ionize the atoms. Thus $\Re \times V$ is low and a higher firing voltage is needed for $\mu = 1$.

Note that the neon plus 0.1% argon gas mixture has a lower firing voltage than does pure neon. This is due to the additional ionization of the Penning reaction discussed in Section 10.4.1. This lower voltage generally makes the Penning mixture preferred so that drive circuit costs can be made lower.

10.4.4 Regions of the I–V Characteristic. This section will cover the details of the I–V characteristic of the gas discharge shown in Fig. 10-5. This discussion will start with the low-current phenomena and progress toward the high-current effects. It should be emphasized that Fig. 10-5 is a steady-state characteristic. A transient discharge will have a significantly different characteristic. The steady-state characteristic is measured by adjusting the power supply and current-limiting resistor to the appropriate values for each point on the curve.

Extreme-Low-Current Region. Figure 10-5 shows that for the lowest currents in the 10^{-8} to 10^{-9} range the characteristic is strongly dependent on priming. These low currents are observed below the firing voltage. They are described by the theory of the feedback model shown in the lower right side of Fig. 10-11. For voltages below firing, the discharge current is proportional to the priming current generated by the external means discussed more fully in Section 10.4.6. This priming current can range over many orders of magnitude. Thus the low-current characteristic can be considerably more variable than shown in Fig. 10-5.

Townsend Discharge Region. Once the firing voltage is exceeded, the current grows sharply, as shown in the upper right corner of Fig. 10-11. In this region the voltage across the gas is constant for a wide range of currents. The Townsend discharge region is sometimes called the self-sustained-current region because, once started, a discharge will continue even if the priming current is removed.

Subnormal-Glow Region. As the discharge current grows, the density of charged particles in the gas volume will eventually reach a point where it will significantly alter the electric field distribution. Under the proper conditions, this can give a negative-resistance region.

Normal-Glow Region. Once the space-charge distribution is fully established, the discharge will configure itself so that the current density on the cathode surface is constant. The variation in current in the normal-glow region occurs by varying the area of the negative glow that fills the cathode so that the constant current density requirement is maintained.

Abnormal-Glow Region. As the discharge current is further increased, a point is reached where the cathode electrode is fully covered with the negative glow. At this point a constant current density can no longer be maintained. The increased current density requires a less efficient electric-field distribution that needs a higher voltage across the gas. This results in a positive resistance in this region.

Arc Region. At very high currents the power dissipation in the discharge becomes so great and the temperatures so high that the cathode material is vaporized. The cathode vapor creates a very-high-pressure gas that can support the much higher currents which generate a much higher degree of field distortion. The temperatures and fields are so high that thermionic emission and field emission become the dominant cathode electron emission mechanisms. These can generate very large currents at voltages as low as 10 volts. These mechanisms generate the very sharp negative-resistance characteristic of the arc.

Because the cathode is severely eroded away by the discharge process in the arc, suitable lifetimes for display devices cannot be achieved. The significance of arcs to plasma displays is that the display system should be designed so that arcs cannot occur, for if they do, the large currents will almost certainly result in destruction of the device.

10.4.5 Spatial Regions of the Normal-Glow Discharge. Figure 10-7 shows the various conditions found in a typical gas discharge that operates in the normal-glow region of Fig. 10-5. Three regions are commonly observed: the negative glow, the Faraday dark space, and the positive column.

Note that there is a high E field region near the cathode where most of the voltage drop occurs. This region will have strong avalanches which will generate the light output associated with the negative glow. The light output will peak near the end of the high field region because that is where the excitation of neutral atoms is the strongest. The shape of the electric field can be determined from the net charge density using Laplace's equation. Thus the field is reduced as one moves from the cathode because there are a large number of ions and very few electrons, creating a positive net charge density.

After the field near the cathode is reduced to near zero, the electron current is still high but the electrons do not have much energy because E is very low, and thus the light generation is very low, resulting in the Faraday dark space.

The next region toward the anode is the positive column. This region contains a plasma of ions and electrons. A simple definition of a plasma is a region with the following properties:

1. The electron density is very nearly equal to the ion density.
2. The electron and ion density is great enough so that the electric field in the region is a low constant value. This is because if an external field is applied to the plasma the electrons and ions are free to move until they reduce the field to the low initial value.
3. Since the electric field is very low, mobility transport of the electron and ions is of less importance. Diffusion is usually dominant. In this case, because the motions of the electrons and the ions are highly coupled, the diffusion is called ambipolar. This means that the electrons have the same diffusion coefficient as the ions. This coefficient is about twice the diffusion coefficient of the ions alone.

As seen in Fig. 10-7, a plasma exists in the positive column because the electron and ion densities are equal and the electric field is near zero. The light from this region can be significant. The atomic-excitation mechanism that yields this light, and the ionization mechanism that maintains the plasma are considerably different from the mechanisms observed in the negative glow. In the positive column, the electric field is too low for a single electron to gain sufficient energy directly from the field to be able to excite or ionize an atom. However, there are a large number of electrons in the plasma of the positive column that are diffusing at a very slow rate. Thus, the electrons will be strongly influenced by coulomb interactions with the other electrons nearby within the plasma. This means that the energies of the electrons are highly coupled, so that a small amount of energy gained by one electron will soon increase the average energy of each of the electrons in the plasma. These electrons will have some distribution of energies which one can usually describe with a single parameter called the electron temperature. This distribution will have a small number of electrons that through many electron-electron coulomb interactions will have gained a large enough energy to excite or ionize an a atom.

In the negative-glow region, the coulomb interactions are insignificant because the electrons move through the region very quickly by mobility transport in the high electric field. Thus, the high electric field in the negative glow gives much more energy to a given electron than does the coulomb interaction of some nearby electron.

Because the excitation mechanisms of the positive column and the negative glow are significantly different, the colors of these two regions are frequently different. This is because a different energy distribution of electrons exists in the two regions and this causes different atomic-energy levels to be excited. This effect is seen most vividly when a gas mixture such as Ne-Xe or Ne-Hg is used. At low currents, the negative glow will be orange and the positive column blue.

10.4.6 Priming. As shown in Fig. 10-11, some initial priming current I_o is necessary to initiate a gas discharge. There are numerous physical mechanisms that can accomplish this. The three most important ones for plasma displays include pilot-cell priming, self-priming, and radioactive priming. Each of these will be discussed in detail below.

Pilot-Cell Priming. The most common priming technique for plasma displays uses a pilot discharge operating near the cell to be primed. This pilot cell will be generating electrons, ions, photons, and metastables which will spill out of the discharge; and by the various reactions occurring in gas discharges, they will create electrons in the cell to be started. Which of these particles is most important depends mostly on the distance between the pilot cell and the primed cell. The electrons and ions will usually follow the electric-field lines and thus will be influential only when the pilot cell is very near the primed cell.

The metastables are not influenced by the field and thus obey the laws of diffusion with the boundary condition that the metastable density at any wall is zero, since any metastables hitting the wall will be de-excited. Also, metastables will be depopulated by the Penning reaction. For most plasma displays, then, the metastables are effective in priming only when the primed cell is near the pilot cell. The metastables can create electrons in the primed cell by either the Penning effect or by hitting the cathode and ejecting an electron by the cathode secondary electron process discussed in section 10.4.1.

The pilot cell will also create a large number of photons, and these can create electrons in the primed cell by photoemission. For most cathodes used in plasma displays, photoemission can only occur for deep ultraviolet photons. These photons are strongly attenuated when passing through glass, and thus there has to be an unobstructed path between the pilot cell and the primed cell for photoemission to be effective. When unobstructed, most of the photons generated by the pilot will travel freely without being absorbed by the gas. Thus the photons can have very-long-range priming abilities, and the distance between the pilot cell and the primed cell can be great.

Self-Priming. Another type of priming frequently used is self-priming. In this case the pilot cell is the same as the primed cell. This is used when the cell must be periodically pulsed. If the period since the last discharge pulse is not too long, there still may be priming particles such as electrons, ions, metastables, or photons in the gas volume. These particles can be used to start the next discharge. For self-priming it is still necessary to prime the cell by some other means when the display system is first turned on for the day. Of course, care must be taken to insure that the critical starting particle decay time between discharges is not exceeded by normal operation, otherwise the cell will possibly not discharge.

Radioactive Priming. One type of priming occasionally used in plasma displays is accomplished by adding a small amount of radioactive gas such as Kr^{85} to the gas mixture or by adding a radioactive Ni^{63} wire to the display. The radioactive particles emitted will have sufficient energy to ionize a large number of atoms. This is attractive because this type of priming does not require any special priming cells or special pulse timing. The amounts of radioactive gas or wire required are so low that the cost is usually negligible. The only major disadvantages of radioactive materials are the health risk associated with handling radioactive materials and the very special care that must accompany their use during manufacturing. The health risk to the user is insignificant since both Kr^{85} and Ni^{63} emit beta particles which cannot penetrate the glass envelope.

10.4.7 Time-Varying Characteristics. The electro-optical characteristics shown in Figs. 10-5, 10-6, and 10-7 are for the steady-state gas discharge. However, pulsed voltages are applied to most plasma displays, so the transient characteristics need consideration. The engineer is frequently concerned with the time for the discharge to grow to steady state so that he can set an appropriate pulsewidth that will insure that the discharge will emit light for the required duration. Also, the decay of the gas discharge (called the afterglow) is significant because it determines the initial densities of priming particles for subsequent discharge pulses. The afterglow characteristics will influence the design of the pulse repetition rate.

Initial Growth of Discharge. Two delay times are generally considered in the discharge growth: statistical delay and formative delay. The actual delay is the sum of the two, which in plasma displays can range from 100 ns to 100 μs.

The statistical delay is due in part to the fact that at least one electron is required to be in the gas volume to start the avalanches rolling. This electron must come from some priming current I_o as shown in Fig. 10-11 and discussed in Section 10.4.2. The value of I_o is frequently very low so that the statistical variations of shot noise become important. In addition, there are certain statistical variations in the size of each electron avalanche and in the number of electrons ejected from the cathode by secondary processes. These statistical delays become quite variable when the loop gain μ of the gas-discharge feedback model seen in Fig. 10-11 is near the value of 1. In this case, a series of avalanches started by a single electron may continue to feedback only one electron for each generation. Such a sequence will terminate if the shot noise statistics ever prevent the cathode ejection of this single electron.

Statistical delay is minimized by increasing the value of the priming current I_o so that shot noise is no longer a factor. It is also significantly reduced by applying a voltage considerably above the firing voltage so that μ is much greater than one. This will insure that many more than a single electron will cause the continued growth of the avalanche sequences.

Once the growth of the discharge has matured beyond the statistical range, one must still wait for the formative delay before the discharge reaches the desired current level. The formative delay is simply the time that it takes for the exponential current growth to develop from the initial current I_o to the final mature current when the statistical delay is neglected. The upper right corner of Fig. 10–11 shows that the discharge-current-growth equation takes the form:

$$I = I_o(A + Be^{at})$$

where I_o is the priming current, A and B are time-independent parameters, a is the positive-growth-rate parameter, and t is time. In plasma display discharges the growth parameter a can vary be-

tween 10^5 and 10^8 s^{-1} depending on how high the applied voltage is above the firing voltage. The formative delay obviously is strongly dependent on the value of a, which depends strongly on the applied voltage. Formative delay is minimized with high applied voltages. The display engineer usually makes a compromise between the amount of applied voltage his circuits can tolerate and the amount of statistical and formative delay he can allow.

Afterglow. The decay of the gas discharge after the applied voltage pulse is removed is frequently called the afterglow. The visible light decays very rapidly, but many of the other particles in the discharge decay much more slowly and influence the priming of subsequent discharges. The mature-gas discharge has created a large population of electrons, ions, photons, and metastables. Most of the photons decay within the first microsecond after the discharge, although new photons can be generated by processes such as electron-ion recombination in the afterglow. Metastables decay at a rate dependent on the speed at which they diffuse to the walls and also on the rate at which the are de-excited by the Penning reaction. For Penning mixtures, Penning de-excitation is the dominant mechanism for determining the metastable decay rate. For 300 torr of neon plus 0.1% argon found in many plasma displays, a decay time-constant of one to two microseconds is typical for metastables.

The decay rate of the ions and electrons depends on whether they are in a plasma state or not. Those outside of the plasma are strongly influenced by the electric fields that are usually still present in the afterglow, and thus they are swept to the wall within a microsecond. The electrons and ions still in the plasma will usually not be influenced by external electric fields, and thus the decay is governed by diffusion and recombination. For typical gas discharges found in plasma displays, the plasma decay-time constant is between 5 and 50 microseconds.

The major influence of the afterglow is on the priming of subsequent discharges. Since priming currents can be very small and the initial afterglow particle densities can be very large, the afterglow can have significant influence on priming for many afterglow decay-time constants after the fall of the voltage pulse. This influence can still be significant for times as long as one millisecond.

10.5 CURRENT-LIMITING TECHNIQUES

As discussed in Section 10.3.3, the gas-discharge electrical characteristic requires that plasma displays incorporate some form of current limiting to restrict the excessive currents that will otherwise destroy the device. There are a number of different ways of accomplishing this current limiting. We will discuss the four techniques that have predominated in the plasma display literature.

Fig. 10-14 shows the two techniques of current limiting that are used in the commercially successful plasma displays. They form the two major divisions of plasma displays: ac displays and dc displays.

10.5.1 DC Current Limiting. The dc display current limiting, shown on the left of Fig. 10-14, simply uses the current-limiting properties of the resistor in the same way as that discussed in Section 10.3.2. The resistor is usually chosen to be in the range of 10 kilohms to

1 megohm and is usually a discrete resistor that is placed external to the gas envelope, even though laboratory displays have been demonstrated with internal resistors.[51,52] In practice, the power source pulse generator produces pulses of one polarity so that the same electode is always the cathode. The current pulse follows the voltage pulse and is also of one polarity.

10.5.2 AC Current Limiting. The ac displays shown on the right of Fig. 10-14 use an internal dielectric-glass layer to limit the current. This dielectric glass forms a small capacitor that is in series with every gas discharge in an ac plasma display. Because the dielectric glass is an excellent insulator, no dc current can flow, so that an ac voltage must be applied to maintain a discharge. This will produce currents that must flow with both polarities. The cathode will alternate on each half cycle between the electrodes. Thus the negative glow will appear to the viewer to be on both electrodes. The current is always of a pulsed nature in marked contrast to the relatively constant current of the dc displays. No external resistor is needed because the buildup of voltage across the dielectric limits the current. In general, for comparable average

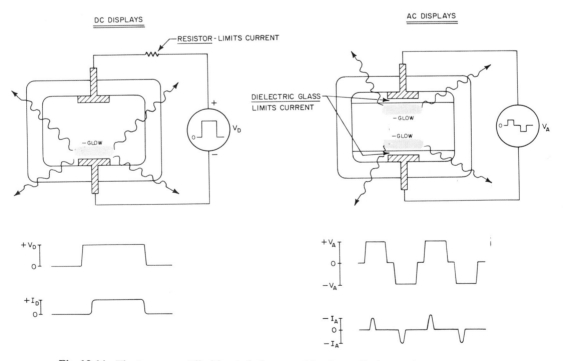

Fig. 10-14. The two current-limiting techniques used in plasma display products in early 1983.

luminous intensity the peak current of the ac display will be greater than that of the dc display because of the low duty cycle of the ac discharge.

As seen in Fig. 10-14, the major difference between ac and dc displays is that the dc electrodes are in intimate contact with the gas and the ac electrodes are insulated from the gas. One could apply an ac voltage to a dc display and it would function properly. However, this arangement would still be considered a dc display because the electrodes are in intimate contact with the gas and a resistor is still required for current limiting.

10.5.3 Other Current-Limiting Techniques.

Two other current-limiting techniques have been used in the laboratory for dc displays. These techniques are not being used in any current products, but they have seen considerable experimentation especially for use in the television displays discussed in Section 10.9.3.

Pulsewidth Current Limiting. The gas-discharge electrical characteristic shown in Fig. 10-5 is for the dc steady-state conditions. If a voltage pulse is applied to a gas cell, a buildup time is required to reach the large currents. Holz realized that the currents could be controlled by pulsing the applied voltage so that the falling edge of the pulse would quench the discharge before the destructive currents would be reached.[53] This technique could be used for dc displays without the need for a current-limiting resistor. As an added bonus, this technique allowed the gas discharge to operated in the memory mode.

One disadvantage of this technique is that the current control is open-loop, in contrast with the closed-loop nature of the other current-limiting techniques. This means that if the gas discharge has different characteristics than anticipated due to such things as manufacturing defects, the current may grow to a destructive level before the end of the voltage pulse.

Abnormal-Glow Current Limiting. Figure 10-5 shows that the abnormal-glow region of the discharge has a positive-resistance region. If the materials and geometry of the discharge are properly chosen, the current levels of the positive-resistance abnormal-glow region can be designed to be low enough to prevent destruction. This kind of dc display can be safely operated with a constant dc supply voltage without a current-limiting resistor. One way to achieve the proper characteristic is to use a graphite cathode.[54,55] The graphite is not used as a resistor but rather as a cathode with a low secondary electron (γ) coefficient that gives the desired abnormal-glow characteristic. A second way to achieve this is to use a helium-based gas mixture at a low gas pressure.[56]

10.6 DC PLASMA DISPLAYS

The left branch of Fig. 10-1 shows the family of dc plasma displays. The first noteworthy point of this family is that none of the commercially successful dc displays currently use the memory feature.* The dc gas discharge is fully capable of exhibiting the memory effect as discussed in section 10.3.2. Of course, memory would be of little value to the segmented displays and the NIXIE displays because of the small number of pixels involved. However, the dc dot matrix and shift displays would benefit from memory operation just as the ac memory dot matrix and shift devices do. Memory can be used to advantage in matrix displays to increase the pixel duty cycle by eliminating the need for scanning the displays. This results in a brighter display, as discussed in Section 5.4.

There must be one bit of memory for each pixel. This means that for a dc display there must be one resistor for each pixel. The gas discharge requires that this resistor be in the range of 10 kilohms to 1 megohm. Unfortunately, it is very difficult to make small resistors in this range of values that will withstand the high voltages and that can easily be incorporated in series with each plasma display pixel.[51,52,57] So, unfortunately, dc memory plasma displays do not appear practical at this time.

The remainder of Section 10.6 will discuss the four major divisions of the dc refreshed devices seen in Figure 10-1.

---

*In a recent development, too late to include in the main text, Thorn EMI Brimar Limited has introduced a dc plasma display product with inherent memory that uses a thick-film resistor in series with each pixel. They have achieved an array of 135 × 150 pixels at a resolution of 11.8 lines per cm (30 lines per inch).

10.6.1 NIXIE Tube. One of the earliest practical gas-discharge display devices was the NIXIE tube developed at Burroughs Corporation in the early 1950's.[32,33,58] This device is shown in Fig. 10-15. It was primarily successful as a numeric indicator but has also seen considerable application as a 14-segment alphanumeric display. The numeric device usually has 10 cathodes and one anode. Each of the ten-wire cathodes is in the shape of one of the ten decimal digits. The cathodes are placed one in front of each other, and a wire-screen anode is placed around them. The anode is typically connected to a 170-volt supply through a 10-kΩ resistor. A digit is selected by switching the appropriate cathode to ground. The negative glow fully covers the wire of the selected cathode. The cathodes not switched to ground do not glow because they are biased to typically 40 volts, which is too high for significant discharge-current flow.

A Penning-mixture gas of neon plus 0.5% argon gas is used at a pressure of 25–50 torr. This is considerably lower than that used in most plasma displays since the distance between the average cathode and the anode was typically 5 to 10 mm. Thus, because of the Paschen relation seen in Fig. 10-13, the pressure must be low to operate at the minimum voltage.

Use of Mercury to Extend Life. The initial life of the first NIXIE tubes was in the hundreds-of-hours range. By adding a small amount of mercury the life could be extended to many tens of thousands of hours.[34] This technique was a key descovery that is still used by all dc plasma display manufacturers today (1983). Mercury reduces a destructive mechanism called sputtering, which erodes away the cathode and thereby alters γ. In NIXIE tubes without mercury, the γ would increase with aging, thereby causing the firing voltage to decrease. This means that the cathodes used most often will eventually have a lower firing voltage than those used less frequently. Thus, after a sufficiently long time, it may not be possible to fire all of the cathodes with a single anode voltage. The display quality, then, will have been degraded, possibly defining the end of its useful life. Also, the material sputtered off the cathode would cause other problems such as coating the inside of the glass envelope like a mirror and coating the other cathodes with material that would change their γ and cause them to glow over only part of the cathode.

The NIXIE tube was the first nonmechanical electronic display device to receive widespread application. Still in use in many instruments today, they are currently being manufactured by National Electronics mostly for the replacement-parts market. They are generally considered obsolete for the following reasons: (1) They utilize the vacuum-tube structure; (2) their display quality is blemished by the fact that the

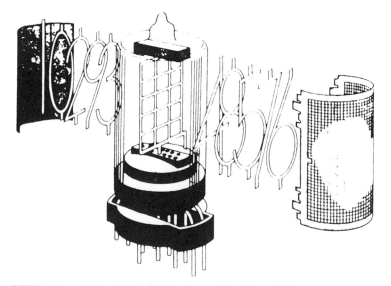

Fig. 10-15. The NIXIE tube display developed by Burroughs. (From "Electronic Numbers" by Alan Sobel. Copyright © 1973 by Scientific American Inc. All rights reserved.)

viewer can see the intervening cathodes, and when the display is rapidly changing digits, the in-and-out motion of the cathode glows is distracting; and (3) they are not matrix addressable and thus require a circuit driver for each digit. In spite of these limitations, the NIXIE tube was a great success story that made practical the concept of the electronic digital readout.

10.6.2 Segmented Discharge Displays. Since the NIXIE tube used the technology of the vacuum tube, it was large and difficult to fabricate. It also required many electrical connections if a large number of digits were needed for the display. Figure 10-16 shows the segmented dc display made by Burroughs that replaced the NIXIE tube.[59] The anode is made of a transparent conductive thin-film material such as tin oxide or indium–tin oxide. The anode electrodes are etched on the front glass plate. The cathodes are screen-printed on a ceramic or glass substrate with thick-film metal, each digit being formed by the seven-segment bar structure. The spacer is typically 0.5 mm thick, and the gas pressure is 60–100 torr of neon-argon Penning mixture with a small amount of mercury added to increase lifetime.

An alternate structure for a similar device is shown in Fig. 10-17. This unit was initially developed by Sperry and is now manufactured and marketed by Beckman. The basic difference of this device is the structure of the cathode substrate, which is a black glass plate molded around the cathode leads that pass through the

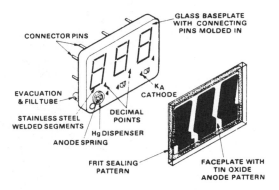

Fig. 10-17. The raised-cathode display made by Beckman. (Courtesy of Beckman Instruments Inc.)

back of the display. The cathode segments are small stainless steel strips spot welded to the substrate leads. This causes the cathodes to be raised slightly off the substrate and allows the cathode glow from one segment to overlap the glow of an adjoining segment, thus giving a more continuous-looking display.

While most applications of segmented dc displays have been for relatively small display areas, very large displays can be achieved by using multiple display panels in a single display. Figure 10-18 shows such a display made by Telegenix. This display has 24 lines of 80 characters with a display area measuring 2.6 X 2.0 meters (102 X 80 inches). It is composed of a mosaic of 240 dc plasma panels, each having eight characters. Each character has 16 segments and is capable of displaying the full range of alphanumerics.

Multiplexing Segmented Displays. Figure 10-16 shows that the cathode substrate of this segmented device requires only seven electrical connections. Each of the cathode contact pads is connected to one of the seven segment cathodes, and each of the corresponding cathodes of all of the digits of the panel are connected in parallel. These parallel connections are accomplished in a metallization layer beneath the gas-discharge area. This metallization is coated by a dielectric glass so that extraneous discharges do not occur. The Fig. 10-16 structure is designed to reduce the total number of circuit drivers required by means of the principles of multiplexing discussed in section 5.3. Plasma displays are ideally suited for multiplexing because of the very large non-linearity associated with gas discharges. This nonlinearity is seen in

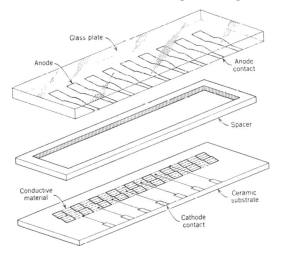

Fig. 10-16. The multiplexed dc segmented display. (From *Electronic Displays* by Sol Sherr, copyright John Wiley & Sons)

Fig. 10-18. Very large plasma display made by Telegenix. (Courtesy of Telegenix Inc. Cherry Hill, New Jersey, USA)

Fig. 10-5. For voltages slightly below the firing voltage, the current may be 5 or more orders of magnitude below the final current for voltages above the firing voltage; thus, the nonlinearity is very strong.

For the Fig. 10-16 device, the multiplexing is accomplished by scanning the anodes one at a time with 170 volts and then placing 0 volts across the selected cathodes which represent the desired digit. The unselected cathodes must have some intermediate voltage, such as 45, to inhibit discharges on these cathodes. The unselected anodes must have a voltage of about 110 volts to reduce extraneous discharges. If instead the unselected anodes were reduced to 0 volts, then these anodes would behave as cathodes to the selected anode, which is at 170 volts.

As with the NIXIE tube, the light from the segmented dc plasma displays comes mostly from the negative glow. The current through the device must be sufficient to cause the glow to cover the full cathode. This means that oper-

ation must be at the upper edge of the normal-glow region and perhaps slightly into the abnormal-glow region of Fig. 10-5. Thus there is a minimum current, below which the full electrode will not be illuminated.

As discussed in Section 10.5.1, a resistor or other impedance is placed in series with a display cell to limit the current. In the multiplexed segmented display of Fig. 10-16 the resistor is usually placed in the cathode lead. This is because when an anode is selected in the scanning process, a number of cathodes will probably be turned on at one time; thus one resistor is available to limit each segment discharge. If the resistor was placed in the anode, then the cathodes would compete for the limited amount of current that the resistor could supply. Thus for digits such as 8, with all of the segments on, most segments would probably be operating in the normal-glow region and the segments would not be fully covered with glow.

For the structure seen in Fig. 10-17, the dis-

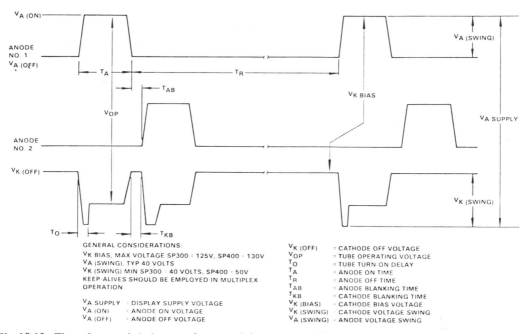

GENERAL CONSIDERATIONS:

V_K BIAS, MAX VOLTAGE SP300 = 125V, SP400 = 130V
V_A (SWING), TYP 40 VOLTS
V_K (SWING) MIN SP300 = 40 VOLTS, SP400 = 50V
KEEP-ALIVES SHOULD BE EMPLOYED IN MULTIPLEX OPERATION

V_A SUPPLY : DISPLAY SUPPLY VOLTAGE
V_A (ON) = ANODE ON VOLTAGE
V_A (OFF) = ANODE OFF VOLTAGE

V_K (OFF) = CATHODE OFF VOLTAGE
V_{OP} = TUBE OPERATING VOLTAGE
T_O = TUBE TURN-ON DELAY
T_A = ANODE ON TIME
T_R = ANODE OFF TIME
T_{AB} = ANODE BLANKING TIME
T_{KB} = CATHODE BLANKING TIME
V_K (BIAS) = CATHODE BIAS VOLTAGE
V_K (SWING) CATHODE VOLTAGE SWING
V_A (SWING) = ANODE VOLTAGE SWING

Fig. 10-19. The voltage and timing waveforms used for segmented displays. (Courtesy of Beckman Instruments Inc.)

play engineer is free to decide to multiplex or not since all of the electrodes come out of the device. Thus the resistors can be placed in either the cathode or the anode lines.

Segmented DC Timing Waveforms. Figure 10-19 shows the typical voltage waveforms applied to multiplexed segmented dc displays. This shows the waveforms for two anodes which correspond to two different digits, and also the cathode voltage for one of the segments in the two digits. The timing of these waveforms can be quite variable depending on the number of multiplexed digits, the refresh rate, and the level of external priming. For instance, the anode off time T_R must be short enough to avoid flicker. In addition, it must not be longer than a few milliseconds or the afterglow will have fully decayed and self-priming will not function. However, this time limitation is not important if a nearby pilot cell is providing sufficient priming.

The anode on time T_A can vary from 100 microseconds to 2 milliseconds. The minimum is set by the uncertainty in the gas-discharge turn-on delay T_O. For a properly primed segment, the statistical and formative delays sum to typically 20 microseconds. If T_A is made shorter than 100 microseconds, then the statis-

tical delay uncertainty will significantly modulate the duty cycle and therefore the brightness of the segment.

The cathode blanking time T_{KB} must be sufficiently long or extraneous discharges will occur between adjacent digits. T_{KB} is typically set at 20 microseconds. Times significantly shorter will not allow the particle densities in the afterglow to sufficiently decay. The extraneous discharge initiated by the excess afterglow particles will occur between the newly pulsed anode and the cathode segments of the previously discharging digit.

The voltage levels defined in Fig. 10-19 are a compromise between minimizing the pulse swing amplitudes and reducing the possibility of extraneous discharges. It is interesting to note that while the total voltage across the display is 180 volts, the address circuit drivers only need to apply pulses of 40 to 50 volts. This is a sufficiently low voltage to be within the range of a large number of integrated circuit drivers.

10.6.3 DC Dot Matrix Displays. Figure 10-1 shows that there are a large number of companies manufacturing dc dot matrix displays. For the purpose of this discussion, dot matrix displays are defined as arrays of dot-shaped pixels controlled by a set of X and Y electrodes that do not utilize the shift-register principle. The shifting displays will be covered in Section 10.6.4 and are considered a separate division of dc plasma displays as seen in Fig. 10-1.

Scanning Technique. The fundamental operation of most dot matrix displays is shown in Fig. 10-20. Each column of cells has a common vertical electrode connected to a scan switch. The seven rows of this display each have common horizontal electrodes which are connected to seven data switches. In a way very similar to the multiplexing of the segmented displays, the display is scanned by closing each of the scan switches one at a time. When a given column scan switch is closed, the appropriate horizontal electrode data switches are closed, causing a discharge in the cells that are to be on for that column of cells. The example shown in Fig. 10-20 has the appropriate switches closed to discharge the leading column of the letter A. The other crosshatched pixels are energized at a different time in the scan sequence. The display is, of course, scanned at a rate high enough to avoid the flicker problem. Note that current-limiting resistors are placed in series with each data switch.

While Fig. 10-20 shows the columns being scanned one at a time, the positions of the scan switches and data switches can be exchanged so that the rows are scanned. This choice is frequently determined by scanning along the axis with the fewest number of electrodes. This maximizes brightness because of the higher duty cycle.

It should be remembered that in actual drive circuit implementation, the scan and data switches do not open and leave the electrodes floating. Another circuit, not shown in Fig. 10-20, is used to bias the electrodes to some voltage level that will not cause extraneous discharges when the switches are open.

Okaya-Oki Display. Figure 10-21 shows the internal structure of a dot matrix display made by Okaya and Oki. The cathodes form the vertical columns which are connected to the scan switches, and the anodes form the rows that are connected to the data switches of Fig. 10-20.

The cathodes are the electrodes closest to the viewer, unlike most other dc plasma displays. The electrodes are made of a comparatively thick sheet metal, which is opaque. Holes are etched in the electrode in the display cell area to allow the light to escape. Since most of the light comes from the negative glow, the front cathode arrangement allows more light to reach the viewer.

The electrode spacing of dot matrix displays is generally much closer than in segmented

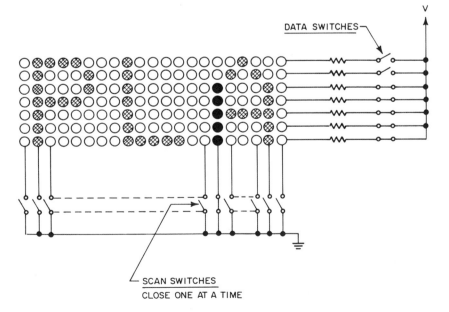

Fig. 10-20. Scanning technique used for matrix displays.

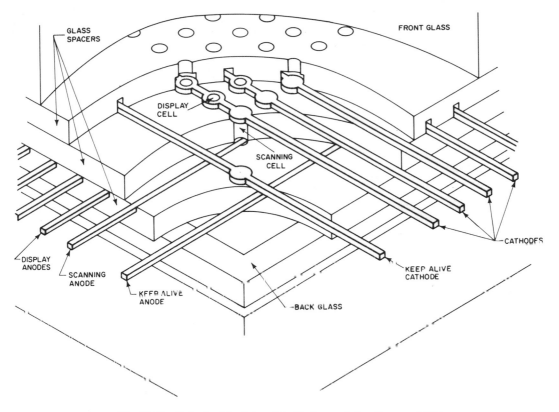

Fig. 10-21. Refreshed dc dot matrix display made by Okaya and Oki.

displays. To prevent extraneous discharges between the adjacent electrodes, a glass spacer sheet is used, having holes positioned at each display cell. This greatly reduces the cell-to-cell interaction, but it also creates a priming problem. The cells are so well isolated that it is difficult to get them to start discharging. This semirandom starting time will influence the duty cycle of the pixel, which will cause significant degradation of the display.

To solve this problem a special system of priming cells is used. There are two types of priming cells: fixed cells and scanning cells. These priming cells are not visible to the viewer since their opaque cathode electrodes do not have holes. The fixed priming cells are located only on the left and right edges of the panel. These cells are discharged constantly by means of the voltage between the keep-alive anode and the keep-alive cathode shown in Fig. 10-21. The scanning priming cells are powered by the potential between the scanning anode and the column cathodes. There is a scanning priming cell for each column of the display. These cells

discharge one at a time as the scan switches which drive the cathodes are closed. This causes the scanning priming cells to follow and prime the scanning display cells. The scanning priming cells are primed by the fixed priming cell when the display scan is at the right or left edge of the display. At other times the scanning priming cells are primed by each other in the scanning sequence.

Matsushita Display. A different means of solving the isolation and priming problems is demonstrated in the Matsushita display shown in Figs. 10-22 and 10-23.[60,61] Here the isolation is achieved with screen-printed partition ribs. This isolates the cells in a column to prevent extraneous discharges. The cells in the row are not isolated with this technique, however. This is not a problem, though, because the display is scanned with the technique discussed in Fig. 10-20, so that at any given time only one column of cells has sufficient voltage to cause a discharge. This means that extraneous discharges along the rows are unlikely.

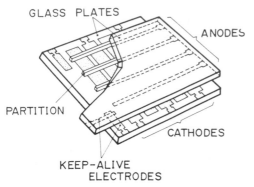

Fig. 10-22. Refreshed dc dot matrix display made by Matsushita. (Reference 61 courtesy Society for Information Display)

The priming problem is solved by having an auxiliary priming discharge for each pixel. Fig. 10-23 shows that this auxiliary discharge area is blocked from view by a black-mask layer and is separated from the display discharge area by a glass barrier. Note that a given anode electrode serves both the auxiliary discharge area and the display discharge area. The panel is scanned with the technique of Fig. 10-20. For each scan period every auxiliary priming discharge is fired once. This is accomplished by closing the data switches for a short period during each longer scan switch closure. This makes the auxiliary discharge emit light for a short period just sufficient to generate the priming particles. If a

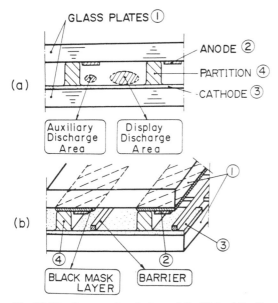

Fig. 10-23. Cross-sectional view of the Matsushita display. (Reference 61 courtesy Society for Information Display)

given pixel is to emit a bright discharge, the appropriate data switch is closed for a longer time so that the auxiliary discharge will bleed over into the display discharge area. Stated another way, all of the data switches close every time the scan switches change from one column to the next. If the data switch closes for a short time, only the auxiliary discharge fires and generates priming particles. If a data switch closes for a long time, then the discharge will develop beyond the auxiliary discharge area into the display discharge area and the viewer will see a much brighter pixel.

10.6.4 Self-Scan™ Display. A major problem with all matrix displays is the high cost of the driver electronics due to the large number of panel electrodes, each requiring circuit drivers. The Self-Scan™ displays developed by Burroughs lower the cost of the electronics by greatly reducing the number of external panel electrodes.[4,41,62,63] This is accomplished by designing the plasma display so that it performs some of the addressing logic internal to the panel. This exploits one of the major attributes of the gas discharge, which is the ability to perform logic functions.

Figure 10-20 shows that the traditional scanning scheme requires both scan switches and data switches. The actual display information flows through the data switches. The scan switches simply close one at a time in a simple sequence. The Self-Scan™ panel performs the function of the scan switch with gas-discharge logic internal to the panel. The scan-switch function can be accomplished with a very low number of external electrodes (4 to 8). Table 10-1 compares the number of circuit drivers required for the Self-Scan™ display and for a standard dot matrix panel. The table shows that the Self-Scan™ offers a very significant economy in electronics cost. Of course, the Self-Scan™ panel cost is generally higher than that of the simple dot matrix panel. The important goal is to lower the total system cost, which is the sum of the panel and the electronics costs.

The remainder of this section will cover the details of the Self-Scan™ structure and operation. Additional details of this technology can be found in an excellent review article by Cola et al.[4]

Table 10-1.   Circuit Drivers Needed for Dot Matrix vs. Self-Scan^{TM} Displays

No. of 5 × 7 Characters	Dot Matrix	No. of Row Drivers	No. of Column Drivers	Total No. of Drivers for Self-Scan^{TM}	No. of Drivers Needed Without Self-Scan^{TM}
1 line of 16 = 16	7 × 96	7	4	11	103
1 line of 40 = 40	7 × 283	7	12	19	290
6 lines of 40 = 240	42 × 283	42	12	54	325
12 lines of 40 = 480	84 × 283	84	12	96	367

Gas-Discharge Switching for the Self-Scan^{TM} Display.   Figure 10-24 shows that each display pixel is broken up into two separate cells: a display cell that the viewer sees and a scan cell, behind the display cell, which the viewer does not see. Separating these two cells is a cathode with a small hole which allows priming particles to pass. The scan-switching action of Fig. 10-20 is accomplished by the scan discharges. These scan cells have the ability to control the state of the display cells by sending active priming particles through the priming hole. At any given time there is only one scan discharge in the scan cell channel. This discharge sequentially moves under each display cell and generates the priming particles which switch the display cell on.

Scan-Discharge Operation.   Figure 10-25 shows how the scan discharge is made to sequentially move from one pixel to the next. This figure shows only the electrodes necessary to make the scan discharge operate properly. Note that except for the reset cathode, every third cathode is connected to a common bus. This connection can be done internal to the panel so that this whole line of scan discharges can be controlled with just five external connections: the three cathode buses, the reset cathode, and the scan anode.

The signals applied to the cathodes are shown on the right side of Fig. 10-25. The discharge sequence is initiated on the left by grounding the reset cathode, ϕ_R, which starts a discharge

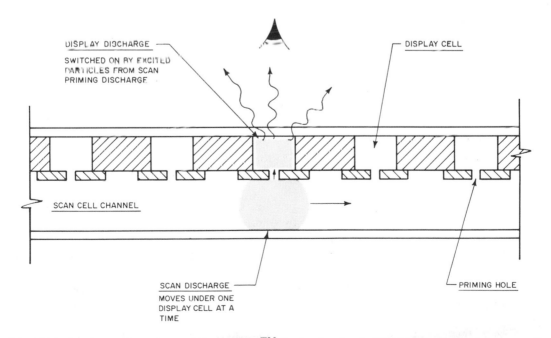

Fig. 10-24. Fundamental operation of the Self-Scan^{TM} display showing the display cells above and the scan cell channel below.

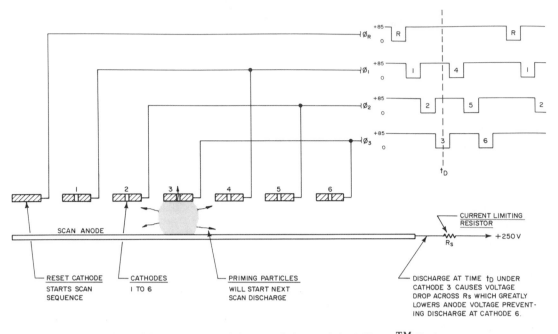

Fig. 10-25. Operation of the scan discharge of the Self-Scan^{TM} display.

between the reset cathode and the scan anode. After this reset period, the ϕ_R voltage is raised back to +85 volts and the voltage ϕ_1 on cathodes 1, 4, 7, ... is reduced to zero. Since 250 volts is across each of these cells, one might expect that discharges would occur in all of them. However, this does not occur because of two fundamental principles:

1. First note that all of the scan cells have a common anode current-limiting resistor. Once one cell starts to conduct, the voltage drop across this resistor will reduce the voltage across all of the other cells, which prevents them from discharging.
2. The cell that receives the most priming particles will conduct first. These particles will be the strongest in the most recently discharged cell.

In the Fig. 10-25 example, at the time of ϕ_1 pulse 1 there will be a discharge only under cathode 1. It will discharge first due to the influence of the priming particles from the reset-cathode discharge, which immediately precedes the cathode-1 discharge. Since cells 4, 7, ... are much farther from the reset cathode at the time of pulse 1, they do not have a chance to dis-

charge before the cathode-1 discharge current reduces their voltage. The next transition of cathode voltages at ϕ_2 pulse 2 will cause the discharge of cathode 1 to prime the discharge of cathode 2. This process will continue to make the single-scan discharge transit one step at a time in one direction across the scan channel. When this discharge reaches the right edge of the scan channel, the scan sequence is re-initiated by pulsing the reset cathode again.

Display-Discharge Operation. Figure 10-26 shows the electrode configuration used for the display discharges. The cathodes are the same electrodes as those in Fig. 10-25. The horizontal-display anodes situated above the cathodes are distinct from the scan anodes of Fig. 10-25. Self-Scan^{TM} displays have a horizontal-scan channel for each display row. Figure 10-26 does not show the four scan cell channels that run horizontally under the cathodes.

Figure 10-26 shows the display-discharge situation for time t_D when the four scan discharges are beneath cathode 3. The switching of the display cells by the scan discharges is accomplished by the priming particles which pass through the priming hole from the scan discharges and by the external data switches.

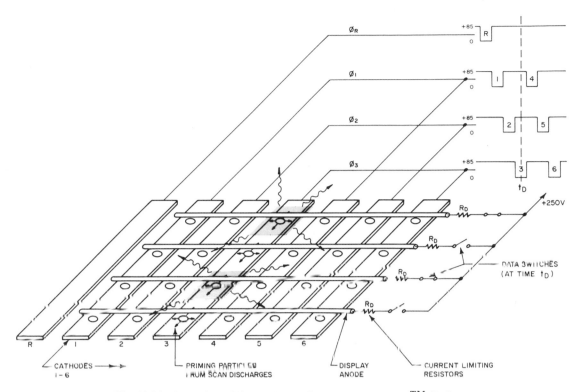

Fig. 10-26. Operation of the display discharge of the Self-Scan^{TM} display.

Note that cathode 3 has priming particles passing through all four holes and that two of these four are selected for discharging by the closed data switches connected to the display anodes. In a manner very similar to that discussed above for the scan discharges, there is a potential conflict, since at time t_D there is 250 volts across the selected cells of both cathode 3 and cathode 6. Only the cells along cathode 3 will discharge, due to the same two fundamental principles discussed above for the scan discharges. The cathode-3 display cells will discharge first, due to the priming particles from the scan discharges, and this discharge will cause a voltage drop across the common display anode resistor R_D which will lower the voltage across the cathode-6 cells and prevent them from discharging.

Self-Scan^{TM} Structures. Figure 10-27 shows the detailed structure of the Self-Scan^{TM} display. Comparing this to Figs. 10-24–10-26, we see that there are seven parallel scan channels corresponding to the vertical dimension of the 5×7-character size. Each of the cathodes has

seven priming holes that operate in parallel. Each of the seven scan anodes has a scan groove which prevents the scan discharges on the various anodes from interacting. Similarly, a ceramic cell sheet with a hole for each display cell is needed to prevent display cell interaction. The reset cathode does not have display cells and has a special shape to reduce its firing voltage, insuring reliable starting. At the left edge of the display there are keep-alive discharges that are fired continuously to provide proper priming for the reset-cathode discharge. Once the reset discharge is primed, the scan discharge sequence provides priming for the rest of the display.

Figure 10-28 shows an improved design of the Self-Scan^{TM} display used for most current production.[62, 63] Note that the center cell sheet is replaced with a greatly simplified glow spacer. Also, the top substrate is grooved for the display anodes. The manufacturing techniques for this type of display can be reasonably low-cost. The scan and display anode grooves are sawed with high-speed diamond saws similar to those used by the semiconductor industry to cut silicon wafers. The priming holes in the cathodes are

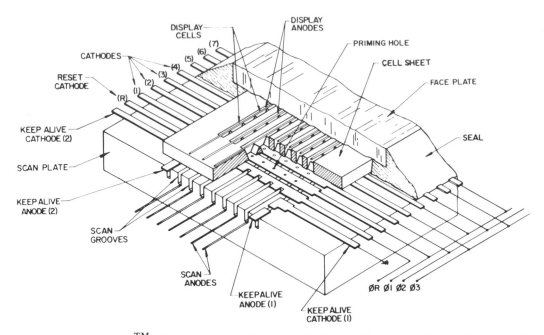

Fig. 10-27. The Self-Scan^{TM}I display made by Burroughs. (Reference 63 courtesy Society for Information Display)

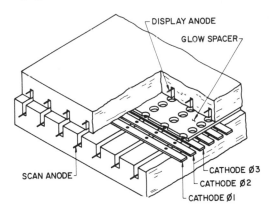

Fig. 10-28. The improved Self-Scan^{TM} I display made by Burroughs. (Reference 63 courtesy Society for Information Display)

machined at high speed with a laser. The glow spacer is a micalike material punched to the desired dimensions. The scan and display anodes are simply wires which are laid in the grooves. These wires have the advantage that they do not tend to open or short like thin- or thick-film electrode structures do.

A third structure, called the Self-Scan^{TM} II, was also developed and produced.[64] This device has only two electrode layers: one for the anodes and one for the cathodes. These elec-

trodes can be fabricated on two separate substrates by screen-printing techniques. A complex isolation spacer is used that allows the scan and the display discharges to be in the same plane between the substrates. The scan channels are etched into the isolation spacer in the regions between the display cell rows. The spacer also has small priming channels for each pixel that connect the scan channels with the display-discharge channels.

Self-Scan^{TM} Electrical Characteristics. The Self-Scan^{TM} display takes full advantage of the priming mechanism of gas discharges to achieve a high degree of multiplexing. Figure 10-29 demonstrates this effect. For this data a normal scan discharge was started with the cathode N held at 0 volts and all other cathodes at 100 volts. This maintains these cells below the firing voltage. The measured current flowing from the cathodes is shown in curve A, which indicates that priming particles from the discharge at cell N are causing small avalanches to occur in the other cells. In the case of the cells closest to cell N, the discharge activity is of sufficient magnitude to distort the field and increase the overall gain of these cells. This lowers the breakdown voltage, which is the effect necessary to achieve

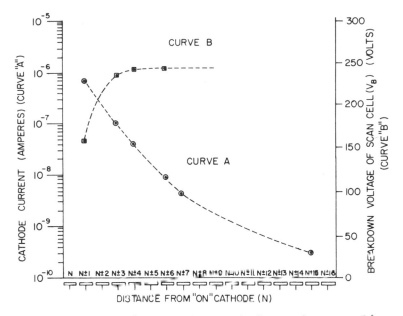

Fig. 10-29. The current and breakdown voltage dependence on the distance of one scan cell from an "on" scan cell at position N. (Reference 4 courtesy of Academic Press)

the scanning of the scan discharges. Note that the cell three cells away from cell N has a much higher breakdown voltage than the cell next to N. This is the effect of fundamental principle number 2.

Figure 10-30 shows how priming from the scan cells affects the I-V characteristic of the display cells. For the scan cell current $I_s = 0$, there is no priming, and the I-V characteristic looks like that seen in Fig. 10-6. As I_s increases, the priming creates the field distortion which lowers the breakdown voltage. Once the breakdown occurs, the current rises and the priming has little influence on the I-V characteristic

because the normal discharge processes cause the field distortion. Characteristics very similar to this can be found for almost all gas discharges where priming influences the breakdown voltage.

Figure 10-31 shows how the remaining excited particles in the afterglow influence the breakdown voltage of display cells. This data was taken by supplying two voltage pulses to the display anodes and by varying the time between the two pulses. The data shows that the breakdown voltage of the second discharge is significantly lowered if the time between the pulses is short. A similar curve could also be drawn for the scan cells. This effect is due to the slow

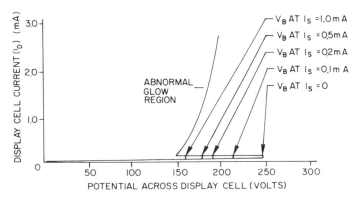

Fig. 10-30. The I-V curve of the display discharge for differing amounts of scan discharge current, I_s. (Reference 4 courtesy of Academic Press)

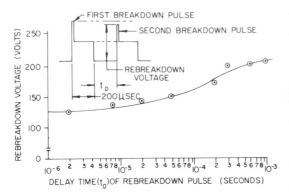

Fig. 10-31. The influences of self-priming on a display cell rebreakdown voltage. (Reference 4 courtesy of Academic Press)

decay of the active priming particles in the afterglow of the discharge, as discussed in Section 10.4.7. In this case this self-priming effect is undesirable because it limits the rate at which the display can be scanned. If firing voltage is again applied to a scan cell before sufficient time has elapsed, then that cell may still have a significant number of priming particles from its own discharge, and it may go on even though it is not primed from a neighboring scan cell. This event would cause the display to misfire because the proper scan sequence would be broken.

From the data in Fig. 10-31, a recovery time of about 200 μs will provide sufficient breakdown voltage recovery. This has the undesirable effect of limiting the scan time. If three phases are used on the scan electrodes and 200 μs is required between discharges, then the cathode pulse width shown in Figs. 10-25 and 10-26 must be at least 100 μs. This effectively limits the number of horizontal cells that can be scanned. If the flicker rate is chosen to be 60 scans per second, then the scan period of 16 ms will only allow 160 cells to be scanned. Since a 5 X 7 character takes a minimum of 6 horizontal dots, such a display would be limited to only 27 characters. One easy way to get around this limitation is to have more than three phases of cathode signals. The only disadvantage to a larger number of phases is the additional circuit cost due to the additional drivers. The additional phases will allow a greater period of time between scan discharges. The maximum number of characters allowed in

the display will be proportional to the increased number of phases. It is common to have seven phases in production displays.

The Self-Scan^{TM} Memory Characteristic. The information on the Self-Scan^{TM} display must be continually refreshed and so it is not a memory display. However, the dc memory effect discussed in Section 10.3.2 is of critical importance to the operation of this device. A very basic premise of Self-Scan^{TM} operation is that a common voltage pulse is applied to a number of cells and that only one of these cells will discharge. This cannot occur without the dc memory effect which allows gas-discharge cells to be in either the discharging or the nondischarging state at a single operating voltage. This is demonstrated by the 500-kΩ, 225-volt loadline characteristic of Fig. 10-6. Here the discharge is stable in either an on or an off state.

Because of the importance of this effect, the design engineer optimizes the gas-discharge-cell geometry to increase the spread between the firing voltage and the minimum discharging voltage. This allows the panel to be operated over a wider range of voltages, increasing the device yield and allowing less critical circuit drivers to be used. Techniques for increasing the voltage spread include operating at higher gas pressures and making the anode grooves narrow.[4]

Self-Scan^{TM} Bar-Graph Display. One rather successful plasma product is the bar-graph display made by Burroughs Corporation. This display utilizes the Self-Scan^{TM} principle. These are digital displays used to present analog information. Figure 10-32 shows the structure of this device. Note that this display requires only seven electrical connections. This is accomplished by using the scan-discharge technique shown in Fig. 10-25. This display has only the scan-discharge area. There is no additional display discharge as there is in the standard Self-Scan^{TM} display. The length of the bar is determined by the number of pixels that the scan discharge is allowed to travel before the scan sequence is re-initiated by re-firing the reset cathode. The same number of electrical connections is required for a longer display having more pixels. The limit to the number

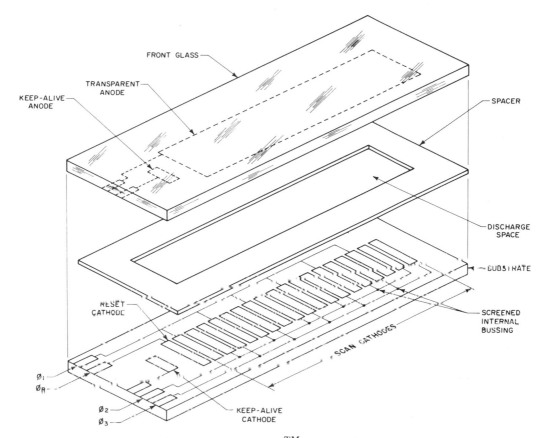

Fig. 10-32. The Self-Scan^{TM} bar-graph display.

of pixels is determined by the usual brightness and flicker constraints common to all refreshed displays. If desired, the pixels can also be arranged in a circle or arc.

Recently Hitachi has developed a bar-graph display that uses a different dc scanning technique.[65,66] They have also demonstrated color by placing phosphors in the device.

10.6.5 Philips DC Display. Philips has recently developed a dc panel with a novel new scanning technique that does not experience duty cycle brightness and flicker limitations as the display is made wider.[67] It also reduces the number of column circuit drivers by nearly a factor of four by means of a gas-discharge logic technique. This design is especially suitable for displays requiring a small number of character lines, each having a large number of characters. For instance, a prototype display was designed with 4 lines of 80 characters having a 5 × 9 dot matrix. This was accomplished with an array of

48 rows and 480 columns. To increase the brightness of this display, the refresh scanning switches are connected to the rows and the data switches are connected to the columns, which is a reversal of the technique shown in Fig. 10-20. For a display of this size, the cost of the 480 column drivers would be rather significant. However, the gas-discharge logic technique requires only 124 column circuit drivers.

Figure 10-33a shows the electrode arrangement for this display. Each pixel is formed at the intersection of a cathode electrode on the bottom substrate and a priming anode electrode on the top substrate. The display anode electrodes are situated between the columns of pixels on the top substrate. The priming anodes perform the priming function and also a selection technique that allows a combination of horizontal and vertical scanning.

The scan sequence is shown on the right side of Fig. 10-33a. It starts at the top of the column at the far left and follows the circuitous route

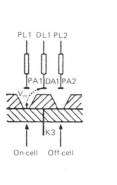

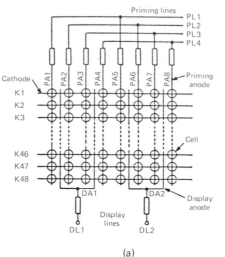

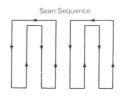

(a)

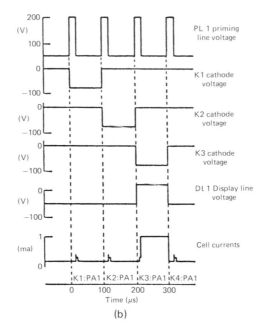

(b)

Fig. 10-33. Refreshed dc dot matrix display developed by Philips. (Reference 67 © 1982 IEEE)

along four columns and then returns to its starting position. This technique is called closed-loop scanning. There is a scan loop for each group of four adjacent columns, and each loop scans simultaneously with the other loops. Closed-loop scanning has the advantage that each pixel is primed by excited particles generated by the neighboring pixel that has just discharged a few microseconds earlier. This means that once the scan loops begin to discharge, they will continue to prime themselves

and therefore eliminate the need for additional pilot priming cells.

The closed-loop scanning sequence is controlled by applying the proper sequence of pulses to the cathodes and the priming anodes. Since every group of four columns has its own scan sequence that operates in parallel with the others, the priming anodes can be bused as shown in Fig. 10-33a. This allows only four address drivers to perform the priming-anode scan job.

One display-anode column electrode is situated between each two priming anodes. A cross-sectional view of this electrode arrangement is shown at far left in Fig. 10-33a. Two of these display-anode electrodes are electrically connected and are driven by a common address driver and current-limiting resistor. The data switches of Fig. 10-20 are used to switch these display anodes. A given pixel can only be addressed when the priming discharge is active in that pixel. Since for a given scan loop there is only one priming discharge at any one time, only one data-switch driver is needed for each scan loop. This trick allows the factor-of-four reduction in column drivers. One might think of each data switch as driving one long column of 192 pixels but with this column cut into four segments of 48 pixels that are rearranged to be parallel to each other.

Figure 10-33b shows the waveforms used for the scan of the first four pixels of the first column. For each pixel period of 100 μs the priming anode is pulsed positive and the selected cathode goes negative. The priming pulse is high for only 20 μs, which is long enough for reliable priming discharge to be initiated. This discharge is not of sufficient intensity to be noticeably visible. If the pixel is to be in the bright state, then the display line voltage is raised for the 100-μs period. This will cause the discharge to continue at a high level after initiated by the priming pulse. For the waveforms of Fig. 10-33b, the cell intersecting K3 is pulsed for the bright state, and the cells intersecting cathodes K1, K2, and K4 are off.

10.7   AC PLASMA DISPLAYS

The fundamental principle of the ac plasma display is to use a capacitor for the current-limiting element, as shown in Fig. 10-14 and discussed in Section 10.5.2.[39, 68] Figure 10-1 shows that a number of successful products use ac plasma displays in a wide range of formats and sizes. Each of these products will be covered in the following text. The reader needing further discussion should consult the other review articles on ac plasma displays.[3, 71-75]

As indicated in Fig. 10-1, ac plasma displays are available in both the memory and the refresh modes. The memory property is unique to plasma displays that use ac current limiting. It is due to the charge that can accumulate on the dielectric walls. Memory means that a refresh scan is not required. Once the image is written, it remains until erased. Although this eliminates flicker and the need for external memory, the most important advantage is that internal memory allows each pixel to operate at a duty cycle of one. Scanned displays operate at a duty cycle of one divided by the number of scanned lines. Because of the saturation effect of gas discharges, the lower duty cycle of scanned displays means lower brightness. So, as a refreshed display scans more lines, it necessarily gets dimmer. Displays with inherent memory retain the same brightness at any size, so that the largest plasma displays on the market are ac with memory.

Figure 10-34 shows the largest plasma display panel commercially available in early 1983. This is manufactured by Photonics and has a display area of 60 by 80 centimeters, giving a 1-meter diagonal with a package thickness of 12 cm (4 inches).[18, 76] The resolution is 19.7 lines per centimeter (50 lines per inch), which yields a 1,212-by-1,596-line display with nearly 2,000,000 pixels. This display is transparent so that a map placed behind the display can easily be seen.

IBM announced the OEM display product shown in Fig. 10-35 in late 1982. This is an array of 960 × 768 electrodes at 28.3 lines per centimeter (72 lines per inch).[77-79] The display area measures 34 by 27.4 centimeters (13.4 by 10.8 inches), giving a 43.7-centimeter (17.2-inch) diagonal. The thickness of the display with electronics is 6.4 centimeters (2.5 inches). This display is used in the IBM 3290 information panel display terminal, introduced in early 1983, which is compatible with the IBM 3270 display terminal product family.

10.7.1 AC Plasma Display Structures

University of Illinois.   Figures 10-36, 10-37, 10-38, and 10-39 show the details of some of the structures used in ac plasma displays. Figure 10-36 shows the original design used by the device inventors at the University of Illinois.[39]

Fig. 10-34. A 1-meter-diagonal ac plasma display with 2 million pixels at 50 lines per inch made by Photonics. (Courtesy of Photonics Technology)

A great deal was learned about the device from this structure; however, the thin glass plates were very fragile and the gas seal had to be done with epoxy, which meant that the gas was easily contaminated. Thus the devices had a very short lifetime. This structure had a center sheet of glass with holes to define the cells. This was necessary to hold the outside pieces of glass apart in order to define the gas gap. Although the cells are isolated from each other in this structure, this is not necessary, as seen below.

Owens-Illinois.   A major advance was the development of the simple structure shown in Fig. 10-37 by Owens-Illinois.[80,81] This was sold under the trade name Digivue™. The substrate glass is typically 6.3 millimeters ($\frac{1}{4}$ inch) thick, and thus the device was not fragile like the University of Illinois structure. Also, all-glass seals were used so that the panel could be baked at high temperatures to drive out contaminants, and thus long lifetimes were achievable. There was no center sheet since the substrate had sufficient rigidity to maintain the gas gap spacing. The individual cells were defined by the electric field generated by the electrodes, and thus there were no critical parts-registration

problems in the manufacturing process since the discharges automatically exist wherever the electrodes intersect. Another advantage of the Owens-Illinois structure over the University of Illinois structure is the reduced operating voltages. This is due to the thinner dielectric glass of the Owens-Illinois structure, which allows a much higher percentage of the voltage applied on the electrodes to appear across the gas. Modern panels with the Owens-Illinois structure are operated with ±90 to 100 volts.

The Owens-Illinois structure has been very successful and is used by all manufacturers of ac memory dot matrix displays. The displays shown in Figs. 10-34 and 10-35 use this structure. The memory capability and the inherent simplicity of this device have made it a tough competitor. Owens-Illinois discontinued manufacture of these displays and sold their plant to other plasma display manufacturers in 1977.

Capillary.   A third structure, shown in Fig. 10-38, was developed and marketed by Control Data Corporation.[82,83] The cells are defined by long, thin glass capillaries that contain the gas. The electrodes are then laminated on both sides of the capillaries. The proponents of the

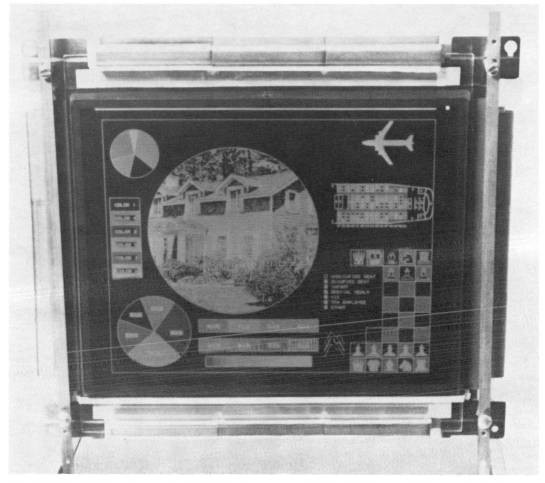

Fig. 10-35. An OEM product made by IBM having an array of 960 × 768 pixels at 72 lines per inch. (Courtesy of International Business Machines Corporation)

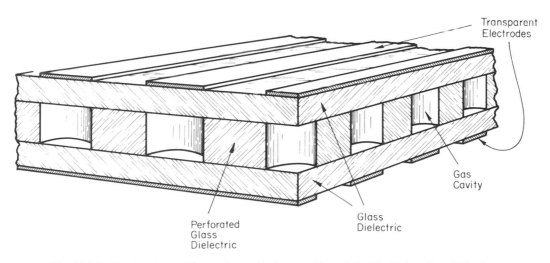

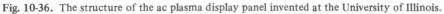

Fig. 10-36. The structure of the ac plasma display panel invented at the University of Illinois.

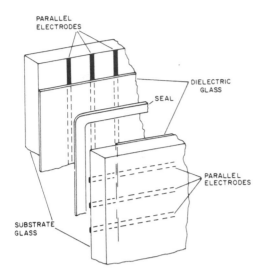

Fig. 10-37. Structure of the ac plasma display panel developed at Owens-Illinois.

The epoxy seals used in production made these devices prone to contamination problems.

Single Substrate. Figure 10-39 shows the single-substrate structure for the ac plasma display.[70,84-86] Note that both the row and the column electrodes are on the same substrate. The fringing fields from these electrodes extend into the glow area and can cause a discharge. This structure has not yet reached commercial production, but it promises displays with higher brightness because there is no front electrode between the discharge and the observer that blocks the light. It may also have less critical fabrication tolerances because, unlike the structure of Fig. 10-37, it does not have a critical distance between the cover glass and the substrate.

structure claimed advantages for making panels as large as one meter on a side. The Owens-Illinois structure has difficulty doing this because the critical tolerance on the gas gap is hard to maintain between the two substrate glass pieces when the dimensions get very large. However, recent results show this to be less of a problem[18,76] (see Fig. 10-34). The capillary structure does not have this problem because the gas gap is defined by the diameter of the capillary.

Panels with the capillary structure were manufactured for a number of years by Control Data but the product line was discontinued in 1978. The problem of handling and sealing a large number of fragile capillaries is substantial.

Details of the Commercial AC Structure. Figure 10-40 shows the details of the ac dot matrix memory panels that have been so successful as products. The substrate material is common float window glass because of its low cost and availability. The electrodes are thin-film gold, $Cr-Cu-Cr$, or Al. They must withstand the firing cycle of the dielectric solder glass which must flow at peak temperatures above $600°C$. Some displays have alterntively used a thin-film dielectric that can be evaporated at comparatively low temperatures.[87,88] The gas is the common Penning mixture for low-voltage operation. The gas pressure is chosen to be lower than atmospheric so that the two substrates do not bow out and distort the gas-gap

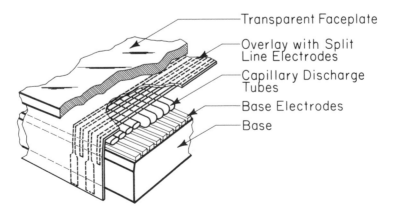

Fig. 10-38. Capillary tube ac plasma display panel developed at Control Data.

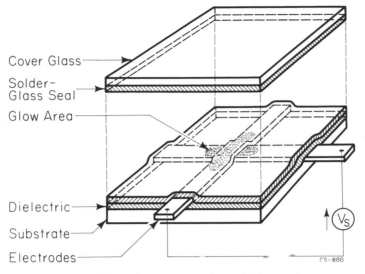

Fig. 10-39. Single-substrate ac plasma display panel.

distance. The gas pressure is made as high as possible so that the gas-gap distance, d, can be made as small as possible and still operate near the minimum of the Paschen curve of Fig. 10-13. The small d allows higher-resolution displays.

MgO Overcoat Layer. One of the most critical parts of this device is the 200-nm thin-film overcoat of MgO that is deposited over the dielectric glass.[89,90] This layer does two things. It is a high γ material, and so from the model of Fig. 10-11 it will have a low firing voltage.[91-94] The γ of MgO is also very stable with time, which will give the devices a very long lifetime.[95-97] MgO is a refractory material that is resistant to sputtering. It achieves very long lifetime without the use of mercury, which is necessary for

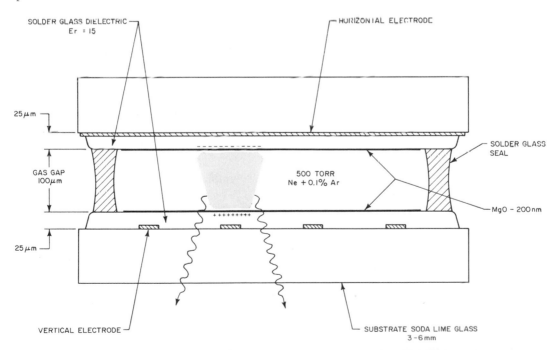

Fig. 10-40. Structural details of a typical ac plasma display product.

all dc displays. Other materials have been developed, but MgO is universally used by the industry.[98]

When done properly, the MgO gives a very uniform and long-life performance. However, it is very susceptible to contamination in the manufacturing process.[99] Once contaminated, the MgO is virtually impossible to clean. Extreme care is necessary to produce commercially acceptable ac displays. This is in contrast to dc displays that are relatively forgiving to contamination.

10.7.2 Electrical Characteristics of the AC Display.

As discussed in Section 10.5.2, an ac voltage called the sustain voltage must be applied to the display to maintain discharging. A charge builds up on the dielectric glass walls as shown in Fig. 10-40. The details of this operation can be understood with the aid of the model shown in Fig. 10-41.[100] This models the capacitors of the dielectric glass C_d and the capacitance of the gas gap C_g. The gas discharge is modeled as a dependent current source. It is dependent on numerous parameters such as the voltage across the gas and the time that the

discharge has had to mature. From the dimensions and the dielectric constant of the dielectric glass shown in Fig. 10-40, it is easy to see that C_d is about 60 times greater than C_g. Thus, to a very good approximation one can assume that with no charge on the dielectric glass wall, the full applied sustain voltage V_s appears across the gas gap between points a and b. The error in this assumption is less than 5%.[100]

When the voltage across the gas gap between points a and b exceeds the firing voltage, the gas-discharge current will begin to flow. This current will deposit charge on the glass dielectric walls. This will charge (or discharge) the capacitors so that the magnitude of the voltage across the gas gap is reduced and the gas discharge is ultimately extinguished. This charge on the glass walls is called wall charge. It is responsible for a voltage component across the gas that is called the wall voltage V_w. This is the voltage that would be across the gas between points a and b if the sustain voltage V_s were zero. The combination of the wall-voltage component plus the sustain-voltage component gives the voltage across the gas gap, which is called the cell voltage V_c. It is always true that:

$$V_c = V_s + V_w$$

Memory Characteristics. If a sustain-voltage waveform of the proper amplitude and proper shape is applied to the panel, it will exhibit bistable memory. Each pixel can then be in either a stable on state or a stable off state. Figure 10-42 shows how the wall voltage and the light output behave for these two states. The sustain pulsewidths are typically 5 to 10 μs at a frequency of 50 kHz. The zero to peak pulse amplitude is in the range of 90 to 100 volts. When a cell is on, it discharges and emits light whenever the sustain waveform first achieves a positive or negative peak. The discharge causes a large transition in the wall voltage. The polarity of the wall voltage is usually drawn so that the voltage across the gas, V_c, can be determined by measuring the distance between the sustain-voltage and the wall-voltage waveforms. In the on state, when the sustain voltage reaches its first positive peak, the wall-voltage component increases V_c,

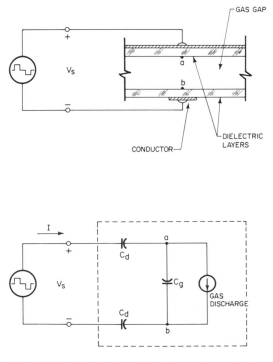

Fig. 10-41. Electrical model of the ac plasma display.

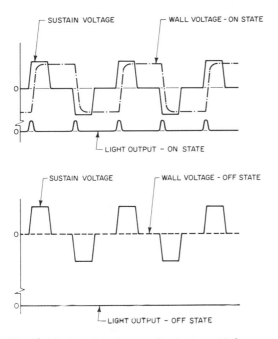

Fig. 10-42. Sustain voltage, wall voltage, and light output for pixels in the on and off states.

so that there is a large enough field across the gas to cause a discharge. The discharge current changes the wall voltage, which reduces V_c until the discharge extinguishes. Now the remaining wall voltage is of the opposite polarity, so that when the sustain voltage is reversed, the high magnitude of V_c will again cause a discharge.

In the off state, there are no discharges and the wall voltage remains at zero. Because of this, $V_c = V_s$, and this V_c is insufficient to cause a discharge large enough to cause a significant change in wall voltage. Thus the wall voltage from a previous strong discharge is needed to achieve sufficient voltage across the gas to cause a new discharge. It follows that because of the action of the wall voltage, an ac plasma cell can be in either the discharging or the nondischarging state with the same sustain voltage being applied.

Hysteresis Effect. As discussed above, an ac plasma pixel can have an on and an off state that both exhibit a memory characteristic. The reason for this memory characteristic is much more complicated to explain than the dc memory effect discussed in Section 10.3.2. At first glance one might consider that the charge stored on the internal dielectric surface is responsible for the memory. This is partially true, because clearly the on state has some charge on the wall and the off state has none. Also, charge stored on a capacitor is a well-recognized form of storage media used in many present-day semiconductor dynamic memories. However, the memory in the ac plasma display cannot be attributed entirely to the charge on the dielectric capacitance, because one can design a device that does not have memory and yet still has charge on the glass wall.

Figure 10-43 shows two characteristics that have been observed in ac plasma pixels. This shows the peak wall voltage as a function of the peak sustain voltage. The hysteresis characteristic on the right side of Fig. 10-43 represents the displays with memory. This is because for a single sustain voltage, two different wall voltages are allowed.

The presence or absence of the memory characteristic depends on device parameters such as gas mixture and on the applied sustain waveforms. The physical mechanisms behind these characteristics have been studied extensively but are beyond the scope of this chapter. The interested reader should consult references on the stability theory for this device;[3,111,112] on techniques for measuring this characteristic;[113-118] and on computer simulations of the discharge.[119-127] Also, these characteristics depend strongly on the geometrical considerations such as electrode width, pitch, and gas gap spacing.[78,79,128,129]

When operating with a memory panel, the sustain voltage must be adjusted to a value between V_f and V_e in Fig. 10-43. For a typical panel with a MgO overcoat the values of V_f and V_e are 100 and 90 volts respectively.

A panel that does not have memory is usually operated at some value well above the voltage V_f in the left-hand characteristic of Fig. 10-43. This is done to assure that all the pixels are uniformly bright.

10.7.3 Addressing. To enter information on an ac plasma display, one must introduce the appropriate address pulses needed to change the wall voltage and state of a cell.[100,101] Figure 10-44 shows the way these address pulses influence the wall voltage.

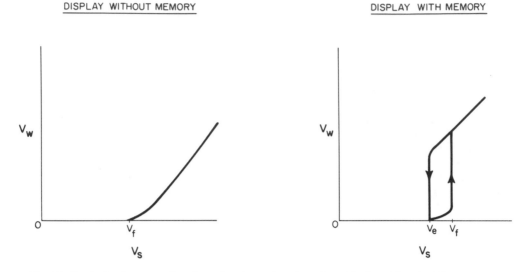

Fig. 10-43. Wall voltage as a function of peak sustain voltage for a display with and without memory.

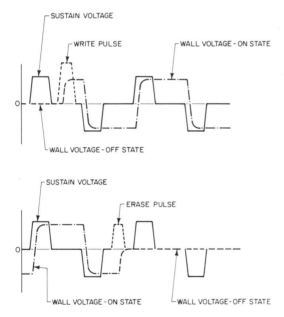

Fig. 10-44. The addressing voltages needed to write and erase a pixel.

The write pulse is shown in the upper part of Fig. 10-44. In this case, the amplitude of the write pulse is adjusted so that the wall voltage goes from zero to the on-state level. The amount of wall-charge transfer is just half of that found in a normal sustain pulse. The amplitude of the write pulse is always greater than the amplitude of the sustain pulse because for zero initial wall voltage, the write pulse must cause a discharge

and the sustain pulse must not. The width of the write pulse is typically 5 μs.

The lower part of Fig. 10-44 shows the erase pulse that should reduce the wall voltage of the selected cell to zero. Like the write pulse, the amplitude and width of this pulse are selected so that only half the amount of wall-voltage change occurs compared to a normal sustain discharge. The amplitude of this pulse is frequently less than the sustain pulse. However, the erase pulse could have an amplitude greater than the sustain pulse if the pulse width is sufficiently short to stop the discharge when exactly half of the normal charge transfer occurs.

As discussed in Section 5.3, a successful X-Y matrix display must have some nonlinearity in the display element that will respond to the full address voltage but not to half of that voltage. The waveforms that accomplish this for the ac plasma display are shown in Fig. 10-45 for a 4 × 4 array of cells. Cell 6 is to be addressed, and all of the other cells in the panel should remain in their initial state. The electrodes B and D that intersect the addressed cell receive address voltages during the addressing period. All other electrodes A and C receive the normal sustain waveform. To determine the voltage that an individual cell receives, one must take the difference between the two electrodes that intersect that cell. For instance, the voltage on

AC PLASMA ADDRESSING VOLTAGES

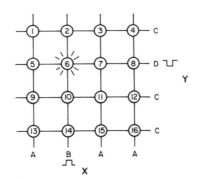

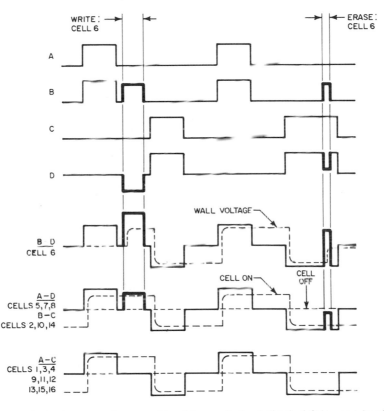

Fig. 10-45. Typical voltage waveforms used to address a single pixel (pixel 6) in a matrix of pixels.

the addressed cell 6 is B – D. This waveform is shown in Fig. 10-45 along with the wall voltages. A write cycle is shown first and next an erase cycle.

In the write cycle, a cell in the off state should be turned on and a cell initially in the on state should remain unperturbed. Also, of course, all of the cells in the panel other than cell 6 should remain unperturbed for both states. The wall voltages for all of these cases are shown in Fig.

10-45. For the B – D waveform during the write period, the addition of the half select pulses of the B and the D waveforms give a high-amplitude pulse that will cause a discharge if the wall voltage is initially zero. Thus the off cell is turned on. The cell that is initially in the on state is not perturbed because the cell voltage V_c is not great enough to cause a discharge at the time of the write pulse. Similarly, the other cells in the panel do not receive a sufficiently large cell voltage during the address pulse to cause a discharge. This is shown in the waveforms for A – D and A – C.

The erase pulse must have characteristics similar to the write pulse in that a perturbation of wall voltage must occur on only cell 6 and only if it is on. The amplitude of the erase pulse has been made the same as the amplitude of the write pulse. This is not necessary but is desirable since the same circuit is usually used for both the write and the erase pulses. The width of the erase pulse is carefully adjusted so that the proper amount of wall voltage is transferred.

The waveforms presented here in Figs. 10-42, 10-44, and 10-45 are commonly used but are not universally accepted. There are numerous other waveforms that may give improved performance in some systems.[17,102-105]

Priming. The ac plasma displays use pilot-cell priming as discussed in Section 10.4.6. The pilot cells are located around the perimeter of the display area. They are frequently called border cells. These cells need to be discharged whenever a pixel in the panel is written. Because of the open cell structure shown in Figs. 10-37 and 10-40, ultraviolet photons generated by the border cells are free to travel to the written pixel and prime it to cause a reliable write operation.

It is frequently helpful to synchronize the border-cell discharge with the write pulse.[106]

10.7.4 Drive-Circuit Considerations. The circuit configuration usually used to generate the waveforms is seen in Fig. 10-46.[107] The circuit for generating the sustain waveform is usually separated from the address drivers. This is done for circuit economy reasons. The address pulsers are a major system cost since there must be one for each display electrode. Thus the amplitude of the address pulses is adjusted so that a minimum amount of voltage is needed. This usually results in lower circuit costs since lower-voltage components can be used. The amplitude of the sustain pulses is frequently considerably greater than the amplitude of the address pulses. Typical values are 90 to 100 volts for the sustain pulses and 60 to 70 volts for the address pulses. Since the two amplitudes are different, it is frequently more economical to have separate circuits for the sustain pulses and the address pulses. This usually means that there is one sustain-voltage generator for the entire display along with address pulse generators for each line. An alternate technique is to use two sustain-voltage generators, one for the X axis and one for the Y axis.

Because of the large number of display lines,

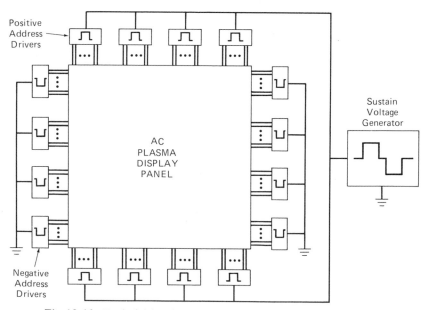

Fig. 10-46. Typical drive circuit configuration for the ac plasma display.

it is frequently desirable to integrate the address drivers. Texas Instruments has developed the SN 75500 positive address driver and the SN 75501 negative address driver for this purpose.[108-110] These devices are available in a 40-pin dual in-line package and are capable of driving 32 display lines with up to 100-volt pulses at currents of up to 20 mA. CMOS shift registers and logic gates are included in each device to reduce the logic level chip interface problem. Devices like these should help to significantly reduce the total system cost of plasma displays.

Figure 10-47 shows the drive circuits for the IBM display shown in Fig. 10-35. Here the address-pulse generators are 40-pin integrated circuits arranged around the perimeter of the circuitboard. These devices have a design very similar to the SN 75500 and SN 75501. Fifty-four of these ICs are used to drive the entire panel. The center part of the circuitboard is occupied by the sustain-voltage generator, the timing and control logic, and some power-supply regulators. One of the advanced features of this display is the on-board microprocessor that automatically optimizes the addressing- and sustain-voltage pulse levels. This eliminates the need for service adjustments as the components age.

The sustain-voltage generator, more frequently called the sustainer, is usually a power circuit that must be capable of rapidly charging and discharging the large capacitance of the plasma panel, and also of conducting the large current of gas discharges when a large number of pixels are on. For instance, for an ac plasma display with the structure of Fig. 10-37 having an array of 512 × 512 pixels with resolution of 23.6 lines per centimeter (60 lines per inch), the capacitance of the entire panel is about 3500 pF.[107] The sustain pulses must rise before the discharge current occurs, otherwise the display will not exhibit good memory characteristics. Thus the rise time of the sustain pulses is in the range of 200 to 400 ns. To charge 3500 pF to 100 volts in 200 ns requires a current of 1.75 A. This is the displacement-current component shown in Fig. 10-48. Within a few hundred nanoseconds of the sustain pulse rise, the discharge current occurs. For this size panel, a reasonable number for the peak discharge current is 40 μA per pixel. Since there are about 250,000 pixels in this panel, the peak discharge current when all of the pixels in the panel are on will be about 10 A.[107] Thus the discharge-current pulse is shown following the displacement-current pulse in Fig. 10-48. The half width of the discharge-current pulse is typically 0.5 μs. Thus the sustainer must be capable of both high voltage and high current.

The address-pulse generators do not have the same stringent requirements. Since in one dimension there are 512 address circuits, a reasonable approximation is that each address driver must supply only one-512th of the current of the

Fig. 10-47. Rear view of the IBM display shown in Fig. 10-35. This shows the integrated-circuit line drivers along the perimeter with the sustain and control logic in the center. (Courtesy of International Business Machines Corporation)

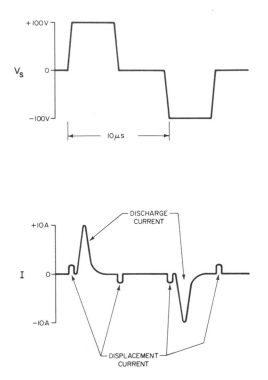

Fig. 10-48. Typical sustain-voltage waveforms and sustainer currents for a 512 × 512 plasma panel with all pixels in the on state.

sustainer. The capacitance of one electrode to all of the other electrodes in the 512 × 512 panel is about 20 pF, making the displacement current only 10 mA. The maximum discharge current for this electrode is about 20 mA based on 40 μA current per pixel and all 512 of the pixels along the electrode in the on state.

10.7.5 Refreshed Segmented AC Plasma Display. As in the case of the refreshed dc displays discussed in Section 10.6.2, it is certainly not necessary to have memory to have a useful display. There are two lines of commerical products that are ac devices but do not utilize the memory characteristic. Figure 10–49 shows the displays marketed by National Cash Register and Nippon Electric Company which use the ac plasma display technology for seven-segment numeric displays.[130,131] Since there are very few display elements in such a display, memory is of little importance. The electrode on the viewer's side is usually made of a transparent conductive material such as tin oxide, and the dielectric glass

covering this electrode is transparent. The back electrode is usually opaque, and its dielectric layer is usually made of black glass to improve the contrast ratio of the display. The two substrates are separated by a layer of glass somewhat like the center sheet seen in Fig. 10-36. The individual segments are isolated from each other by this layer.

Distributed-Glow Characteristic. The ac seven-segment displays have a couple of advantages over their counterpart dc displays. First, the glow covers the electrode of the ac device more uniformly and may thus give a more appealing display. The reason for this is that the current limiting of the ac device is accomplished by the capacitor, which is distributed over the face of the electrode. In the dc case, the current limiting for the electrode is accomplished by the external lumped resistor. All areas of the dc electrode compete with each other for the current that the single resistor can supply. Thus, some of the areas will get more of the current than others and will be correspondingly brighter. In the ac case, each part of the electrode has its own capacitor to limit the current. As soon as that capacitor is discharged, that area will stop discharging. Since all areas of the electrode have the same capacitance, the whole electrode will have the same brightness. This phenomena works so well that it is useful to make ac plasma cells with a large area such as 1 × 3 cm. When transparencies are placed in front of these large areas, they can display a message that can be changed depending on the needs of the user.

NEC uses the distributed-glow characteristic for making large billboard displays suitable for outdoor viewing. Each pixel is a large-area ac discharge. At high sustain frequencies, the high brightness of these displays makes them sunlight readable.

Wide Operational-Temperature Range. The second advantage of the ac device over the dc device is a wider operating-temperature range for the ac display. The basic reason for this is that the dc device uses mercury and the ac device does not. At too low a temperature the vapor pressure of the mercury is too low, and it will not act as an anti-sputtering agent as

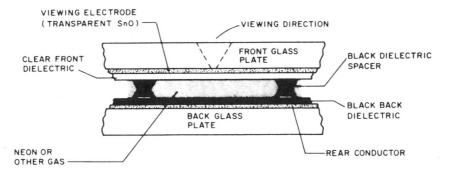

Fig. 10-49. Structure of the refreshed ac plasma display manufactured by NCR. (Reference 130 © 1972 IEEE)

discussed in Section 10.6.1. Thus the dc device operated at low temperature for prolonged periods will have low lifetime. The ac devices achieve long lifetime without the aid of mercury. The temperature range of gas discharges is generally very large, with NCR specifying operating temperature between −40°C and +85°C. The low temperature end of this range could probably be extended considerably without trouble. Since the ac devices operating outside of the memory range are very tolerant to applied voltage variations, any variations in parameters caused by temperature are unimportant. Thus, the temperature limits of the ac device are usually determined by the temperature limits of the drive circuits.

The brightness of these ac displays is proportional to the drive frequency. Designers can expect a proportionality factor of about 3 fL/kHz. Brightnesses as high as 500 fL are specified. Such a brightness would require a drive frequency of 167 kHz, which is very reasonable for these devices.

Multiplexing. The ac seven-segment device is multiplexed in a way very similar to that used in the dc devices as discussed in Section 10.6.2. The differences are that no current-limiting resistors are needed and an ac voltage is applied instead of a dc voltage. Thus, the ac device with the same number of digits will have the same number of external connections. The discharges of the segments are all isolated from each other by the center spacer and they cannot prime each other. Thus, only self-priming can be used as discussed in Section 10.4.6. The segments that are to remain off must be discharged at a low

rate to maintain sufficient active particles. These priming discharges reduce the contrast ratio slightly, but it is generally not objectionable.

10.7.6 Refreshed Dot Matrix AC Plasma Display.

Another ac plasma display which operates outside of the memory range is a scanned dot matrix panel made by Nippon Electric Company,[132] Matrix displays as large as 160 X 280 lines have been made. Since the panel has no memory, it must be scanned with the same technique shown in Fig. 10-20. The switching on the horizontal electrodes must be accomplished by external circuits in a way very similar to the dc panel shown in Fig. 10-21.

Because it is a scanned display, there is a problem with brightness, as discussed in Section 5.3, due to the low duty cycle. To combat this problem, everything possible is done to increase the light output. Pure neon gas is used for this display because it is brighter than the Penning gases. A small penalty of higher required voltages for pure neon is worth the brightness benefit. A transparent electrode on the faceplate is used to allow more light to be emitted. The biggest increase is achieved by making the sustain frequency 500 kHz. The memory panels cannot operate at a sustain frequency much greater than 50 kHz before they lose significant memory margin. However, this limitation is unimportant for refreshed mode. Since brightness is roughly proportional to sustain frequency, the 500-kHz rate gives a factor of 10 improvement in brightness.

When all of this is done, the panel has reasonable brightness if the scanning is accomplished along the shortest dimension of the matrix. For instance, for the 160 X 280 panel mentioned above there are 280 column electrodes and 160 row electrodes. The scanning is done along the rows so that the duty cycle is higher and the display is brighter. The scan switches connected to the column electrodes in Fig. 10-20 are connected to the row electrodes in this display.

As with the seven-segment refreshed ac displays, the addressing is accomplished by selectively applying the sustainer to the electrodes that intersect the cell to be lit. Because this sustain voltage is very high, the discharge activity of one cell tends to bleed over to the neighboring cell. To prevent this, barriers are placed between the cells to reduce this interaction. These barriers also act as the center spacer that determines the discharge gap spacing.

10.7.7 AC Shift Panels.

One disadvantage of dot matrix displays is that a circuit driver is needed for each of the many electrodes, and this circuitry can result in a very significant amount of the system cost, as discussed in Section 5.7. In most cases the circuits cost more than the display panel itself. One way to reduce the circuit cost is to use some sort of shifting technique similar to that used in the dc Self-Scan[TM] panels discussed in Section 10.6.4. Figure 10-1 shows that Fujitsu has a product line that employs a shift technique for the ac memory display.[42,133]

It is worth noting that the ac shift technology has memory whereas the dc shift technology (Self-Scan[TM]) does not. This means that the ac display will not have flicker or have brightness problems for larger numbers of scanned lines. In a very recent development, the engineers at Burroughs have eliminated this limitation by introducing their Self-Scan[TM] memory display discussed in Section 10.8.1 and shown on Fig. 10-1 as an ac–dc hybrid.

Figures 10-50 and 10-51 show the electrode configuration of the Fujitsu ac self-shift product.[134-138] The electrode geometry is commonly referred to as the meander electrode pattern. Display data is entered into the W electrodes, which can then be shifted to the left when the appropriate waveforms are applied to the X and Y electrodes. Except for the W electrodes, all of the other electrodes are bused together so that the X and Y electrodes require only four drivers for the entire panel: X1, X2,

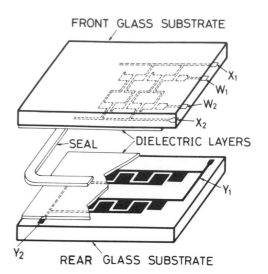

Fig. 10-50. Exploded view of the meander electrode ac plasma self-shift panel manufactured by Fujitsu. (Reference 137 courtesy Society for Information Display)

Y1, and Y2. This device requires one W electrode driver for each display row and the above-mentioned four drivers for all of the display columns. Thus, as with the Burroughs Self-Scan[TM] display, there is a very significant reduction in the number of column drivers and no reduction in the number of row drivers.

This arrangement is especially economical for a single line of characters where very few address drivers are needed for the horizontal-row electrodes. Of course, the shift technique allows a much slower panel update time and less random access flexibility than conventional matrix addressing. However, many small displays do not require this flexibility, so shift is an excellent compromise. A major challenge with all shift techniques is to hold the increased costs of manufacturing the shift panel below the savings realized by the reduced number of drive circuits. Because this display has memory, the entire display is simultaneously shifted across the screen as it is entered. This means that, as desired, the display shifting can be stopped and the image will remain in place, or it can even be reversed.

There are two physical mechanisms that can be used to accomplish the shifting action in the ac self-shift panels. One is the priming coupling, and the other is wall-charge coupling. Both of these mechanisms are in use in different Fujitsu products. The priming coupling was developed

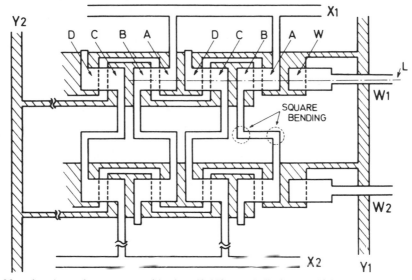

Fig. 10-51. Meander electrode pattern used in the self-shift panel. (Reference 137 courtesy Society for Information Display)

first and utilizes priming particles from one cell to reduce the firing voltage of a neighboring cell which will then go on.[42,139] This is very similar to the mechanism used in the dc Self-Scan[TM]. The wall-charge coupling uses appropriate applied waveforms to transfer wall charge from one cell to its neighbor.[140,141] This technique offers a much faster shifting speed over the priming coupling technique.

AC Shift with Priming Coupling. The details of the priming coupling shift technique are shown in Figs. 10-51, 10-52, and 10-53. Figure 10-52 shows the edge view of the shift panel and the locations of the discharges W, A, B, C, and D. The appropriate waveforms are applied

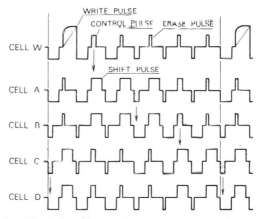

Fig. 10-53. Waveforms that appear across the discharges shown in Figs. 10-51 and 10-52 for the priming coupling technique. (Reference 137 courtesy Society for Information Display)

to the electrodes W, X1, X2, Y1, and Y2 to give the cell waveforms seen in Fig. 10-53. For instance, the cell A waveform is obtained by subtracting the Y1 waveform from the X1 waveform.

Figure 10-53 shows the various types of pulses that are used to control the shifting operation. The write pulse appears only on W cells and it controls the input of information to the shift register. If a cell is in the on state, the control pulse will cause a discharge that will generate the priming particles necessary to reduce the firing voltage of the next cell which will turn it on. This next primed cell will have a normal-

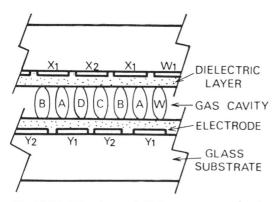

Fig. 10-52. Side view and discharge sequence for the self-shift panel. (Reference 137 courtesy Society for Information Display)

width sustain pulse called a shift pulse. The erase pulse is applied to reduce any further discharging which would cause adverse priming.

The shift sequence is started by the application of a write pulse on electrode W1 if the pixel should be turned on. All cells other than the W cells continually receive the sequence of (1) control pulse, (2) erase pulses, (3) shift pulse, and (4) sustain pulses. The phase of this sequence is delayed for each successive cell so that the information is shifted from one cell to the next as indicated by the downward arrows in Fig. 10-53.

Note that after each shift pulse, a pair of ordinary sustain pulses are needed to stabilize the discharge sequence so that it is completely in the on state before the next control pulse is applied. The erase waveform is comprised of narrow pulses of the normal sustain amplitude that are too narrow to transfer the full on-state wall voltage. Once the full display is shifted to the desired position, a different set of waveforms is applied that sustains the pixels and holds the image in the memory mode.

For 25-μs sustain cycles it takes 450 μs for one complete pixel shift cycle. This means that for a line of 40 5 $\times$ 7 characters, 108 ms will be taken to shift in the entire line, which is a rate of 370 characters per second.[134] This is quite acceptable for many display applications.

The electrode geometry of the meander pattern shown in Figs. 10-50 and 10-51 has a number of interesting characteristics. The pattern has the advantage that all of the busing is accomplished in the outer edges of the glass substrates without the need for complicated electrical crossovers. The absence of crossovers very significantly reduces the cost of panel fabrication. Figure 10-51 shows that the column electrodes have regions of square bending between the pixels. This arrangement greatly reduces the adverse interaction between cells in adjacent rows and therefore eliminates the need for glass barriers to isolate the rows. Although the meander pattern is quite complex and requires very careful alignment between the two substrates, Fujitsu has products with pixel resolutions as high as 19.7 lines per centimeter (50 lines per inch) and containing 256 7 $\times$ 9 font characters.[142]

AC Shift with Wall-Charge Coupling. The first practical device that used the wall-charge coupling technique is shown in Figs. 10-54 and 10-55. This was first developed at National Electronics and further developed at National Cash Register.[140,141,143] This device is distinguished by having the display electrodes on the two substrates parallel to each other. Just as with the meander shift panel, the display information is entered through the write electrodes and is shifted to the left. Because of the parallel electrode structure, glass barriers are necessary to create discharge channels that define the rows of the display.

The waveforms that are applied to the electrodes of this display are shown in Fig. 10-55. When comparing this figure to the priming coupling waveform Fig. 10-53, one should remember that Fig. 10-55 shows the electrode voltages whereas Fig. 10-53 shows the voltage across the various pixels. The wall-charge coupling waveforms are considerably more simple. If a write pulse is applied to electrode W at time T1, the resulting discharge will leave a positive wall charge on the dielectric over the first Y1 electrode. This wall charge will add to the pulse on X1 at time T2 to cause a discharge that will deposit wall charge on the X1 dielectric. This process continues and causes the display information to be shifted to the left. When the shifted discharge reaches the end of panel, the erase electrode is pulsed to neutralize the wall charge on the last X2 dielectric.

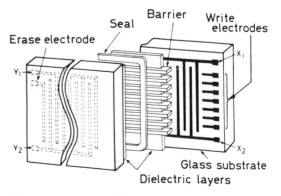

Fig. 10-54. Structure for the wall-charge coupling ac shift panel developed by National Electronics and NCR. (Reference 144 © 1982 IEEE)

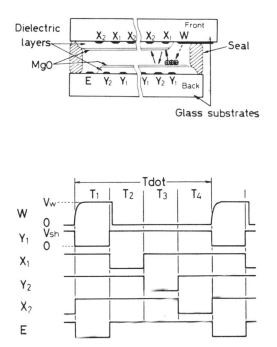

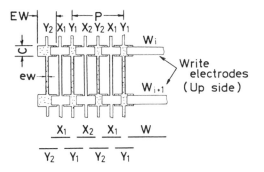

Fig. 10-56. Electrode structure developed by Fujitsu for use in a wall-charge coupling ac shift display. This structure eliminates the need for the glass barriers shown in Fig. 10-54. (Reference 144 © 1982 IEEE)

Fig. 10-55. Side view and waveforms applied to the electrodes for the wall-charge coupling technique. (Reference 144 © 1982 IEEE)

The wall-charge coupling mechanism has much stronger coupling than the priming coupling mechanism. This gives wall-charge coupling a significant advantage in shifting speed. The fundamental shift period Tdot shown in Fig. 10-55 is typically 60 μs. This allows a 5 × 7 character shift rate of 2,700 characters per second. This is more than 7 times faster than the priming coupling mechanism. With this technique, a panel with rows of 40 characters can be fully updated in 15 ms.

Fujitsu has developed a wall-charge coupling shift panel that does not require the internal glass barriers seen in Fig. 10-54. The new device shown in Fig. 10-56 uses a special electrode structure that confines the discharges to channels without the use of glass barriers.[144,145] By designing the electrode width EW to be 190 μm and the width ew to be 50 μm, the shifting discharges will follow the wide electrode channels from left to right without significant interaction along the vertical direction. This electric-field type of confinement is sufficiently good to allow pixel resolutions of 19.7 lines per centi-

meter (50 pixels per inch) with shift speeds of 2,500 characters per second.

Other AC Shift Technologies. There are numerous other ways of achieving shift operation in ac plasma displays. One can achieve shift operation in a standard memory dot matrix ac plasma panel having the structure of Fig. 10-37 by making the appropriate external electrode connections.[146,139] For a practical display it is desirable to make these connections directly on the display panel, as shown in Fig. 10-57. This display, which was developed by IBM, requires metal paste crossovers, which increase the panel fabrication costs.[147,148] However, the alignment of the two substrates is not as critical as for the meander electrode pattern. Experimental panels have achieved resolutions of 19.7 pixel per centimeter (50 pixels per inch) at a shift speed of 360 7 × 9 characters per second.

Bell Labs has achieved shifting in a panel with the structure of Fig. 10-37 through the use of wall-charge coupling.[149-151] This technique offers the advantage of a higher shifting rate of 600 7 × 9 characters per second. In addition, it requires only two electrodes per display pixel, in contrast to the three electrodes per pixel for the Fig. 10-57 display.

Although the shift technique greatly reduces the number of drive circuits for the columns, the schemes discussed above still require one circuit for every row. This large number of circuits can be considerably reduced by designing the system to do the shifting in both the x and y

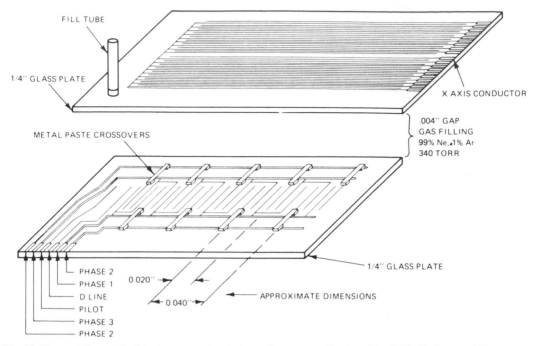

FILL TUBE

1/4" GLASS PLATE

X AXIS CONDUCTOR

METAL PASTE CROSSOVERS

.004" GAP
GAS FILLING
99% Ne,.1% Ar
340 TORR

1/4" GLASS PLATE

PHASE 2
PHASE 1
D LINE
PILOT
PHASE 3
PHASE 2

0.020"

0.040"

APPROXIMATE DIMENSIONS

Fig. 10-57. ac shift panel with the conventional electrode structure developed by IBM. (Reference 148 courtesy Society for Information Display)

directions.[133,152] This is achieved at the expense of display speed. The technique uses seven row drivers to horizontally shift in a complete line of 5 × 7 characters along the top of the panel. When the top character line is complete, the whole line is shifted vertically down along with the rest of the information on the panel. The shifting operation is arranged so that when the top line is shifted in horizontally, the rest of the information below this line does not shift.

Shift Caveat. The manufacturing yield of shift panels is very significantly reduced because every cell in a shift panel must be operational. This is true because if one cell is bad it will break a link in the shift register and the rest of the display line will not be addressed. Of course, a bad display line is an unacceptable failure that will cause a rejection of the panel. This 100% pixel yield requirement is severe compared to matrix panels where a limited number of bad pixels are acceptable. Thus, without careful processing of the panels, low panel yield will make the increased panel costs greater than the savings in circuits. This is one reason why shift panels seem more suited to smaller displays; to achieve 100% pixel yield in a large area would be very difficult.

10.8 HYBRID AC–DC PLASMA DISPLAYS

Recently two new displays were developed that utilize both the ac and the dc current-limiting techniques. In these displays, some of the gas discharges are ac and some are dc. It is the intention of the designers to utilize the best features of both techniques to make a higher-performance display. One of these displays, the Burroughs Self-Scan[TM] Memory plasma display, has made it to the marketplace and is shown in the family tree of Fig. 10-1 as a combination of the dc Self-Scan[TM] and the ac memory displays.[153] The other panel has been developed by Sony and is projected to be a product soon.[154]

10.8.1 Self-Scan[TM] Memory Panel. The standard Self-Scan[TM] display has the advantage of a low number of drive circuits due to the clever internal gas-discharge logic technique discussed in Section 10.6.4. It has the disadvantage of being a refreshed display which does not have memory. This limits its performance in brightness and in the maximum number of pixels per row. The Self-Scan[TM] memory display obviates these limitations by adding memory to the Self-Scan[TM] feature.

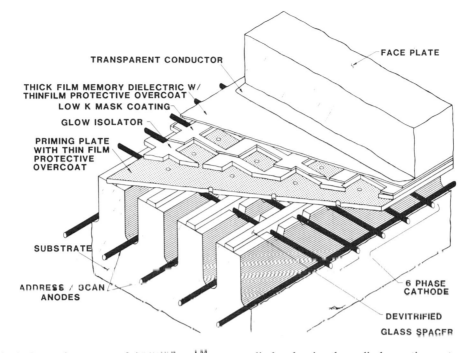

TRANSPARENT CONDUCTOR

FACE PLATE

THICK FILM MEMORY DIELECTRIC W/
THINFILM PROTECTIVE OVERCOAT

LOW K MASK COATING

GLOW ISOLATOR

PRIMING PLATE
WITH THIN FILM
PROTECTIVE
OVERCOAT

SUBSTRATE

ADDRESS / SCAN
ANODES

6 PHASE
CATHODE

DEVITRIFIED
GLASS SPACER

Fig. 10-58. Internal structure of the Self-Scan[TM] memory display showing the ac display section on top and the dc address section below. (Courtesy of Burroughs Corporation)

Figures 10-58, 10-59, and 10-60 show the structure of the Self-Scan[TM] memory display.[153] This display can be described as a dc Self-Scan[TM] structure that is placed behind an ac memory display. The viewer sees only the ac part of the display. The pixel addressing is accomplished by the dc Self-Scan[TM] section, and the pixel memory function is the job of the ac section. The memory function gives this display the attributes of high brightness at any display size with freedom from flicker. The Self-Scan[TM] function reduces the number of circuit drivers, which should reduce total system cost.

The device structure is shown in Fig. 10-58. The ac portion of the display is the structure above the priming plate. The two ac electrodes are formed by the priming plate and the transparent conductor on the faceplate. Both of these electrodes are solid and cover the entire area of the display panel. This requires a dielectric glow isolator sheet that separates the two electrodes, defining and isolating the pixels. The transparent conductor is coated with the standard ac dielectric glass and a thin-film protective overcoat material such as MgO. The priming plate has no dielectric glass but only the thin-film

protective overcoat. Unlike other ac displays, this display uses the capacitance of dielectric on only one of the two electrodes to limit the current. This is a perfectly acceptable technique for ac current limiting. These two ac electrodes are connected to an ac sustain-voltage generator that supplies power to the pixels and maintains them in either the on or the off state, as shown in Fig. 10-59.

The dc Self-Scan[TM] section is found below the priming plate. This operates similarly to the scan discharge section of a standard Self-Scan[TM] display as shown in Fig. 10-25. An important difference is that the scan does not operate continuously but only when addressing is required. Also, only selected scan anodes are discharged, depending on which areas of the panel are being addressed. This technique saves power and extends display lifetime. The cathode wires are held in place with a devitrified glass spacer. This prevents the cathodes from shorting with the priming plate.

Writing a pixel in this display proceeds as follows. First, the appropriate scan discharge is moved along the scan channel until it is under the addressed pixel. This corresponds to the time

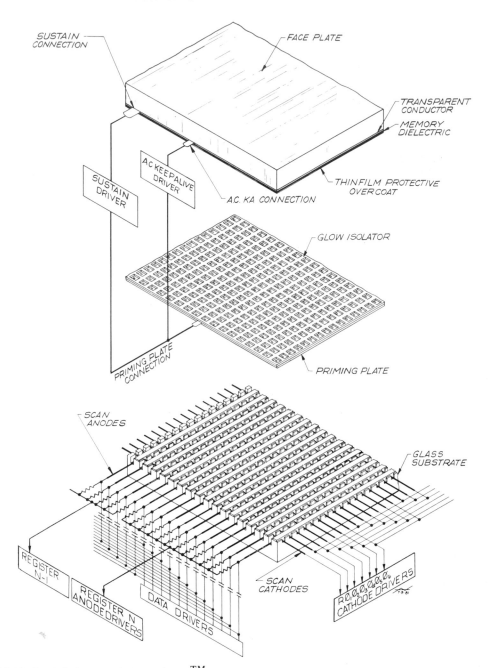

Fig. 10-59. Exploded view of the Self-Scan^{TM} memory display showing the external addressing electronics. (Courtesy of Burroughs Corporation)

1 condition in Fig. 10-60. While the scan discharge is active under this pixel, the scan anode is pulsed negative for about 1 microsecond while the sustainer signal is on the positive half of its cycle. The scan discharge is interrupted and electrons from the scan glow flow through the small hole in the priming plate into the ac section of the panel as shown at time 2 of Fig. 10-60. These electrons change the wall voltage of the ac section dielectric, causing the pixel to be written.

A pixel is erased in a similar manner. The scan glow is interrupted when the sustainer signal is at its midpoint value and when the ac-section dielectric is charged positively. Electrons from

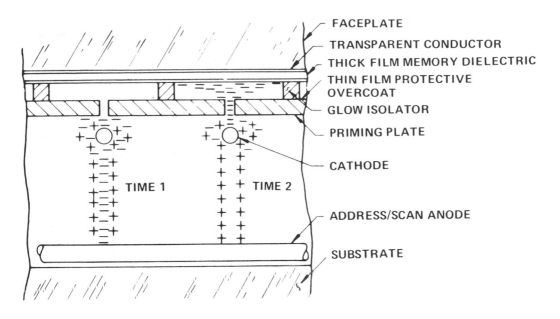

FACEPLATE
TRANSPARENT CONDUCTOR
THICK FILM MEMORY DIELECTRIC
THIN FILM PROTECTIVE OVERCOAT
GLOW ISOLATOR
PRIMING PLATE
CATHODE
ADDRESS/SCAN ANODE
SUBSTRATE

Fig. 10-60. Side view of the Self-Scan™ memory display manufactured by Burroughs. (Reference 153 courtesy Society for Information Display)

the scan glow move through the priming hole, neutralizing the positive charge and returning the ac section to a quiescent off condition.

Figure 10-59 shows that the number of active circuit drivers for the scan anodes can be reduced by a resistor capacitor decode scheme. The anode drivers are connected to the buses of current-limiting resistors for the scan anodes. These drivers supply the power for the scan discharge. Only one of these anode drivers is active at any one time, and so the scan discharges are active only in this selected group. When the scan discharge is under the desired pixel, one of the data drivers is pulsed negative for 1 μs. This couples through the capacitor to interrupt the scan discharge so that the electrons are sent through the priming hole as shown in time 2 of Fig. 10-60. This technique significantly reduces the number of active address drivers needed along the rows to the sum of the number of anode drivers and data drivers. An additional capacitor is needed for each row, but this is a relatively low-cost part. Of course, the columns of this display are bused together as for the normal Self-Scan™ display. Figure 10-59 shows a 6-phase scanning technique that requires only 7 cathode drivers for addressing all of the display columns.

10.8.2 Sony AC-DC Hybrid. Figures 10-61 through 10-64 show the ac–dc panel developed by Sony.[154] This display takes quite a different approach from the Burroughs panel discussed just above. Here the display discharge is dc and the addressing discharge is ac. This display does not have internal memory and must be refreshed.

Figure 10-61 shows the superb resolution characteristics of this display, which demonstrates a 1,024 $\times$ 512-line display with a resolution of 50 lines per centimeter (127 lines per inch). The display area measures 10 $\times$ 20 centimeters (4 $\times$ 8 inches). This is the highest-resolution plasma display demonstrated up to early 1983. The character capacity of this display can be appreciated by noting that the characters in the lower right corner of Fig. 10-61 have a 5 $\times$ 7 dot matrix.

The structure of this display is shown in Fig. 10-62. It is basically a refreshed dc display that uses glass barrier ribs to isolate the discharges. What makes this panel unique is the ac trigger electrode and dielectric layer on the lower substrate. This ac electrode is used to address and prime the display as described below.

The right side of Fig. 10-63 shows the addressing waveforms. If a pixel is to emit light, then the discharge is initiated with appropriate pulses

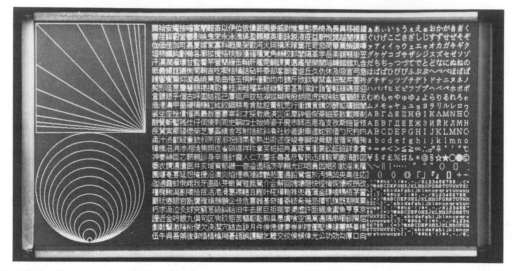

Fig. 10-61. Plasma display developed by Sony having an array of 512 × 1,024 pixels at 127 lines per inch. (Courtesy of Sony Corporation)

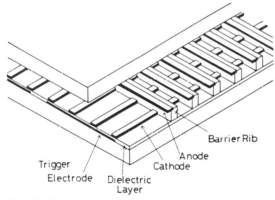

Fig. 10-62. Internal structure of the Sony display showing the dc anodes and cathodes and the ac trigger electrodes. (Reference 154 courtesy of Society for Information Display)

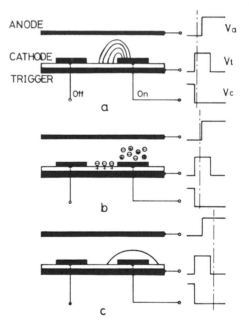

Fig. 10-63. Electrodes, waveforms, and discharge activity for addressing the Sony plasma display. (Reference 154 courtesy of Society for Information Display)

that cause an ac discharge between the cathode and the trigger electrode as shown in Fig. 10-63a. The wall charge builds up on the dielectric layer and eventually extinguishes the ac discharge as shown in Fig. 10-63b. This ac trigger discharge generates a large number of priming particles that will turn on the dc discharge between the cathode and the anode as shown in Fig. 10-63c. This dc discharge then emits a significant amount of light that is used for the display output.

The ac trigger pulse has an amplitude of 180 volts. This causes a strong ac discharge that does a very good job of priming the dc discharge. This greatly reduces the statistical and the formative delay times of the dc discharge and therefore

allows for faster scan rates. Also, the dc address voltages do not need to be as large, since a large voltage is not needed to make the delay time short. The anode pulse is 30 volts and the cathode pulse is 50 volts. These low voltages allow the utilization of lower-cost integrated circuits.

Figure 10-64 shows a multiplexing technique used to reduce the number of cathode circuit

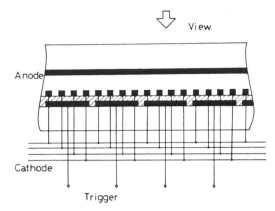

Fig. 10-64. Electrode-connection technique for doing internal-discharge logic in the Sony display. (Reference 154 courtesy of Society for Information Display)

drivers. By busing every fourth cathode and designing separate trigger electrodes under each group of four cathodes, the selection of a discharge on one cathode can be accomplished by pulsing one of the trigger electrodes and also pulsing one of the cathode groups. For this 16-cathode example, only 4 cathode drivers and 4 trigger drivers are needed. Of course, the circuit-driver savings becomes much more significant for larger numbers of display lines. For instance, 512 lines might use 32 cathode drivers and 16 trigger drivers. This technique effectively allows the gas discharge to act as an "and" gate between the cathode input and the trigger input. This internal-logic "and"-gate concept had been previously developed for the ac plasma panel.[155,156] These designs allow a factor-of-ten reduction in the number of circuit drivers. None of these techniques have reached commercial production in early 1983.

10.9 IMAGE DISPLAYS

Up to this point, the discussion has been on message-type displays in which each pixel need only be either on or off. This section discusses the plasma displays that can be used for image displays. These require that the pixels exhibit gray scale and sometimes color. Virtually all of the displays covered in the previous sections have been successful in the marketplace. None of the displays discussed in this section have appeared on the market. This fact is a commentary on the difficulty of achieving flat image displays

since there are very few flat image displays of any technology on the market. This is in spite of the great potential market for flat-panel television. The CRT is a formidable competitor, especially in the area of image displays.

The remainder of this section will cover gray-scale and color techniques which have been demonstrated with both ac and dc displays. Gray scale can be achieved by modulating the discharge current or duty cycle. Color plasma displays can be achieved with different gas mixtures and by exciting phosphors with the gas discharge. Considerable success has been achieved at combining gray scale and color in plasma displays for real-time color television.

10.9.1 Gray Scale

Self-Scan[TM] *Gray Scale.* Gray scale can be achieved in both ac and dc plasma displays. It has seen more success with the dc panels, especially with the Self-Scan[TM] displays. The first high-quality real-time television displays were achieved by Zenith and Bell Labs with a standard Self-Scan[TM] panel using special drive electronics.[43,44] Earlier plasma displays showed real-time television with considerably less quality.[30,157] Gray scale can be achieved by analog modulation of the discharge current and by duty cycle modulation of a constant discharge current. Since the Self-Scan[TM] technology requires that the panel be scanned, it was natural to use this scanning in a scanned television mode. The major difference is that instead of scanning horizontally as in Fig. 10-20, the panel is rotated 90 degrees and scanned vertically. Each vertical display anode is then modulated in parallel so that a horizontal display line is discharged during the 63-microsecond television horizontal-display period. The scan rate of the panel is the same as the vertical scan rate of the television image.

Figure 10-65 shows a Self-Scan[TM] display operating with a real-time television signal. The display image compared very favorably with that of a standard television receiver tuned for the same color and brightness. Its high quality generated great excitement. The performance of this display, however, fell far short of what is easily obtainable with a CRT. The brightness was only 10 to 20 foot-lamberts, and the lum-

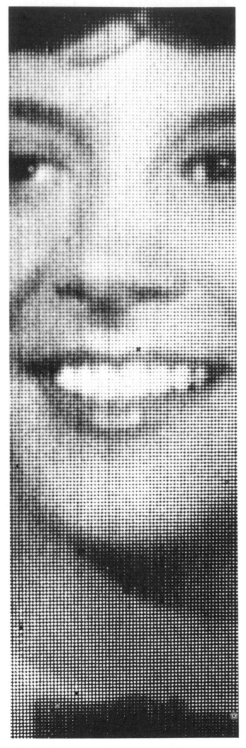

Fig. 10-65. Real-time, full-gray-scale television picture on a standard dc Self-Scan[TM] panel. (Reference 12 courtesy Society for Information Display)

inous efficiency was between 0.1 and 0.2 1m/W. The color of black and orange was unacceptable even for monochrome displays, but there is potential for correcting this through phosphors as discussed below. The circuits required for this type of display were quite extensive and complicated, making the cost of this display system much more expensive than the CRT. In spite of all these difficulties, the very good visual quality of the image on the Self-Scan[TM] panel encouraged many other researchers to continue work in the area.

AC Plasma Gray Scale. AC plasma displays have not yet received the same attention as d-c displays concerning image displays. This is probably because of the early incorrect perception that the intensity of ac panels is not easily modulated. The ac memory displays have only two intensities: on and off. Although limited success was achieved in an ac plasma display with three intensity levels,[69,158] the voltage margins achievable were very small, thus making the technique impractical. Therefore, only duty cycle and spacial modulation have received any significant attention. However, there is no reason why duty cycle modulation will not work perfectly well for gray scale in ac memory displays, as discussed below. In fact, the memory can be considered an advantage since the duty cycle of the brightest pixels can be one. This will give a higher brightness than will a refreshed display.

Figure 10-66 shows a 512 X 512 ac plasma display operating as a real-time television display.[159,160] This was operating in the memory mode and used duty cycle modulation to achieve the gray scale. Each cell in the panel is left on for a fraction of the vertical scan period, depending on the required brightness. This means that each cell must be turned on and off once each frame period. These speeds require that the vertical electrodes be addressed in parallel so that a horizontal display line is addressed at once. This is very similar to the requirement for the Self-Scan[TM] panels. However, unlike the Self-Scan,[TM] the ac display requires that the horizontal electrodes have a circuit driver for each line. This technique requires a frame memory that is used to remember the intensities of the

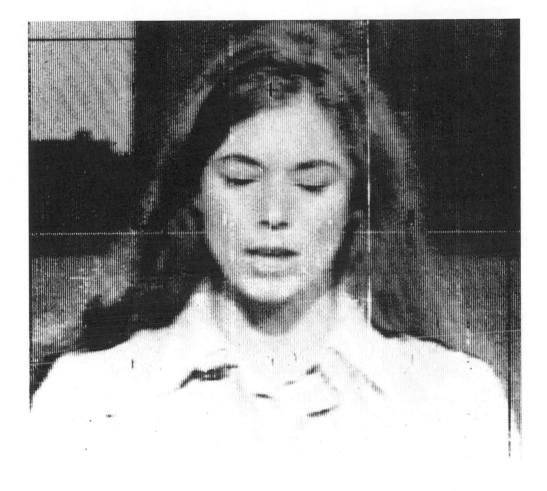

Fig. 10-66. Real-time, full-gray-scale television picture on a standard ac plasma display. (Reference 12 courtesy Society for Information Display)

pixels. This memory is continuously scanned to determine the proper times to turn a cell on and off. The visual quality of this display was good, but like the Self-Scan,[TM] it suffers from brightness, luminous efficiency, and circuit deficiencies that would make it unfavorable compared with the CRT.

Other monochrome displays that have demonstrated gray scale include a novel dc device that uses a constricted glow discharge,[161,162] and a new dc memory television display that uses graphite cathodes for current limiting.[55]

Pseudo Gray Scale. The high resolution of the ac plasma display makes spacial modulation an attractive way to do gray scale. This technique is similar to the halftone images used in the

printing industry. The technique is applicable to any bi-level displays which have sufficient resolution to trade for gray scale. One technique that has received considerable attention is ordered dither, developed at Bell Labs.[163–170] Figure 10-67 shows an image that uses this algorithm in a 512×512 ac plasma display. The major advantage of this technique is that the algorithm is very simple and requires only a small amount of hardware or software. It can be implemented on an existing display designed for use as a computer terminal. An attractive application for this technique is to transmit images in a facsimile mode that does not require much additional hardware to the computer terminal.

The ordered dither technique does not give the highest-quality image achievable on a bi-

Fig. 10-67. Pseudo gray scale on a standard bi-level ac plasma display using the ordered dither technique. (Courtesy of Drew White)

level display. With more complicated algorithms that adapt themselves depending on image, a measureable amount of improvement can be achieved.[163,165,171,172] These algorithms take a large amount of hardware or software. Ordered dither does not need to adapt to the image and so is easy to implement. This simplicity makes it possible to do real-time television displays using the dither algorithm.[173,174] If the image does not move too rapidly, a surprisingly good television reproduction can be achieved with a standard commercially available 512 X 512 panel that allows random access to any pixel every 20 microseconds. The image quality does not equal that of Figs. 10-65 and 10-66. The amount of hardware is modest, however, making it a good potential solution for applications not requiring superior image quality.

10.9.2 Color Plasma Displays.

Almost all plasma displays on the market have the neon-orange color. Other colors can be achieved by using different gas mixtures or by introducing phosphors into the display.

Gas Mixtures with Different Colors. Changing the gas mixture is the easiest technique for achieving other colors. The problem with this is that the other inert gas mixtures do not have nearly the luminous efficiency of neon.[175] This means that to achieve the same brightness as neon, the gas discharge must have at least ten times the drive current or duty cycle. This has the undesirable effect of greatly increasing the drive power and causing a proportional decrease in display lifetime.

One exception is a mercury-based argon gas mixture that exhibits a blue color with efficiencies and brightnesses comparable to neon.[176] This mixture is of diminished practical value because the efficiencies and brightness are strongly dependent on mercury pressure, which is strongly dependent on temperature. Such mixtures can change brightness by a factor of five between 20°C and 60°C.

Recently Nighan and Ferrar have demonstrated high-brightness and high-efficiency green and blue emissions in an ac display filled with a noninert gas mixture.[177,178] These experiments showed very wide emission bands from the fluorescence of excimer molecules. These are molecules that are semistable in the excited state but that do not exist in the absence of the discharge. They are typically composed of xenon and a halogen or oxygen. These gases have been studied extensively for use in high-pressure gas lasers. While a $Ne-Xe-Cl_2$ showed bright efficient emissions in the blue that were comparable to the standard $Ne-Xe$ Penning mixture, it is unlikely that this mixture will be practical because of the very high chemical reacitivity of Cl_2. A second mixture studied was $Xe-O_2$, which gave comparably bright green emissions. This gas has a much higher probability of being chemically compatible with the other materials in plasma displays. While these results are encouraging, considerable work needs to be done before practical long-lived devices are available.

Color with Electron-Excited Phosphors. Good color can be achieved by adding phosphors to the panel structure. There is at least one commercial product which uses a gas-discharge-excited phosphor, and there will undoubtably be more. Such phosphors can be excited either by ultraviolet photons[52] or by electrons.[179] For electron excitation, the phosphor is placed on the anode of the discharge, where there is a large flux of electrons. The energy of such electrons is very low, with an absolute maximum being the ionization energy of the gas used, which for neon is 21 electron volts. These electrons have energies many orders of magnitude lower than those typically used in CRTs. Most common phosphers will, therefore, be unsuitable. Only one known phosphor will give the required efficiencies with low-energy electrons, and that is $ZnO:Zn$.[179] This phosphor is a green color and has yielded a 30-fL brightness in a Self-Scan™ panel operated at 1% duty cycle.

Operation in dc displays requires that the phosphor be electrically conductive. Otherwise, it would charge up, and the electrons would no longer strike it. The solution, commonly used in the CRT, of placing a conductive film of aluminum over the phosphor will not work here, because even the thinnest films will completely attenuate the low-energy electrons from the gas discharge. $ZnO:Zn$ has the necessary conductivity.

The color of $ZnO:Zn$ has an interesting characteristic. When neon gas is used, the orange light of the neon combines with the green light of the $ZnO:Zn$ to give a white light. This is potentially useful for black-and-white image displays.[179]

Color with UV-Excited Phosphors. Photoluminescent phosphors have been studied extensively for use in plasma displays.[52,180-190] These are excited by the ultraviolet light which is generated by the gas discharge. The gas used is usually not neon but a mixture (frequently containing xenon) that does not generate very much visible light but is optimized for ultraviolet emission. The spectra of these gases are rich in lines and bands in the range from 100 to 200 nm. A number of phosphors have been developed for ultraviolet excitation. Good efficiencies and brightnesses have been achieved for all three primary colors. In many cases the efficiencies and brightnesses of the phosphor system are considerably greater than those of neon gas. Table 10-2 shows some of these phosphors along with the brightnesses and efficiencies achieved. These phosphors must be designed differently than CRT phosphors since they have to be sensitive to the ultraviolet radiation generated by the gas discharge.

Phosphor Lifetime in Plasma Displays. Phosphors in plasma displays face two fundamental lifetime problems that have slowed their application. First, the efficiency of the phosphor can be reduced by the gas discharge, and second, the firing voltage of the gas discharge can be altered by the decomposition of phosphors. The excited particles in the gas discharge can have sufficient energy to poison the phosphor and reduce its light output. This problem is especially severe when the phosphors are placed at the cathode of the gas discharge. In this configuration, the ions from the discharge can sputter away the phosphor and cause significant damage to the phosphor. In dc displays, material sputtered off the cathode can coat the phosphor with a film that will block the UV radiation. The drift in the firing voltage of the gas discharge is caused by the changing secondary electron γ coefficient of the phosphor cathode surface.

It was initially obvious that the phosphor should not be placed at the cathode if lifetimes greater than a few hundred hours were desired. In dc displays, the anode might be a good place since ion sputtering cannot occur there. This increases lifetime considerably at the expense of brightness and efficiency, since most of the ultraviolet light is generated in the cathode fall region near the cathode. In ac displays, one does

not have this option since the cathode and anode are exchanged every half sustain cycle. Here one can put bars or donuts of phosphor around or to the side of the discharge area.[180] Even this, however, does not work well enough to make marketable devices. A more recent approach, shown in Fig. 10-68, uses a single-substrate ac plasma panel and places the phosphor on the cover plate away from the discharge. This achieves good coupling with the UV light, which gives the good luminous efficiencies shown in Fig. 10-69. This three-color panel has been aged 2,000 hours without degradation.[182] This technology holds considerable promise for the future.

Even with proper placement of the phosphor in the most benign region of the discharge, degradation still occurs, so discharge-resistant phosphors are being developed. The first thing that might be considered is coating the phosphor with a refractory material that is sputter-resistant. This has had some success,[183] but the major problem is finding a protective material that will transmit the deep ultraviolet light. Even if such a material is found, the problem of coating the irregularly shaped phosphor particles with a pinhole-free film is formidable.

The more reasonable approach to increased lifetime is to make the phosphor more rugged. Success in this area will continue to increase as interest in color plasma displays continues.

One indication of the possibilities is the one commercial product currently on the market. This is a green dc plasma display made by Okaya and Oki in Japan and marketed in the United States by IEE. This panel is the refreshed dc dot matrix type with the structure shown in Fig. 10-21. The phosphor is placed on the front glass so that UV light from the negative glow of the adjacent cathode electrode will efficiently excite the phosphor. This requires careful pro-

Table 10-2   Some UV-Excited Phosphors Used in Color Plasma Displays

Phosphor Materials			Chromaticity Coordinates		Relative Luminance	Reference
Red	Green	Blue	u	v		
$Y_2O_3:Eu$	$Zn_2SiO_4:Mn$	$CaWO_4:Pb$	.18	.35	0.58	190
$Y_2O_3:Eu$	$Zn_2SiO_4:Mn$	$Y_2SiO_5:Ce$	.18	.33	1.0	196
$YP_{.65}V_{.35}O_4:Eu$	$Zn_2SiO_4:Mn$	$YP_{.85}V_{.15}O_4$	.18	.33	.83	195, 201
$(Y,Gd)BO_3:Eu$	$BaAl_{12}O_{19}:Mn$	$BaMgAl_{14}O_{23}:Eu$	.19	.31	1.7	21

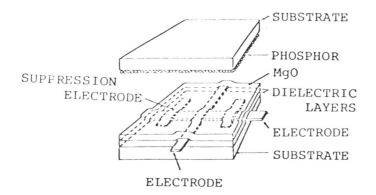

Fig. 10-68. Color ac plasma display that used the single-substrate structure to extend phosphor lifetime, developed by Fujitsu. (Reference 142 courtesy Society for Information Display)

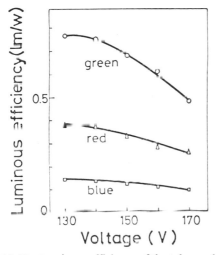

Fig. 10-69. Luminous efficiency of the color ac plasma display. (Reference 142 courtesy Society for Information Display)

cessing to minimize cathode sputtering, which will deposit cathode material on the phosphor and cut off the UV excitation. This arrangement gives the display reasonable lifetimes. The manufacturer specifies that the display half-intensity lifetime is 30,000 hours. This is an impressive result which suggests that we may see more green plasma displays in future marketplaces.

One problem with using phosphors in plasma displays is that the contrast ratio is reduced because of the phosphor's high ambient-light reflectivity. Thus, the high-brightness phosphors shown in Table 10-2 may not have a higher contrast ratio than the phosphor-free neon panel when the ambient light is considered.

10.9.3 Color Television. With the color and gray-scale abilities discussed above, good-looking television displays have been made with gas discharges. The major problems of achieving brightness and luminous efficiencies equivalent to the CRT have not been solved, but steady progress is being made, especially in the research labs of Japan. A number of excellent review articles have covered the state of the art prior to 1976.[12,191–193]

None of the displays discussed in this section are currently on the market, nor are any of them likely to be before 1990. Because of the very large market that a flat television display will have, a technology that achieves the required performance will be regarded as highly proprietary. This means that the displays that appear in the literature and are discussed below are likely to be the ones that do not work to the required specification. However, the steady progress of the performance of the research displays in the literature indicates that there is an excellent chance that practical flat-panel television using gas discharges will be a reality.

Table 10-3 lists some of the research television displays that have appeared in the literature. A large number of techniques have been tried. All of the color-television plasma displays have been dc, probably because of the early difficulty in getting good lifetime in color ac plasma panels. This may change in the future, since recent results show that color ac plasma panels with good lifetime and high resolution can be achieved.[182] The UV light that excites the phosphor has come from the negative glow or the positive column. Some of the panels have used memory to increase the duty cycle of the display elements to obtain greater brightness. The memory displays have

Table 10-3. Research Color-Television Plasma Displays

Negative Glow

Sony 1973 (References 197, 198; Fig. 10-71)
Hitachi 1974 (References 194, 195; Fig. 10-70)
Japan Broadcasting (NHK) 1974 (Reference 196)
Japan Broadcasting (NHK) 1979 (References 20, 21; Figs. 10-76, 10-77)

Positive Column

Philips 1972 (References 52, 181)
Hitachi 1976 (References 201, 56)
Hitachi 1977 (References 199, 200; Figs. 10-74, 10-75)
Japan Broadcasting (NHK) 1978 (References 21, 202)
AEG-Telefunken 1980 (Reference 203)

been developed using all of the different dc current-limiting techniques discussed in Section 10.5.

Duty Cycle Considerations. The cathode-ray tube pixel generally is excited by the electron beam for only 100–200 nanoseconds for each 33-ms display frame. Since its duty cycle is so low, the peak brightness of the CRT phosphor must be very large to achieve reasonable brightness. Plasma displays would saturate at such high brightness, and so their duty cycle must be greater. This is accomplished in two ways. First, all of the cells in a horizontal display line are discharged during the 63-microsecond horizontal scan period. This requires that a circuit driver be placed on each vertical electrode. This, therefore, allows a duty cycle of 1/525 for standard television. Second, duty cycles approaching one can be achieved if each pixel has memory. This will allow the brightest pixels to be on for the entire 33-ms frame period. The required peak brightness will be very close to the average brightness. This low peak brightness is a potential advantage not only because of the brightness-saturation problem but because many gas discharges lose efficiency as they are driven harder, so that the memory panels may have an efficiency advantage.

Negative-Glow Television Displays. All of the alphanumeric and graphic plasma displays on the market today utilize the light from the negative-glow region of the discharge shown in Fig.

10-7. The early television plasma displays utilized this same phenomenon. Figure 10-70 shows a display developed by Hitachi[194,195] which places the phosphor very near the cathode to achieve maximum flux of UV light. This display was capable of Self-Scan™ operation so that the number of circuit drivers in one dimension could be drastically reduced, as discussed in Section 10.6.4. This display gave good-quality pictures, but with an average area luminance of only 5 fL at a luminous efficiency of only 0.05 lm/W. There were also serious life problems because the phosphor was so close to the cathode.

Figure 10-71 shows a display developed by Sony that is of the matrix type.[197,198] This structure is considerably simpler than that in Fig. 10-70, utilizing only thick-film deposition techniques. However, the savings in the panel cost may be displaced by the increased cost of the driver circuits, because the matrix approach requires a driver for each line in both dimensions, almost double the number required for Self-Scan™ displays. The phosphors were placed near the anode to prevent sputtering degradation. Glass barriers and barrier electrodes were used to prevent the spreading of the discharge from one cell to the next. The brightness and efficiency of this panel were comparable to those of the Hitachi display of Fig. 10-70.

Positive-Column Television Displays. One possible solution to the brightness and efficiency problem is to use UV light generated by the positive-column region shown in Fig. 10-7. The common fluorescent lamp used in room lighting achieves efficiencies of 80 lm/W by using phosphors excited by UV light from a gas-discharge positive column. It was reasoned[52,181] that by using the Paschen firing voltage dependence on pd, discussed in Section 10.4.3, the size of the fluorescent tube could be greatly reduced by simply increasing the gas pressure. This sounds easier to do than it is in reality, but significant advances have been made through use of the positive column.

The possibilities of this technique are illustrated in a display cell developed by Zenith as shown in Fig. 10-72.[188] This cell achieved a time-averaged spot brightness of 360 fL at 3.4 lm/W with a green phosphor. These are the kinds of

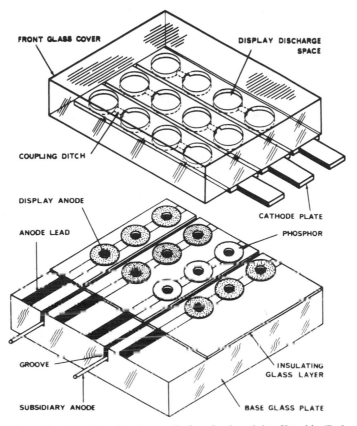

Fig. 10-70. Cathode-glow color-television dc plasma display developed by Hitachi. (Reference 195 © 1975 IEEE)

numbers needed for a practical television display. This design has a Self-Scan™-type structure. The positive column is formed along the front phosphor-coated groove between the front anode wire and the metal ribbon cathode. The back anode wire forms the scanning discharges needed for the Self-Scan™ effect. The luminance and efficiency as a function of instantaneous drive current is seen in Fig. 10-73. The luminance is time-averaged with a duty cycle of 1.8×10^{-3}. This clearly demonstrates the classic saturation effect of gas discharges and shows that there is a tradeoff between brightness and efficiency. Note that this device is capable of exceedingly high peak luminance. For instance, for the 5.6-lm/W data point in Fig. 10-73, dividing the average 200 fL by the duty cycle gives a peak luminance of 110,000 fL.

The Zenith device was filled with 100-torr He with mercury. Because of the high pressure of mercury needed, the cell had to be operated at 86°C to achieve the excellent results. This greatly

reduces the probability that this device will reach the marketplace. However, these results are a great encouragement for further research.

A three-color postive-column display developed by Hitachi is shown in Fig. 10-74.[199, 200] The major new feature that this display adds is memory. As discussed above, this allows the duty cycle of the discharge to approach unity, which in turn allows for considerably lower peak currents. Figure 10-73 demonstrates that there is a great efficiency advantage to operating at lower currents. This display achieved a spot luminance of 1750 fL in the green at 0.87 lm/W and area luminance of 105 fL for white at 0.29 lm/W. This was achieved with a xenon gas using no mercury, so that there are no significant temperature limitations. The addressing is done by simple matrix techniques as shown in Fig. 10-75. Note that each pixel has two discharge cells, the front one for display and the auxiliary discharge for priming. Since the postive column requires rather high voltages and a long delay

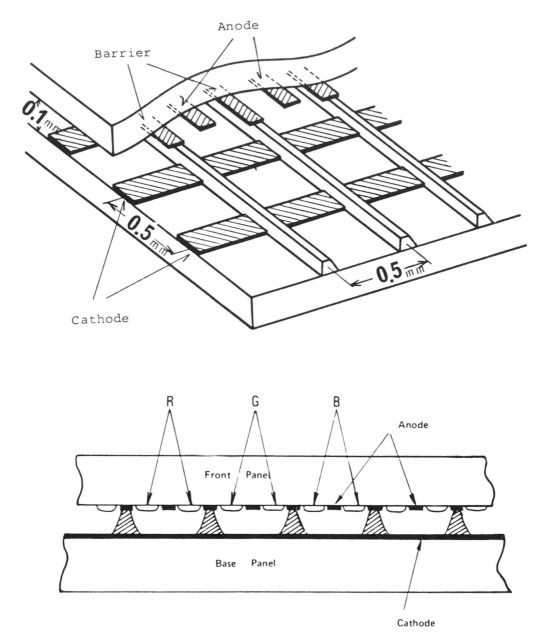

Fig. 10-71. Cathode-glow color-television dc plasma display developed by Sony. (Reference 198 © 1975 IEEE)

time to build up, good priming is necessary for a practical display. In this device, this is achieved by having a discharge active in each pixel at all times. The display discharge has high luminance, and the auxiliary discharge has very low luminance. Note that to achieve memory in this device, a 1–2-Mohm resistor is needed for each pixel. These might present fabrication problems, as discussed at the beginning of Section 10.6.

NHK Color Television. The most impressive-looking plasma display demonstrating color television as of early 1983 is shown in Fig. 10.76. This was developed by the Japanese Broadcasting Corporation (NHK).[20,21] This shows an array of 240×320 pixels with a diagonal of 16 inches and a pitch of 1 mm. It gives a very good color image comparable to that of a CRT with similar resolution. This is a refreshed dc

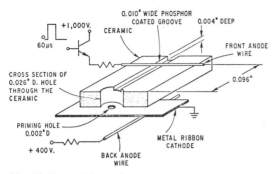

Fig. 10-72. Positive-column color-television display with high luminous efficiency and brightness developed by Zenith. (Reference 188 courtesy Society for Information Display)

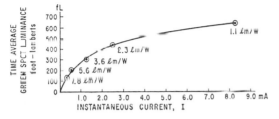

Fig. 10-73. The time-averaged spot luminance as a function of current for the color-TV display of Fig. 10-72. Also plotted are the luminous efficiencies for each data point. (Reference 188 courtesy Society for Information Display)

display without memory. The phosphor is excited with UV light from the negative glow. This display achieved an area white luminance of 7 fL at a white luminous efficiency of 0.05 lm/W.

Figure 10-77 shows the structure of this display. It is basically a Self-Scan™ structure with phosphor dots placed near the display anodes. This display is not operated like a Self-Scan™ display but rather like a refreshed dc dot matrix display as shown in Fig. 10-20. The scan anode discharge is not used for addressing but rather for priming the display discharge. Good priming is necessary to minimize the statistical and formative delays that otherwise would force this display to scan at too low a rate.

This display uses a special drive technique to increase the brightness by doubling the duty cycle of each pixel. This is effectively done by dividing the panel in two and then scanning the upper and lower halves of the panel with separate sets of address electronics. Thus a row in the upper half can be discharging at the same time as another row in the lower half, thus doubling

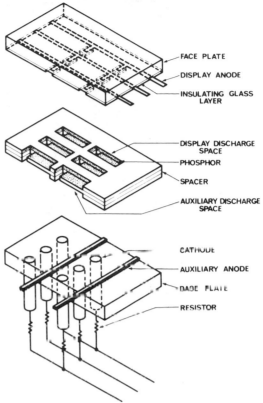

Fig. 10-74. Positive-column color-television display with memory developed by Hitachi. (Reference 199 courtesy Society for Information Display)

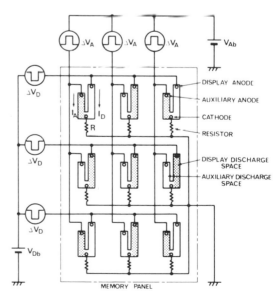

Fig. 10-75. Addressing technique used in the Hitachi display Fig. 10-74. (Reference 199 courtesy Society for Information Display)

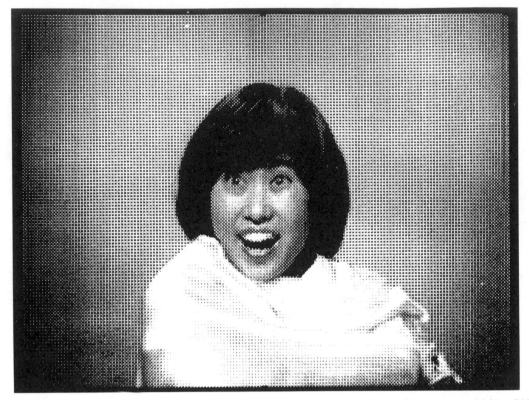

Fig. 10-76. Real-time full-color television display developed by NHK. This display has an array of 240 × 320 pixels and a 16-inch diagonal. (Reference 21 courtesy of NHK)

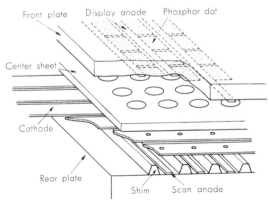

Fig. 10-77. Structure of the NHK color-television display shown in Fig. 10-76. (Reference 21 courtesy of NHK)

the duty cycle. This technique doubles the required number of column drivers, and it requires a memory to store a display frame.

10.10 HYBRID PLASMA-CRT

The gas-discharge and the CRT technologies have been combined recently to achieve a high-performance display. Figure 10-78 shows one such display developed by Siemens.[204] A similar display developed by Lucitron is shown in Fig. 10-79.[205,206] These devices use the gas discharge as a matrix-addressable source of electrons. The electrons are then accelerated to 4 kV, where they hit a phosphor screen that emits light like other CRTs. This display is capable of gray scale and full color. Perhaps the most impressive feature is the high brightness and high luminous efficiency that can be achieved with this new flat-panel technology.

This display has high brightness because of the exceedingly large currents that the gas discharge can supply. Currents in the milliamp range can easily be delivered to the phosphor, which means that phosphor life will be the major brightness-limiting factor.

The luminous efficiency of this device will approach that of the CRT. Siemens claims 6 lumens per watt at 60 foot-lamberts.[204] This high efficiency occurs because the fundamental light-emitting mechanism is the same as that of the CRT.

To operate properly, this device must have a strong gas discharge in the rear section with roughly 200 volts between the control plate

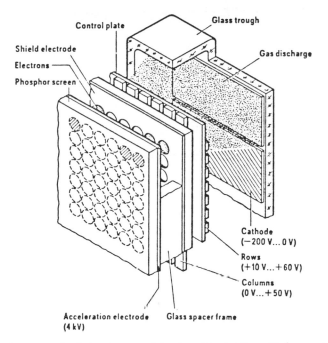

Fig. 10-78. Plasma electron excited phosphor panel developed by Siemens. (Reference 204 © 1982 IEEE)

and the cathode; and at the same time it must maintain the phosphor screen at +4 kV above the control plate without getting a gas discharge or arc in the front section, which would cause the whole screen to light up in an uncontrollable manner. These requirements are designed with the aid of the Paschen curve seen in Fig. 10-13. This shows how the firing voltage depends on the product, pd, of the gas pressure and the gas-gap distance. The breakdown-voltage require-

ments are achieved by optimizing the interelectrode spacings.

This display is operated at a pressure close to 1 torr with an inert gas such as helium or neon. The distance between the shield electrode and the acceleration electrode in Fig. 10-78 is 1 mm. This gap is operated at a pd of 0.1 torr · cm, which is on the far left edge of the Paschen curve in Fig. 10-13. Extrapolating the neon data gives a breakdown voltage of roughly 8 kV. Thus the

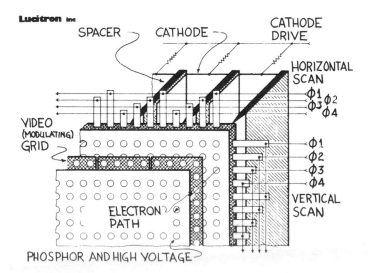

Fig. 10-79. Structure of the gas-electron-phosphor display developed by Lucitron. (Courtesy of Lucitron® Inc.)

applied 4 kV is well below the firing voltage for this gap.

The distance between the control plate and the cathode in Fig. 10-78 is 25 mm. This means that pd for this gap is 2.5 torr · cm. This is very near the minium firing voltage for the neon data in Fig. 10-13, which means that the applied −200 volts can cause a strong discharge. Also, this 25-mm gap allows the discharge to spread over a large area of the cathode, which gives high currents at low cathode current density.

The device developed by Lucitron (Fig. 10-79) uses a similar gas-discharge-excited-phosphor concept as the Siemens panel of Fig. 10-78, but there are a number of significant differences. The Lucitron panel uses the internal-logic ability of the gas discharge to reduce the number of external connections. Figure 10-79 shows that the row and column electrodes are internally bused to four drive phases. By using a technique very similar to that show in Fig. 10-25 for the Self-Scan^{TM} display, the discharge is internally scanned from one pixel to another. Unlike the Self-Scan^{TM} however, this device can scan in both the horizontal and the vertical directions, which gives an even greater reduction in external circuit connections.

The Lucitron panel also has a modulating grid between the scan electrodes and the high-voltage phosphor electrode. The video information is applied to this grid to modulate the electron current. Note that there is one modulating grid for each cathode electrode. This allows a parallel scan technique to be used. Each cathode-grid combination can continuously support one lit pixel. This means that at any given time the number of pixels lit in the display equals the number of cathode-grid combinations. This allows an additional level of multiplexing that greatly reduces the necessary scan rate and increases the duty cycle of each pixel.

The width of the modulating grids and the cathodes is 2.5 centimeters (one inch). The spacers between the cathodes make the Lucitron structure self-supporting. This allows very large panels (4 ft^2) to be made without the very heavy faceplate necessary to hold back the atmospheric pressure. The Siemens display in Fig. 10-78 does not have these spacers and requires a thick faceplate even for a small panel.

Although both of these devices are in the early development stages, they have achieved impressive displays. Table 10-4 shows the specifications for the two panels. The less complicated Siemens structure allows much higher resolutions. The Lucitron panel requires an order of magnitude fewer electrical connections and a much larger area display.

Both of these devices should be capable of a full-color image with suitable phosphors just like other CRTs. Although color television has not yet been demonstrated in 1983, it inevitably will be shortly. This technology has great potential for being a commercially successful flat-panel color-television display.

10.11 FABRICATION OF PLASMA DISPLAYS

Considerable care must be used in the fabrication of plasma displays in order to achieve good yields and good lifetimes.[207, 208] The number one concern is to maintain cleanliness in the gas and on discharge surfaces. It is very easy to contaminate the gas mixture and the cathode surfaces unless the necessary precautions are observed.[99, 74] This means that a glass and metal structure that can withstand a high-temperature bakeout is essential to achieving the required cleanliness. The sealing process must be done with a high-temperature glass frit process. Some

Table 10-4. Specifications for Siemens and Lucitron Hybrid Plasma-CRT Displays

Mfr.	Display Area Width, Inches	Height, Inches	Resolution Vert. (LPI)	Horiz. (LPI)	No. of Columns	No. of Rows	No. of External Connections
Siemens	9.0	7.0	79.6	63.6	720	448	1,168
Lucitron	27.6	20.5	18.6	12.5	512	256	96

manufacturers have tried epoxy seals, but these panels all have very poor lifetime. The requirement for high-temperature processing means that materials must be carefully chosen to match expansion coefficients. The choice of metal and glass type is almost always determined by the expansion and high-temperature compatibility requirements during manufacture. Fortunately, gas discharges operate properly with a wide range of metal and glass types.

The substrate glass is usually soda-lime window glass since it is readily available in large flat sheets and has an expansion coefficient compatible with the sealing solder glass.

The electrodes can be deposited by a number of techniques. These include screen-printed thick films, photosensitive thick films,[209] evaporated[210,211] or sputtered thin films, and transparent electrodes such as indium-tin oxide. Electrodes that do not require deposition on a substrate include etched sheet metal[212] or taut wires held in place by seal glass. In general, for low-resolution panels, the screen-printed electrodes have a considerably lower cost than thin-film processing. The thin films are usually required at the higher resolutions.

Plasma displays frequently require the application of a dielectric-glass layer at some stage in the processing. For ac displays, the dielectric layer is usually lead solder glass, although evaporated dielectrics can be used.[87] In dc displays, dielectric layers are used for crossovers and as discharge barriers. Dielectrics are useful in preventing the dc discharge from following along the electrode connecting the display electrode to the outside of the panel. These dielectric glasses can be made black by use of suitable dyes to increase contrast. They can also be made milky translucent, which is useful for back projection images on ac panels. The lead-based dielectric-glass frit is either screen-printed or sprayed on the substrate and is fired at between 500°C and 650°C depending on the solder glass composition and uniformity required.

After application of the dielectric layer in ac panels 200 to 300 nm of MgO is deposited, usually by electron-beam evaporation. This film is sputter-resistant and greatly increases device lifetime. It also has a high secondary emission

coefficient (γ) that reduces the firing voltage of the panel. It is highly susceptible to contamination, however, so it is applied just before the panel is sealed.

Some displays insert a small radioactive Ni^{63} wire at the appropriate spot to prime the discharge for initial startup, as discussed in Section 10.4.6. The alternative technique is to use a small amount of Kr^{85} in the fill gas.

There is always some sort of spacer material placed between the two substrates that determines the cathode-anode distance. Spacers can take the form of metal or glass rods or sheets with the necessary temperature and expansion characteristics. Spacers can be screen-printed or etched. Some spacers can be quite complex, such as the Self-Scan[TM] II spacer sheet that is made from the photosensitive Corning Fotoform material.[64]

Once all of the electrode and dielectric layers are on the substrate, the two substrates are sealed together, and a glass-filling tubulation is sealed to one of the substrates. This is accomplished by another solder glass that has a lower softening temperature than the dielectric glasses used in earlier processing. These glasses usually are fired at between 390°C and 480°C.

DC plasma displays usually insert an ampule of mercury just before sealing that can later be opened after the panel is completely sealed off. An alternate technique is to insert a metal ring containing a mercury compound. After the panel is sealed, a RF induction heater is used to heat the ring to drive out the mercury. Other manufacturers simply inject mercury directly into the panel before sealing. This mercury is essential for increasing the device lifetime, as discussed in Section 10.6.1.

Once all but the tubulation has been sealed, the panel is placed on a vacuum system and evacuated. During evacuation it is baked at a temperature of between 300°C and 400°C. This is necessary to drive out contaminants that are adsorbed on the surfaces and absorbed in the glass structure. The major culprit is H_2O, which is always absorbed by glass structures. After the bakeout, the panel is filled with a high-purity neon gas mixture and then the tubulation is tiped off to seal the panel completely.

During the evacuation, the panels are carefully checked for leaks. Since impurity levels as low as 100 ppm can significantly alter the firing voltage of plasma displays, care must be taken. The gas volume of most plasma displays is usually very small, and the device lifetime is very long. Thus, leak rates below the detectability of many systems may be the life-limiting factor of some panels with marginal seals.

After sealing, the panels are tested and aged. The aging process allows the gas discharge to sputter clean the cathode surfaces of the display. In a well-designed materials system, the burn-in process will occur in the first day or so, and from then on the firing voltages will be reasonably stable with time. This aging process can usually be accelerated by operating the panel at higher voltage and current levels. In some cases, open electrodes discovered at the test stage can be repaired by making connection to the opposite edge of the panel.[213]

One problem that limits the resolution of many potential designs is the hysteresis effect of glass.[214] During any of the high-temperature processing steps, the glass undergoes considerable expansion. The problem is that when the glass cools, it does not have the same dimensions that it did before the high-temperature process. This causes problems with any structures that must be carefully aligned. Problems with this effect can be controlled by annealing the glass substrate before the electrode patterns are deposited, using the same temperature profile that will be required for the subsequent processing.

Edge connections to the panel are made by a number of techniques. Pins can be soldered to some of the screen-printed electrode materials. The designs that use internal sheet-metal or pin electrodes simply bring the electrodes out and solder them directly to the circuitboard. Thin-film electrodes are usually connected by a pressure contact to a flexible circuit strip held in place by a spring clip. In this arrangement it is useful to have a rubber gasket to seal out the reactive gases that can destroy the pressure contact in time.

10.12 FUTURE OF PLASMA DISPLAYS

This chapter has shown that plasma displays have been commercially successful over a remarkably wide range of display sizes and device technologies. Plasma displays enjoy the dominant market position in large flat-panel display technologies in 1983. But what does the future hold for this technology? What follows is an editorial on its prospects.

It seems clear that plasma displays will maintain this dominant position in large flat-panel vectorgraphic displays in the near future for a number of reasons. Currently, no competing technologies can claim such a complete set of flat-panel attributes as is found in Section 10.1. For instance, it will be very difficult to develop any type of matrix display with a simpler structure than that of the ac plasma display in Fig. 10-37, and one that has the desirable feature of internal memory to boot. Memory is usually necessary to have large matrix displays with sufficient brightness. The recent product announcements by IBM of the large computer terminal display shown in Fig. 10-35 will surely enhance plasma's credibility as a viable technology in the marketplace.

The smaller plasma displays such as the segmented displays will continue to receive severe competition from the other technologies such as light-emitting diodes, vacuum fluorescence, electroluminescence, and liquid crystals. Plasma's continued success in this market is not assured.

Advances will continue in the research and development of image and color-television displays. However, these are not likely to see large-scale commercial production for many years. Plasma displays must first prove to be competitive with the CRT in the large vectorgraphic displays where technological advances of color and gray scale are not as critical.

The major problem with all large flat-panel displays is cost. To successfully displace an existing technology, there must be a very significant cost-performance ratio improvement. As of 1983, plasma displays are not yet cost-competitive with the cathode-ray tube, and therefore plasma displays are number two in large computer displays, since flatness is not as important as cost for most of this market. Plasma finds most of its applications in places where the CRT does not fit or where cost is a secondary factor. Plasma is now struggling to lower its cost to capture a larger share of the market from the CRT.

On the other hand, any technology that tries to overtake the solid number-two position of plasma in large vectorgraphic displays will have to outperform it at a lower cost. This is not likely, since the major cost factors of device processing and drive circuits are shared by all matrix technologies. As discussed below, plasma displays have a considerable head start and also unexplored potential in these two important cost areas.

The materials cost of the display device is usually negligible. It is the processing costs that determine the cost of the device. It follows that the simpler the structure, the lower the processing costs. Display technologies that require a large number of processing steps, such as those that use a thin-film transistor for each pixel, will have a very difficult time in a head-on competition with the simple plasma displays. Such technologies may, however, be viable in areas which plasma can't touch, such as very-low-power passive displays.

The thin-film ac electroluminescent displays offer a structure that is roughly as simple as the simplest plasma panels. However, EL technology will have to struggle to match the volume-production capability of the existing plasma-display-manufacturing facilities in order to get its cost down. Even if they succeed in doing this, EL devices can only hope to equal the device costs of the plasma displays.

For most large flat-panel matrix displays, the drive electronics are the major cost, because of the large number of drivers required. Integrating these drivers will go a long way toward reducing these costs. However, it is difficult for even the most sophisticated integration to lower the cost to a level competitive with the CRT. This is because the typical matrix display requires roughly 100 times as many electrode drivers as the CRT!

Plasma displays have demonstrated a number of commercial products that significantly reduce the electronics cost and volume by performing much of the logic and decoding function with gas-discharge logic located in the display device. Such techniques can reduce the number of circuit drivers by factors of ten and more. Plasma is the only flat-panel technology with existing products that use this internal-logic ability. While great progress has been made

already, the internal-logic ability has a lot of unexplored potential. The rate of new ideas in this area has not yet started to decline, as is evidenced by the new internal-logic devices introduced in the three years prior to this writing (1980–1983) and which are shown in Figs. 10-33, 10-56, 10-58, 10-64, and 10-79. The gas discharge is so rich in phenomena conducive to switching that there is considerable room for new invention. The goal is to drastically reduce the electronics costs without significantly increasing the device costs.

Internal logic is plasma's major hope for undercutting the cost of the CRT. The other large flat-panel matrix technologies can only rely on circuit integration to lower the circuit costs. If circuit integration is successful for the others, then it will also be successful for plasma. The gas-discharge logic gives plasma displays a second potential road to success.

It is sometimes said that plasma is a mature display technology, with the implication that most major advances have already occurred. While its long history certainly qualifies plasma technology as mature, the rate of significant new developments in the research labs and in the plasma marketplace in the last few years makes this field as dynamic and as exciting as ever. The tremendous flexibility of the gas discharge as a light emitter with many electrical properties highly suitable for matrix displays will insure a great amount of invention, innovation, and market success in the future.

REFERENCES

1. Sobel, A. "Gas-Discharge Displays: The State of the Art." *IEEE Trans. Electron. Devices*, Vol. ED-24, pp. 835–847, July 1977. Also appears in 1976 Biennial Display Conference, Conf. Rec., New York, pp. 99–109.
2. Slottow, H. G. "Plasma Displays." *IEEE Trans. Electron. Devices*, Vol. ED-23, pp. 760–772, July 1976.
3. Criscimagna, T. N., and Pleshko, P. "AC Plasma Display." Chapter 3 in *Topics in Applied Physics*, Vol. 40, Display Devices, Springer-Verlag, Berlin, pp. 91–150, 1980.
4. Cola, R., et al. "Gas Discharge Panels with Internal Line Sequencing ("Self-Scan" displays)." *Advances in Image Pick-up and Display*, Vol. 3, Academic Press, New York, pp. 83–170, 1975.
5. Jackson, R. N., and Johnson, K. E. "Gas Discharge Displays: A Critical Review." *Advances in*

Electronics and Electron Physics, Vol. 35, Academic Press, New York, pp. 191–267, 1974.

6. Weston, G. F. "Plasma Panel Displays." *Scientific Instruments*, Vol. 8, pp. 981–991, December 1975.

7. Bylander, E. G. *Electronic Displays*, Chapter 4: "Gas Discharge Displays." McGraw-Hill, New York, 1980.

8. Maloney, T. C. "Plasma Displays." Electro/81, Session Record, paper 18/2, New York, 1981.

9. Boardman, C. M., and Deschamps, J. "Plasma Display Panels Have Come of Age." *Displays*, July 1982, pp. 135–146.

10. Sasaki, A., and Takagi, T. "Display-Device Research and Development in Japan." *IEEE Trans. Electron. Devices,* Vol. ED-20, pp. 925–933, November 1973.

11. Baur, G., and Meier, G. "Progress in Display-Device Development in Western Europe." *IEEE Trans. Electron. Devices*, Vol. ED-20, pp. 934–940, November 1973.

12. Chodil, G. "Gas Discharge Displays for Flat-Panel." *Proc. SID*, Vol. 17, pp. 14–22, 1976.

13. Kojima, T. "Recent Flat-Panel Display Developments in Japan." 1980 SID Int. Symposium, San Diego, pp. 22–23.

14. Deschamps, J. "What Does the Future Hold for the Plasma Panel?" Proc. 1981 European Display Research Conference, Munich, pp. 177–182.

15. Burke, R. W.; Hoehn, H. J.; and Fein, M. E. "Optical Characteristics of AC Plasma Panels." 1974 SID Int. Symposium, San Diego, pp. 104–107.

16. Crisp, M. D.; Hinson, D. C.; and Siegel, J. I. "Luminous Efficiency of a Digivue® Display/Memory Panel." 1974 IEEE Conf. Display Devices, Conf. Rec., New York, pp. 27–29. Also, appears in *IEEE Trans. Electron. Devices*, Vol. ED-22, pp. 681–685, September 1975.

17. Criscimagna, T. N.; Beidl, J. R.; and Trushell, J. B. "Write and Erase Waveforms for High Resolution AC Plasma Display Panels." 1980 Biennial Display Research Conference, Conf. Rec., Cherry Hill, N.J., pp. 31–39. Also appears in *Proc. SID*, Vol. 22, pp. 204–212, 1981, and *IEEE Trans. Electron Devices*, Vol. ED-28, pp. 630–638, June 1981.

18. Willis, D. R.; Johnson, R. L.; Ernsthausen, R. E.; and Wedding, D. R. "Large Area Displays." Proc. 1981 European Display Research Conference, Munich, pp. 191–194.

19. Schermerhorn, J. D. "Color and Contrast Modifications in AC Plasma Display/Memory Products." 1982 SID Int. Symposium, San Diego, pp. 168–169.

20. Kojima, T.; Toyonaga, R.; Sakai, T.; Tajima, T.; Sega, S.; Kuriyama, T.; Koike J.; and Murakami, H. "Sixteen-Inch Gas-Discharge Display Panel with 2-Lines-at-a-Time Driving." *Proc. SID*, Vol. 20, pp. 153–158, 1979.

21. Image Devices Research Group. "Gas-Discharge

Panels for Color TV Display." NHK Technical Monograph, No. 28, pp. 1–47, March 1979.

22. Benjamin, P. *A History of Electricity*. Wiley, New York, 1898.

23. Schroeder, H. *History of Electric Light*. Smithsonian Institution, Harrison, NJ, 1923.

24. Meyer, H. W. *A History of Electricity and Magnetism*. MIT Press, Cambridge, Mass., 1971.

25. Faraday, M. *Experimental Researches in Electricity*, Vol. 1, pp. 490–496. Taylor and Francis, 1839, also Dover, 1965.

26. Magie, W. F. *A Source Book in Physics*. McGraw-Hill, New York, 1935.

27. Phillips, U. J. *Early Radio Wave Detectors*. Peter Peregrinus Ltd., London, 1980.

28. Elwell, C. F. *The Poulsen Arc Generator*. Ernest Benn Limited, London, 1923.

29. Miller, S. C., and Fink, D. G. *Neon Signs*. McGraw-Hill, New York and London, 1935.

30. Gray, F.; Horton, J. W.; and Mathes, R. C. "The Production and Utilization of Television Signals." *Bell Sys. Tech. J.*, Vol. 6, pp. 560–603, October 1927.

31. Bacon, R. C., and Pollard, J. R. "The Dekatron: A New Cold Cathode Counting Tube." *Electronic Engineering*, Vol. 22, pp. 173–177, 1950.

32. Hampel, H. J. U.S. Patent 2,874,320, 1954.

33. McLoughlin, N.; Reaney, D.; and Turner, A. W. "The DIGITRON: A Cold-Cathode Character Display Tube." *Electronic Engineering*, Vol. 32, pp. 140–143, 1960.

34. McCauley, J. H. U.S. Patent 2,991,387, 1961.

35. Maynard, F. B.; Carluccio, J.; and Poelstra, W. G. "Grid Switched Gas Tube for Display Presentation." *Electronics*, Vol. 29, pp. 154–156, August 1956.

36. Findeisen, B. "Flat Display Device vs Gas-Discharge Unit." *Electronic Equipment*, Vol. 4, pp. 16–18, October 1956.

37. Moore, D. W. "Gas Discharge X-Y Display Panel." Winter Conference on Military Electronics, Professional Group on Military Electronics, pp. 1-8 to 1-13, 1962.

38. Lear Siegler Inc. "Development of Experimental Gas Discharge Display." Quarterly Progress Reports #2, 3, 4, 5, 6, and 7, Contact #NOBSR-89201 Bu Ships, August 1963–June 1965.

39. Bitzer, D. L., and Slottow, H. G. "The Plasma Display Panel–A Digitally Addressable Display with Inherent Memory." Presented at the 1966 Fall Joint Computer Conf., Washington, D.C., AFIPS Conf. Proc., Vol. 29, p. 541, 1966.

40. Nolan, J. F. "Gas Discharge Display Panel." Conf. Digest, 1969 Int. Elect. Dev. Meeting, Washington D.C.

41. Holz, G. E. "The Primed Gas Discharge Cell– A Cost and Capability Improvement for Gas Discharge Matrix Displays." *Proc. SID*, Vol. 13, pp. 2–5, 1972.

42. Umeda, S., and Hirose, T., "Self-Shift Plasma

Display." 1972 SID Int. Symposium, San Francisco, pp. 38–39.

43. Chen, Y. S., and Fukui, H. "A Field-Interlaced Real-Time Gas-Discharge Flat-Panel Display with Gray-Scale." 1972 IEEE Conf. Display Devices, Con. Rec., pp. 70–76; also *IEEE Trans. Electron. Devices*, Vol. ED-20, pp. 1092–1098, November 1973.

44. Chodil, G. J.; De Jule, M. C.; and Markin, J. "Good Quality TV Pictures Using a Gas-Discharge Panel." 1972 IEEE Conf. Display Devices, Con. Rec., pp. 77–81; also *IEEE Trans. Electron. Devices*, Vol. ED-20, 1098–1102, November 1973.

45. Cobine, J. D. *Gaseous Conductors*. Dover, New York, 1958.

46. von Engle, A. *Ionized Gases*. Oxford University Press, Clarendon, London and New York, 1965.

47. Loeb, L. B. *Basic Processes of Gaseous Electronics*, University of California Press, Berkeley, 1961.

48. Meek, J. M., and Craggs, J. D. *Electrical Break down of Gases*. Wiley, New York, 1978.

49. Kruithof, A. A., and Penning, F. M. "Determination of the Townsend Ionization Coefficient α for Mixtures of Neon and Argon." *Physica*, Vol. 4, pp. 430–449, June 1937.

50. Nighan, W. L. "Basic Kinetic Processes in Neon Gas Discharge Displays." *Proc. SID*, Vol. 22, pp. 199–204, 1981.

51. Smith, J. "DC Gas-Discharge Display Panels with Internal Memory." *IEEE Trans. Electron. Devices*, Vol. ED-20, pp. 1103–1108, November 1973.

52. van Houten, S.; Jackson, R. N.; and Weston, G. F.; "DC Gas Discharge Display Panels." *Proc. SID*, Vol. 13, pp. 43–51, 1st quarter 1972, and p. 113, 2nd quarter 1972.

53. Holz, G. E. "Pulsed Gas Discharged Display with Memory." 1972 SID Int. Symposium, San Francisco, pp. 36–37.

54. Smith, J., and Johnson, K. E., "Experimental Storage Display Panels Using DC Gas Discharges Without Resistors." 1974 IEEE Conf. Display Devices, Conf. Rec., New York, pp. 110–115. Also appears in *IEEE Trans. Electron. Devices*, Vol. ED-22, pp. 642–649, September 1975.

55. Murakami, H.; Sega, S.; and Tajima, T. "An Experimental TV Display Using a Gas-Discharge Panel with Internal Memory." *Proc. SID*, Vol. 21, pp. 327–332, 1980.

56. Okamoto, Y., and Mizushima, M. "A Positive-Column Display Memory Panel Without Current-Limiting Resistors for Color TV Display." *IEEE Trans. Electron. Devices*, Vol. ED-27, pp. 1778–1783, September 1980.

57. Walters, F. "DC Gas Discharge Storage Displays." 1974 SID Int. Symposium, San Diego, pp. 126–127.

58. Weston, G. F. Chapter 9 in *Cold Cathode Glow Discharge Tubes*. ILIFFE Books Ltd., London, 1968.

59. Maloney, T. C. "Panaplex II^{TM}: A Low Cost Digital Display." 1972 IEEE Conference on Display Devices, Conference Report, pp. 19–21.

60. Akutsu, H., and Nakagawa, Y. "Scanning in a DC Plasma Panel." 1981 SID Int. Symposium, New York, pp. 166–167.

61. Akutsu, H., and Nakagawa, Y. "A dc Plasma Display Panel Unit with Higher Reliability and Simpler Construction." *Proc. SID*, Vol. 23, pp. 61–65, 1982.

62. Miller, D.; Ogle, J.; Cola, R.; Caras, B.; and Maloney, T. "An Improved Performance Self-Scan 1^R Panel Design." 1980 SID Int. Symposium, San Diego, pp. 146–147.

63. Miller, D.; Ogle, J.; Cola, R.; Caras, B.; and Maloney, T. "An Improved Performance Self-Scan I Panel Design." *Proc. SID*, Vol. 22, pp. 159–163, 1981.

64. Miller, D. E., and Cola, R. A. "Self-Scan II Panel Displays, A New Family of Flat Display Devices." 1976 Biennial Display Conference, Conf. Rec., New York, p. 38, 1976.

65. Okamoto, Y.; Mitani, E.; Sumioka, A.; and Okabe, T.; "A Color Bar-Graph Display Panel Using DC Gas Discharge." *Proc. SID*, Vol. 21, pp. 321–324, 1980.

66. Miani, E., and Okamoto, Y. "A New Self-Scanning Method for Producing High Luminance in a DC Gas-Discharge Bar-Graph Display Panel." *IEEE Trans. Electron. Devices*, Vol. ED-29, pp. 1745–1748, November 1982.

67. Smith, J. "A Gas Discharge Display for Compact Desk-Top Word Processors." 1980 Biennial Display Research Conference, Conf. Rec., Cherry Hill, N.J., pp. 79–82. Also appears in *IEEE Trans. Electron Devices*, Vol. ED-29, pp. 174–178, February 1982.

68. Wilson, R. H. "A Capacitively Coupled Bistable Gas Discharge Cell for Computer Controlled Display." Coordinated Science Laboratory Report R-303, University of Illinois, June 1966.

69. Petty, W. E., and Slottow, H. G. "Multiple States and Variable Intensity in the Plasma Display Panel." *IEEE Trans. Electron. Devices*, Vol. ED-ED-18, pp. 654–658, September 1971.

70. Dick, G. W. "Single Substrate AC Plasma Display." 1974 SID Int. Symposium, San Diego, pp. 124–125.

71. Reisman, A., and Park, K. C. "AC Gas Discharge Panels: Some General Considerations." *IBM J. Res. Dev.*, Vol. 22, pp. 589–595, November 1978.

72. Pleshko, P. "AC Plasma Display Technology Overview." *Proc. SID*, Vol. 20, pp. 127–130, 1979.

73. Pleshko, P. "AC Plasma Display Device Technology: An Overview." *Proc. SID*, Vol. 21, pp. 93–100, 1980.

74. Greeson, Jr., J. "AC Plasma Technologies and Characteristics." 1980 SID Int. Symposium Sem-

inar Lecture Notes, Vol. 2, pp. 101–123, San Diego, May 1980.

75. Miwa, H.; Uozumi, T.; Nakayama, N.; Umeda, S.; Furuta, H.; and Hayashi, H. "Plasma Display– New Interactive Display Terminal." *Proc. SID*, Vol. 14, pp. 34–38, 1973.

76. Soper, T. J.; Ernsthausen, R. E.; Wedding, D. K.; and Willis, D. R. "High Resolution Meter-Size Display Technology." 1982 SID Int. Symposium, San Diego, pp. 162–163.

77. Pleshko, P.; Smith, G. W.; Thompson, D. R.; and Vecchiarelli, N. "The Characteristics and Performance of an Experimental AC Plasma 960 × 768 Line Panel and Electronics Assembly." 1981 IEDM, p. 299.

78. Lorenzen, J. "Design Optimization of a 960 × 768 Line AC Plasma Display Panel." 1982 Int. Display Research Conference, Conf. Rec., Cherry Hill, N.J., pp. 107–110.

79. Hairabedian, B. "Empirical Relation of Margin to Gap and Line-Width in AC Plasma Panel." 1982 Int. Display Research Conference, Conf. Rec., Cherry Hill, N.J., pp. 111–114.

80. Nolan, J. "Gas Discharge Display Panel." Presented at the 1969 International Electron Devices Meeting, Washington, D.C.

81. Hoehn, H. J., and Martel, R. A. "A 60 Line Per Inch Plasma Display Panel." *IEEE Trans. Electron. Devices*, Vol. ED-18, pp. 659–663, September 1971.

82. Mayer, W. N., and Bonin, R. V. "Tubular AC Plasma Panels." 1972 IEEE Conf. Display Devices, Conf. Rec., New York, pp. 15–18.

83. Strom, R. "32-Inch Graphic Plasma Display Module." 1974 SID Int. Symposium, San Diego, pp. 122–123.

84. Dick, G. W., and Biazzo, M. R. "A Planar Single-Substrate AC Plasma Display." *IEEE Trans. Electron. Devices*, Vol. ED-23, pp. 429–437, April 1976.

85. Sato, S.; Yamamoto, H.; Shirouchi, Y.; Iemori, T.; Nakayama, N.; and Morita, I. "Surface-Discharge-Type Plasma Display Panel." *IEEE Trans. Electron. Devices*, Vol. ED-23, pp. 328–331, March 1976.

86. Dick, G. W., and Biazzo, M. R. "A Planar Single Substrate AC Plasma Display Without Vias." 1978 Biennial Display Conference, Conf. Rec., Cherry Hill, N.J., pp. 80–84. Also appears in *IEEE Trans. Electron. Devices*, Vol. ED-26, pp. 1168–1172, August 1979.

87. Park, K. C., and Weitzman, E. J., "E-Beam Evaporated Glass and MgO Layers for Gas Panel Fabrication." *IBM J. Res. Dev.*, Vol. 22, pp. 607–612, November 1978.

88. O'Hanlon, J. F.; Park, K. C.; Reisman, A.; Havreluk, R.; and Cahill, J. G. "Electrical and Optical Characteristics of Evaporable-Glass-Dielectric AC Gas Display Panels." *IBM J. Res. Dev.*, Vol. 22, pp. 613–621, November 1978.

89. Urade, T.; Iemori, T.; Nakayama, N.; and Morita, I. "Considerations of Protecting Layer in AC Plasma Panels." 1974 IEEE Conf. Display Devices, Conf. Rec., New York, pp. 30–33. Also appears in *IEEE Trans. Electron. Devices*, Vol. ED-23, pp. 313–318, March 1976.

90. Byrum, B. "Surface Aging Mechanisms of AC Plasma Display Panels." *IEEE Trans. Electron. Devices*, Vol. ED-22, pp. 685–691, September 1975.

91. Uchiike, H.; Miura, K.; Nakayama, N.; Shinoda, T.; and Fukushima, Y. "Secondary Electron Emission Characteristics of Dielectric Materials in AC-Operated Plasma Display Panels." *IEEE Trans. Electron. Devices*, Vol. ED-23, pp. 1211–1217, November 1976.

92. Chou, N. J., and Sahni, O. "Comments on 'Secondary Emission Characteristics of Dielectric Materials in AC-Operated Plasma Display Panels.'" *IEEE Trans. Electron. Devices,* Vol. ED-25, pp. 60–61, January 1978.

93. Aboelfotoh, M. O., and Lorenzen, J. A., "Influence of Secondary-Electron Emissions from MgO Surfaces on Voltage-Breakdown Curves in Penning Mixtures for Insulated-Electrode Discharges." *J. Applied Phys.*, Vol. 48, pp. 4754–4759, Nov. 1977.

94. Andoh, S.; Murase, K.; Umeda, S.; and Nakayama, N. "Discharge Time-Lag in a Plasma Display-Selection of Protective Layer (γ Surface)." *IEEE Trans. Electron. Devices*, Vol. ED-23, pp. 319–324, March 1976.

95. Abeolfotoh, M. O., and Sahni, O. "Aging Characteristics of AC Plasma Display Panels." *Proc. SID*, Vol. 22, pp. 219–227, 1981, also appears in *IEEE Trans. Electron. Devices*, Vol. ED-28, p. 645, 1981.

96. Pleshko, P. "AC Plasma Display Aging Model and Lifetime Calculations." *Proc. SID*, Vol. 22, pp. 228–232, 1981.

97. Aboelfotoh, M. O. "On the Stability of the Operating Voltages of AC Plasma Display Panel." *IEEE Trans. Electron. Devices*, Vol. ED-29, pp. 247–254, February 1982.

98. Shinoda, T.; Uchiike, H.; and Andoh, S. "Low Voltage Operated Plasma-Display Panels." *IEEE Trans. Electron. Devices*, Vol. ED-26, pp. 1163–1167, August 1979.

99. Ahearn, W. E., and Sahni, O. "Effect of Reactive Gas Dopants on the MgO Surface in AC Plasma Display Panels." *IBM J. Res. Dev.*, Vol. 22, pp. 622–625, November 1978.

100. Johnson, R. L.; Bitzer, D. L.; and Slottow, H. G. "The Device Characteristics of the Plasma Display Element." *IEEE Trans. Electron. Devices*, Vol. ED-18, pp. 642–649, September 1971.

101. Criscimagna, T. N. "Additive Pulses Turn Cells On Reliably." *Electro-Optics System Design*, August 1971, pp. 34–36.

102. Umeda, S.; Murase, R.; Andoh, S.; and Naka-

yama, N. "A Highly Stabilized AC Plasma Display." *IEEE Trans. Electron. Devices*, Vol. ED-23, pp. 324–328, March 1976.

103. Kleen, B. G., Lamoureux, W. R.; and Pearson, K. A. "The Design of a Versatile Gas Panel Subsystem." *Proc. SID*, Vol. 20, pp. 139–145, 1979.

104. Criscimagna, T. N. "Low Voltage Circuits for Plasma Display Panel." *Proc. SID*, Vol. 17, pp. 124–129, 1976.

105. Lyons, N.; Bitzer, D.; and Tucker, P. "Improving the Memory Margin in Plasma Display Panels." 1980 SID Int. Symposium, San Diego, pp. 88–89.

106. Ngo, P. "Dynamic Keep-Alive to Improve Plasma Panel Write and Sustain Margins." *IEEE Trans. Electron. Devices*, Vol. ED-22, pp. 676–681, September 1975.

107. Johnson, W. E., and Schmersal, L. J. "A Quarter-Million-Element AC Plasma Display with Memory." *Proc. SID*, Vol. 13, pp. 56 60, 1972.

108. Schermerhorn, J. D. "Plasma Display/Memory Monitor Requirements and Characteristics." 1980 SID Int. Symposium, San Diego, pp. 80–81.

109. Curran, P. "Plasma Display/Memory Monitor Electronics." 1980 SID Int. Symposium, San Diego, pp. 82–83.

110. Spencer, J. "The High Voltage IC and Its Future." 1982 SID Int. Symposium, San Diego, pp. 256–257.

111. Slottow, H. G., and Petty, W. D. "Stability of Discharge Series in the Plasma Display Panel." *IEEE Trans. Electron Devices*, Vol. ED-18, pp. 650–654, September 1971.

112. Slottow, H. G. "The Voltage Transfer Curve and Stability Criteria in the Theory of the AC Plasma Displays." *IEEE Trans. on Electron. Devices*, Vol. ED-24, pp. 848–852, July 1977.

113. Schlig, E. S., and Stilwell, Jr., G. R. "Characterization of Voltage and Charge Transfer in AC Gas Discharge Displays." *IBM J. Res. Dev.*, Vol. 22, pp. 634–640, November 1978.

114. Weber, L. F. "Measurement of Wall Charge and Capacitance Variation for a Single Cell in the AC Plasma Display Panel." 1976 Biennial Display Conference, Conf. Rec., New York, pp. 39–44. Also appears in *IEEE Trans. Electron. Devices*, Vol. ED-24, pp. 864–869, July 1977.

115. Weber, L. F. "Measurement of a Plasma in the AC Plasma Display Panel Using RF Capacitance and Microwave Techniques." IEEE Trans. Electron Devices, Vol. ED-24, pp. 859–864, July 1977.

116. Weber, L. F., and Weikart, G. Scott, "A Real-Time Curve Tracer for the AC Plasma Display Panel." *IEEE Trans. Electron. Devices*, Vol. ED-26, pp. 1156–1163, August 1979.

117. Ngo, P. "Charge Spreading and Its Effect on AC Plasma Panel Operating Margins." 1976 Biennial Display Conference, Conf. Rec., New York, pp.

118–120. Also appears in *IEEE Trans. Electron. Devices*, Vol. ED-24, pp. 870–872, July 1977.

118. Sahni, O., and Jones, W. P. "Spatial Distribution of Wall Charge Density in AC Plasma Display Panels." *IEEE Trans. Electron. Devices*, Vol. ED-25, pp. 223–226, March 1979.

119. Sahni, O., and Lanza, C. "Origin of the Bistable Voltage Margin in the AC Plasma Display Panel." *IEEE Trans. Electron. Devices*, Vol. ED-24, pp. 853–859, July 1977.

120. Lanza, C., and Sahni, O. "Numerical Calculation of the Characteristics of an Isolated AC Gas Discharge Display Panel Cell." *IBM J. Res. Dev.*, Vol. 22, pp. 641–646, November 1978.

121. Sahni, O., and Aboelfotoh, M. O. "The Pressure Dependence of the Bistable Voltage Margin of an AC Plasma Panel Cell." *Proc. SID*, Vol. 22, pp. 212–218, 1981. Also appears in *IEEE Trans. Electron. Devices*, Vol. ED-28, p 638, 1981.

122. Sahni, O., and Lanza, C. "Importance of the Secondary Electron Emission Coefficient on E/Po for Paschen Breakdown Curves in AC Plasma Panels." *J. Appl. Phys.*, Vol. 47, p. 1337, 1976.

123. Sahni, O., and Lanza, C., "Influence of the Secondary Electron Emission Coefficient of Argon on Paschen Breakdown Curves in AC Plasma Panels for Neon + 0.1% Argon Mixture." *J. Appl. Phys.*, Vol. 47, p. 5107, 1976.

124. Lanza, C. "Analysis of an AC Gas Display Panel." *IBM J. Res. Dev.*, Vol. 18, pp. 232–43, May 1974.

125. Sahni, O.; Lanza, C.; and Howard, W. E. "One-Dimensional Numerical Simulation of AC Discharges in a High-Pressure Mixture of Ne + 0.1% Ar Confined to a Narrow Gap between Insulated Metal Electrodes." *J. Appl. Phys.*, Vol. 49, p. 2365, 1978.

126. Sahni, O., and Lanza, C. "Failure of Paschen's Scaling Law for Ne – 0.1% Ar Mixtures at High Pressures." *J. Appl. Phys.*, Vol. 52, p. 196, 1981.

127. Weber, L. F. "Theory of Memory Effects for the AC Plasma Display Panel." 1976 SID Int. Symposium, Beverly Hills, pp. 82–83.

128. Nakayama, N., and Andoh, S. "Design of a Plasma Display Panel." *Proc. SID*, Vol. 13, pp. 61–66, 1972.

129. O'Hanlon, J. F. "A Phenomenological Study of AC Gas Panels Fabricated with Vacuum-Deposited Dielectric Layers." *IBM J. Res. Dev.*, Vol. 22, pp. 626–633, November 1978.

130. Coleman, W. E.; Gaur, J. P.; Hoskinson, J. H.; Janning, J. L.; and Smith, R. C. "An All-Thick-Film AC Plasma Line Display." 1972 IEEE Conf. Display Devices, Conf. Rec., New York, pp. 22–28.

131. Coleman, W. E.; Gaur, J. P.; Hoskinson, J. H.; Janning, J. L.; and Smith, R. C. "A Time-Shared Plasma Line Display." *Proc. SID*, Vol. 13, pp. 26–34, 1972.

132. Yano, A.; Inomata, I.; and Iwakawa, T. "Plasma Display." NEC (Nippon Electric Co.), Res. Devel., No. 30, pp. 54–63, 1973.

133. Sato, S.; Yamaguchi, H.; and Umeda, S. "Self-Shift Character Display." *Fujitsu Scientific and Technical Journal*, Vol. 11, No. 2, pp. 81–98, June 1975.

134. Andoh, S.; Oki, K.; Yoshikawa, K.; Miyashita, Y.; Shinoda, T.; Sato, S.; and Sugimoto, Y. "Self-Shift Plasma Display Panels with Meander Electrodes or Meander Channels." *IEEE Trans. Electron. Devices*, Vol. ED-25, pp. 1145–1151, September 1978.

135. Andoh, S.; Yoshikawa, K.; and Ohtsuki, O. "New Type Meander-Electrode AC PDP." 1978 Biennial Display Conference, Conf. Rec., Cherry Hill, N.J., pp. 69–72.

136. Murase, K.; Ishizaki, H.; Yamaguchi, H.; Kawada, T.; and Kashiwara, H. "A Multirow Drive Scheme for Self-Shift Display Panels." *Proc. SID*, Vol. 19, pp. 85–89, 1978.

137. Andoh, S.; Oki, K.; and Yoshikawa, K.; "Self-Shift PDP with Meander Electronics." 1977 SID Int. Symposium, Boston, pp. 78–79.

138. Murase, K.; Yamaguchi, H.; and Ueda, Y. "A Partial Shift Technique for Self-Shift PDPs." SID Int. Symposium, San Francisco, pp. 42–43.

139. Jones, L. M., and Johnson, R. L. "A Parallel Self-Shift Technique for Plasma Display/Memory Panels." *IEEE Trans. Electron. Devices,* Vol. ED-22, pp. 235–239, May 1975.

140. Coleman, W. E. and Graycroft, D. G. "A Serial Input Plasma Charge Transfer Display Device." Presented at 1975 SID Symposium, Washington D.C., Conference Digest, pp. 114–115.

141. Coleman, W. E., and Graycroft, D. G. "A Plasma Charge Transfer Device—A Plasma Display Incorporating Address and Memory Within the Panel." *Proc. SID*, Vol. 17, pp. 169–175, 1976.

142. Andoh, S., and Murase, K. "Trends in AC Plasma Displays." Presented at 1981 SID Int. Symposium, New York, Conference Digest, pp. 160–161.

143. Coleman, W. E., and Hall, S. W. "Device Characteristics of the Plasma Charge Transfer Shift Display." 1980 Biennial Display Research Conference, Conf. Rec., Cherry Hill, N.J., pp. 64–71. Also appears in *IEEE Trans. Electron. Devices*, Vol. ED-28, pp. 673–679, June 1981.

144. Yamaguchi, H.; Yoshikawa, K.; and Miyashita, Y.; "High Speed and High Resolution Self-Shift Plasma Display." 1982 Int. Display Research Conference, Conf. Rec., Cherry Hill, N.J., pp. 80–83.

145. Sato, S.; Wakitani, M.; Miyashita, Y.; Miura, S.; Murase, K.; and Umeda, S. "High Resolution, High Speed, Self Shift Plasma Display." IEEE Electron Device Lett., Vol. EDL-3, pp. 301–302, October 1982.

146. Schermerhorn, J. D. "A Shift-Logic Plasma Display/Memory Unit." 1975 SID Int. Symposium, Washington D.C., pp. 112–113.

147. Beidl, J. R.; Campagna, F. C.; and Criscimagna, T. N. "High Resolution Shift Panel." 1978 SID Int. Symposium, San Francisco, Conference Digest, pp. 40–41.

148. Beidl, J. R.; Campagna, F. C.; and Criscimagna, T. N. "A High-Resolution Shift Panel," Proc. SID, Vol. 20, pp. 147–151, 1979.

149. Ngo, P. D. T. "Electron Transport Self-Shift Display." 1980 SID Int. Symposium, San Diego, pp. 84–85.

150. Ngo, P. "Charge Transport in an AC Plasma Panel." *Proc. SID*, Vol. 22, pp. 233–239, 1981.

151. Ngo, P., and Maliszewski, S. R. "Electron Transport Self-Shift Display Using an AC Plasma Panel." *Proc. SID*, Vol. 22, pp. 240–246, 1981.

152. Weikart, G. S. "Independent Subsection Shift and a New Simplified Write Technique for Self-Shift AC Plasma Panels." *IEEE Trans. Electron. Devices*, Vol. ED-22, pp. 663–668, September 1975.

153. Holz, G.; Ogle, J.; Andreadakis, N.; Siegel, J.; and Maloney, T. "A Self-Scan Memory Plasma Display Panel." 1981 SID Int. Symposium, New York, pp. 168–169.

154. Amano, Y.; Yoshida, K.; and Shionoya, T. "A High-Resolution DC Plasma Display Panel." 1982 SID Int. Symposium, San Diego, pp. 160–161.

155. Schermerhorn, J. D., and Miller, J. W. V. "Discharge-Logic Drive Schemes." *IEEE Trans. Electron. Devices*, Vol. ED-22, pp. 669–673, September 1975.

156. Criscimagna, T. N.; Beidl, F. R.; Steinmetz, M.; and Hevesi, J. "Coupled-Matrix, Threshold-Logic AC Plasma Display Panel." *Proc. SID*, Vol. 17, pp. 176–179, 1976.

157. de Boer, T. J. "An Experimental 4000 Picture-Element Gas Discharge TV Display Panel." Ninth Nat. Symp. on Inf. Display, Los Angeles, May 1968, p. 193.

158. Arora, B. M.; Bitzer, D. L.; and Slottow, H. G. "Ternary Operation of the Plasma Display Cell." *IEEE Trans. Electron. Devices*, Vol. ED-20, pp. 78–81, January 1973.

159. Anderson, B. C., and Fowler, V. J. "AC Plasma Panel TV Display with 64 Discrete Intensity Levels." 1974 SID Symp. Digest, pp. 28–29.

160. Kurahashi, K.; Tottori, H.; Isogai, F.; and Tsuruta, "Plasma Display with Gray Scale." 1973 SID Int. Symposium, New York, pp. 72–73.

161. Hori, H.; Kasahara, K.; and Inoue, K. "A New Gas-Discharge Display Device Using Through-Hole Enhancement." 1970 IEEE Conf. Display Devices, Conf. Rec., pp. 140–143.

162. Hori, H.; Kasahara, K.; and Inoue, K. "A Picture-Display Panel Using a Constricted-Glow Dis-

charge." *IEEE Trans. Electron. Devices*, Vol. ED-21, pp. 372–376, June 1974.

163. Judice, C. N., and Slusky, R. D. "Processing Images for Bilevel Digital Displays." *Advances in Image Pickup and Display*, Vol. 4, pp. 157–229, Academic Press, New York, 1981.

164. Judice, C. N.; Jarvis, J. F.; and Ninke, W. H. "Using Ordered Dither to Display Continuous Tone Pictures on an AC Plasma Panel." *Proc. SID*, Vol. 15, pp. 161–169, 1974.

165. Jarvis, J. F.; Judice, C. N.; and Ninke, W. H. "A Survey of Techniques for the Display of Continuous Tone Pictures on Bilevel Displays." *Computer Graphics Image Processing*, Vol. 5, pp. 13–40, 1976.

166. Judice, C. N.; Jarvis, J. F.; and Ninke, W. H. "Bi-Level Rendition of Continuous-Tone Pictures on an AC Plasma Panel." 1974 IEEE Conf. Display Devices, Conf. Rec., New York, pp. 89–93.

167. Netravali, A. N. and E. G. Bowen, "Display of Dithered Images." *Proc. SID*, Vol. 22, pp. 185–190, 1981.

168. Judice, C. N.; White, A. B.; and Johnson, R. L. "Transmission and Storage of Dither Coded Images Using 2-D Pattern Matching," *Proc. SID*, Vol. 17, pp. 85–91, 1976.

169. Judice, C. N. "Data Reduction of Dither Coded Images by Bit Interleaving," *Proc. SID*, Vol. 17, pp. 92–101, 1976.

170. White, A. B. "Video Imaging on the Plasma Display Panel." Coordinated Science Laboratory Report R-677, University of Illinois, Masters Thesis, April 1975.

171. Stucki, P. "Comparison and Optimization of Computer-Generated Digital Halftone Pictures." 1975 SID Int. Symposium, Washingon D.C., pp. 34–35.

172. Floyd, R., and Steinberg, L. "An Adaptive Algorithm for Spatial Grey Scale." 1975 SID Int. Symposium, Washington D.C., pp. 36–37. Also appears in *Proc. SID*, Vol. 17, pp. 75–77, 1976.

173. Judice, C. N. "Dithervision—A Collection of Techniques for Displaying Continuous Tone Still-And-Animated Pictures on Bilevel Displays." 1975 Symp. Digest, SID, pp. 32–33.

174. White, A. B.; Johnson, R. L.; and Judice, C. N. "Animated Dither Images on the ac Plasma Panel." Proc. 1976 Biennial Display Conference, pp. 35–37.

175. Ahearn, W. E., and Sahni, O. "The Dependence of the Spectral and Electrical Properties of ac Plasma Panels on the Choice and Purity of the Gas Mixture." 1978 SID Int. Symposium, San Francisco, pp. 44–45.

176. Sahni, O. "Blue Color ac Plasma Panel." 1980 SID Int. Symposium, San Diego, pp. 86–87.

177. Ferrar, C. M., and Nighan, W. L. "Excimer Fluorescence for Display Applications." 1982 Int. Display Research Conference, Conf. Rec., Cherry Hill, N.J., pp. 101–106.

178. Nighan, W. L., and Ferrar, C. M. "Excimer Fluorescence for Plasma Displays." *Appl. Phys. Lett.*, Vol. 40, pp. 223–224, 1982.

179. Krupka, D. C.; Chen, Y.; and Fukui, H. "On the Use of Phosphors Excited by Low-Energy Electrons in a Gas Discharge, Flat Panel Displays." *Proc. IEEE*, Vol. 61, pp. 1025–1029, July 1973.

180. Hoehn, H. J., and Martel, R. A. "Recent Developments on Three-Color Plasma Panels." *IEEE Trans. Electron. Devices,* Vol. ED-20, pp. 1078–1081, November 1973.

181. van Gelder, Z., and Mattheij, M. M. M. P. "Principles and Techniques in Multicolor dc Gas Discharge Displays." *Proc. IEEE*, Vol. 61, pp. 1019–1024, July 1973.

182. Shinoda, T.; Miyashita, Y.; Sugimoto, Y.; and Yoshikawa, K. "Characteristics of Surface-Discharge Color ac-Plasma Display Panels." 1981 SID Int. Symposium, New York, pp. 164–165.

183. Yamashita, H.; Andoh, S.; and Shinoda, T. "A Green ac Plasma Display." 1976 SID Int. Symposium, Beverly Hills, pp. 80–81.

184. Koike, J.; Kojima, T.; Toyonaga, R.; Kagami, A.; Hase, T.; and Inaho, S. "Phosphors for Gas-Discharge Color Display Panels." 1980 SID Int. Symposium, San Diego, pp. 150–151.

185. Kaufman, R. G. "Lamp Phosphors and Color Gamut in Positive Column Gas-Discharge Displays: The State of the Arts." 1976 Biennial Display Conference, Conf. Rec., New York, pp. 45–46. Also appears in *IEEE Trans. Electron. Devices*, Vol. ED-24, pp. 884–890, July 1977.

186. Yokozawa, M. "Nitrogen Gas-Mixture for Color Plasma Display Panel." Proc. 1981 European Display Research Conference, Munich, pp. 183–186.

187. Kamegaya, T.; Matsuzaki, H.; and Yokozawa, M. "Basic Study on the Gas-Discharge Panel for Luminescent Color Display." *IEEE Trans. Electron. Devices*, Vol. ED-25, pp. 1094–1100, September 1978.

188. DeJule, M. C., and Chodil, G. J. "High-Efficiency, High-Luminance Gas-Discharge Cells for TV Displays." 1975 SID Symp. Digest, pp. 56–57.

189. Forman, J. "Phosphor Color in Gas Discharge Panel Displays." *Proc. SID*, Vol. 13, pp. 14–20, 1972.

190. Brown, F. H., and Zayac, M. T. "A Multicolor Gas-Discharge Display Panel." *Proc. SID*, Vol. 13, pp. 52–55, 1972.

191. Markin, J. "Some Factors Affecting the Development of Flat-Panel Video Displays." *Proc. SID*, Vol. 17, pp. 2, 1976.

192. Van Raalte, J. A. "Matrix TV Displays: Systems and Circuit Problems." *Proc. SID*, Vol. 17, pp. 8–13, 1976.

193. Sobel, A. "Summary: New Techniques in Video Displays." *Proc. SID*, Vol. 17, pp. 56–61, 1976.

194. Fukushima, M.; Murayama, S.; and Kaji, T. "Color-TV Display Using a Flat Gas-Discharge Panel." 1974 Symp. Digest, SID, pp. 120–121.

195. Fukushima, M.; Murayama, S.; Kaji, T.; and Mikoshiba, S. "A Flat Gas-Discharge Panel TV with Good Color Saturation." *IEEE Trans. Electron. Devices*, Vol. ED-22, pp. 657–662, September 1975.

196. Ohishi, I.; Kojima, T.; Ikeda, H.; Toyonaga, R.; Murakami, H.; Koike, J.; and Tajima, T. "An Experimental Real-Time Color TV Display with a dc Gas Discharge Panel." 1974 IEEE Conf. Display Devices, Conf. Rec., New York, pp. 110–115. Also appears in *IEEE Trans. Electron. Devices*, Vol. ED-22, pp. 650–662, September 1975.

197. Amano, Y. "A New Flat Panel TV Display System." 1973 Int. Electron Devices Meet., Tech. Dig., pp. 196–197.

198. Amano, Y. "A Flat-Panel Color TV Display System." 1974 Conf. Display Devices and Systems, Conf. Rec., pp. 99–102. Also appears in *IEEE Trans. Electron. Devices*, Vol. ED-22, pp. 1–7, January 1975.

199. Mikoshiba, S.; Shinada, S.; Takano, H.; and Fukushima, M. "A New Positive Column Discharge Memory Panel for TV Display." 1977 SID Int. Symp. Digest, pp. 80–81.

200. Mikoshiba, S.; Shinada, S.; Takano, H.; and Fukushima, M. "A Positive Column Discharge Memory Panel for Color TV Display." *IEEE Trans. Electron. Devices*, Vol. ED-26, pp. 1177–1181, August 1979.

201. Okamoto, Y., and Mizushima, M. "A New dc Gas-Discharge Display with Internal Memory." *Japan J. Appl. Phys.*, Vol. 15, pp. 719–720, 1976; and "A New dc Gas Discharge Panel with Internal Memory for Color Television Display." *IEEE Trans. Electron. Devices*, Vol. ED-25, pp. 8–16, 1978.

202. Kaneko, R.; Kamegaya, T.; Yokozawa, M.; Matsuzaki, H.; and Suzuki, S. "Color TV Display Using 10″ Planar Positive-Column Discharge Panel." 1978 SID Int. Symposium, San Francisco, pp. 46–47.

203. Schiekel, M. F., and Sussenbach, H. "DC Pulsed Multicolor Plasma Display." 1980 SID Int. Symposium, San Diego, pp. 148–149.

204. Schauer, A. "A Plasma Electron Excited Phosphor Flat Panel Display." 1982 IEDM, pp. 304–307.

205. DeJule, M.; Sobel, A.; and Markin, J. "A Gas-Electron-Phosphor (GEP) Flat-Panel Display." 1983 SID Int. Symp., Philadelphia, pp. 134–135.

206. Glaser, D.; DeJule, M.; Whelchel, C. J.; Stone, C. S.; Sobel, A.; and Markin, J. "The Flatscreen Display: Construction and Circuitry." 1983 SID Int. Symp., Philadelphia, pp. 136–137.

207. Hoehn, H. J. "Plasma Display/Memory Panel Fabrication." 1977 SID Int. Symposium, Boston, pp. 18–19.

208. Reisman, A.; Berkenblit, M.; and Chan, S. A. "Single-Cycle Gas Panel Assembly." *IBM J. Res. Dev.*, Vol. 22, pp. 596–600, November 1978.

209. Jolley, J. E. "Photosensitive Thick Film Compositions for Fabricating Fine Line ac Plasma Display Electrodes." 1978 Biennial Display Research Conference, Conf. Rec., Cherry Hill, N.J., pp. 85–87.

210. Hammer, R. "Obtaining Gas Panel Metallization Patterns by Vacuum Deposition Through Rib-supported Mask Structures." *IBM J. Res. Dev.*, Vol. 22, pp. 601–606, November 1978.

211. Brusic, V.; d'Heurle, F. M.; MacInnes, R. D.; Alessandrini, E. I.; Angilello, J.; Dempsey, J. J.; and Sampogna, M. "Al-Cu Alloy for Gas Panels." *IBM J. Res. Dev.*, Vol. 22, pp. 647–657, November 1978.

212. de Vries, G. H. F. "Thick-Film Technology for High Resolution dc Gas Discharge Panels." 1982 SID Int. Symposium, San Diego, pp. 164–165.

213. Pleshko, P.; Greeson, Jr., J. C.; and Pearson, K. A. "Repair of ac Plasma Display Panels." *Proc. SID*, Vol. 21, pp. 333–340, 1980.

214. Perry, C. H. "Mechanical Changes in Plasma Display Substrates During Thermal Processing." Electro-Chemical Society Meeting, Los Angeles, Extended Abstracts, Vol. 79-2, pp. 783–784, October 1979.

11

NONEMISSIVE DISPLAYS

P. ANDREW PENZ, *Texas Instruments*

11.1 INTRODUCTION

11.1.1 General Characteristics of Nonemissive Displays. A nonemissive display is a device that presents information by changing the ambient-light pattern at the observer, or in some special cases by altering the transmission of externally generated light through the device. The term passive display is also used for a nonemissive display to distinguish it from active or light-emitting technologies. Externally generated light usually results in a projection display. Unfortunately, there is no commonly accepted name for the technologies which are the subject of this chapter, other than terms that contrast them with light-generating technologies. The lack of a forceful-sounding name has not prevented nonemissive displays from playing an important role in information display. As will be shown, they are becoming dominant in applications that require battery operation and sunlight viewing.

Electromagnetic radiation is described by an electric-field vector which (1) propagates in a given direction $\vec{k}/|\vec{k}|$ where $|\vec{k}| = 2\pi/\lambda$ and λ is the wave-length, (2) has a magnitude or amplitude, (3) has an orientation in the plane perpendicular to the propagation direction referred to as the polarization, and (4) has a sinusoidal dependence on spacial coordinates (phase factor $\vec{k} \cdot \vec{r} - \omega t$). The energy in the wave is proportional to the square of the amplitude. The polarization of the field in the plane can be random, linear, elliptical, circular, etc. The type of polarization in a medium is determined by the phase factor in the medium and the incident polarization. The radiation can be modulated by changing the amplitude of the wave (absorp-

tion), the direction of the wave propagation (scattering or refraction), or the polarization of the wave. Reorientation of the electric field in the plane perpendicular to the direction of propagation cannot be detected by the unaided eye. Detection of the reorientation requires the use of polarizers. Reorientation of the polarization is generally produced by a birefringent media.

The term *birefringence* refers to the phenomenon of light traveling with different velocities in crystalline materials depending on the propagation direction and the orientation of the polarization relative to the crystalline axis. The velocity of light in a material medium depends on the distortion produced in the medium by the electric field as measured by the dielectric constant ϵ. The index of refraction—the velocity of light in a vacuum (c) divided by the velocity in the medium—is given by $\sqrt{\epsilon}$. Since the distribution of electrons in a crystalline material is anisotropic, it follows that the dielectric constant and the index of refraction will be anisotropic. For uniaxial crystals, there is one axis for which light when polarized parallel to the axis travels with an extraordinary index n_e and when polarized perpendicular to the axis travels with an ordinary index n_o. The birefringence of the medium is defined as the difference in the indices: $n_e - n_o$. This number is sufficient to characterize the birefringent materials to be discussed here. For further reading on the topic of birefringence, consult Reference 1.

Consider light propagating in a uniaxial material in a direction perpendicular to the unique axis, or the optical axis. We arrange incident linear polarization so that, initially ($z = 0$), the electric field has equal components along the optic axis and perpendicular to the optic axis. What happens to the polarization at a position z further into the sample for a free-space wavelength λ? The phase of the extra-ordinary wave will be given by:

$$2\pi n_0/\lambda - \omega t$$

while the phase of the ordinary wave will be

$$2\pi n_e z/\lambda - \omega t$$

Generally the two waves will have unequal phases, and elliptically polarized light will result from a phase change equal to:

$$\frac{2\pi(n_e - n_o)z}{\lambda}$$

For particular values of z such that the phase difference is $\pi/2$, the light will be circularly polarized, etc. This illustrates how propagation in birefringent media can produce polarization changes. Since the phase difference depends on the wavelength of light, these polarization changes generally are reduced to color changes by a polarizer/analyzer pair.

The simple case analyzed above turns out to be much more complicated in the general case. Fortunately, the general case is of little interest in display technology. In fact, the use of bire-fringence to achieve a color display is not used to any great extent. The most important liquid-crystal display, the twisted nematic effect, in-volves the use of a birefringent material but is a wavelength-independent phenomenon to be described below. In this effect and in other birefringence effects, the contrast can depend dramatically on viewing angle since the effec-tive thickness and the effective birefringence both depend on viewing angle.

The quality factors for judging the human perception of a display were discussed in Chap. 2. The definition of brightness ratio was given in Eq. 5-2 and of contrast ratio in Eq. 5-4. Both PCR and DBR are independent of ambient light-ing unless a transillumination technique is used. For the majority of commercial nonemissive displays, transillumination is important only at very low light levels and then only in some applications, primarily watches. When trans-illumination is not used, the independence of DBR on ambient can give a misleading repre-sentation of quality. At very low light levels, the display becomes unreadable even through the DBR remains constant. This is the reason for backlighting is some watches. For the rest of the chapter, it will be assumed that transillumi-nation is not used.

In order to measure the brightness of a non-emissive electronic display relative to this well-known standard, it is traditional in the non-emissive display field to use a highly reflective white layer (MgO) as the surround reflectance (R_s in Eq. 2-3). The reflectance of MgO is nearly 100% as opposed to the 18% of an aver-age scene. Since MgO is always used as a non-emissive standard of brightness, the brightness ratios quoted are always less than unity and, with polarizers used on the display, considerably less then unity.

The question of anisotropic viewing char-acteristics makes it important to specify how numbers like PCR and DBR are measured. Both the incident ambient light and the characteristics of the detector are important. Does the inci-dent light come from a point source, is it colli-mated, or does it come uniformly from a hemi-sphere of 2π steradians around the display? What is the aperture of the incident light? Since one cannot usually control the char-acteristics of ambient light, it is important to known how a given display behaves under a variety of conditions. For instance, a scattering display will require a light source in correct relation to the display and to the eye so that the display will scatter light into the eye. Nat-urally, the position and aperture of the detector are important, too. The general question of display characterization was covered in Chaps. 3 and 4. In fact, few standards exist in the non-emissive industry. It is important to understand, therefore, under what viewing conditions (viewing cone) the contrast ratio of a display is measured.

Nonemissive displays normally are viewed in reflection. This means that there must be a reflecting medium behind the display. Its func-tion is to return light to the viewer so that he can perceive changes. Thus, a reflection display is actually a double-transmission effect. Light can be modulated on the way to the mirror and/or on the way from the mirror to the observer. Nonemissive displays can work in transmission, but this requires some source of light behind the display. Transmissive systems that use projection optics obviously do not belong in a book on flat panels. Thus, this chapter will concentrate on reflective displays.

Nonemissive displays have two basic advantages over light-emitting displays: they require much less power to operate and they do not wash out in high ambients. The term washout refers to the loss of contrast in a light-emitting display when the ambient light reflected off the background becomes comparable with the light from the active area. Since nonemissive displays modulate ambient light, their contrast ratio is not sensitive to ambient level and washout is not a problem. All-ambient viewing of passive displays can be achieved by providing a low-brightness light source to illuminate the display in the dark. The high sensitivity of the dark-adapted eye permits the low brightness and, consequently, low power demands for the light source.

The main reason for the emergence of nonemissive displays in the 1970s has been the fact that they require very little power to operate *and* that the semiconductor industry has come up with products that operate off small batteries, e.g., electronic watches and calculators. As discussed in Section 5.2.5, the generation of a watt of optical power requires significantly more than a watt in the power supply. The power saved by a nonemissive display can be used to dramatically extend the battery life in a given device. For some applications, e.g., watches, a continuous display of data becomes possible as compared to data on demand. Longer battery life can be traded off for smaller batteries, which can lead to styling advantages, e.g., the slim LCD calculators. Thus, the system advantages of nonemissive displays have given them a dominant position in portable applications which, generally, also require sunlight-readable displays.

Nonemissive displays have some generic disadvantages when compared with light-emitting displays, especially in large-information-content flat-panel displays. The most basic disadvantage is that passive displays have a limited, and in some cases nonexistent, matrix-addressing capability. As has been discussed earlier in this book, matrix addressing requires that the electro-optic characteristic of the material being addressed have a threshold property. For high information content, this characteristic must be exceedingly sharp. Several nonemissive displays lack the threshold property. The others have a limited sharpness, especially when viewed under large-aperture conditions. It follows that the number of lines which can be multiplexed is limited. The commercial maximum was 7 pixel lines (or 1 character line) for LCDs in 1980 and 32 pixel lines (4 character lines) in 1982, with not much promise for the 100 pixel lines necessary for graphics applications at acceptable contrast and viewing conditions.

The limited capability of passive displays for x-y matrix drive is related in part to their basic light-modulating character. The maximum amount of light that a nonemissive display can generate at the eye is the ambient-light level, whereas the minimum is limited by stray reflections to a finite amount. Thus, the dynamic range of a nonemissive display is severely limited as compared to emissive displays in an office ambient. In fact, the contrast ratio of a passive display is generally expressed on a linear scale, whereas the brightness characteristic of an active display is often presented on a logarithmic scale. This means that the change in the electro-optic characteristics as a function of voltage for emissive displays is fundamentally larger/sharper than for passive displays. It follows from what has been said above about multiplexing that intrinsic-drive passive displays will always be less applicable to high-information-content flat-panel displays than their light-emitting cousins. The limited information content of nonemissive displays leads to the conclusion that color will not be an important issue for intrinsic addressing. Typically, color can only be achieved by sacrificing pixel counts, which already are small for nonemissive displays. The lack of color is a second disadvantage compared to emissive displays. An exception is transilluminated displays, where color can be achieved by filtering for very restricted formats where a warning message is needed. Another exception is extrinsic addressing where a much larger pixel count is possible.

The limitations of passive displays in an intrinsic matrix-addressing application does not mean that they are automatically excluded from high-information-content flat-panel applications. The limitations arise from the fact that a given pixel must retain a high average contrast for a time much longer than the time available for its address in a matrix application. It follows that a switch in series with each pixel can permit flat-panel high-information-content passive displays. An array of such switches has been called an active matrix, a term proposed and actively pursued by Brody. Brody[2] has made an eloquent defense of the proposal that any nonmemory flat-panel technology, be it emissive or nonemissive, will require such an active matrix. A number of workers in the passive display field, including this author, agree with Brody. The type of active matrix is open to question, but the need for it for large information displays appears to be clear.

The third disadvantage of nonemissive displays is their lack of absolute brightness. The only commercial passive displays appear dull, and it is well known that people like bright, colorful displays. Active displays fulfill this human-factor need well, but at a significant cost in power. This does not mean, however, that a bright, colorful, high-information-content display cannot be achieved. The examples of Eastman and Land in photography indicate that nonemissive color pictures can be made. The challenge for the remainder of the century will be to make electro-optic effects mimic color pictures.

The remainder of this chapter will be devoted to the science of various nonemissive electro-optic technologies. These will include liquid crystals, electrochromics, electrophoretics, dipolar suspensions, and magnetic and mechanical phenomena. For a graphic listing of the technologies to be covered, see Table 11-1.

11.1.2 History. The history of commerically important nonemissive displays begins with a series of papers by Heilmeier and co-workers in 1968 on two liquid crystal displays, the dynamic scattering mode,[3] and the dichroic

dye mode[4] (guest-host effect). Other nonemissive display technologies came into existence in the same time frame, namely electrochromic and electrophoretic displays, but they have yet to become widely used in products. There was another simultaneous technological development which was to be very influential. This was large-scale integration of semiconductors, which made possible computations of great complexity in a portable format. The combination of LSI, LCDs, and LEDs led to the portable calculator and digital watches as we know them today.

Initially, the calculator/watch market was dominated by LEDs because of a lack of contrast and reliability in the LCD dynamic scattering mode. In 1970 a new dielectric effect known as the twisted nematic LCD was discovered.[5,6] The LCD has high contrast and very low power consumption and was soon made to be reliable by consumer standards. For a discussion of standards, see Reference 7. By the mid 1970s the twisted nematic LCD had become dominant in portable displays of low and intermediate information content. LCD research in the early 1980s concentrated on increasing the information content and size of LCDs to achieve full-page alphanumeric and graphic displays, as well as to extend the reliability into automotive and industrial markets.

The other nonemissive technologies discussed in this chapter continue to be worked on in the laboratories of the world because of three drawbacks to the twisted nematic LCD: limited information content, limited color capability, and poor brightness. These other technologies so far have not found a combination of advantages and reliability relative to LCDs to make them a strong factor in the nonemissive marketplace.

11.1.3 Definitions and Acronyms

1. Alignment: the orientation of the liquid-crystal director at a planar surface; also used as the process for achieving the orientation.
2. Anisotropy: a variation with angle of some material property such as dielectric

constant; typical of crystalline solids and found in the special class of liquids known as liquid crystals.

3. Cholesteric: the liquid-crystal phase characterized by the director adopting a spiral-staircase spacial structure, formed from molecules with left or right handed structure (see Fig. 11-2).

4. DAP: Deformation of Aligned Phase.

5. Dichroic dye: a dye whose absorption characteristics are different depending on whether the electric field of the light is parallel or perpendicular to the dye axis; pleochroic also used.

6. Dielectric constant: a number (ϵ) which measures the amount of dielectric distortion produced in a medium by an applied electric field.

7. Dielectric anisotropy: the difference between the dielectric constant with the applied electric field parallel to the director and the dielectric constant with the applied field perpendicular to the director ($\epsilon_\parallel - \epsilon_\perp$).

8. Director: a unit vector which is parallel to the average local orientation of liquid-crystal molecules.

9. Disclination: the liquid-crystal defect structure which occurs at a line where there is a discontinuous change in director orientation.

10. DSM: Dynamic Scattering Mode.

11. ECD: Electrochromic Display.

12. Elastic constant (K): the liquid-crystal constant which describes the linear relationship between an applied torque and the induced director orientational distortion.

13. Electro-optic characteristic: the plot of optical response of a display as a function of voltage.

14. EPD: Electrophoretic Display.

15. FLAD: Fluorescence-Activated Display.

16. Homeotropic alignment: the alignment for which the director at a surface is everywhere perpendicular to the surface.

17. Homogeneous alignment: the alignment for which the director at a surface is everywhere parallel to the surface and parallel to a given direction in the surface.

18. Index of refraction: the ratio of the speed of light in a vacuum to the speed of light in the material of interest; equal to the square root of the dielectric constant at the optical frequency of interest.

19. ITO (Indium–Tin Oxide): transparent electrode.

20. Nematic: the liquid-crystal phase characterized by the molecules being roughly parallel everywhere but the centers of mass not being correlated; the least-ordered form of liquid crystal (see Fig. 11-1).

21. LC: Liquid Crystal.

22. LCD: Liquid-Crystal Display.

23. LSI: Large-Scale Integration.

24. n_e: the extraordinary index of refraction; describes the index of refraction for light polarized parallel to the director.

25. n_o: the ordinary index of refraction; describes the index of refraction for light polarized perpendicular to the director.

26. Order parameter: a measure of the degree of molecular order as given by the thermal average of $1/2 (3 \cos^2 \phi - 1)$ where ϕ is the fluctuation angle referenced to the mean orientation.

27. PLZT: Lead, Lanthanum, Zirconium Titanate.

28. S: liquid-crystal order parameter ($0 \leqslant S \leqslant 1$).

29. Smectic: a liquid-crystal phase where the molecules are ordered into planes (see Fig. 11-3).

30. Texture: the optical appearance of a thin layer of liquid crystal; used to characterize the crystalline nature of a display.

31. TFT: Thin-Film Transistor.

32. TN: Twisted Nematic.

33. T_{NI}: temperature at which a liquid crystal changes from nematic to isotropic.

34. V_{th}: threshold voltage.

11.2 THE LIQUID-CRYSTAL PHASE

11.2.1 Liquid-Crystal Symmetry.
The term liquid crystal[8], when heard for the first time disturbs most people as being a contradiction

in terms. A liquid is generally thought to be without long-range order. Crystals possess long-range order. For approximately one hundred years, however, some organic materials have been known to melt into a stable, ordered liquid state before melting again into an isotropic liquid state. The molecules which demonstrate the ordered liquid or liquid-crystal state tend to be flattened rods which are quite rigid. An example, the nematic phase, is shown in Fig. 11-1. The two benzene rings in the molecule give it flatness and length. The bond between the rings does not permit much bending, so that the molecule is rigid.

If the shape of the liquid-crystal molecule is represented by a cigar, the order in the nematic liquid-crystal state can be indicated by the array of cigars in Fig. 11-1. The cigars line up with their axes parallel, but their center of masses unordered. The molecules are free to translate with respect to each other, so the phase has liquid properties. The material will flow to fill a container, but as one might expect, the viscosity is different if the shear rate is parallel or perpendicular to the molecular orientation. It is customary to indicate the direction in which

the molecules point by a unit vector called the director. The symmetry associated with the director makes the liquid ordered. Thus, a liquid phase can be ordered, i.e., liquid crystal does make sense after all as a descriptive term.

The molecular array shown in Fig. 11-1 represents an ideal case since the phase is present at $300°K$ where a great deal of thermal fluctuation is occurring. The molecules undergo considerable agitation, but on the average they do point in a given direction. The deviation of the molecular motion from the mean direction is measured by a thermal average known as the order parameter S. When the solid first melts into the nematic phase, S is about 0.7. As the temperature increases, S decreases slightly until the nematic-isotropic temperature (T_{NI}) is approached. Then S falls rapidly to zero at T_{NI} and is zero in the isotropic state. Perfect order as indicated in Fig. 11-1 would correspond to $S = 1$.

There are other liquid-crystal phases besides nematic. Figure 11-2 shows an idealized pic-

MBBA

$$CH_3 - O - \langle \rangle -CH=N- \langle \rangle -C_4H_9$$

T < 20°C	:	Crystalline Solid
20°C < T < 47°C	:	Liquid Crystal (Nematic)
47°C < T	:	Isotropic Liquid

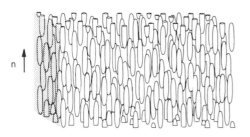

Fig. 11-1. The Nematic Phase: A nematic liquid crystal is composed of long, flat molecules such as methoxy benzylidene butylanalyine (MBBA). The hexagons in the molecular structure indicate benzene rings. The crystalline, liquid crystalline, and isotropic liquid temperature ranges are indicated. A schematic representation of the molecular order is shown with each ellipse portraying an individual molecule. Brownian motion spoils the perfect alignment indicated in the figure.

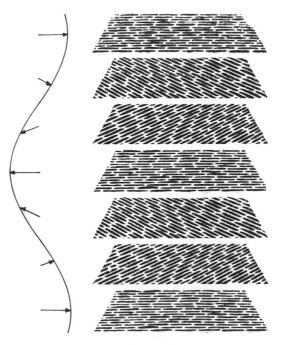

Fig. 11-2. The Cholesteric Phase: The three-dimensional structure of the cholesteric phase is shown, again in an idealized drawing. Each plane has a nematic orientation, but the direction of the director changes from plane to plane so that a spiral staircase (helix) structure is produced.

ture of the cholesteric phase. Note that each plane in the figure looks like a plane in the nematic phase, but that the direction of the molecular axis in one plane twists smoothly in going to the next. This circular-staircase structure is characteristic of the cholesteric phase, which derives its name from the fact that many derivatives of the molecule cholesterol form cholesteric phases. The pitch p is defined as the distance along the helix over which the molecular orientation is repeated. When the pitch matches the wavelength of incident electromagnetic radiation, strong Bragg reflection will occur. With white light incident, strong color play occurs and is the basis for the popular Mood Rings of the mid 1970s.

Another liquid-crystal phase, actually a family of phases, is referred to as smectic. A schematic diagram of one smectic[8] type is shown in Fig. 11-3. The basic characteristic of the smectic microstructure is an arrangement of molecules in layers. The layers can flow with respect to each other and the molecules can move in the layers, so the phase has fluid properties. The layered structure and the orientational order in the layers make this phase a liquid crystal, but more "crystalline" than the nematic or cholesteric phase. A typical melting characteristic of a smectic would be crystalline to smectic to nematic to isotropic liquid. Finally, there are materials which form anisotropic liquid structures when dissolved in water. Detergent/water mixtures are typical of this phase, which is known as lyotropic.[8]

11.2.2 Liquid-Crystal Materials.

Since most commercial devices rely on the nematic liquid-crystal phase, this section will concentrate on the materials that adopt this phase. As shown in Fig. 11-1, the basic feature of nematic materials is the double-ring structure. The materials are broken into classes depending on the groups which join the rings. The $CH \equiv N$ group shown in Fig. 11-1 is known as the Schiff base linkage, and this class is therefore known as Schiff base materials. When the linkage is a single bond, the double-benzene-ring structure is known as biphenyl. Other common linkages are ester and azoxy. There are other ring structures which form liquid crystals. Cyclohexane (single bonds

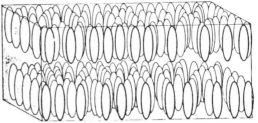

Fig. 11-3. The Smectic Phase: The molecules in the smectic liquid-crystal phase are arranged in layers. The layers are free to move relative to each other, giving the phase its fluid characteristic. The orientation order within the layers can take several forms, leading to several distinct smectic phases.

all around the carbon ring) and pyrimidene (nitrogen in the ring) materials are now used in commercial devices. Three ring structures are also used.

The end groups on the double-ring structure are used to provide variability in the dielectric properties and to provide components for mixtures. The material shown in Fig. 11-1 was the first pure material to be liquid crystalline at room temperature. The liquid-crystalline temperature range of 20 to 47°C clearly is not sufficient for commercial devices. This range can be extended by mixing materials from the same class (or other classes) to form a eutectic. For instance, a molecule with a methoxy group (CH_3), Fig. 11-1, can be combined with a similar molecule containing an ethoxy group to produce the mixture used in the first commercial devices (-10 to 60°C). Mixtures in 1983 were available which spaned -30°C to +80°C and 0°C to 100°C.

The difference in dielectric polarizabilities for electric fields parallel to the director and perpendicular to the director is important for device reasons. The molecule in Fig. 11-1 has a dielectric constant parallel to the molecule ($\epsilon_\parallel$) less than the dielectric constant perpendicular to the molecular ($\epsilon_\perp$). The material is said to possess negative dielectric anisotropy ($\epsilon_\parallel - \epsilon_\perp < 0$). The most common LCD requires a molecule with positive dielectric anisotrophy ($\epsilon_\parallel - \epsilon_\perp > 0$). This is accomplished by making one of the end groups on the double-ring structure into a $C \equiv N$ group. The triple bond is very polarizable parallel to the bond and leads to $\epsilon_\parallel \gg \epsilon_\perp$.

It is beyond the scope of this chapter to discuss the specific application/material choices used in commercial devices. Naturally, many mixtures are considered proprietary. Most materials are supplied by a few large chemical/drug houses. As a rule, Schiff base and azoxy materials were used in the early devices (before 1974) and have been replaced with biphenyl cyclohexane, and ester materials for reasons which include stability to water and UV radiation.

11.2.3 Physics of Liquid-Crystal Displays.

The mechanical properties of the liquid-crystal phase are specified by a compressability common to isotropic liquids and by unique elastic constants relating distortions in the director vector field to torques necessary to produce them. Since the director is an orientational parameter, the appropriate mechanical description is in terms of angle, torque, and angular momentum. Consider the distortions shown in Fig. 11-4. They are called splay, twist, and bend distortions and are measured by the vector differential operations listed below the schematics in Fig. 11-4. The material tries to oppose these distortions, and the relation between the distortions and the torques are linear. The constants of proportionality are the three elastic constants. K_{11}, K_{22}, and K_{33} are the splay, twist, and bend material parameters. They are all remarkably small ($\sim 10^{-12}$ N). This means that it requires very little torque to distort a nematic liquid crystal. If the distortion is of amplitude θ over a gap d, the torque is on the order of $K(\pi/d)^2 \theta$, where K is one of the three elastic constants.

Two other important properties of liquid-crystal materials for display applications are their low- and high-frequency electric polarizabilities: the dielectric constant and the index of refraction, respectively. In particular, the dielectric constant anisotropy leads to a torque which will tend to reorient the director in an applied electric field so that the largest dielectric constant is along the field. If the initial angle between the applied field and the director is $90 - \theta$, where θ is small, the torque is $(\epsilon_{\parallel} - \epsilon_{\perp}) \epsilon_O E_0^2 \theta$. E_0 is the applied field strength, and ϵ_O is the permittivity of free space—an SI constant. This torque can work against the elas-

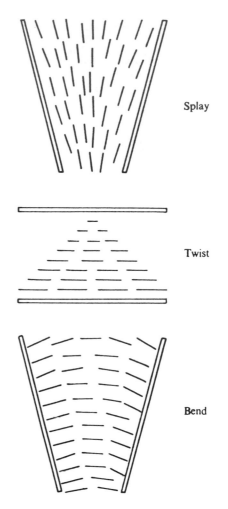

Fig. 11-4. Liquid-Crystal Deformation Structures: The three examples of liquid-crystal deformation are splay, twist, and bend. The molecules are now represented by short lines. Splay involves the divergence operation of vector calculus, while twist and bend involve the curl. The three-dimensional twist structure is indicated in the figure by the decrease in the line length.

tic torque described above to produce a net distortion. Now the high-frequency polarizability comes into play. The birefringence of the distortion can be detected by the eye, i.e., one has a display.

Consider the device structure indicated in Fig. 11-5. Here a nematic liquid crystal is contained between two electrodes which apply a voltage V_0 across a gap d. The surfaces of the electrodes have been treated so that the nematic director is almost parallel to the electrodes at zero voltage. Now consider what happens if

FIELD EFFECT - PARALLEL CELL

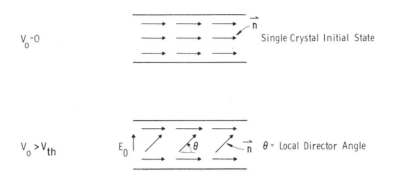

Fig. 11-5. Liquid-Crystal Field-Effect Geometry: The cross section of a liquid-crystal display is shown in field-free ($V_0 = 0$) and energized ($V_0 \geqslant V_{th}$) states. A single-crystal structure is produced by wall orientation so that the director lies in the plane of the cell and in the plane of the figure. Above threshold voltage, the director in the center of the cell is tilted by an angle θ which measures the amplitude of the distortion.

there is a fluctuation of amplitude θ in the director pattern and a voltage is applied. Thermal fluctuations are significant in liquid crystals—the director equivalent of Brownian motion can be detected in a polarizing microscope. For the present discussion the magnitude of θ is not important. Since thermal fluctuations will produce a distribution of θ in the sample, the director experiences an electric torque $(\epsilon_\parallel - \epsilon_\perp)\epsilon_O (V_0/d)^2 \theta$. The distortion is opposed by the elastic torque $K_{11}(\pi/d)^2\theta$. If the electric torque is greater (less) than the elastic torque, the distortion will grow (be damped). The distortion will be in unstable equilibrium when the two torques are equal, which means that:

$$V_{th}^2 = E_{th}^2\, d^2 = \pi K_{11}/(\epsilon_\parallel - \epsilon_\perp)\,\epsilon_O$$

Note that the fluctuation amplitude has canceled out. This means that there is a threshold voltage, V_{th}, in the problem. For different thickness samples, the distortion begins at the same voltage, independent of the electric field. Below V_{th}, fluctuations are damped, and above V_{th} they grow. Because K_{11} is so small, V_{th} is generally in the 1-to-4-V range. This is ideal for integrated-circuit operation.

Another reason for the commercial success of LCDs is the fact that the materials have a large birefringence relative to crystalline solids. The birefringence is on the order of 0.1 to 0.3 for typical materials. This means that liquid crystals can produce large modulations of light, i.e., the display can have good contrast from a variety of angles. The net effect is that the orientation of the director can be changed by low-frequency electric fields and detected visually.

Flow in a liquid crystal is described by a shear rate and a viscous stress which are related by a viscosity tensor. Tensor representations are generally required to fully describe the mechanical and electrical relationships in liquid crystals due to the anisotropic nature of the medium. While these representations are beyond the scope of this chapter, two basic principles are involved in the flow case. First, there is a traditional frictional effect which opposes the relative motion of two adjacent fluid layers. The second viscosity principle is that when a shear rate is present, there is also a torque which acts to reorientate the director. This principle is unique to liquid crystals. By reciprocity, director reorientation will involve fluid motion and, therefore, viscous loss. This means that director reorientations are damped by viscous effects.

The time for a distortion to disappear is typically $\eta d^2/K$ where η is an average viscosity and K an average elastic constant. The time for an electric field E_0, to produce a distortion is typically $\eta d^2/E_0^2$. For rapid response, one wants low viscosity and thin samples. E_0 and K are related by the threshold condition and so cannot strongly influence the response times. Most commercial devices have response times in the 100-ms range at room temperature. Since viscosity coefficients typically decrease by a factor of two for every 10°C rise in temperature, the response times are temperature dependent. Typically, at low temperatures, i.e., 0°C, response times become long (>1 sec) even though the material is still in the liquid-crystalline state. Low-temperature properties of LCDs are usually limited by response times rather than by the crystalline to nematic temperature. During the early 1980s, commercial liquid-crystal blends appeared with very low viscosities, and so LCDs began to appear in outdoor applications such as gasoline pumps and automobiles.

Since the eye has a response time on the order of 30 ms, changes in LCD data appear to have a slight fade in/fade out appearance at room temperature. To avoid this smear in real-time picture displays (TV), the thickness of the display must be decreased to the 3 μm region. Here sample fabrication becomes a significant problem since it is very difficult to hold a 10% tolerance on the cavity thickness over the whole display area. Clearly, lower-viscosity materials will be required for direct-view, smear-free displays.

Note that the expression for the dielectric torque acting on a liquid crystal is proportional to the square of the applied field. This is because the dielectric polarization is induced by the applied field which then acts to twist the polarization it induced. The square dependence means that the sign of the torque does not change when the field changes sign—as with ac waveform drive. Since the driving torque is proportional to the square of the applied electric field, the liquid crystal responds to the RMS value of the applied field. The RMS characteristic has an important practical advantage: the display can be operated with ac or dc fields. DC drive has a big disadvantage in liquids: It can cause ion migration and produce unwanted electrochemical effects at the electrodes. Electrochemical effects can lead to a serious lifetime problem in LCDs. They are now avoided by operating with audio-frequency-drive signals.

The power drawn by a typical square centimeter of liquid-crystal display can now be esti-mated. For a dielectric effect, the current produced by a voltage of amplitude V_0 on a material of average dielectric constant of ϵ and a material thickness d, at an angular frequency of ω is $\omega \epsilon \epsilon_o V_0^2 / d$ for a unit surface area. This number turns out to be in the 1 μW/cm^2 region for a typical liquid-crystal display. That is, field-effect LCDs are not just low-voltage, they are also very-low-power. This fact has made LCDs dominant in portable applications.

As one might imagine from Fig. 11-1, impurity ions in a liquid-crystal solvent can travel more easily along the director than across the director. This means that the conductivity parallel ($\sigma_\parallel$) to the director is greater than the conductivity perpendicular ($\sigma_\perp$), typically by as much as 50%. Naturally, the relationship between the current density and the applied electric field involves a conductivity tensor.

Up to this point the figures in this chapter have depicted single-crystal liquid crystals. Just as with solid crystals, however, single crystals are the exception to the polycrystalline rule. A bottle of material in the liquid-crystal phase will appear milky due to the scattering from the "grain" boundaries called disclinations. Since a disclination defines the sharp change in orientation of a birefringent material, one expects strong optical coupling, i.e., scattering. The name nematic comes from the Greek word for line, which is the structure of disclinations in that phase. The twist structure of the cholesteric phase makes its disclination structure even more complicated than for nematics, with herringbone textures common in thin samples. Most LCDs employ a single-crystal state. Methods for achieving this initial condition are referrred to as alignment processes.

11.2.4 Liquid-Crystal Alignment. For flat-panel displays the geometry is typically a sandwich of the type shown in Fig. 11-5. For reasons of speed and contrast, the thickness of the liquid-crystal layer is in the 10-μm range. Boundary conditions at the surfaces and the elastic properties of the material will determine the director pattern throughout the cavity. There are two types of boundary or alignment conditions: homeotropic and homogeneous. The homeotropic case occurs when the director is perpendicular to the surface, and the homogeneous case occurs when the director is in the plane of the surface as in Fig. 11-5. The homeotropic boundary condition is relatively straightforward to produce. The surfaces of interest are treated with surfactants which bond to the surface and trail long hydrocarbon chains off the surface into the cavity. The liquid-crystal molecules align parallel to these chains and orthogonal to the surface.

The homogeneous orientation is more difficult to produce and is the dominant commercial orientation. Note that a single crystal requires that the molecules not only lie in the plane of the surface, but also adopt a single direction in that surface. It is this double requirement which makes the homogeneous case more subtle. There are two basic techniques for producing this alignment: mechanical rubbing and vacuum evaporation. The rubbing method has been known for some time and is the simpler of the two. One merely rubs the surface to be oriented with a soft fabric.[8] Cotton is a typical example. Several strokes in a single direction are made on each of two surfaces. The surfaces are made into a sandwich structure with the rubbing directions antiparallel. When the liquid crystal is injected into the cavity, very often the director will be aligned parallel to the rubbing direction and a single crystal results. The "very often" caveat means that the method is not always that simple. The surface being rubbed, the cloth, and the liquid-crystal type can be important to the success of the method. Some surfaces are traditionally difficult to orient by rubbing. In particular, indium–tin oxide (ITO) transparent electrodes belong to this class. One method for improving the rubbing operation is to treat the inorganic electrode with an organic surfactant before rubbing.[9] The details of this process are generally considered proprietary by most manufacturers since it is the most important step in the device fabrication.

The other surface alignment technique was discovered by Janning in 1970.[10] He found that evaporating an inorganic dielectric such as SiO_x would produce orientation when the angle between the beam and the surface was small, e.g.,

5 degrees. In the SiO_x formula, x is some number less than 2, indicating that the compound is not stoichiometric. The evaporation alignment is neither pure homeotropic nor homogeneous. The director ends up at an angle of approximately 30 degrees with respect to the surface in the beam/surface normal plane. This has a basic disadvantage for some flat-panel applications, as will be discussed below. It is possible to produce low-tilt angle alignment with a double-evaporation process, but this increases the cost of the process considerably. Since rubbing usually produces low-tilt angle alignment and since it is less expensive than evaporation alignment, rubbing is the process used in the majority of low-tilt angle applications.

The multilayer sandwich structure of a typical LCD is shown in Fig. 11-6. It is assumed here that polarizers are required to detect the distortion in the liquid crystal. The blocking layers in Fig. 11-5 are usually some form of SiO_2 applied to improve alignment of the ITO and to protect against ionic contaminants in the glass substrate. Assuming that the display requires polarizers and is viewed in reflection, there are fourteen layers in the display: polarizer, adhesive, glass, transparent conductor, blocking layer, alignment layer, liquid crystal, alignment layer, blocking layer, transparent conductor, glass, polarizer adhesive, polarizer, and mirror. Then there are adhesives to join the glass substrates and to seal the hole through which the liquid crystal is filled. Assembly of this sandwich is a bit of an art. The hole for filling liquid-crystal material can be in one of the substrates, as shown, or more commonly is in an opening in the adhesive ring which joins the two substrates.

Each layer must perform its function without interfering with the others. The layers tend to be interactive, a feature that now will be illustrated by the alignment layer/perimeter seal processes. Schiff base and azoxy materials hydrolize with the moisture in the air and rapidly lose their liquid-crystal nature. These types of materials must be sealed hermetically to achieve long-life operation. This means a glass frit for the perimeter seal and consequent high temperatures ~480°C. Most rubbing pro-

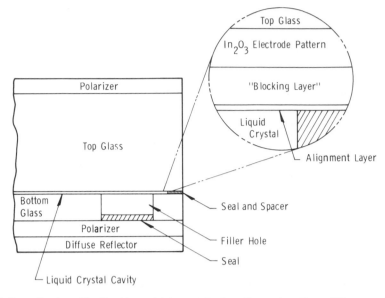

Fig. 11-6. LCD Cross Section: The liquid crystal is contained as the center of a multilayer sandwich. The glass substrates are coated on the interior side with In_2O_3 transparent electrodes to apply local electric fields (see blow-up of liquid-crystal cavity). A dielectric (blocking) layer and an alignment layer are present on both interior substrate surfaces. Polarizers and a reflector complete the planar structure with perimeter and fill hole seals indicated.

cesses will not produce good alignment after exposure to these temperatures, but SiO_x evaporation will work. Thus, most displays made with frit seals have SiO_x alignment, while epoxy seals usually mean rubbing alignment.

In the early 1970s there was considerable disagreement over which type of process was adequate for consumer products. Since then, more stable liquid-crystal materials have become available, e.g., biphenyls, cyclohexanes, and esters, permitting reasonable life in epoxy-sealed packages. Since the rubbing alignment/epoxy seal is cheaper, and is capable of multiplex address, it has become the standard LCD fabrication technique for watch/calculator/game applications. Displays exposed to more severe temperature/humidity environments (e.g., for automobiles) may still require the hermetic seal of the frit/evaporated alignment process.

11.3   LCD CONFIGURATIONS

Although the electro-optic effect shown in Fig. 11-5 constitutes a display, it is not a high-quality one. The molecular reorientation, in conjunction with external polarizers, produces interference colors through the mechanism out-lined in Section 11.1.1. Interference colors are not high-contrast, however, and the colors are not saturated. The design of Fig. 11-5 is not commercially viable. During the late 1960s and early 1970s several new display geometries were invented. The dynamic scattering mode (DSM), the dichroic dye mode, the cholesteric/nematic display, the twisted nematic mode, the fluorescence-activated display (FLAD), the smectic LCD, and disclination LCDs will be discussed. The majority of the discussion will center on the twisted nematic mode since it is the dominant technology at present.

11.3.1 Dynamic Scattering LCD. Dynamic scattering was invented by Heilmeier, Zanoni, and Barton and was announced in 1968.[3] The display operates by scattering light at dislocations in a turbulent structure. The precursor of the turbulent structure is shown in cross section in Fig. 11-7. Note the domain pattern as opposed to the uniform distortion in Fig. 11-5. The domains are not especially strong scatterers, but at higher voltages dislocations in the pattern produce much stronger scattering. Unfortunately, the scattering is at small angles, i.e.,

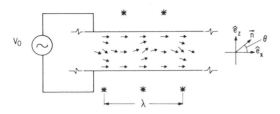

Fig. 11-7. Dynamic Scattering Mode: The time-independent liquid-crystal structure which precedes dynamic scattering is shown in cross section. A periodic distortion of wavelength refracts light into a series of domain lines above and below the sample. The director pattern is accompanied by an array of antiparallel vortices.

forward scattering. The display does not have wide-angle (large-aperture) viewing when the sources of light are restricted to a few directions. For a comparison of the PCR and DBR of dynamic scattering displays with other displays, see Table 11-1. The small-angle scattering relates to the fact that the liquid-crystal cavity is restricted to greater than 6 to 8 μm for practical reasons. Visible light has wavelengths much smaller than this and therefore cannot be scattered at large angles.

The physics behind the DSM is more complicated than for the other displays which operate in the dielectric/elastic fashion discussed earlier. The DSM mostly occurs in materials with negative dielectric anisotropy. In an applied electric field, one expects that the director will tend to align perpendicular to the field. For the homogeneous boundary condition in Fig. 11-7, there is no dielectric torque to tilt the director out of the electrode plane. The domain structure is therefore unexpected. Helfrich[11] was responsible for the model which explains this mode. He showed how vortex flow in the sample

could be set up by space charge produced by the applied electric field. The shear rates involved in the vortex motion produce a torque on the director via the shear torque interaction described above. The sign of the torque is such that the distortion grows above a threshold voltage. This leads to more space charge which leads to larger shear rates, etc., until nonlinearities limit the growth. The period of the distortion can be explained by rigorously solving the two-dimensional boundary value problem.[12]

Dynamic scattering has a threshold voltage of approximately 6 V and operates well at voltages over 10 V. Space charge is essential to the device, and the pure liquid is doped with ionic salts to provide the conductivity. Since the voltage and current requirements are larger for the DSM than for the field effect in Fig. 11-5, the power necessary to operate DSM is two orders of magnitude greater than for a field effect.

11.3.2 Dichroic Dye LCD. In 1968 the workers at RCA announced another liquid-crystal display, the dichroic dye display.[4] This is an absorbing rather than a scattering mode. It takes advantage of the fact that dye molecules will align parallel to the director when dissolved in a liquid crystal. Thus, the name guest-host display has also been used. For this display to work, the dye must be dichroic. The absorption coefficient of the dye depends on whether the electric field of the incident light is aligned parallel or perpendicular to the long axis of the dye. Typically, the absorption constant parallel to the axis is much greater than the perpendicular constant. Consider the liquid-crystal orientation of Fig. 11-5 with a dye dissolved in the liquid crystal. In the off state, the dye

Table 11-1. Non-emitting Display Comparison

Display	PCR*	DBR*	Angle Dependence	V_{on}	Power or Charge/cm^2
TN/LCD	6–8	0.25	Significant	3 VRMS	0.5 μW
DSM/LCD	5	0.5	Very significant	12 VRMS	10.0 μW
DYE/LCD	4	0.5	Small	10 VRMS	2.0 μW
ECD	5	0.8	Small	2 V dc	5.0 mC
EPD	5	0.65	Small	50 V dc	0.1 μC

*Measured at best viewing angle, typical data for room ambients.

will absorb incident unpolarized light, i.e., the components with electric field parallel to the director will be absorbed. In the on state, this absorption will be less. Since light polarized perpendicular to the plane of the figure will not be absorbed, however, such a display would have an unacceptably low contrast. The contrast can be enhanced by the use of a single polarizer which is oriented so that it absorbs light polarized perpendicular to the dye in the liquid crystal.

As originally proposed, the dichroic dye display did not meet with much commercial success. The appearance of the display is whitish/gray characters on a colored background. This contrast arrangement has not met with consumer acceptance for watch/calculator application due to the fact that the majority of the display is dark. Also, there was a problem getting dyes to order sufficiently well in the liquid-crystal matrix so that the on state really did not absorb. Another problem was obtaining sufficient solubility so that the off state really did absorb. Before these problems were solved, another mode was invented that solved these problems by using two polarizers: the twisted nematic. Nevertheless, the problems have continued to receive a great deal of attention during the late 1970s and early 1980s, and displays are beginning to be marketed using a variation on the dichroic dye mode to be discussed below.

11.3.3 Cholesteric-Nematic LCD. The cholesteric-nematic phase change display takes advantage of the fact that an applied electric field in a positive dielectric anisotropy cholesteric will change the strongly scattering state of zero applied field into a clear state at high field. The scattering "off" state results from a homeotropic surface orientation which in turn forces the helix axis to lie in the display plane. Since the helix axis orientation can point in random directions in the plane, a complicated "fingerprint" pattern results. The term fingerprint is used because the disclinations have a fingerprint appearance. This condition results when the cholesteric pitch is much less than the cavity spacing. This pattern also scatters light and does it considerably more effectively than dynamic scattering. When the electric field is applied,

the scattering sites gradually disappear as the nematic phase is induced. At high enough voltage, the texture is completely homeotropic. While the scattering is stronger for the cholesteric-nematic phase transformation, this comes at the cost of considerably higher fields necessary to switch the device. Voltages are $2\times$ to $10\times$ greater than for a nematic field effect. This is because the formulas describing the electric-field effects have an additional length, the pitch of the helix p in addition to the dimension of the cavity. The elastic energy is much higher since $p \ll d$ while the voltage is still applied across the cavity width so the electric field remains the same. This leads to higher threshold voltages, an undesirable feature. In addition, the general preference for an absorbing display has limited the commercial application of this display. For other cholesteric displays see References 14 to 17.

White and Taylor[18] realized that the cholesteric structure could be combined with the guest-host concept to produce a dye display which would not need a polarizer as in the old RCA geometry.[4] The electromagnetic normal modes in a cholesteric are different than for a nematic, and this leads to the absorption of all the light even though the dye molecules are still parallel to the host molecular axis. The display still has a white-on-color format. The fact that polarizers are not needed makes this display attractive for harsh temperature application, e.g., automobiles. There are still problems in achieving sufficiently large values of the order parameter S, in achieving sufficient solubility, and in the lack of significant multiplex capability.

11.3.4 Deformation of Aligned-Phase LCD. In 1971 Schiekel and Fahrenschon[19] reported a two-polarizer display which uses a nematic material and a homeotropic-type boundary condition. The deformation of aligned phase (DAP) mode requires a negative dielectric anisotropy material. The director is initially oriented homeotropically and is tilted perpendicular to the applied electric field as shown in Fig. 11-8. Dynamic scattering is avoided by operating at a frequency above the space-charge relaxation frequency, requiring the use of exceedingly pure

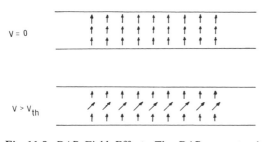

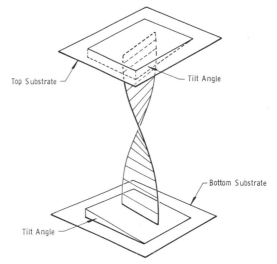

Fig. 11-8. DAP Field Effect: The DAP geometry is shown in cross section. The initial homeotropic alignment is distorted into a uniform bend pattern by the electric field/negative dielectric anisotropy torque.

materials. When the sample is placed between crossed polarizers, it does not transmit light in the off state. When the voltage is applied, the molecules tilt and light is transmitted. Unfortunately, this does not work out to be a high-contrast display when viewed in reflection in ambient light. If the display is viewed in transmission with collimated light, i.e., a very narrow viewing angle, a more satisfactory display is produced. The electro-optic threshold characteristics of the DAP effect tends to be very sharp, and this has important flat-panel advantages as will be discussed below.

11.3.5 Twisted Nematic LCD. In 1970 two groups,[5,6] independently developed a two-polarizer display which has become the overwhelming success in LCD technology. As the name twisted nematic implies, the display geometry produces a 90-degree twist in a nematic material, as shown in Fig. 11-9. The twist results from homogeneous boundary conditions on the two surfaces being oriented at 90 degrees to each other. The electromagnetic modes of this inhomogeneous birefringent material are quite complex. For a particular twist pitch and birefringence range, however, visible light is transmitted through the material in a unique manner. If the light is incident on the sample polarized either parallel or perpendicular to the director at the surface, the light is rotated 90 degrees and emerges plane-polarized from the opposite side. The reader is warned that this is a unique behavior and does not yield to typical birefringent analysis. The key is that the birefringent axis is constantly changing in space.

Fig. 11-9. The Twist Structure: The twisted nematic structure is produced by surface alignment layers being twisted by 90 degrees with respect to each other. The director pattern is a uniformly twisted structure as shown. This give rise to a twisted birefringent material—a structure unique to liquid crystals. The alignment is homogeneous with a slight tilt to insure that the molecules all turn in the same direction above threshold.

The display contains crossed polarizers which are aligned with the alignment axis (see Fig. 11-10). Ambient light incident on the display is polarized by the first polarizer and then is twisted 90 degrees by the twisted nematic.

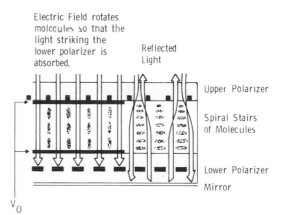

Fig. 11-10. The Twisted Nematic LCD: The result of the twisted nematic structure combined with crossed polarizers is shown in cross section. In the field-free region, the polarized light incident on the liquid crystal is twisted to a gray off state. Where the voltage is over threshold, the twist structure is destroyed. The initial polarized light is absorbed by the bottom polarizer, and the on state is dark.

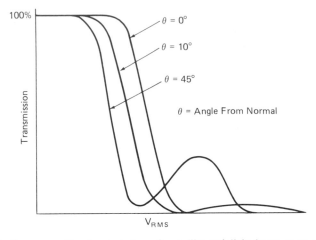

Fig. 11-11. Electro-optic Response: The transmission of a collimated light beam at an angle θ relative to the normal of a twisted nematic display is plotted as a function of the applied voltage. Note that the lowest voltage at which transmission begins to decrease for a given θ rises as θ decreases. This can be understood by realizing that the transmission is minimum in an angular region around the director orientation at the center of the cell. As the director tilts up, the transmission decreases at larger values of θ, (not same θ as in 11.5).

The light emerging from the liquid crystal is now incident on the second polarizer in an orientation such that the light is not absorbed. It travels to the mirror, is reflected, and passes again through the whole system with little attenuation. When a voltage is applied, the director tilts up, thereby undoing the twist structure. When the polarized light now propagates through the liquid crystal, it is not twisted and therefore is absorbed by the second crossed polarizer. The appearance of the display is dark segments on a metallic gray field, i.e., correct for easy observation. The mode is a nematic field effect and so operates at low voltage and very low power. The transmission of a twisted nematic display as a function of voltage and angle of view from the normal is shown in Fig. 11-11. Note that the viewing angle does influence the electro-optic characteristic. Even so, the field of view of the twisted nematic display is considerably larger than DSM, and its contrast is equal to or better than dichroic dye displays. For these reasons, the twisted nematic display has found wide consumer acceptance and in the early 1980s was manufactured at running rates exceeding 10 M per month worldwide.

11.3.6 Fluorescent LCD. As useful as the twisted nematic LCD is, it is better described as functional rather than attractive. The loss in brightness associated with the polarizers gives a generally dull appearance. One way to increase the apparent brightness of a display is to backlight it. In 1977 Baur and Greubel[20] announced a particularly ingenious method of collecting ambient light and directing it through a twisted nematic display. They called it the fluorescence-activated display (FLAD). A cross-section diagram of the display system is shown in Fig. 11-12. The twisted nematic part of the structure is similar to that in Fig. 11-10 with the exception that the polarizers are parallel rather than crossed. When the display is on, one sees through the activated segments to the fluorescent plastic behind the display. Since the plastic can be quite thin, the display system remains flat.

The key to the FLAD is the fluorescent dye contained in the plastic. The dye absorbs light in the blue end of the spectrum and re-emits the light at longer wavelengths. By using the right dye in the plastic, energy in the blue (where the eye is not sensitive) is converted into energy in the green or orange, where the eye is very sensitive. The re-emitted light is radiated uniformly over a 4π steradian solid angle, and so most gets trapped in the plastic light pipe. It travels to the end of the substrate or surface defects, where it escapes. This results in a large brightness gain. The effect is already used commercially in draftsmen's drawing triangles. In the FLAD application, the plastic is grooved to spoil the total in-

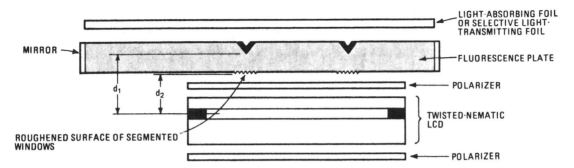

Fig. 11-12. Fluorescence-Activated Display: A fluorescent plastic is grooved in the places where the display in front of the plastic is to present information. This produces a bright 1888 pattern in the plastic used for a clock. The twisted nematic display, with parallel polarizers, acts as a light shutter to let the observer see selected segments of the 1888 pattern. The display appears to be emitting light.

ternal reflection behind the display segments. Local "light" sources are produced from ambient light. When the display is turned on, one sees a green or orange glow very similar to that of a light-emitting display. The display is especially effective when viewed on axis, since parallax between the plastic ridges and the display segments cuts down off-axis viewing.

11.3.7 Smectic LCD. The high degree of order present in smectic liquid crystals means higher viscosities and much longer response times at reasonable voltages than for nematic liquid crystals. Thus, there are no practical smectic displays driven by electric fields alone. In 1973 Kahn[21] discovered that a smectic liquid crystal could be altered between scattering and nonscattering states by the application of a heat pulse to raise the smectic into the nematic phase at the pixel in question. As the nematic phase cooled, it was subjected to an electric field to align the molecules if a clear smectic state was called for. If the pixel was to be scattering in texture, the electric field was not applied. The display mode has memory, i.e., the scattering and clear states of the smectic are stable with time. This memory feature has attractive potential for matrix addressing as will be discussed below.

11.3.8 Disclination Display Modes. As discussed above, most liquid-crystal structures used for displays originate from a single-crystal structure. In 1980, however, two modes were invented that specifically took advantage of carefully con-

structed defects in the liquid crystal to produce a memory-type operating characteristic. Boyd, et al.[22] demonstrated a dye in a nematic host that had a bistable configuration. A high-tilt angle alignment method was employed on both surfaces of the cell, but the cell was assembled so that the director orientations at the surfaces were not parallel but at an angle of approximately 60 degrees. The director can assume two topologically inequivalent configurations with these boundary conditions. One has the average orientation of the director nearly vertical, the other nearly horizontal. The horizontal state involves a disclination at the center of the cell, where the director pattern must accommodate the nonparallel director contours coming from the surfaces. To switch between the configurations requires moving the disclination, and so they are bistable. The states are distinguished optically by the addition of a dye which absorbs strongly in the horizontal state but less so in the vertical state. The contrast of this display is inferior to the twisted nematic mode, and the electrical switching requires nontraditional electrodes which are more difficult to fabricate than the row/column format.

Also in 1980, Berreman and Heffner[23] invented a bistable display which employed the cholesteric liquid-crystal phase. Again, high-tilt angle homogeneous alignment is employed. The liquid crystal is blended so that the pitch of the cholesteric is approximately equal to the thickness of the liquid-crystal layer. Under these conditions there are two stable configurations: an untwisted state which has a nematic-like structure and a state having a 360-degree twist about the normal to the surface. One detects differences in the two states using crossed polarizers, using the fact that the 360-degree-twist state depolarizes light. Switching between the two states is accomplished by using voltage pulses of less amplitude for the untwisted state and more amplitude for the twisted state. As with the bistable nematic effect, the bistable cholesteric effect lacks the strong contrast of the twisted nematic display.

11.4   INTRINSIC MATRIX ADDRESSING OF LCDS

Now that the relevant electro-optic modes in liquid crystals have been discussed, it remains to demonstrate how these effects can be used to produce flat-panel displays. As usual, it comes down to a matter of addressing. Several schemes will be discussed. They are the intrinsic matrix-address techniques of fast scanning, two-frequency, hysteresis, and pointer. System considerations include temperature, power supply, reliability, and cost.

11.4.1 Fast-Scan Matrix Addressing. The standard electrode structure for addressing a dot matrix LCD format is shown in Fig. 11-13. A linear array in the horizontal (x) direction is fabricated on one surface of the display, and a second linear array in the vertical (y) direction is fabricated on the other surface. Scan pulses are applied sequentially to the rows, and data pulses are fed in on the columns in a fashion correlated with the scan pulses. The question becomes what form are these pulses to take. If the information on the display is not going to flicker to the observer, the frame rate at which the pulses are applied to the whole display must be greater than 30 Hz. If N lines are scanned, the frequency

(a)

Fig. 11-13. Dot Matrix LCDs: An alphanumeric format in a portable computer is addressed by an x-y array of electrodes. The display is a 16×80 array unfolded to mimic a 8×160 array. The left-hand side of the display being addressed by the first eight scan electrodes and then the right-hand side by the second eight scan electrodes. Similar displays were commercially available in 1982. The viewing angle problem can be understood by comparing Fig. 11-13(a) with Fig. 11-13(b). Figure 11-13(a) was taken under ideal conditions. Figure 11-13(b) demonstrates the background turning on because of the half select voltages at 1/16 duty cycle.

(b)

Fig. 11-13. (*Continued*)

seen by individual dots will have components $\simeq Nx$ 30 Hz. These rates are much faster than the response of typical displays (1/100 ms). What will be the averae liquid-crystal distortion?

Kmetz[24] was first to point out that LCDs would respond to pulses of this sort in a root-mean-square sense. The average orientation of the liquid-crystal director in a static scene would be the angle produced by the RMS voltage applied to that pixel over the whole frame cycle. The orientation would not be simply that produced by the voltage at the time when the pixel was addressed because of the slow response times.

To illustrate this point, consider the following drive scheme (Fig. 11-14). The scan pulses are V_S in amplitude and the data pulses are $+V_D$ or $-V_D$ depending on whether the pixel is to be off or on. If N rows are being scanned and the pixel under consideration is supposed to be off ($V_S - V_D$ during selected period), the mean squared voltage across the pixel is given by:

$$V_{OFF}^2 = (V_S - V_D)^2 \frac{1}{N} + V_D^2 \left(1 - \frac{1}{N}\right)$$

Similarly, the mean squared voltage across an "on" pixel is given by:

$$V_{ON}^2 = (V_S + V_D)^2 \frac{1}{N} + V_D^2 \left(1 - \frac{1}{N}\right)$$

For the pixel to be, in fact, off requires that $V_{OFF} \leqslant V_{th}$. The on voltage is broken into two parts relative to the threshold voltage and the voltage above threshold Δ: $V_{ON} = V_{th} + \Delta \simeq V_{OFF} + \Delta$. Note that V_{OFF} and V_{ON} both approach V_D^2 for large N (for V_D and V_S given). This means that as N increases, Δ becomes smaller and smaller, or in other words the contrast gets poorer and poorer, as shown in Fig. 11-11. If

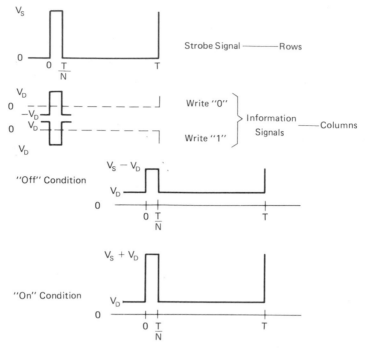

Liquid-crystal matrix-drive waveforms.

Fig. 11-14. Strobe and Data Signals: The pulse waveforms used by Alt and Pleshko to analyze LCD matrix addressing are shown. The strobe signals have amplitude V_S and have a pulse width of T/N. The data signals have amplitude V_D and are in phase with the strobe signal for the "0" state and 180 degrees out of phase for the "1" or written state. A given pixel sees only one strobe signal during one addressing period, but it sees a constant stream of data. The net waveforms across an "off" and an "on" pixel are also shown in the figure. Note that as N increases, both waveforms tend to RMS amplitude of V_D. This is the electrical manifestation of the LCD matrix-address problem.

one picks a contrast required for an application which sets a required value Δ_R from the electro-optic characteristic, Alt and Pleshko[25] have shown that the maximum number of lines that can be scanned for Δ_R is

$$N_{MAX} = \frac{(1+P)^2 + 1^2}{(1-P)^2 + 1}, \quad P \gg 1 \frac{1}{P^2}$$

where $P = \Delta_R / V_{th}$. V_S and V_D are adjusted to achieve this limit, with V_D going to $0.707\ V_{th}$ and V_S rising as $\sqrt{N}\ V_{th}$ for large N.

The algebra of fast-scan matrix addressing reflects the physical fact that the more lines scanned, the less time any given pixel is told what to do. This means less voltage above threshold and less contrast. The formula shows that the sharper the electro-optic characteristic, the more lines that can be addressed. The key is to reduce the width of the voltage step between threshold and saturation while keeping V_{th} as

low as possible to permit ease of driving from batteries.

There are two basic techniques used to sharpen the threshold characteristic. Ths surest method is to reduce the angle of the display viewing and illumination. The shape of the electro-optic characteristic depends on these conditions. Since birefringent materials are anisotropic by definition, the threshold can be sharpened significantly by restricting viewing. Additional benefit comes from backlighting the display with collimated light. Unfortunately, both of these methods seriously degrade the fundamental advantages of the traditional LCD: low power and wide viewing angles. The DAP[26] and twisted nematic[27] displays can be operated with a large multiplex capability (up to 200 lines), but only under these limitations.

A second method of increasing the sharpness is to optimize the liquid-crystal material constants. This is the most common approach and is the most closely guarded for obvious reasons. Material vendors and display manufacturers are constantly mixing various materials from a given class and now are mixing different classes of materials. There are some physical models to follow, but the process is more of an art than a science.[28-30]

The bottom line on fast-scan multiplexing is that a limited amount of information can be presented in reflection. Sixteen scan lines became commercially available in the early 1980s in the twisted nematic mode. If two sets of column drivers are used, one set feeding data to the top sixteen pixel rows and the other set feeding the bottom sixteen pixel rows, four alphanumeric-character rows can be addressed with reasonable viewing characteristics. Recent analysis indicates that 20 to 30 scanned pixel lines will be the limit for reflective displays.[28]

11.4.2 Two-Frequency Addressing.

Up to this point, little has been said about the frequency dependence of the liquid-crystal material constants. In 1971, two groups[31,32] realized that the frequency of the drive signals was a variable that could be used to increase the number of scanned lines using the space-charge relaxation frequency associated with the DSM. The DSM is observed above a threshold voltage at low frequencies, but the same voltage at frequencies above the space-charge relaxation frequency will not disturb the sample. To take advantage of this fact, the strobe signals are at one frequency and the data signals at a second frequency on the other side of the critical frequency from the first. Generally, the high-frequency signal is the one that turns off a pixel, i.e., reinforces the off state. The low-frequency signal turns the pixel on. The amplitudes of the two signals are adjusted so that the strobe signal by itself will activate a pixel, whereas the strobe plus a high-frequency data pulse will leave the pixel off.

The two-frequency addressing scheme removes the threshold voltage limit on the scan and data signals. For a large number of scanned lines, however, the pixels still have very little time over which to be activated. To achieve a sufficient net RMS voltage, the absolute voltages must become large. Since the two frequencies are separated by a few orders of magnitude, the high-frequency pulses draw a significant amount of power. These are two fundamental limits to this addressing technology.

The twisted nematic display can also be driven by a two-frequency scheme provided a special material is used which has dispersion in the dielectric anisotropy: positive at low frequency and negative at high frequency.[33] These materials do exist, and successful prototypes have been fabricated. The crossover frequency between positive and negative dielectric anisotropy (and the space-charge relaxation frequency) is temperature-sensitive. This means that operation over a wide temperature range will require temperature-controlled frequencies, a significant electronics problem. In the early 1980s no commercial displays had appeared due to the voltage, power, and temperature drawbacks.

11.4.3 Hysteresis Multiplex Addressing.

Liquid-crystal display modes with memory can be driven in a qualitatively different fashion than nonmemory displays such as the twisted nematic. Memory means that a display remains in either the "on" or "off" state without power. The technique is basically to write each line of the array sequentially, going on to the next line only when the present line has achieved the required contrast. The written lines remain written, so flicker is not a factor. The 30-ms frame requirement becomes irrelevant. This in turn removes the requirement for a very sharp threshold. The drawback is that it may take a very long time to write a large display. If one pixel takes 100 ms to write, a 100 X 100 array will take 10 seconds to write using line-at-a-time addressing. This is unacceptable for most applications.

Workers at Nippon Electric[34] have developed a novel scheme for addressing a variation of the cholesteric-nematic phase transition. They construct the display so that the pitch of the cholesteric is almost the same length as the cell thickness, as opposed to the cell described in Section 11.3.3 which had the pitch very much smaller than the cavity thickness. The off state of the Nippon Electric display is clear. When a voltage is applied to the positive dielectric anisotropy material, the texture converts to the strong scattering fingerprint structure. Further increase in voltage to a critical voltage V_H produces the

The Nippon workers[34] invented a unique method of using the temporal characteristics to achieve as many as 100 lines of matrix addressing. The drawback, as discussed above, was that the panel takes several seconds to address. The contrast of the panel was also weak, as is the case with many scattering display modes. The bistable nematic[22] and bistable cholesteric[23] effects were also invented to take advantage of the lack of a refresh requirement in hysteretic display modes. These displays were in the research phase in the early 1980s.

11.4.4 Thermal Addressing. Another memory display is the smectic mode driven by both heat and electric fields. Two groups have produced large-information-content displays using this effect by generating local ohmic heat in the scan lines. Typically, a pulse of current is delivered to an individual scan line, the magnitude of the current being sufficient to heat and melt the material from the smectic phase into the nematic phase. The electric fields are then applied on those data lines which are to be written clear. The field-free regions are scattering [36] or absorbing.[37] The absorbing display is produced by dissolving a dye in the smectic liquid crystal. In the disordered smectic texture it absorbs; in the homeotropic texture it is clear. Several hundred scan lines can be addressed in this fashion. The problem with the thermally addressed displays is the power required to generate the heat. The power is measured in watts, and the drivers required to generate hundreds of milliamperes of current are expensive. This type of design had only reached the prototype stage in 1983.

11.4.5 Other Intrinsic Addressing Modes. For completeness, several other intrinsic addressing schemes should be mentioned. The first method uses data pulses which depend on the states of *all* the pixels in the chosen data column instead of the conventional scheme where the data voltage at any instant depends on the desired state of the element being addressed. A theoretical study[38] has been performed using completely general strobe and data waveforms in order to determine the absolute optimum for RMS responding matrixes with arbitrary patterns. It was found that for some cases the Alt and Pleshko

clear, bulk homeotropic structure which must result with a positive dielectric anisotropy material. The unique features of this design are that the electro-optic characteristic has a pronounced hysteresis,[35] and that the display has an unusual time-dependence. Once the pixel under consideration is driven to the high-voltage clear state, it can return to the initial, nonscattering state, or to a scattering state, depending on the pulse sequence.

limits[25] can be exceeded, but the improvement is only significant for matrixes with two rows.

Flat-panel displays need not have an x-y grid format, and for some applications this format is not desirable. In particular, rate-of-change information is more readily read from a pointer or an analog display. Analog display watches, e.g., the traditional format with a long and short hand, appeared commercially in 1978. The reader is referred to a recent review article[39] for further information.

11.4.6 System Considerations for Intrinsic Matrix Addressing. Direct matrix addressing of any device depends on the details of the electro-optic characteristic. In the case of LCDs, this characteristic tends to be temperature-sensitive, which means that a given design may not operate over the entire liquid-crystal range. This problem can be attacked in a variety of ways. First, the liquid-crystal material can be blended to minimize the temperature dependence. This is the approach taken in many calculator numeric displays. Second, one can take advantage of the natural temperature-dependence of some component in the power supply to match the temperature-dependence of the display.[40] A third possible technique is to measure some property of the display, say the capacitance, and regulate the drive voltage according to the known temperature-dependence of that property.[41] Fourth, one can control the temperature of the display by heating.[42] If the temperature range is to be wide and the duty cycle is to be anything less than $\frac{1}{7}$, blending is not sufficient, and one of the last three methods must be employed.

A second important system consideration is the power supply voltage. The trend in LCDs has been toward lower and lower voltages, driven by the cost of batteries. Since low duty cycle operation by direct methods requires increasingly higher voltages,[25] a tradeoff must be made. The low threshold voltage of the twisted nematic mode has made it a favorite over higher-voltage dynamic scattering mode and cholesteric-nematic devices. In environments where large voltages are readily available, e.g., automobiles and line-operated devices, this contrast is removed. One can expect other displays to emerge to compete with the twisted nematic in this regime.

There is also the question of humidity. Water can destroy polarizers, degrade liquid-crystal resistivity, and in some cases destroy the liquid-crystal material altogether. A frit seal process is much more tolerant to humidity than an epoxy seal, so frit seals may be necessary in high-humidity applications. Since polarizers are also humidity-sensitive, one can expect dichroic dye displays to show strength in high-humidity environments, e.g., marine applications.

After the viewability of any display technology has been established, the next most important system consideration is usually cost, especially for the consumer or commercial markets. A potentially important technology for lowering the cost of glass-substrate LCDs is the use of plastic substrates.[43,44] There are distinct manufacturing advantages to moving a flexible film as in the graphic arts industry. This is opposed to a process flow which must consider a rigid, brittle substrate such as glass. These plastic advantages should lead to lower cost and may lead to a viewing-angle improvement due to the very thin substrates. The basic problem to be overcome in melding the polymer and liquid technologies is to come up with a design that packages a good solvent such as a liquid crystal in a plastic which will not react with the liquid crystal and not have deleterious birefringent characteristics.

11.5 EXTRINSIC MATRIX ADDRESSING OF LCDs

The problems with intrinsic matrix addressing of LCDs have been covered. Basically, one can have only a few lines of alphanumeric information and maintain the traditional benefits of the LCD: low power, low voltage, large viewing angle, and sunlight readability. As discussed above, compromising one or more of these advantages can

be allowed to gain a larger degree of intrinsic addressing. To keep all the advantages, however, and get a large information content too, requires an extrinsic matrix of some sort. The various designs will now be described, with fabrication difficulties that often go far beyond those of making a simple LCD. The key, of course, is to make an extrinsic array that does not become more expensive than portable products can afford in cost, energy, weight, etc.

11.5.1 Ferroelectric Addressing. The hysteresis loop of ferroelectric materials can be used as the active switch in an LC/PLZT sandwich display.[45,46] Since the nature of this and other active-switch matrices are covered elsewhere in this book, the operation of the switches will not be discussed in detail. A diagram of the display in perspective and cross section is in Fig. 11-15. A ferroelectric wafer (7/65/35 PLZT) is addressed by an array of row electrodes on one side and column electrodes on the other. On the side in contact with the liquid crystal a third array of electrodes is deposited. These feed-through or distribution electrodes couple the charge in the ceramic to the liquid crystal. A fourth electrode is the transparent conductor on the other side of the liquid crystal. The device operates as follows: The ceramic material is written with data by a series of microsecond pulses on the x y electrodes while the transparent conductor is held at a high potential to prevent charge coupling to the liquid crystal. The data is stored as poled regions of the ceramic. Once the data is completely entered, the y array and the transparent conductor are taken to ground potential. This couples the charge to the liquid crystal via the isolated distribution electrodes and also resets the ceramic for the next series of pulses. The local electric fields across the liquid-crystal layer stimulate the display action for the remainder of the frame. Since the write part of the frame can be made much shorter than the read or display time, each pixel is effectively connected to its data source for the whole frame. The pixels see unity duty cycle, and the lack of sharpness of the liquid-crystal electro-optic characteristic is therefore not relevant. This is, of course, the basis of the whole active-matrix concept.

Prototypes of a 32 × 32 device were fabri-

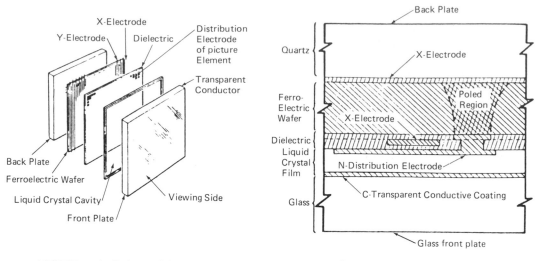

LC/PLZT matrix display sandwich.   Cross-section of display picture element

Fig. 11-15. PLZT/LCD Extrinsic Matrix: The ferroelectric active matrix is shown in perspective and in cross section. A PLZT ceramic is used to address the liquid crystal via the poled region that is created by an x-y array on the PLZT. The PLZT used here is of a different design than PLZT used directly as a display medium as in Fig. 11-23.

cated at Rockwell in the early 1970s. The biggest drawbacks to the design were the high voltages (up to 120 V) and a cross-coupling produced by hysteresis creep in the ceramic. Another problem was the high cost of the ceramic substrates, which must be sliced from a large block and then lapped and polished.

11.5.2 Varistor Addressing. In 1978 Castleberry[47] used the breakdown phenomenon in a varistor material to address a dichroic dye display. The ZnO substrate is an insulator below 60 V and a conductor above this voltage. The I-V characteristic is very steep. When the varistor is placed in series with a liquid-crystal pixel, the LC pixel has zero voltage across it until the varistor breaks down. The pixel then charges to a high potential which is sufficient to keep the pixel on for the rest of the frame. The effective voltage on the cell is not the RMS voltage as in the fast-scan addressing technique, but rather the voltage applied to the pixel during the write period only.

Figure 11-16 shows a cross section of the varistor/liquid crystal design. The design is somewhat simpler than the ferroelectric design of the previous section, since only three electrodes are required instead of four. The holes

through the dielectric layers on either side of the ZnO represent a significant technical challenge, since the varistor capacitance must be small with respect to the liquid-crystal capacitance. Since the substrate is not optically clear, the display mode is restricted to a scattering or dye display.

In 1982 a 180 × 252 matrix was reported[48] over a 5 × 7-inch area. At these small pixel sizes, the General Electric workers reported a significant uniformity problem in the threshold voltage of the varistor due to the finite grain size of the ZnO relative to the pixel state. This was thought to be a fundamental problem for larger arrays at higher pixel densities.

Varistor electrical characteristics can also be produced by a metal insulator/metal (MIM) structure where the insulator is Ta_2O_5. Baraff, et al,[49] reported this structure in 1980 and demonstrated a twisted nematic device driven at 1/100 duty cycle. The structure is especially interesting because of its ease of fabrication with respect to the ZnO ceramic.

11.5.3 Thin-Film Transistor Addressing. Varistor and ferroelectric addressing schemes make use of the nonlinear impedance characteristics of two terminal devices. Three terminal devices, or switches, can also be fabricated using thin-film or

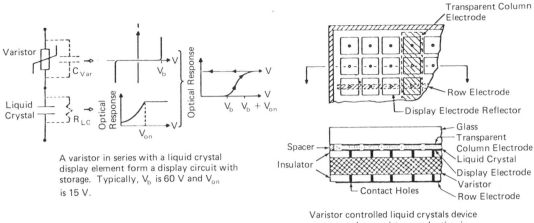

A varistor in series with a liquid crystal display element form a display circuit with storage. Typically, V_b is 60 V and V_{on} is 15 V.

Varistor controlled liquid crystals device structure where varistor conduction is through the thickness of the wafer.

Fig. 11-16. Varistor Extrinsic Matrix: The electrical, electro-optic, and spatial characteristics of a varistor active matrix are shown. An individual varistor is fabricated in series with each pixel. Its sharp electrical characteristic gives the system a sharp electro-optic characteristic.

bulk semiconductors. Semiconductor switches are unequaled for low-power applications and therefore are a natural for display addressing. The design of a semiconductor switch, typically a MOS field-effect transistor, for a liquid-crystal application is shown in Fig. 11-17. A transistor is addressed by an x-y matrix of leads. When the transistor gate is turned on, charge is transferred to the liquid-crystal capacitor and to an auxiliary storage capacitor. This charge operates the display for the rest of the frame time. Since the liquid crystal is a lossy dielectric, additional charge on the auxiliary capacitor is used to keep the pixel on for the whole frame when the gate electrode is turned off to isolate the pixel from the drive circuitry.

While the design is beautiful conceptually, it is a nontrivial effort to fabricate such an array in practice. Consider the case of thin-film transistors (TFT). Early work on TFTs concentrated on the compound semiconductor CdSe because of the high mobilities that were achievable in thin-film form. Recent CdSe work is covered in References 50 and 51. More recently, silicon TFTs have been fabricated in amorphous, polycrystalline, and laser recrystallized form because of the enormous world experience with silicon processing. An excellent review article on the state of the art in 1982 with respect to silicon TFTs has been published by Lakatos.[52] Advantages of TFTs are that they can be fabricated on transparent substrates (thus allowing use of the high-contrast twisted nematic display), can be made in a much larger area than bulk silicon slices, and, hopefully, can be made cheaper than bulk silicon.[2]

In the early 1980s the state of the art was well past demonstrating that individual TFTs could be made with adequate characteristics. The problem had become a developmental one of achieving uniform switching characteristics over large arrays of transistors at a sufficiently low cost. One technological barrier was achieving dielectric layers of high electrical and mechanical quality using low-temperature processing. Another barrier involved fabricating uniform-mobility semiconductor material without a high concentration of traps, etc. Another problem involved the very high impedance of a liquid-crystal pixel at high spacial densities, e.g., greater than 30 lines per inch. The resistive and reactive impedance characteristics of the semiconductor switch become more important at these densities. In 1982 CdSe arrays had been used in the laboratory to address 250 × 250 arrays over an area of 5 × 5 inches.[53] Silicon arrays were smaller in size at that time.

Details such as those mentioned above have prevented the commercial flourishing of TFT-addressed liquid-crystal arrays. The conceptual simplicity and the example of large-scale integration insure that this approach will continue

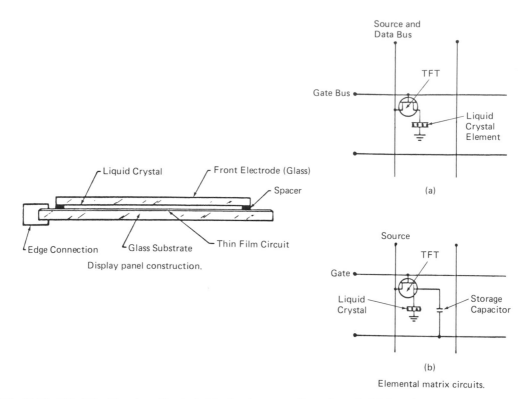

Fig. 11-17. Thin-Film Transistor Extrinsic Matrix: An array of transistors is fabricated on glass using thin film technology. The substrate is cheap, can be made "arbitrarily" large, and is transparent. Again, there is an active element for each pixel, and the *x-y* array addresses the transistors rather than the liquid crystal.

to be followed. Note that the ultimate TFT (or bulk silicon) display could include shift registers to address the 2N leads in the N^2 pixel array. This would mean a few interconnections to the display instead of the 500 or more connections needed for a standard x-y matrix. For very large information content at reasonable cost, this integration will be essential.

11.5.4 Bulk Silicon Addressing. The best understood semiconductor material is, of course, silicon. This makes the material a natural for active matrix addressing. There has been a great deal of work invested in R&D on this technology with outstanding technical success.[54-57] The basic pixel memory element for bulk silicon and TFT devices are very similar, although the processing methods are obviously different. The details of both processes are, naturally, closely held secrets. It is reasonable to assume that the design rules for the MOS devices are different from standard LSI design rules. In display technology the silicon process must produce a very

high yield ($\geqslant 99.7\%$) of relatively simple transistor structures over an area the size of the whole slice. In LSI fabrication, very complex structures are fabricated, and low yields can often be tolerated due to the small size of the finished device and its high functionality.

The use of silicon as one substrate for a liquid-crystal display means that the display mode used cannot have a rear polarizer, e.g., dynamic scattering, or a dye display. It also means that large-area displays must be fabricated from a number of silicon slices which were at 100-mm size in standard production in the early 1980s. A picture of a video display built at Hughes is shown in Fig. 11-18. The 2.5-inch-diagonal display has 30,625 elements. The display can be run at video frame rates and, obviously, is approaching the resolution necessary for television (black-and-white). Defect-free devices have been reported using silicon transistors, while TFT devices are not as far advanced. Recently Matsushita[56] and Toshiba[57] have shown a prototype portable TV using the bulk silicon to address the dynamic

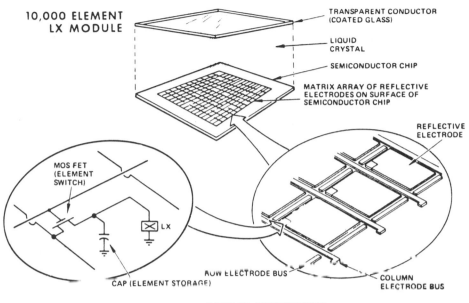

DISPLAY CONSTRUCTION

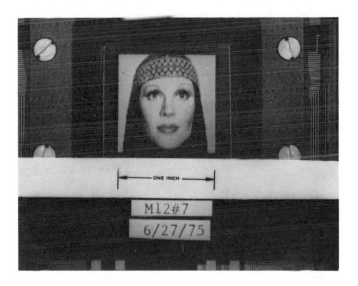

Fig. 11-18. Silicon Extrinsic Matrix: An array of transistors is fabricated on a single-crystal silicon wafer. The circuitry is similar to the TFT design with the liquid crystal in contact with the transistors. Silicon transistor action is much better understood and more reliable that TFT operation, but the substrate is opaque, expensive, and limited in area. An example of a picture displayed on a Hughes array is also shown. (John Gunther)

scattering mode. Suwa Seikosha[58] (Epson) has developed a portable TV with bulk silicon addressing the dye mode LCD. Daini Seikosha[59] has integrated shift registers on the edge of the bulk silicon array to solve the large number of interconnects to the 2N leads from a simple N X N array.

From the large amount of work going on with bulk silicon addressing of LCDs, especially in Japan, it is reasonable to forecast portable LCD TVs becoming available commercially in the mid 1980s. Whether they gain wide acceptance, however, depends on two problems this approach seems to have generically. First of all, the limited size and the limited contrast, both due to the bulk silicon substrate, will make the display distinctly inferior to a CRT display that is the existing standard. Second, the cost of the display alone will also be significant due to the silicon fabrication cost. Whether these drawbacks limit the market for these technological masterpieces will be one of the most interesting display-market questions in the 1980s.

11.6 ELECTROCHROMIC DISPLAYS

11.6.1 Introduction. An electrochromic display (ECD) is a battery which has one of the electrodes serving a display function. A battery stores electrical energy by changing it into chemical energy via electrochemical reactions at both electrodes. In an ECD the display electrode changes from transparent to absorbing. This reaction is termed electrochromic and is of interest for displays if the reaction is sufficiently reversible.

A typical electrochemical cell is shown in Fig. 11-19. Electrons are supplied to the battery by some conductor, usually transparent in the case of the working, or visible, electrode. The electrical conductor is covered by an electrochemically active material, e.g., WO_3, which changes color. The electrochromic material is also in contact with an electrolyte, an ionic conductor but an electronic insulator. The electrolyte's role is to supply electrochemically active ions to the electrodes without providing a path for electrons to short-circuit the chemical energy. The cell is

completed with a second electronic conductor/electrochemical material structure called the counterelectrode. This electrode can work on the same or a different electrochemical couple as the working electrode.

For a typical WO_3 ECD coloring cycle the following reactions take place:

Working electrode:
$$xH^+ + xe^- + WO_3 \rightarrow H_xWO_3$$

Counterelectrode:
$$xPdH \rightarrow xPd + xH^+ + xe^-$$

At the working electrode, a quantity x of hydrogen ions (from the electrolyte) and a quantity x of electrons (from the transparent conductor) are electrochemically injected into the WO_3 (clear) to form a tungsten bronze H_xWO_3 (blue). The lack of stoichiometry ($x < 1$) in this reaction will be explained below. Since hydrogen ions are being removed from the electrolyte, some positive ion must be added to the electrolyte by the counterelectrode reaction. Since Pd is well known for its ability to store hydrogen, the PdH system is often chosen for research devices. Charge neutrality is accomplished when x electrons leave the cell via the counterelectrode's electronic conductor.

The voltages and rates at which electrochemical reactions occur are the subject of the discipline called electrochemistry. This area has an old and honored position in science and is documented in numerous texts, for instance, Reference 60. For this reason the subject of voltages and rates will be covered briefly here. A battery is not a capacitor which stores energy as the electric field in a dielectric. A battery stores energy chemically, often at a unique potential related to the free energies of the two half-cell reactions occurring at two electrodes. This voltage is always in the $<3V$ range, which makes ECDs an ideal match for semiconductor electronics.

The current/voltage characteristics of an ECD are more complex than the ohmic-type behavior of an LCD. Four extra complications are present in an ECD due to the fact that current is carried by both ions and electrons. First, current in the electrolyte is carried by relatively massive ions. Normally, the driving force is not electrical. In

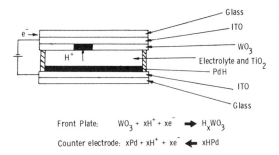

Front Plate: $WO_3 + xH^+ + xe^- \Rightarrow H_xWO_3$

Counter electrode: $xPd + xH^+ + xe^- \Leftarrow xHPd$

Fig. 11-19. WO_3 ECD: A tungsten oxide ECD is shown in cross section. The working electrode is a WO_3 layer on ITO (indium–tin oxide) transparent conductor. The counterelectrode is a Pd layer charged with atomic hydrogen. The cell operates by electrochemically injecting hydrogen into the WO_3 layer to cause a blue absorbing layer.

most batteries the ionic conductivity of the electrolyte is so high that most of the electric field is at the electrode/electrolyte interfaces. This means that ion motion in the bulk of the electrolyte must be accomplished by diffusion, driven by concentration gradients. Since diffusion coefficients decrease with decreasing temperature, electrochromic displays can be expected to slow down as they get cold just as LCDs do. Diffusion of ions in the electrode is an additional mass transfer process in the WO_3 ECD. Secondly, ionic concentrations (activities) also enter into the free energy of a half-cell reaction. The potential of a cell under charge or discharge will differ from the open-circuit potential.

The third complication is the electron transfer problem. Once the ion has been delivered to the electrode, it must react with the electron and possibly with other material at the electrode. While this is energetically favorable, an energy barrier will also exist due to the liquid/solid interface. This barrier will determine the electron transfer rates. The ion may be complex with the solvent or the electrode before reaching the final energy state. This, again, will have a rate effect. The kinetics of electrochemical processes is a complex subject. For further reading, see Reference 60.

Finally, there is the question of reversibility of the charge/discharge process. The type of battery which cycles numerous times is termed secondary vs. primary for the one-shot application, e.g., a flashlight battery. A good secondary battery will cycle a few thousand times before losing so much active chemical concentration that the device is no longer useful. This cycle life is simply not useful for most electronic display applications. Something more like a million cycles is the absolute minimum. ECD prototypes have been reported with as large as 10 million useful cycles, although only one product, a digital watch, had made it to the marketplace by 1982. One key to highly reversible devices is to avoid side reactions, i.e., other chemical reactions than the display reaction. The cleanliness of the system now becomes a big factor as it was with LCDs. Also, the drive conditions will have to be more carefully controlled than for LCDs. An ECD can be overdriven to generate extra chemicals and/or "burn" the electrochromic layer. Either of these will produce failure.

If ECDs seem to be delicate as compared to LCDs, why is there still interest in the ECD, which has been known[61] in its WO_3 form as long as LCDs? The answer is a simple one: an ECD can be made to look significantly better than any known LCD. The WO_3 display consists of strong blue segments on a bright white background (Fig. 11-20). The key here is the bright background (BR > .5) as compared with the gray twisted nematic background. The second advantage is that the contrast does not depend on angle of observation for an ECD—it looks much like ink on paper. These two factors give a strong human-factors advantage.

The various ECD technologies will now be reviewed. They are broken into three categories: nonstoichiometric materials, deposition effects, and miscellaneous. The section on ECD concludes with the flat-panel system problem.

11.6.2 Nonstoichiometric ECDs. The tungsten oxide ECD has been worked on since the late 1960s when Deb invented the display at American Cyanamid.[61] As described above, a thin film of amorphous WO_3 colors was electrochemically converted to a tungsten bronze H_xWO_3. The x in the formula indicates that the bronze structure is not stoichiometric and will take a value from 0.3 to 0.4 for a typical display application. Obviously, the valence state of the W atom must vary between +6 (clear) and +5 for the bronze to form. The fractional value of x indicates that the hydrogen must influence a number of tungsten atoms so that the net effect in the solid is a lack in integer valence states in the formula. Faughnan and Crandall[62] have interpreted the optical absorption by an intervalency charge transfer model.

The bronze structure has an electrical potential which depends on the value of x. In the classical electrochemical terminology, the battery is of the concentration type. The potential of a written pixel will be positive with respect to an unwritten electrode by as much as 1 V. This has serious implications for intrinsic multiplex operation. The x-y grid electrodes short large numbers of pixels together—in this case batteries

Fig. 11-20. ECD: A $3\frac{1}{2}$-digit WO_3 ECD prototype is pictured compared with a standard twisted nematic LCD. The ECD on the bottom is much more visible at 40 degrees off axis. The ECD has blue digits on a white background. It can be viewed from all angles, a feature which makes it superior to the twisted nematic LCD for some applications. (Jeffrey Sampsell)

charged to various voltages. The addressing electrodes will permit the written pixels to discharge their charge to the unwritten pixels. The display will rapidly deteriorate to a uniform blue. Thus, it appears that WO_3 type ECDs cannot be intrinsically multiplexed by first-principle arguments.

As a class, most ECDs have the quality of memory. Once the display is written, it can be put on open circuit and left for extended periods without losing its data. The battery nature of the display is seen again: a battery stores charge. The question of how much charge varies from ECD to ECD. For the WO_3 system, acceptable contrast can be achieved with 5 mC/cm^2 of active area. This charge must be supplied only when the display is changed. Whether it represents a power problem depends on how often the display is switched. For a watch application, continuous display of seconds is precluded from battery-drain and switchability considerations, but hours and minutes seem to be achievable.

The art of making a WO_3 ECD is less well understood than for the LCD. WO_3 films can be deposited by a variety of techniques: evaporation, sputtering, thermal oxidation, and anodization. Since it is important that the film be amorphous, post-deposition heat treating

cannot be used lest it recrystallize the film. The W-to-O ratio of the virgin film seems to be important, and the role of water in the film also seems to be crucial. There is a lore about depositing "good" films which seems to indicate that humid days are better than dry ones. These processes are naturally considered highly proprietary.

The electrolyte in an ECD can be as important as the electroactive film. For WO_3 displays, the classical water-based electrolyte systems are out. Water dissolves WO_3. This leaves organic, semisolid, and solid electrolytes. Again, the role of water in trace quantities is an important question. Since the understanding of organic electrolytes is not as well established in the electrochemical community, the progress has not been as rapid as was originally hoped.

A second type of nonstoichiometric ECD has been announced recently by scientists at Bell Laboratories.[63] This display makes use of anodic iridium oxide films (AIROF) which are stable in aqueous and nonaqueous electrolytes. The coloration may occur via insertion-extraction of hydroxide ions rather than protons in the WO_3 display. Initial results indicate an increased reliability over the WO_3 system, but the commercial reality has not yet been demonstrated. The temperature-insensitive response time of this system makes it attractive for outdoor applications.

11.6.3 Deposition ECDs. Electrochromic displays can also be made by depositing an absorbing layer on an electrode. In 1973 workers at Philips[64] announced that an insoluble purple compound, heptyl viologen bromide $((C_7H_{15})_2 (NC_5H_4)_2 Br)$, could be deposited and removed electrochemically from an electrode in a display. The mechanism is essentially reversible electroplating and is distinct from the nonstoichiometric mechanisms discussed above. The display has memory, but its voltage/current characteristic is more typical of the standard battery, i.e., the potential of a written electrode is the same regardless of how much viologen is deposited. The electrochemistry can be made to operate in well-known aqueous electrolytes.

The early viologen ECDs used gold electrodes due to their excellent chemical stability. Purple on a reflective metal background, however, is not an attractive display. Device work turned to transparent electrodes to get purple-on-white appearance. The reversibility of the viologen reaction on tin oxide electrodes is a major problem. To get some idea of the difficulty, consider what must happen at the display electrode. First, the electron must transfer from the electrode to the radical ion, and then the colored molecule must deposit on the electrode. This deposition requires active bonding sites on the electrode, i.e., the electrode must remain scrupulously clean during the entire life of the device.[65] Then the colored molecule must bleach and redissolve on erase. The molecule cannot get stuck. Unfortunately, staining was an all too common occurrence, especially in regions of high electric field. As a result, work on this display had been discontinued at most laboratories in the late 1970s.

There is at least one exception to this statement. Workers at IBM, Ltd. in England have combined the viologen ECD and the silicon active matrix technology to produce a 64 × 64-pixel array.[66] They report less than 1-ms response times, an impressive achievement. Another variation on the viologen theme has also been developed at IBM by Arellano, et al.[67]

The colored species in this display decays rapidly and thus eliminates the multiplex difficulty mentioned above. A benefit of this design is multiplex capability, but at greatly increased power consumption.

Other types of deposition ECDs are, of course, possible. Recent work in France has concentrated on the silver system.[68]

11.6.4 Other Electrochromic Mechanisms. If small organic molecules exhibit electrochromic action, it is reasonable to look for electrochromism in larger aggregates. Such behavior has indeed been observed. Nicholson and Galiardi[69] have investigated thin films of rare-earth phthalocyanines and demonstrated rapid color changes at remarkably low temperatures. The films are sufficiently absorbing in the off state so that the devices must be observed in transmission. Workers at IBM have demonstrated electrochromic operation in cross-linked donor-bound polymer films.[70] Kaufman[71] has given a review of these systems. The conclusion is that it is still too early to tell about the commercial future of the large organic molecule systems. Given the current interest in electroactive polymers, e.g., conducting polyacetylene, it does not seem beyond the realm of possibility that other display phenomena may be discovered in these long-chain systems.

11.7 COLLOIDAL DISPLAYS

11.7.1 Introduction. In the late 1960s and early 1970s displays using colloidal suspensions were demonstrated.[72-74] Before discussing these displays, some of the properties of colloidal suspensions will be reviewed. The field of colloidal chemistry is, of course, another time-honored discipline and is well documented.[75] The terms colloidal suspension or dispersion refer to a mixture of solid particles in a liquid. They imply more than rocks in a stream or a true solution. Colloidal refers to solid/liquid systems where the interfacial forces between the liquid and the solid play a dominant role in the rheology and stability of the system. Under appropriate conditions, the particles can be made to remain in suspension for long periods of time, i.e., they do not flocculate to form macroscopic clumps and/or settle from the liquid under gravity. The size of the particles is in the 5-μm or smaller regime. The magnitude of the interfacial and inertial forces on the particles is crucial to the determination of colloidal stability.

Colloidal suspensions are electrically active in many cases. The particles can have a net charge in some cases, while in others the particles can have a dipole moment associated with a long particle axis. In the net charge case, the liquid must surround the particle with a charge of the opposite sign to preserve electrical neutrality, i.e., a double layer exists. The charge at the solid/liquid surface will experience a force due to the external electric fields. This is a classical boundary value problem which relates electrophoresis (particle translation in an electric field with zero fluid velocity) and streaming potential (electric potential produced by a pressure head with fluid flow) to the dielectric constant, viscosity, conductivity, and Zeta potential (related to net charge). Particle translation has, of course, potential for display applications. In the case of asymmetric particles, an external electrical field will exert a torque on the particles, much as in the case of nematic liquid crystals. The rotation tends to produce an ordered structure in the colloid case rather than reorienting an already oriented material in the liquid-crystal case. The colloid orientation is opposed by thermal fluctuations.

The key to all colloidal displays is the stability of the suspension. The particles must remain independent of each other for the lifetime of the device. This is not achieved easily. The particles are attracted to each other by a Van der Walls-type potential ($\simeq r^{-6}$). The finer the particles, the closer together they can get and the stronger the attractive force. Thus, the particle size can be too small for stability reasons.

Repulsive forces can be electrostatic and/or steric. The colloidal particles all acquire the same charge, so there is a repulsive interaction between them. The repulsion is complicated by the screening effect of oppositely charged ions which surround the particles. The net effect is measured by an electrical potential called the Zeta potential. This is a measure of the electrical potential of the double layer and must be above a critical value (25 mV) to achieve stability.

The second repulsive force has to do with surfactant materials attached to the particles, usually intentionally. Long-chain molecules can prevent the particles from approaching each other closer than twice the length of the chain. This technique can be used to appreciably increase the lifetime of suspensions and is currently being studied extensively.

Gravitational settlement can be minimized by making the density of the particles close to the density of the liquid. Unfortunately, the best particles from the electrophoretic point of view are ones which strongly scatter light. This means that the particles should have a large index of refraction, which necessarily correlates with high density. This in turn means that the density match is impossible with uncoated particles. Composite particles and organic coatings have been used to lower the net density of the particle.

The question of ionic purity of the liquid is critically important. Extra ionic material leads to conduction current in the device, which does not contribute to electrophoresis or particle rotation. This current is parasitic and can cause serious deterioration of the display by promoting unwanted electrochemical reactions at the electrodes. To get pure liquids, it is generally necessary to use low-dielectric-constant liquids. High-dielectric-constant liquids bind impurities tightly and are subject to chemical and electrochemical reactions. This is unfortunate, since large-dielectric constants mean large electrophoretic mobility. There is a tradeoff between speed of response and stability—an effect well known in display technology. References 76 and 77 are good reviews of electrophoretic physics and chemistry for the reader who requires more complete information.

11.7.2 Electrophoretic Displays (EPD). The electrophoretic display (EPD) is shown in cross section in Fig. 11-21. A suspension of TiO_2 particles in a black fluid forms the center of what by now is a familiar glass/transparent conducting electrode sandwich structure. The operation of the device is deceptively simple. The particles are positively charged and their color is white. As the field is applied, the particles migrate to the negative electrodes of the seven-segment digit. As they "deposit" on the segments, the

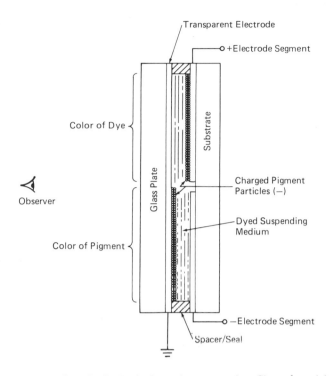

Fig. 11-21. EPD Cell: An electrophoretic display is shown in cross section. Charged particles are suspended in an opaque medium. Transparent electrodes apply a dc field across the colloidal suspension, causing the particles to translate to the addressed electrode. The white particles vs. dye background make a high-contrast, wide-viewing-angle display.

numbers appear as a bright white against the black of the surrounding fluid. The particles stick to the electrodes, so that the display has an inherent memory capability. The electric-field strengths required to achieve reasonable contrast at reasonable response times tend to require high voltages. Typical voltages are in the 25- to 100-V range.

The appearance of this display is excellent. Since the TiO_2 particles are good scatterers, the display has a broad viewing angle. It is also bright, since polarizers are not required. The contrast can be colored by suspending white pigment particles in a colored liquid. It is also possible to use colored pigment particles and to change the color of the segments by having two colored pigments of opposite charge.

The simplicity of Fig. 11-21 disguises some problems. First of all, the particles deposited on the electrodes are very densely packed. This can lead to agglomeration of the particles when the display is erased, i.e., particles are erased in groups rather than individually. Since the display is operated in a direct-current mode, there

is the possibility for unwanted electrochemistry. This is especially true when the ionic contamination leads to significant ionic currents, as opposed to electrophoretic currents. It should be remembered that TiO_2 has been extensively studied specifically for its electrochemical activity. In spite of these difficulties, EPDs have been reported to switch in excess of one million cycles.[78]

The drive requirements of EPDs are significantly more difficult than ECDs and LCDs because of the higher voltages involved. The first devices required direct drive, but recently a threshold characteristic has been developed.[78] Workers at Matsushita have developed a 32 × 184 x-y matrix which requires 12 seconds to address. While the threshold characteristic exists, it tends to be time-dependent due to the stability problems mentioned above. Another approach was suggested by workers at Philips.[79] A third "grid"-type electrode can be used to make a triode-type structure. At the present time, however, there has been minimal commercial use of EPDs. Work continues at several labs, however, due to the display's superior aesthetics. For other reviews of recent work see References 80 and 81.

11.7.3 Dipolar Suspension Displays.   A suspension of dichroic particles has been shown by Marks[72] to have interesting display properties. In the zero external electric-field state, the particles are randomly oriented as shown in Fig. 11-22. The particles are specially fabricated to be long fibers so that their electromagnetic interaction with light is anisotropic. If the particles are herapathite, they absorb light polarized along the long axis. The randomly oriented particles will absorb unpolarized light, leading to a dark off state depending on the thickness of the suspension. As an electric field is applied, the particles are rotated so that the long axis is parallel to the electric field. In the on state, the particles do not absorb as much. A bright on state vs. a dark off state is achieved. The electric fields required to achieve rotation are lower than those required to achieve translation, so the voltages tend to be somewhat lower than for EPDs. More importantly, the suspensions should tend to be more stable since the particles need not deposit. To avoid deposition, the applied electric fields are ac in the 1-kHz range. This also avoids electrochemical effects as in the liquid-crystal case.

The dipolar suspension display has not achieved significant commercial impact in the

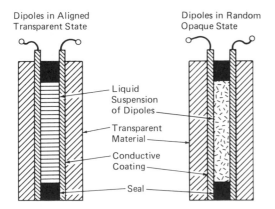

Fig. 11-22. Dipole Suspension Display: The on and off states of the dipole suspension display are shown in cross section. In the off state, the dipoles are randomly aligned and so absorb light. In the on state, the particles align parallel to the field and so transmit light.

early 1980s. The voltages are still significant, and the lack of a threshold in simple-device geometries[82] remains another serious problem. An active-matrix approach would solve the latter difficulty and would, of course, be applicable to EPDs as well.

11.8   ELECTROACTIVE SOLIDS

11.8.1 Introduction.   Solids can interact with light by a number of effects. Electric forces are employed in the Pockels and Kerr effects where the birefringence is proportional to the first and second powers of the electric fields, respectively. Magnetic materials rotate the plane of polarization by the Faraday effect. Electric fields can also produce scattering and physical deformations by electrostriction. While these effects exist in isotropic materials, it is only in ferroelectric and ferromagnetic materials that the magnitudes of the optical coupling become large enough to be useful for displays.

Cooperative effects in solids constitute another extremely large discipline. Since ferroelectric and ferromagnetic effects are well documented,[83] only the aspects relevant to displays will be covered here. The ceramic materials which support these states are usually anisotropic in crystal structure. Their crystal structure together with electric and magnetic susceptibilities lead to long-range ferroelectric and ferromagnetic order, respectively. Usually, the ordered structure does not extend uniformly over the whole sample, but is present in an array or in domains. Often the material will have a random distribution of these domains, so that the sample does not have a net dipole moment. Application of an appropriate field aligns the domains, often leading to induced-field vs. applied-field characteristics which have hysteresis. Since critical temperatures for the ordered structures exist, temperature can also be used as a switching parameter. The optical properties of the field-switchable domain structure can be detected, usually by crossed polarizers. Thus the display capability. Since the reorientation is an electronic/lattice deformation in a solid, the switching times are much faster than the viscous response times in LCD, ECD, and EPD devices.

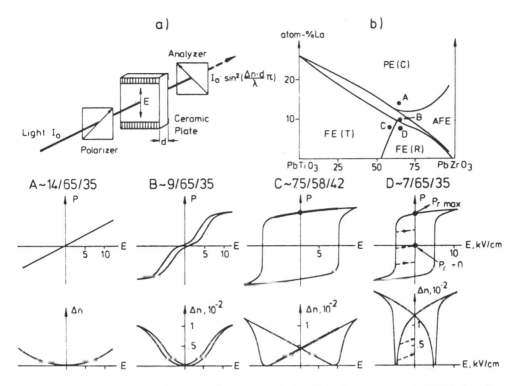

Fig. 11-23. PLZT Display Characteristics: The figure shows the PLZT display geometry, the PLZT phase diagram, and the electrical and electro-optic characteristics of various phases of the material. The material is limited in its application by the small birefringence achievable and the large voltages required.

11.8.2 Ferroelectric Displays. Ferroelectric displays became technologically feasible in the early 1970s when Haertling[84] developed an inexpensive method for producing optically clear (Pb, La) (Zr, Ti)O_3 ceramic materials, abbreviated to PLZT. By varying the ratios of the metals, various ferroelectric states can be produced with different crystal structures and polarization characteristics. An example from a review article by Hardtl[85] is shown in Fig. 11-23. Naturally, the characteristics of a given device will depend on the crystal structure. Note, however, that the magnitudes of the birefringence changes are on the order of 10^{-2} or less, and it generally takes voltages in the 50- to 100-V range to achieve these changes. The small change in index anisotropy means that electrically controlled birefringence and scattering must be viewed in transmission with narrow-aperture optics. As was discussed above, this makes it almost impossible for a nonemissive display to be considered a flat panel. For a liquid crystal, by comparison, optical changes are more than an order of magnitude larger at voltages which are more than an order of magnitude less. These ferroelectric system disadvantages more than outweigh the solid-state and rapid-response advantages in all but a very few applications. Flash-protection goggles for military pilots are one such exceptional application.

11.8.3 Ferromagnetic Displays. Hill and Schmidt have recently reported a thin-film iron-garnet display.[86] Using a gadolinium gallium garnet (GGG) as a substrate, islands of iron bismuth substituted garnet are grown on the GGG surface. Since the islands are in the 10- to 100-μm range, they form single magnetic domains, with the magnetization normal to the thin film as shown in Fig. 11-24. By properly constructing the island material, changes of Faraday rotation as large as 45 degrees can be achieved between the magnetization up vs. down states. Once again the display must be viewed in transmission. The large electro-optic effect may allow a relatively flat, diffuse rear light source which can be engineered into a "flat" panel geometry. An x-y matrix-addressed display is shown in Fig. 11-25.

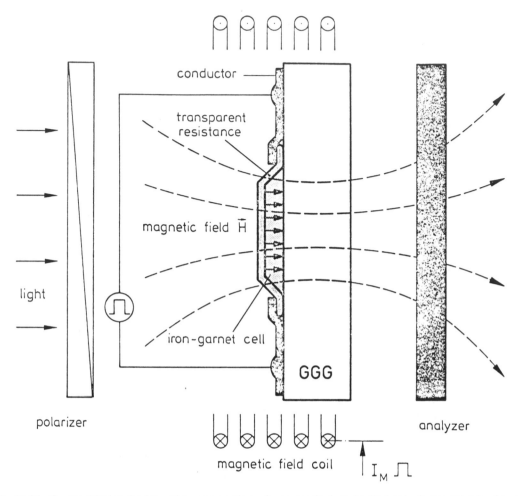

Fig. 11-24. Garnet Display Design: The cross section of a magnetic domain display is shown. Islands of iron-garnet are grown on a GGG substrate. The birefringence of the islands can be switched by heat and magnetic fields. This means that the Faraday effect can be switched locally and a display can be produced.

The display consists of a 64 × 64 array with a pitch of 60 μm. Switching is accomplished by local heating from electrical pulses in the presence of an externally applied magnetic field. Since 30- to 40°C temperature changes above room temperature can be achieved in 10 ms, and since a well-defined switching threshold exists, flicker-free real-time data can be presented.

As usual, there are disadvantages to the iron-garnet display, just as there are to all the displays discussed in this chapter. The device operates best in monochromatic light. The cost of GGG substrates is nontrivial, and the maximum size of the substrates is 2 inches in diameter in the early 1980s. The maximum size of a single domain island is limited also and at present is so small that magnification is necessary. This implies projection optics and therefore nonflat-panel geometries at the 60-μm pitch size. Since the technology is relatively new, commercial feasibility will not be tested for some time. One factor will be the success of magnetic bubble memories, which should drive down the cost of the substrate material.

11.8.4 Surface Deformation Displays. The surface of a solid can be changed from flat to wavy by external forces. Electric fields cause certain PLZT ceramics to deform on the surface, leading to the FERICON display.[87] A similar effect induced in plastics is used in the display known as the Ruticon.[88] The modulation produced by these structures is not, however, very strong. Since Schlieren optics are required to

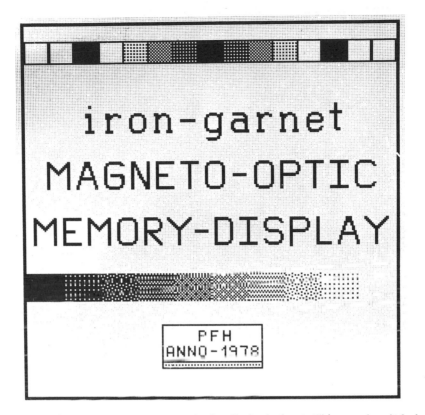

Fig. 11-25. Garnet Display: A prototype garnet projection display is shown. This array is switched electrically by heat generated by currents in the array electrodes.

give contrast, these devices are not at present applicable to the flat-panel geometry.

11.9 ELECTROMECHANICAL DISPLAYS

11.9.1 Introduction. Electromechanical displays are, of course, the original electronic displays, e.g., the galvanometer and the rotating numbers on a gas pump. Recently, attempts have been made to engineer this type of mechanical motion into high-density, flat-panel displays. Mechanical here means the motion of macroscopic structures and, as such, includes the topics discussed in this section as well as the section on suspension-based displays. The colloidal section was broken out as a separate topic because of the problems associated with that metastable formation. The technologies remaining to be discussed are cantilever foil-based displays, rotating magnetic balls, and glavanometer displays.

11.9.2 Minielectronic Shutters. Electrostatic and ferroelectric forces can be used to physically move thin plastic foils. Bruneel and Micheron[89]

used corona-charged polypropylene film as an electret which was switched mechanically between two electrodes by electrostatic interactions. In one position the film blocked light from a source behind the array, and in the other position it permitted the light to reach the viewer in front.

Toda and Osaka[90] used polyvinylidene fluoride (PVF_2) as a bimorph cantilever to achieve a similar mechanical bending motion. The PVF_2 is ferroelectric so that electric fields applied across the film will produce the display action shown in Fig. 11-26. The amplitude of the cantilever motion can be on the order of 3 mm for 60 V. There have not been any announced plans to commercially exploit this concept. For completeness, other shutter technologies should be mentioned, although they are for a projection system and thus not relevant to flat panels. Workers at Westinghouse have developed a process for etching single-crystal silicon into an array of "toadstools" which deflect under electrostatic forces.[91] A polymer membrane will accomplish a similar function.[92]

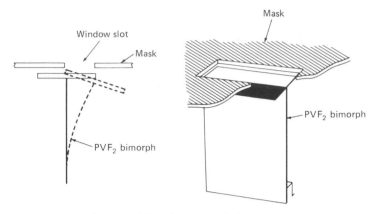

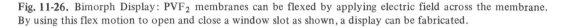

Structure of slotted-mask type display element.

Fig. 11-26. Bimorph Display: PVF$_2$ membranes can be flexed by applying electric field across the membrane. By using this flex motion to open and close a window slot as shown, a display can be fabricated.

11.9.3 Rotating Ball Displays. Consider a magnetic sphere where the north magnetic hemisphere is black and the south magnetic hemisphere is white. The ball can be oriented up vs. down by the appropriate magnetic polarity field, and the magnitude of the field will depend on the manner in which the particle is suspended. The rotating magnetic ball concept has been demonstrated by Lee.[93] Particles on the order of 10 to 200 μm are encapsulated in a plastic with a lubricant around the particle to permit low-friction rotation. By properly choosing the magnetic medium, rotation can be achieved by seven gauss fields. The display is addressed in a matrix fashion by currents which generate the required fields. Typical pixel sizes are 1 mm; that is, each pixel contains a large number of magnetic spheres. The display requires considerable energy to drive (50 mJ for 1.7 mm spot), but the inherent memory capability can be used to minimize power in a semistatic-type application.

Sheridan and Berkowitz[94] have reported on an electrostatic version of the rotating ball display. The Gyricon consists of dielectric balls, black on one side and white on the other, encapsulated in a plastic with a liquid as a lubricant. The structure of the Gyricon is shown in Fig. 11-27. The balls have an anisotropic double layer due to the different hemispheres (black on white). This layer leads to a torque in an applied electric field. Fifty volts is required to switch the balls in 200 ms.

Neither the magnetic nor the electric ball

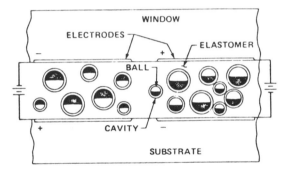

Fig. 11-27. Gyricon Display: Dielectric spheres can be made to rotate by an applied electric field. One hemisphere is made absorbing and the other reflecting.

display has achieved commercial reality. This is due in part to the limited contrast caused by the filling-factor restriction. Also, the displays do not seem to offer any particular multiplex drive advantages.

11.9.4 Galvanometer Displays. When large-area displays are needed, the size of the galvanometer mechanism is not a drawback. Peprnik[95] has reported on billboard-type alphanumeric displays using this system.

11.10   CONCLUSION

This chapter has discussed the various technologies that modulate light: liquid crystals, electrochromic reactions, colloidal suspensions, ferroelectric and ferromagnetic solids, and mini-mechanics. While significant development work has gone into all these electro-optic effects, the twisted nematic LCD has emerged in the 1970s as the overwhelming commercial flat-panel display for nonemissive applications. Granted the applications so far are of limited information content relative to light-emitting displays. Non-emitting technologies which cannot compete in these low-information-content applications, however, must be considered suspect in larger-information-content applications where the requirements become increasingly severe. Certainly any competitor will have to possess most of the advantages of the LCDs, plus beat them in their basic weak points: lack of strong optical appearance, lack of a high degree of direct multi-plex capability, and relatively slow response.

To understand why LCDs are dominant, it is useful to compare them to each competitor in turn (see Table 11-2). ECDs are significantly better-looking, but the multiplex problem is worse. There is also the basic problem that ECDs involve a chemical rather than a physical mechanism. A chemical reaction produces a much higher energy-density effect. The electric field at the double layer is very high compared to the uniform, lower electric field in an LCD. The ordered structure of the liquid-crystal phase permits the lower stress environment and will ultimately give the LCD a basic reliability superior to that of the ECD in all applications requiring very large numbers of switching cycles. ECDs will compete with LCDs in large-area, relatively low-information-content displays with relatively static data.

Colloidal systems are very similar to ECDs in their advantages and disadvantages. They look better than LCDs because they employ a direct absorption, but stability and multiplexability as well as high voltages are a problem. The electro-phoretic mechanism is physical, but the display is run under conditions where electrochemical effects can become important. The colloidal system cannot compete with the liquid-crystal phase for stability for very basic reasons. EPDs will be limited to the same applications as ECDs but with the additional handicap of high voltages.

Ferroelectric and ferromagnetic effects are very similar to LCDs in their basic interaction with light via birefringence. This means that these solid-state mechanisms cannot lead to better-looking displays than LCDs and may produce significantly inferior displays due to the small birefringence and the lack of the wave-

Table 11-2. Classification of Nonemitting Information Displays

	LCD			ECD	
Scattering	*Absorption*	*Phase/polarization*		*Deposition*	*Ion insertion*
Dynamic scattering	Dichroic dye	Deform. align. phase		Viologen	WO$_3$
Cholesteric/nematic	Thermal smectic	Twisted nematic		Silver	IrO$_2$
Thermal smectic		Fluorescence (FLAD)			Pthyalocyanine
					Polymer

	Colloidal		Electroactive solids			Electromechanical
Deposition	*Realignment*		*Ferroelectric*	*Surface*	*Ferromagnetic*	Foil shutter
TiO$_2$	Herapathite		PLZT	Fericon	GGG	Rotating ball
				Ruticon		Galvanometer

length-insensitive structure. Being a physical mechanism, the ferro-displays can be expected to be reliable in terms of switching. The basic advantage of rapid switching speeds may find applications in real-time data presentation. The high multiplexing capacity reported for the garnet display will be interesting if the pixel size can be significantly increased.

Minimechanical structures have a cultural problem. They are mechanical devices in an increasingly solid-state-device world. This has not bothered the CRT, but the minimechanical devices have not demonstrated any basic market advantages either. Lack of significant optical modulation and fabrication problems combine to make these devices a laboratory curiosity with respect to direct-view flat-panel displays. An important exception to this last observation is the galvanometer display for very large area status signs.

Thus, LCDs are not perfect, but they have four advantages: the liquid-crystal order, the large and "twistable" birefringence, the physical mechanism, and the enormous lead in the marketplace. Barring a radically new electro-optic effect which has these advantages and more, the author expects LCD technology to continue to dominate nonemissive flat-panel displays in the 1980s. Large information content will probably be achieved by solving the addressing problem with an extrinsic matrix rather than by improving the electro-optic characteristics. The company that can combine LCD and extrinsic matrix technologies will have a basic advantage in the late-1980s marketplace. Initially, the displays will have too slow a response for real-time presentation, but materials improvements will eventually solve this problem. The solution to full-color displays remains to be found.

Note added in proof: In section 11.4.1 on fast-scan multiplexing, it was stated that 20 to 30 scanned pixel lines would be the limit for this addressing technology. It has recently become apparent that 100 or more scanned lines are possible if the total display area is filled with electrodes (full xy matrix rather than a 5 × 7 dot character/no graphics format) and if the user is willing to read a significantly lower effective brightness ratio. With a full array,

there are two contrast states: select and nonselect. Voltage is applied everywhere in the array so that the relatively bright zero voltage state is not available to the viewer. As the duty cycle decreases, the brightness of the nonselect state, relative to the zero voltage state continually decreases. The important human factors point is that the contrast between the select and the nonselect state at 1/100 duty cycle is large enough to be legible even at the lower brightness ratio of the nonselect state. The whole display looks very "muddy" due to the dark background, but the darkness is fairly angle independent as is the darker select state. The result will be 25 × 80 character LCD displays with graphics capability in the marketplace by 1985.

REFERENCES

1. Bloss, F. D. *An Introduction to the Methods of Optical Crystallography.* New York: Holt, Rinehart and Winston, 1961.
2. Brody, T. P., "When–If Ever–Will the CRT be replaced by a Flat Display Panel?" *Microelectronics 11*, 5, 1980.
3. Heilmeier, G. H.; Zanoni, L. A.; and Barton, L. A. "Dynamic Scattering in Nematic Liquid Crystals." *Appl. Phys. Letters 13*, 46, 1968.
4. Heilmeier, G. H., and Zanoni, L. A. "Guest-Host Interactions in Nematic Liquid Crystals. A New Electro-optic Effect." *Appl. Phys. Letters 13*, 91, 1968.
5. Fergason, J. "Display Devices Utilizing Liquid Crystal Light Modulation." U.S. Patent #3,731,986, 1973.
6. Schadt, M., and Helfrich, W. "Voltage-Dependent Optical Activity of a Twisted Nematic Liquid Crystal." *Appl. Phys. Letters 18*, 127, 1971.
7. Castellano, J. A. "Reliability and Standards in USA." *Mol. Cryst. and Liq. Cryst. 63*, 265, 1981.
8. Chandrasekhar, S. *Liquid Crystals.* Cambridge: Cambridge University Press, 1977.
9. Kahn, F. J.; Taylor, G. N.; and Schanhorn, H. "Surface-Produced Alignment of Liquid Crystals." *Proc. IEEE 61*, 823, 1973.
10. Janning, J. L. "Thin Film Surface Orientation for Liquid Crystals." *Appl. Phys. Letters 21*, 173, 1972.
11. Helfrich, W. "Conduction-Induced Alignment of Nematic Liquid Crystals: Basic Model and Stability Considerations." *J. Chem. Phys. 51*, 4092, 1969.
12. Penz, P. A. "Electrohydrodynamic Wavelengths and Response Rates for a Nematic Liquid Crystal." *Phys. Rev. 10A*, 1300, 1974.
13. Wysocki, J. J.; Adams, J.; and Haas, W. "Electric-

Field-Induced Phase Change in Cholesteric Liquid Crystals." *Phys. Rev. Letters 20*, 1024, 1968.

14. Meyer, R. B. "Effects of Electric and Magnetic Fields on the Structure of Cholesteric Liquid Crystals." *Appl. Phys. Letters 12*, 281, 1968.

15. Heilmeier, G. H., and Goldmacher, J. E. "A New Electric-Field-Controlled Reflective Optical Storage Effect in Mixed-Liquid Crystal Systems." *Appl. Phys. Letters 13*, 132, 1968.

16. Kahn, F. J. "Electric-Field-Induced Color Changes and Pitch Dilation in Cholesteric Liquid Crystals." *Phys. Rev. Letters 24*, 209, 1970.

17. Ohtsuka, T.; Tsukamoto, M.; and Tsuchiya, M. "AC Electric-Field-Induced Cholesteric-Nematic Phase Transition in Mixed Liquid Crystal Films." *Jap. J. Appl. Phys. 12*, 22, 1973.

18. White, D. L., and Taylor, G. N. "New Absorptive Mode Reflective Liquid-Crystal Display Device." *J. Appl. Phys. 45*, 4718, 1974.

19. Schiekel, M. F., and Fahrenschon, K. "Deformation of Nematic Liquid Crystals with Vertical Orientation in Electrical Fields." *Appl. Phys. Letters 19*, 391, 1971.

20. Baur, G., and Greubel, W. "Fluorescence-Activated Liquid Crystal Display." *Appl. Phys. Letters 31*, 4, 1977.

21. Kahn, F. J. "IR-Laser-Addressed Thermo-optic Smectic Liquid-Crystal Storage Displays." *Appl. Phys. Letters 22*, 111, 1973.

22. Boyd, G. D.; Cheng, J.; and Ngo, P. D. T. "Liquid-Crystal Orientational Bistability and Nematic Storage Effects." *Appl. Phys. Letters 36*, 556, 1980.

23. Berreman, D. W., and Heffner, W. R. "New Bistable Cholesteric Liquid-Crystal Display." *Appl. Phys. Letters 37*, 109, 1980.

24. Kmetz, A. R. "Liquid Crystal Display Prospects in Perspective." *IEEE Trans. Elect. Devices 20*, 954, 1973.

25. Alt, P. M., and Pleshko, P. "Scanning Limitations of Liquid-Crystal Displays." *IEEE Trans. Elect. Devices 21*, 146, 1974.

26. Robert, J. "TV Image with Liquid Crystal Display." *IEEE Trans. Elect. Devices 26*, 1128, 1979.

27. Kaneko, E.; Kawakami, H.; and Hanmura, H. "Liquid Crystal TV Display." 1978 SID Int. Symp. Digest, p. 92.

28. Kahn, F. J. "Multiplexing Limits of Twisted Nematic Liquid Crystal Displays and Implications for the Future of High Information Content LCDs." In *the Physics and Chemistry of Liquid Crystal Devices*, edited by G. J. Sprokel. New York: Plenum, 1980, p. 79.

29. van Doorn, C. Z.; deKlerk, J.; and Gerritsma, C. J. "Influence of the Device Parameters on the Performance of Twisted Nematic Liquid Crystal Matrix Displays." In *The Physics and Chemistry of Liquid Crystal Devices*, edited by G. J. Sprokel. New York: Plenum, 1980, p. 95.

30. Schradt, M., and Gerber, P. R. "Physical Properties, Multiplexability, Bending and Electro-optical

Performance in TN-LCDs." 1981 SID Int. Symp. Digest, p. 80.

31. Wild, P. J., and Nehring, J. "Turn-on Time Reduction and Contrast Enhancement in Matrix-Addressed Liquid-Crystal Light Values." *Appl. Phys. Letters 19*, 335, 1971.

32. Stein, C. R., and Kashnow, R. A. "A Two-Frequency Coincidence Addressing Scheme for Nematic-Liquid-Crystal Displays." *Appl. Phys. Letters 19*, 343, 1971.

33. Bucher, H. K.; Klingbiel, R. T.; and Van Meter, J. P. "Frequency-Addressed Liquid Crystal Field Effect." *Appl. Phys. Letters 25*, 186, 1974.

34. Tani, C.; Ogawa, F.; Naemura, S.; Ueno, T.; and Saito, F. "Storage-Type Liquid-Crystal Matrix Display." *Proc. SID 21*, 71, 1980.

35. Greubel, W. "Bistability Behavior of Texture in Cholesteric Liquid Crystals in an Electric Field." *Appl. Phys. Letters 25*, 5, 1974

36. Hareng, M.; LeBerre, S.; Hehlen, R.; and Perbet, J. N. "Flat Matrix Addressed Smectic Liquid Crystal Display." 1981 SID Int. Symp. Digest, p. 106.

37. Lu, S.; Davies, D. H.; Albert, R.; Chung, D.; Hochbawn, A.; and Chung, C. "Thermally-Addressed Pleochroic Dye Switching Liquid Crystal Display." 1982 SID Int. Symp. Digest, p. 238.

38. Greubel, W., and Kmetz, A. R. "Ultimate Limits for Matrix Addressing of RMS-Responding Liquid-Crystal Displays." *IEEE Trans. Elect. Devices 26*, 795, 1979.

39. Penz, P. A. "Analog Displays—The Other Approach to LCDs." In *the Physics and Chemistry of Liquid Crystal Devices*, edited by G. J. Sprokel. New York: Plenum, 1980, p. 119.

40. Gruebel, R. L.; Marks, B. W.; Noble, R. T.; Penz, P. A.; and Surtani, K. H. "A Radial Format LCD/ Semiconductor System for Analog Watch Applications." *IEEE Trans. Elect. Devices 26*, 1134, 1979.

41. Hilsum, C.; Holden, R. J.; and Raynes, E. P. "A Novel Method of Temperature Compensation From Multiplexed Liquid Crystal Displays." *Elect. Letters 14*, 430, 1978.

42. Smith, G. W.; Kaplit, M.; and Hayden, D. B. "An Automotive Instrument Panel Employing Liquid Crystal Displays." Int. Expo. S.A.E. 770274, 1977.

43. Penz, P. A.; Surtani, K.; Wen, W. Y.; Johnson, M. R.; Kane, D. W.; Sanders, L. W.; Culley, B. G.; and Fish, J. G. "Plastic Substrate LCD." 1981 SID Int. Symp. Digest, p. 116.

44. Takahashi, S.; Shimokawa, O.; Inoue, H.; Uehara, K.; Hirose, T.; and Kikuyama, A. "A Liquid Crystal Display Panel Using Polymer Films." 1981 SID Int. Symp. Digest, p. 86.

45. Grabmaier, J. G.; Gruebel, W. F.; and Kruger, H. H. "Liquid Crystal Matrix Displays Using Additional Solid Layers for Suppression of Parasite Currents." *Mol. Cryst. and Liq. Cryst. 15*, 95, 1971.

46. Tannas, L. E., and York, P. K. "Liquid Crystal/Ferroelectric Matrix Addressable Display." *Ferroelectrics 10*, 19, 1976.

47. Castleberry, D. E. "Varistor-Controlled Liquid Crystal Displays." *IEEE Trans. Elect. Devices 26*, 1123, 1979.

48. Castleberry, D. E.; Becker, C. A.; and Levinson, L. M. "A 5 × 7″ Varistor-Controlled LC Matrix Display." 1982 SID Int. Symp. Digest, p. 246.

49. Baraff, D.; Boynton, R.; Gribbon, B.; Long, J. R.; MacLawrin, B.; Miner, C; Serinken, N.; Streater, R.; and Westwood, W. "The Application of Metal-Insulator-Metal Nonlinear Devices in Multiplexed Liquid Crystal Displays." 1980 SID Int. Symp. Digest, p. 200.

50. Luo, F.; Hester, W. A.; and Brody, T. P. "Alphanumeric and Video Performance of 6″ × 6″ 30 Lines-per-Inch Thin-Film Transistor–Liquid Crystal Display Panel." 1978 SID Int. Symp. Digest, p. 94.

51. Erskine, J. C., and Cserhati, A. "Cadmium Selenide Thin-Film Transistors." *J. Vac. Sci. Technol. 15*, 1823, 1978.

52. Lakatos, A. I. "Promise and Challenge of Thin-Film Silicon Approaches to Active Matrices." *IEEE Trans. Elect. Devices 30*, 525, 1983.

53. Luo, F., and Hoesley, D. "Hybrid Processed TFT Matrix Circuits for Flat Display Panels." 1982 SID Int. Symp. Digest, p. 46.

54. Lipton, L. T.; Stephens, C. P.; Lloyd, R. B.; Shields, S. E.; Toth, A. G.; and Tsai, R. C. "A 2.5″ Diagonal, High Contrast, Dynamic Scattering Liquid Crystal Matrix Display with Video Drivers." 1978 SID Int. Symp. Digest, p. 96.

55. Goede, W. "Status of New Display Technology in Japan." *Proc. Soc. Photo-Optical Inst. Eng. 199*, 53, 1979.

56. Yoshiyama, M.; Matsuo, T.; Kawasaki, K.; Tatsuta, H.; and Ishihara, T. "Small Liquid Crystal Television Display." *Mol. Cryst. and Liq. Cryst. 68*, 247, 1981.

57. Kasahara, K.; Yanagisawa, T.; Sakai, K.; Adachi, T.; Inoue, K.; Tsutsumi, T.; and Horl, H. "A Liquid-Crystal Display Panel Using an MOS Array with Gate-Bus Drivers." *IEEE Trans. Elect. Devices 28*, 744, 1981.

58. Hosokawa, M.; Oguchi, K.; Ikeda, M.; Yazawa, S.; and Endo, K. "Dichroic Guest-Host Active Matrix Video Display." 1981 SID Int. Symp. Digest, p. 114.

59. Ymasaki, T.; Kawahara, Y.; Motte, S.; Kamamori, H.; and Nakamura, J. "A Liquid Crystal TV Display panel with Drivers." 1982 SID Int. Symp. Digest, p. 48.

60. Lingane, J. J. "Electro-analytical Chemistry." New York: Interscience, 1958.

61. Deb, S. K. "A Novel Electrophotographic System." *Appl. Opt. Suppl. 3*, 192, 1969.

62. Faughnan, B. W., and Crandall, R. S. "Electrochromism in WO_3 Amorphous Films." *RCA Rev. 36*, 177, 1975.

63. Gottesfeld, S.; McIntyre, J. D. E.; Beni, G.; and Shay, J. L. "Electrochromism in Anodic Iridium Oxide Films." *Appl. Phys. Letters 33*, 208, 1978.

64. Schoot, C. J.; Ponje, J. J.; Van Darn, H. T.; Van Doorn, R. A.; and Bolwijn, P. T. "New Electrochromic Memory Display." *Appl. Phys. Letters 23*, 64, 1973.

65. Jasinski, R. "The Viologen Electrochromic Process." Conf. Record 1978 Biennial Disp. Res. Conf., p. 26.

66. Barclay, D. C.; Bird, C. L.; Kirkman, D. H.; Martin, D. H.; and Moth, F. T. "An Integrated Electrochromic Data Display." 1980 SID Int. Symp. Digest, p. 124.

67. Arellano, A.; Keller, G. S.; Melz, P. J.; Shattuk, M. D.; and Wilbur, C. V. "A Refreshed Matrix-Addressed Electrochromic Display." 1978 SID Int. Symp. Digest, p. 22.

68. Duchene, J.; Meyer, R.; and Delapierre, G. "Electrolytic Display," *IEEE Trans. Elect. Devices 26*, 1243, 1979.

69. Nicholson, M. M., and Galiardi, R. V. "A Multicolor Electrochromic Display," 1978 SID Int. Symp. Digest, p. 24.

70. Kaufman, F. B., and Engler, E. M. "Solid-State Spectroelectrochemistry of Cross-Linked Donor Bound Polymer Films." *J. Am. Chem. Soc. 101*, 547, 1979.

71. Kaufman, F. B. "New Organic Materials for Use as Transducers in Electrochromic Display Devices." Conf. Record 1978 Biennial Disp. Res. Conf., p. 23.

72. Marks, A. M. "Electro-optic Characteristics of Dipole Suspensions." *Appl. Optics 8*, 1397, 1969.

73. Evans, P. F.; Lees, H. D.; Maltz, M. S.; and Dailey, J. L. "Color Display Device." U.S. Patent #3,612,758 1971.

74. Ota, I.; Ohnishi, J.; and Yoshiyama, M. "Electrophoretic Image Display (EPID) Panel." *Proc. IEEE 61*, 832, 1973.

75. Mysels, K. J. "Introduction to Colloid Chemistry." New York: Interscience, 1959.

76. Chiang, A. "Electrophoretic Display Technologies and Characteristics." SID 1980 Symp. Seminar Notes, *1*, 137.

77. Fitzhenry-Ritz, B. "Optical Properties of Electrophoretic Image Displays." *IEEE Trans. Elect. Devices 28*, 726, 1981.

78. Ota, I.; Tuskamoto, M.; and Ohtsuka, T. "Developments in Electrophoretic Displays," *Proc. SID 18*, 243, 1977.

79. Singer, B., and Dalisa, A. L. "An X-Y Addressable Electrophoretic Display." *Proc. SID 18*, 255, 1977.

80. Special Issue on Electrophoretic and Other Particle Type Displays, *Proc. SID 18*, Nos. 3 and 4, 1977.

81. Novotny, V., and Hooper, M. A. "Optical and Electrical Characterization of Electrophoretic Displays." *J. Electrochem. Soc. 126*, 2211, 1979.

82. Marks, A. M. "An XY Raster Utilizing a Dipolar Electro-optical Medium." *Proc. SID 11*, 2, 1970.

83. Ashcroft, N. W., and Mermin, D. *Solid State Physics*. New York: Holt, Rinehart and Winston, 1976.

84. Haertling, G. H. "Improved Hot-Pressed Electro-optic Ceramics in the (Pb, La) (Zr, Ti) O_3 System." *J. Am. Cer. Soc. 54*, 303, 1971.

85. Hardtl, K. H. "Ferroelectric Displays." In *Nonemissive Electro-optic Displays*, edited by A. R. Kmetz and F. K. Von Willisen. New York: Plenum Press, 1976, p. 241.

86. Hill, B., and Schmidt, K. P. "Thin-Film Iron-Garnet Display Components." 1979 SID Int. Symp. Digest, p. 80.

87. Land, C. E., and Smith, W. D. "Reflective-Mode Ferroelectric Image Storage and Display Devices," *Appl. Phys. Letters 23*, 57, 1973.

88. Sheridan, N. "The Ruticon Family of Erasable Image Recording Devices." *IEEE Trans. Elect. Devices 19*, 1003, 1972.

89. Bruneel, J. L., and Micheron, F. "Optical Display Device Using Bistable Electrets," *Appl. Phys. Letters 30*, 382, 1977.

90. Toda, M., and Osaka, S. "Application of PVF_2 Bimorph Cantilener Elements to Display Devices." *Proc. SID 19*, 35, 1978.

91. Thomas, R. N.; Guldberg, J.; Nathanson, H. C.; and Malinberg, P. R. "The Mirror Matrix Tube: A Novel Light-Valve for Projection Displays." 1975 SID Int. Symp. Digest, p. 28.

92. Hornbeck, L. J.; Carlo, J. T.; Cowens, M. W.; Osterberg, P. M.; and DeSmith, M. M. "Deformable Mirror Projection Display," 1980 SID Int. Symp. Digest, p. 228.

93. Lee, L. L. "Fabrication of Magnetic Particles Displays," *Proc. SID 18*, 283, 1977.

94. Sheridan, N. K. and Berkowitz, M. A. "A Twisting Ball Display." *Proc. SID 18*, 289, 1977.

95. Peprnik, H. O. "A Light-Reflecting Electromagnetic Display." 1974 SID Int. Symp. Digest, p. 136.

INDEX